# CRC HANDBOOK OF
# AVIAN BODY MASSES

EDITED BY
## JOHN B. DUNNING, JR.
DEPARTMENT OF ZOOLOGY
AND INSTITUTE OF ECOLOGY
UNIVERSITY OF GEORGIA
ATHENS, GEORGIA

CRC Press
Boca Raton   Ann Arbor   London   Tokyo

**Library of Congress Cataloging-in-Publication Data**

Dunning, John B. (John Barnard)
    CRC handbook of avian body masses / by John B. Dunning, Jr.
        p.   cm.
    Includes bibliographical references (p. ) and index.
    ISBN 0-8493-4258-9
    1. Birds—Size—Tables.   2. Body size—Tables.   I. Title.   II. Title: Handbook of avian body masses.
QL697.D86   1993
598—dc20                                                                                                92-20884
                                                                                                          CIP

© 1993 by CRC Press, Inc.

International Standard Book Number 0-8493-4258-9

Library of Congress Card Number 92-20884
Printed in the United States  3  4  5  6  7  8  9  0
Printed on acid-free paper

To my parents, for their support.

# EDITOR

**John B. Dunning, Jr., Ph.D.,** is Assistant Research Scientist at the Institute of Ecology, University of Georgia, Athens, Georgia.

Dr. Dunning graduated in 1978 from Kent State University, Kent, Ohio, with a B.S. degree in biology (summa cum laude), and obtained his Ph.D. in ecology from the Department of Ecology and Evolutionary Biology, University of Arizona, Tucson, Arizona in 1986.

Dr. Dunning is a member of the American Ornithologists' Union, the Cooper Ornithological Society, the Wilson Ornithological Society, the Ecological Society of America, the American Society of Naturalists, the Association of Field Ornithologists, and the Western Bird Banding Association.

Among other awards, Dr. Dunning received Elective Member status in the American Ornithologists' Union in 1990. During his college career, he received the Kent State University Distinguished Scholar Award, and a National Science Foundation Predoctoral Fellowship from 1978–1981.

Dr. Dunning has received research grants from the National Science Foundation, the U.S. Department of Energy, the U.S. Forest Service, among other grants. His research is on the community and population ecology of North American sparrows. In particular, he has focused on the impacts that human land-use change has had on bird populations across large spatial scales. Dr. Dunning has published 23 papers, presented lectures, and has contributed to 17 annual meetings that have been based on his research. He is a member of Phi Beta Kappa and Beta Beta Beta honoraries.

# TABLE OF CONTENTS

Part I

# PART I

## BODY MASSES OF BIRDS OF THE WORLD

# INTRODUCTION

## John B. Dunning, Jr.

Many studies in avian biology require estimates of body size for the species being investigated. Ecological and physiological studies, for instance, often report measures of body size as baseline descriptive statistics when a large number of species are being compared. For this purpose, adult body mass is often the best single estimator of avian body size (Rising and Somers 1989). Body masses are also necessary for allometric scaling of metabolic processes (Calder 1984), and for some community structure analyses, such as calculating the total biomass of consumers supported by local resources ( Schluter and Repasky 1991, Wiens and Dyer 1975). For many other types of avian research, masses are extremely useful statistics (Clark 1979).

In spite of their utility, body masses are often difficult to locate even for relatively common species, however (Dunning 1984, 1985). Gleaning the relevant data from varied and some-times obscure publications is time-consuming. Weights are routinely recorded in banding operations and during museum assession of specimens, however, gathering data from these unpublished sources is also a difficult process.

This handbook is a compilation of data on body masses for 6283 species of birds, and is intended to alleviate the difficulty with which avian masses can be found. The handbook includes the largest and most complete sample of body masses located for each species from published or unpublished sources. Taxonomy and sequence of the handbook follow the fifth edition of Clements' (1991) checklist of birds of the world.

## SOURCES OF DATA

The data presented here were gathered primarily by searching the published literature. Clark's unpublished bibliography of body mass references and Brough's (1983) compilation were of great value in this effort. Because mass data were irregularly reported in papers published prior to 1960, I searched systematically for articles reporting mass data in the volumes of the major bird journals published from 1960 to 1990. Journals searched system-atically included *Ibis, Auk, Wilson Bulletin, Condor, Emu, Corella, Notornis, Journal of Field Ornithology,* and a number of smaller journals devoted primarily to bird banding. I also searched for individual articles published in these journals prior to 1960 when I encountered a citation that suggested that mass data might be found in a specific article. A wide variety of other journals were searched less systematically. Coverage of English-language literature published in the Western Hemisphere was more complete than coverage of Eastern Hemi-sphere journals, or journals published in French, German, or other languages. This bias reflects my access to the relevant literature, but probably does not reflect the distribution of published data. Mass data were located in articles covering a wide number of topics, including breeding biology, morphological variation, community structure, and physiology.

In addition, I surveyed regional handbooks and compilations such as Ali and Ripley (1968–1974) for India, Cramp and Simmons (1977–1988) and Baver and Glutz (1966) for Western Europe and North Africa, Brown et al. (1982) and subsequent volumes for Africa, and Clench and Leberman (1978) for eastern North America. Faunal surveys that focused on particular groups of birds provided extensive data for those groups, including Isler and Isler (1987) for tanagers, Turner and Rose (1989) for swallows, and Palmer (1976, 1988) for North American waterfowl and hawks, respectively. Finally, Brough's (1983) list of masses for over 2000 species worldwide provided both data and an extensive list of references, especially for Eurasian and African species.

Unpublished data were requested when possible from museum collections, banding operations, and researchers. I visited the collections of several research museums to record data from specimen labels, especially the collections at the University of Arizona, Louisiana State University, and the Philadelphia Academy of Natural Sciences. Numerous other museums responded to requests for information (see Acknowledgments). Unpublished banding data were solicited primarily for North American species. Much of this data was reported in Dunning (1984). A few unpublished dissertations were consulted, especially for Latin American species (e.g., Binford 1968, Weske 1972).

## DATA COLLECTION

As in Dunning (1984), I included the "best available sample" that I located for each species. The best available sample was defined as the data from the single source that had (1) the largest available sample size, or (2) the most complete descriptive statistics, if sample size from several sources were equivalent. I avoided combining data from different sources if possible because data taken in different seasons, at different locations, or using different techniques increases the heterogeneity of variance found within the sample, reducing the value of the composite mean (Sokal and Rohlf 1973). I did not follow this rule when all samples were very small. If no source provided masses of at least ten individuals, I lumped data from several sources to increase the overall sample size. I considered the increased heterogeneity that might result from this practice preferable to reporting means based on the mass of only one or two individuals.

I preferred data from breeding birds, however, the lack of published information for many species required the liberal use of data from other seasons. If ages were reported in the original source, I only used masses of adult birds. I present separate means for males and females for sexually dimorphic species. Dimorphism was assessed by testing for significant differences between sexes with a Student's t test when this was possible. A statistical test was not possible in many cases due to incomplete statistical information in the original source. In these instances I report separate means for males and females if the means differ by more than 10%. A single mean is presented for species that were not judged to be sexually dimorphic by these criteria, and for species in which the sex of the weighed individuals was not indicated in the original source.

Many species of birds vary in size across their geographic range. My ability to express this geographic variation is limited by the lack of adequate data for most species. However, I have reported several means when suitable samples were available from different locations throughout the range of geographically variable species. Such samples are identified by the collecting location, or the subspecific identity of the samples, whichever were reported by the original source. I did not assign samples to subspecies if the original source did not do so, since subspecies' ranges and validity are poorly studied in most birds.

## DEFINITIONS

The table was compiled using as much of the following information as was available in the selected source(s): sample size, mean, standard deviation, range, sex, collecting season, and location. Species names and order follow Clements (1991), which is based on Sibley and Monroe (1990), except in the treatment of some orders and families (e.g., Parulidae). Scientific names were modified in a few cases to reflect recent taxonomic suggestions (e.g., Robbins and Ridgely 1991). Species are numbered in the table using a numbering system following Clements (1991). Thus gaps in the species numbers in the table indicate species in Clements' list for which no data were found.

All means and ranges are given in grams. Standard deviation, when available, is given for samples that include masses of ten or more individuals. When necessary, standard deviation was calculated from standard error.

The following codes are used to describe the sex of the samples: M = male, F = female, B = data for both sexes lumped into a single mean, U = sex of weighed individuals not given in original source. To describe the collecting season, I used the following codes: B = breeding, PB = postbreeding, S = spring migration, F = fall migration, M = spring and fall migration combined, W = winter, Y = year round. In many sources, the data were not collected during a time period that fit neatly into any of the above categories (i.e., April through November). No collecting season is indicated in the table for these species.

All sources are numbered in the Literature Cited section, including sources of unpublished data such as banding programs and museum collections. Data in the table are referenced to the original source by citing the relevant numbers from the Literature Cited section.

## LIMITATIONS AND USES OF MASS DATA

A variety of topics in avian research require a measure of a bird's size. Size can be measured most accurately through multivariate analysis of several mensural characteristics (Freeman and Jackson 1990), however, the data for such an approach are rarely available, especially when a large number of species are being studied. In recent comparative studies, body mass has been shown to be the single most accurate univariate measure of size in birds. Rising and Somers (1989) determined that body mass correlated more strongly with overall body size (as estimated by multivariate analysis) than did body length, wing chord, or other commonly-measured linear characteristic. Freeman and Jackson (1990) showed that body mass and tarsal length were the two best univariate measures of overall body size. Tarsal length is a less desirable measure, at least in part because its measurement is less standardized and can be more subject to measurement error than are other linear measurements of birds (Lougheed et al. 1991). In addition, tarsal lengths are published for even fewer bird species than body masses.

Body mass is a measure of the overall size of an organism, including both the organism's structural framework and its nutrient reserves (Clark 1979, Piersma and Davidson 1991). The structural size of an organism is the relatively invariant mass associated with skeletal elements and internal organs; while the nutrient reserves are the protein and fat stores that vary with an organism's condition. Studies of structural size can be done most accurately using multivariate approaches to estimate the size of skeletal elements (Freeman and Jackson 1990, Piersma and Davidson 1991) or using skeletal volume (Moser and Rusch 1988). Both techniques require sacrifice of the study organisms, however, and therefore cannot be used in long-term studies of live birds. Representative samples of skeletal masses, linear measurements, or volumes are even less available in the published literature than are body masses for most birds.

Although body mass can be a useful measure of overall body size, the use of large number of masses collected from a variety of sources, such as those presented here, may not be justified in all comparative studies. The body mass of birds varies according to time of day, season, sex, and throughout many species' geographic range (Clark 1979). Such variation may reduce the validity of comparing mass samples collected from different sources. Researchers contemplating using the data in this handbook are urged to consider the effect of such variation on their study. When questions arise, the original source of the data should be consulted.

## A CALL FOR DATA

Although this compilation contains data for a large proportion of the world's avifauna, I anticipate that a more complete compilation will be possible in the future. About 35% of the world's birds are not included in this handbook, while a significant number of the species that are included have sample sizes of less than 10 birds. In particular, the birds of China, Japan, Southeast Asia, and islands of the south Pacific and Indian oceans are poorly represented in the table. Better data probably exist for many species, therefore, I expect that this handbook will need to be updated in the future. This updating would improve the handbook in several ways. First, my search of the literature was not exhaustive, and significant papers were probably missed. This is especially true for papers that were not published in English. Second, museum collections and banding operations are a rich, but underutilized, source of unpublished data. Finally, new research published after 1991 will undoubtedly include improved samples for many species, as well as data for species not included in this handbook. For all these reasons, I expect future editions of the handbook with additional data will be necessary. Researchers with data that would be an improvement over those published here are encouraged to contact the author.

## ACKNOWLEDGMENTS

I was supported by NSF grant BRS-8817950, and a postdoctoral fellowship at the Institute of Ecology, University of Georgia, during the final compilation of the database, and the preparation of the manuscript for this handbook. I thank Nora Mays, Rick Bowers, and Eugene Howard for help in creating the database from which the manuscript was prepared. Rick Bowers also provided extensive advice on programming. James Clements provided a copy of his checklist on which the database was organized. George Clark, Steve Russell, and especially Charles Collins helped locate many sources of data. I also thank Ron Pulliam for his patience during the preparation of this manuscript.

In the course of this compilation, I consulted the libraries of the University of Arizona, University of Georgia, Kent State University (Kent, Ohio, USA), University of Canterbury (Christchurch, New Zealand), MacQuarie University (Sydney, Australia), the Cleveland Museum of Natural History (Cleveland, Ohio, USA), the Alexander Library of the Edward Grey Institute (Oxford, UK), and the Josselyn Van Tyne Memorial Library of the University of Michigan (Ann Arbor, Michigan, USA). Jon Fisher of the Western Foundation of Vertebrate Zoology (Los Angeles, California, USA) and Peter Smallwood of the University of Arizona provided copies of articles to which I had no access.

I also thank the curators and directors of the following collections and institutions who provided me with data: University of Arizona, Royal Ontario Museum (Canada), University of Alaska, Texas Cooperative Wildlife Collections (Texas A & M University), Louisiana State University, Philadelphia Academy of Natural Sciences, Cowan Vertebrate Museum (University of British Columbia, Canada), California State University-Sacramento, University of Oklahoma, University of Washington, University of Minnesota, Delaware Museum of Natural History, Denver Museum of Natural History, Field Museum of Natural History (Chicago, Illinois), Yale Peabody Museum, Museum of Vertebrate Zoology (University of California-Berkeley), Western Foundation of Vertebrate Zoology (Los Angeles, California, USA), the National History Museum (Bulawayo, Zimbabwe), the New Zealand National Museum (Wellington, New Zealand), the Los Angeles County Museum (California, USA), the U.S. National Museum, British Museum of Natural History, and the San Diego Museum of Natural History.

The following observatories and organizations provided unpublished data from their research programs: Long Point Bird Observatory (Ontario, Canada), Point Reyes Bird

Observatory (California, USA), Manomet Bird Observatory (Massachusetts, USA), Braddock Bay Raptor Research (New York, USA), Cape May Bird Observatory and the Cape May Raptor Banding Project (New Jersey, USA), Hastings Reservation (California, USA), National Audubon Society, Powdermill Nature Center (Pennsylvania, USA), the Canadian Wildlife Service, and the U.S. Fish and Wildlife Service. Data from captive birds were provided by The Zoological Society of Philadelphia (Pennsylvania, USA), Milwaukee County Zoological Park (Wisconsin, USA), Kobi Oji Zoo (Kobe, Japan), New York Zoological Society (New York, USA), and the Audubon Zoological Gardens (New Orleans, Louisiana, USA).

This project benefited from the cooperation of a large number of ornithologists. Approximately a third of the data in Dunning (1984) was previously unpublished and provided by researchers, especially the members of the Western Bird Banding Association. These data have been included in the current handbook when these samples were still the best available. Thus, I am indebted to the people who contributed to the previous compilation. The current handbook contains less unpublished data than the 1984 compilation, probably because I had less contact with researchers with unpublished data from areas other than North America. Extensive unpublished data were provided by John Blake and Bette Loiselle (Panama and Costa Rica), Doug Stotz (Brazil), and John Fitzpatrick (Peru). Charles Collins provided me with his unpublished compilation of mass data for swifts. Robert Sutton provided me with sources of data for Jamaican birds, while John Innes put me in touch with sources of New Zealand data.

In addition to the individuals cited above, I thank the following people for providing unpublished data and advice: Keith Arnold, Jon Atwood, Yves Aubry, Tom Bancroft, Luis Baptista, Jon Barlow, Carl Barrentine, John Bates, Range Bayer, Craig Benkman, Byron Berger, Richard Bierregaard, Linda Birch, Sharon Birks, Robin Bjork, Charles Blake, John Blake, David Blockstein, E. Clark Bloom, Peter Bloom, Jeffrey Bouton, Rick Bowers, Roger Boyd, Dawn Breese, H. Brieschke, Kay Burk, Bill Calder, Peter Cannell, Richard Cannings, David Capin, Charles Chase, Jane Church, Roger Clapp, George Clark, William Clark, Mario Cohn-Haft, Charles Collins, Charles Corchran, Malcolm Coulter, Humphrey Crick, Alexander Cruz, Christian Dau, Dave DeSante, James Dick, Edward Diebold, George Divoky, Robert Duncan, Erica Dunn, Bruce Eichhorst, Susan Elbin, John Fitzpatrick, Charles Francis, Daniel Gibson, Robert Gill, David Goldstein, Walter Graul, Ken Graupman, John Groves, Joseph Grzybowski. Jeremy Hatch, Rod Hay, Sue Heath, Paul Hendricks, Joanna Hill, Larry Hood, Robert Hosea, C. Stuart Houston, Peter Hubbell, Steve Hubbell, George Hunt, John Innes, Ross James, Joseph Jehl, Kent Jenson, Kent Johnson, Ned Johnson, Virginia Johnson, Jim Jolly, Philip Kahl, Kenn Kaufman, Paul Kerlinger, Brina Kessel, Lloyd Kiff, Greg Lasley, Peter Lawson, Calvin Lensink, Fred Lohrer, Bette Loiselle, Annarie Lyles, Carl Marti, Thomas Martin, Harold Mayfield, H. Elliott McClure, Bonnie McKinney, Barbara McKnight, Martin McNicholl, David Melville, Rita Mesquita, L. Richard Mewaldt, Alex Middleton, George Millikan, Burt Monroe, R.J. Moorhouse, P.J. Mundy, Koichi Murata, Nancy Newfield, David Niles, Robert Norton, Gary Nuechterlein, Yoshika Oniki, Jose Ottenwalder, James Otto, George Page, Dennis Paulson, Donald Payne, Margaret Peterson, Lisa Petit, Kenneth Prescott, Trevor Price, Ronald Pulliam, Elizabeth Pullman, Michael Putnam, William Radke, John Ratti, Amadeo Rea, Roland Redmond, Kerry Reese, Robert Ricklefs, Mark Robbins, Sievert Rohwer, Roland Roth, Ruth Russell, Steve Russell. Andrew Sanders, S.D. Schemnitz, Dolph Schluter, Gary Schnell, Donald Schroeder, Chris Schultz, Peter Shannon, David Shepherd, Jay Sheppard, Doug Stotz, Joseph Strauch, Robert Sutton, John Terborgh, Bernie Tershy, Betsy Trent Thomas, Robert Tintle, Pepper Trail, Jolan Truan, Joan Tweit, Robert Tweit, J. Van Remsen, Jerry Verner, D.L. Walker, George Wallace, Philip Walters, Richard Weatherly, Elizabeth Webb, John Weske, Pamela Williams, Kerry-Jayne Wilson, David Wingate, D. Scott Wood, John Woods, Richard Zusi.

## Body Masses of World Birds

| Species | Sex | N | Mean | Std dev | Range | Sn | Location | Number |
|---|---|---|---|---|---|---|---|---|
| | | | | | ORDER: TINAMIFORMES | | FAMILY: TINAMIDAE | |
| Tinamus tao | U | | 2000.0 | | | | Peru 623 | 1.0 |
| Tinamus solitarius | M | | | | 1200.0–1500.0 | | Brazil | 2.0 |
| | F | | | | 1300.0–1800.0 | | 564 | |
| Tinamus osgoodi | M | 1 | 1285.0 | | | | Peru 193 | 3.0 |
| Tinamus major | B | 10 | 1052.0 | 163.10 | 824.0–1337.0 | | Belize 510 | 4.0 |
| Tinamus guttatus | U | | 600.0 | | | | Peru 623 | 5.0 |
| Nothocercus bonapartei | B | 8 | 763.0 | | 455.0–1050.0 | | Panama; Columbia 243, 244, 387 | 6.0 |
| Crypturellus cinereus | U | | 450.0 | | | | Peru 623 | 10.0 |
| Crypturellus soui | M | 6 | 198.0 | 26.91 | 165.0–204.0 | | Belize | 11.0 |
| | F | 7 | 235.0 | 17.12 | 213.0–268.0 | | 510 | |
| Crypturellus obsoletus | F | 3 | 482.0 | | 357.0–600.0 | | Brazil; Peru 37, 193 | 13.0 |
| Crypturellus cinnamomeus | B | 14 | 419.0 | 17.96 | 401.0–448.0 | | Mexico; Belize 274, 457, 510 | 14.0 |
| Crypturellus undulatus | U | | 540.0 | | | | Peru 623 | 15.0 |
| Crypturellus transfasciatus | B | 6 | 283.0 | | | | Peru 672 | 16.0 |
| Crypturellus strigulosus | M | 1 | 390.0 | | | | Bolivia 22 | 17.0 |
| Crypturellus boucardi | M | 22 | 418.0 | 20.50 | 375.0–464.0 | | Belize | 18.0 |
| | F | 18 | 468.0 | 43.84 | 374.0–526.0 | | 510 | |
| Crypturellus noctivagus | M | 1 | 800.0 | | | | Brazil 37 | 22.0 |
| Crypturellus atrocapillus | M | 2 | 453.0 | | 450.0–455.0 | | Peru 193 | 23.0 |
| Crypturellus variegatus | U | 4 | 384.0 | | | | Brazil 45 | 24.0 |
| Crypturellus bartletti | U | 6 | 241.0 | | | | Peru 193 | 26.0 |
| Crypturellus parvirostris | U | 2 | 212.0 | | 200.0–225.0 | | Brazil 37, 610a | 27.0 |
| Crypturellus tataupa | U | 7 | 264.0 | | 184.0–421.0 | | 37, 610, 668 | 29.0 |
| Rhynchotus rufescens | U | 1 | 900.0 | | | | Brazil 37 | 30.0 |

## Body Masses of World Birds (continued)

| Species | Sex | N | Mean | Std dev | Range | Sn | Location | Number |
|---------|-----|---|------|---------|-------|----|----------|--------|
| Nothoprocta perdicaria | U | 4 | 458.0 | | | | Chile 685 | 36.0 |
| Nothura maculosa | F | 1 | 300.0 | | | | Brazil 37 | 40.0 |
| Eudromia elegans | M | 1 | 660.0 | | | | Argentina 411 | 44.0 |

**ORDER: RHEIFORMES                    FAMILY: RHEIDAE**

| Species | Sex | N | Mean | Std dev | Range | Sn | Location | Number |
|---------|-----|---|------|---------|-------|----|----------|--------|
| Rhea americana | U | | 23000.0 | | 10500.0–40000.0 | | 78 | 48.0 |

**ORDER: STRUTHIONIFORMES              FAMILY: STRUTHIONIDAE**

| Species | Sex | N | Mean | Std dev | Range | Sn | Location | Number |
|---------|-----|---|------|---------|-------|----|----------|--------|
| Struthio camelus | U | 20 | | | 63000.0–104000.0 | | domesticated 115 | 50.0 |

**ORDER: CASUARIIFORMES                FAMILY: DROMICEIDAE**

| Species | Sex | N | Mean | Std dev | Range | Sn | Location | Number |
|---------|-----|---|------|---------|-------|----|----------|--------|
| Dromaius novaehollandiae | U | 10 | 31160.0 | | | | 141, 254 | 51.0 |

**ORDER: CASUARIIFORMES                FAMILY: CASUARIIDAE**

| Species | Sex | N | Mean | Std dev | Range | Sn | Location | Number |
|---------|-----|---|------|---------|-------|----|----------|--------|
| Casuarius casuarius | U | | 44000.0 | | 29200.0–58500.0 | | 78 | 52.0 |

**ORDER: DINORNITHIFORMES              FAMILY: APTERYGIDAE**

| Species | Sex | N | Mean | Std dev | Range | Sn | Location | Number |
|---------|-----|---|------|---------|-------|----|----------|--------|
| Apteryx australis | M | 15 | 2120.0 | | 1720.0–2730.0 | | New Zealand | 55.0 |
| | F | 31 | 2540.0 | | 2060.0–3850.0 | | 102 | |
| Apteryx owenii | M | 61 | 1135.0 | 119.00 | 880.0–1356.0 | | New Zealand | 56.0 |
| | F | 41 | 1351.0 | 164.00 | 1000.0  1400.0 | | 308 | |

**ORDER: PODICIPEDIFORMES              FAMILY: PODICIPEDIDAE**

| Species | Sex | N | Mean | Std dev | Range | Sn | Location | Number |
|---------|-----|---|------|---------|-------|----|----------|--------|
| Rollandia rolland | M | 2 | 232.0 | | 225.0–239.0 | | Brazil | 58.0 |
| | F | 1 | 265.0 | | | | 37 | |
| Tachybaptus ruficollis | B | 50 | 201.0 | | 91.0–290.0 | Y | Britain 258 | 60.0 |
| Tachybaptus novaehollandiae | M | 2 | 220.0 | | 199.0–242.0 | | | 61.0 |
| | F | 1 | 189.0 | | | | 217, 506 | |
| Tachybaptus pelzelnii | M | 1 | 172.0 | | | | Madagascar | 62.0 |
| | F | | | | | | 39 | |
| Tachybaptus dominicus | M | 12 | 129.0 | 11.40 | | | Panama | 64.0 |
| | F | 12 | 116.0 | 11.80 | | | 244 | |
| Podilymbus podiceps | B | 33 | 442.0 | 65.90 | 343.0–551.0 | B | Wisconsin, USA 442 | 65.0 |

## Body Masses of World Birds (continued)

| Species | Sex | N | Mean | Std dev | Range | Sn | Location | Number |
|---|---|---|---|---|---|---|---|---|
| Podilymbus gigas | M | 2 | 830.0 | | 804.0–856.0 | | Guatemala | 66.0 |
| | F | 2 | 568.0 | | 552.0–584.0 | | 329 | |
| Poliocephalus poliocephalus | M | 14 | 258.0 | 30.20 | 202.0–311.0 | | | 67.0 |
| | F | 14 | 223.0 | 27.60 | 190  –276.0 | | 609 | |
| Poliocephalus rufopectus | B | 15 | 249.0 | | 170.0–291.0 | | New Zealand | 68.0 |
| | | | | | | | 609 | |
| Podiceps grisegena | B | 6 | 1023.0 | | 743.0–1270.0 | | Alaska, USA | 70.0 |
| | | | | | | | 445 | |
| Podiceps cristatus | M | 7 | 738.0 | | 596.0–813.0 | | Netherlands | 71.0 |
| | F | 4 | 609.0 | | 568.0–686.0 | | 115 | |
| Podiceps auritus | U | 47 | 453.0 | | 327.0–528.0 | W | Maryland, USA | 72.0 |
| | | | | | | | 263 | |
| Podiceps nigricollis | U | 8 | 292.0 | | 218.0–375.0 | F | SW USA; Mexico | 73.0 |
| | | | | | | | 48, 594 | |
| Podiceps occipitalis | F | 4 | 334.0 | | 278.0–410.0 | | | 75.0 |
| | | | | | | | 607 | |
| Podiceps gallardoi | B | 4 | 575.0 | | 420.0–740.0 | | | 77.0 |
| | | | | | 420.0–740.0 | | 608 | |
| Aechmophorus occidentalis | U | 13 | 1477.0 | | 795.0–1818.0 | W | Washington, USA | 78.0 |
| | | | | | | | 445 | |

### ORDER: SPHENISCIFORMES            FAMILY: SPHENISCIDAE

| Species | Sex | N | Mean | Std dev | Range | Sn | Location | Number |
|---|---|---|---|---|---|---|---|---|
| Aptenodytes patagonicus | U | 5 | 13220.0 | 470.00 | | PB | Possession Is. | 80.0 |
| | | | | | | | 98 | |
| Aptenodytes forsteri | U | | 34000.0 | | 22700.0–45300.0 | | | 81.0 |
| | | | | | | | 78 | |
| Pygoscelis papua | M | 32 | 6400.0 | 700.00 | | B | S. Georgia Is. | 82.0 |
| | F | 32 | 5500.0 | 650.00 | | | 605 | |
| Pygoscelis adeliae | M | 15 | 5000.0 | | | B | Ross Island | 83.0 |
| | F | 10 | 4700.0 | | | | 4 | |
| Pygoscelis antarctica | U | 44 | 4150.0 | | | | S. Shetland Is. | 84.0 |
| | | | | | | | 124 | |
| Eudyptes chrysocome | M | | 2700.0 | | | | MacQuire Is. | 85.0 |
| | F | | 2300.0 | | | | 268 | |
| Eudyptes pachyrhynchus | U | 208 | 3500.0 | | 2000.0–5950.0 | | | 86.0 |
| | | | | | | | 658 | |
| Eudyptes chrysolophus | U | | 3900.0 | | | | | 89.0 |
| | | | | | | | 49 | |
| Eudyptes schlegeli | M | | 4500.0 | | | | MacQuire Is. | 90.0 |
| | F | | 4000.0 | | | | 268 | |
| Eudyptula minor | B | 3195 | 1105.0 | 110.00 | | Br | Tasmania | 92.0 |
| | | | | | | | 324 | |
| Eudyptula albosignata | U | 18 | 1327.0 | 176.00 | | Br | New Zealand | 93.0 |
| | | | | | | | 324 | |

## Body Masses of World Birds (continued)

| Species | Sex | N | Mean | Std dev | Range | Sn | Location | Number |
|---|---|---|---|---|---|---|---|---|
| Spheniscus demersus | M | 127 | 3310.0 | 260.00 | | | South Africa | 94.0 |
| | F | 127 | 2960.0 | 310.00 | | | 172 | |
| Spheniscus humboldti | U | | 5000.0 | | | | | 95.0 |
| | | | | | | | 171 | |
| Spheniscus magellanicus | U | | 4500.0 | | | | Brazil | 96.0 |
| | | | | | | | 564 | |
| Spheniscus mendiculus | U | | 2500.0 | | | | | 97.0 |
| | | | | | | | 582 | |

### ORDER: PROCELLARIIFORMES      FAMILY: DIOMEDEIDAE

| Species | Sex | N | Mean | Std dev | Range | Sn | Location | Number |
|---|---|---|---|---|---|---|---|---|
| Diomedea exulans | M | 3 | 8400.0 | | 8000.0–8600.0 | | MacQuarie Is. | 98.0 |
| | F | 3 | 6900.0 | | 6400.0–7500.0 | | 629 | |
| Diomedea amsterdamensis | B | 33 | 6270.0 | 780.00 | 4800.0–8000.0 | B | Amsterdam Is. | 99.0 |
| | | | | | | | 309 | |
| Diomedea epomophora | M | | 8200.0 | | | | estimated | 100.0 |
| | | | | | | | 486 | |
| Diomedea irrorata | M | 7 | 3750.0 | | 3110.0–4390.0 | | | 101.0 |
| | F | 13 | 3040.0 | | 2700.0–3380.0 | | 78 | |
| Diomedea nigripes | U | 306 | 3148.0 | | | | | 103.0 |
| | | | | | | | 657 | |
| Diomedea immutabilis | M | 233 | 3230.0 | 61.40 | 2400.0–4100.0 | B | Midway Island | 104.0 |
| | F | 134 | 2853.0 | 73.40 | 1900.0–3600.0 | | 204 | |
| Diomedea melanophris | M | 132 | 3922.0 | | | | S. Georgia Is. | 105.0 |
| | F | 94 | 3206.0 | | | | 468 | |
| Diomedea cauta | U | | 4100.0 | | | | | 106.0 |
| | | | | | | | 3 | |
| Diomedea chrysostoma | M | 133 | 3751.0 | | | | S. Georgia Is. | 107.0 |
| | F | 95 | 3264.0 | | | | 468 | |
| Diomedea chlororhynchos | B | 16 | 2460.0 | 240.00 | 2130.0–2800.0 | B | Kerguelen Is. | 108.0 |
| | | | | | | | 665 | |
| Phoebetria fusca | B | 176 | 2500.0 | | 2100.0–3400.0 | | Marion Is. | 110.0 |
| | | | | | | | 42 | |
| Phoebetria palpebrata | B | 10 | 2785.0 | | 2050.0–3200.0 | | | 111.0 |
| | | | | | | | 42, 671 | |

### ORDER: PROCELLARIFORMES      FAMILY: PROCELLARIIDAE

| Species | Sex | N | Mean | Std dev | Range | Sn | Location | Number |
|---|---|---|---|---|---|---|---|---|
| Macronectes giganteus | M | 9 | 5190.0 | | 4650.0–5300.0 | | Possession Is. | 112.0 |
| | F | 7 | 3944.0 | | 3350.0–4900.0 | | 115 | |
| Macronectes halli | U | | 4000.0 | | | | | 113.0 |
| | | | | | | | 173 | |
| Fulmarus glacialis | M | 16 | 609.0 | 77.90 | 485.0–727.0 | W | Oregon, USA | 114.0 |
| | F | 29 | 479.0 | 50.00 | 395.0–582.0 | | 236 | |
| Fulmarus glacialoides | U | | 1000.0 | | | | | 115.0 |
| | | | | | | | 3 | |

## Body Masses of World Birds (continued)

| Species | Sex | N | Mean | Std dev | Range | Sn | Location | Number |
|---|---|---|---|---|---|---|---|---|
| Thalassoica antarctica | M | 11 | 813.0 | | | B | Antarctica | 116.0 |
| | F | 5 | 696.0 | | | | 389 | |
| Daption capense | U | 17 | 428.0 | | 375.0–500.0 | B | | 117.0 |
| | | | | | | | 445 | |
| Pagodroma nivea | B | 52 | 268.0 | 35.90 | 202.0–322.0 | B | Antarctica | 118.0 |
| | | | | | | | 515 | |
| Lugensa brevirostris | U | 126 | 357.0 | 43.20 | 255.0–451.0 | | Marion Is. | 120.0 |
| | | | | | | | 532 | |
| Pterodroma inexpectata | U | 89 | 316.0 | 32.10 | 247.0–441.0 | B | New Zealand | 128.0 |
| | | | | | | | 486 | |
| Pterodroma hypoleuca | B | 168 | 176.0 | 13.00 | 150.0–220.0 | B | Hawaiian Is. | 129.0 |
| | | | | | | | 242 | |
| Pterodroma leucoptera | M | 1 | 125.0 | | | | New Zealand | 130.0 |
| | | | | | | | 248a | |
| Pterodroma cookii | M | 3 | 164.0 | | 112.0–250.0 | | New Zealand | 131.0 |
| | F | 7 | 193.0 | | 149.0–241.0 | | 363 | |
| Pterodroma pycrofti | M | 16 | 153.0 | | | B | New Zealand | 132.0 |
| | | | | | | | 175 | |
| Pterodroma brevipes | B | 20 | 136.0 | | 116.0–158.0 | | Fiji | 133.0 |
| | | | | | | | 653a | |
| Pterodroma alba | U | | 272.0 | | | | Christmas Is. | 136.0 |
| | | | | | | | 488 | |
| Pterodroma arminjoniana | U | 1 | 161.0 | | | | Australia | 138.0 |
| | | | | | | | 641 | |
| Pterodroma phaeopygia | B | 38 | 434.0 | 52.90 | | B | Hawaiian Is. | 139.0 |
| | | | | | | | 566 | |
| Pterodroma ultima | U | 13 | 360.0 | | 325.0–377.0 | | Pitcairn Is. | 143.0 |
| | | | | | | | 677 | |
| Pterodroma macroptera gouldii | M | 56 | 560.0 | | | B | New Zealand | 145.0 |
| | F | 28 | 505.0 | | | | 276 | |
| Pterodroma lessoni | B | 18 | 698.0 | 58.60 | 580.0–810.0 | B | Kerguelen Is. | 147.0 |
| | | | | | | | 665 | |
| Pterodroma mollis | B | 85 | 312.0 | 34.70 | 250.0–380.0 | B | Marion Is. | 150.0 |
| | | | | | | | 532 | |
| Pterodroma incerta | U | | 520.0 | | | | | 151.0 |
| | | | | | | | 3 | |
| Pterodroma cahow | U | 1 | 246.0 | | | | Bermuda | 152.0 |
| | | | | | | | 681a | |
| Pterodroma hasitata | U | 1 | 278.0 | | | | Dominican Repub. | 153.0 |
| | | | | | | | 441a | |
| Halobaena caerulea | U | 215 | 202.0 | 17.50 | 163.0–251.0 | B | Marion Is. | 154.0 |
| | | | | | | | 208 | |
| Pachyptila vittata | U | 29 | 196.0 | | | | | 155.0 |
| | | | | | | | 486 | |

## Body Masses of World Birds (continued)

| Species | Sex | N | Mean | Std dev | Range | Sn | Location | Number |
|---|---|---|---|---|---|---|---|---|
| Pachyptila salvini | B | 153 | 164.0 | | 130.0–210.0 | B | Crozet Is. 64 | 156.0 |
| Pachyptila desolata | U | 118 | 147.0 | 14.10 | 115.0–183.0 | B | Kerguelen Is. 665 | 157.0 |
| Pachyptila belcheri | U | 66 | 145.0 | 12.50 | 118.0–180.0 | B | Kerguelen Is. 665 | 158.0 |
| Pachyptila turtur | U | 54 | 137.0 | | 110.0–175.0 | B | Crozet Is. 64 | 159.0 |
| Pachyptila crassirostris | M | 1 | 102.0 | | | | Australia 372 | 160.0 |
| Bulweria bulwerii | U | 191 | 99.0 | 13.80 | 78.0–130.0 | B | Hawaiian Is. 242 | 161.0 |
| Procellaria aequinoctialis | U | 87 | 1213.0 | 134.00 | 980.0–1885.0 | B | Crozet Is. 310 | 163.0 |
| Procellaria parkinsoni | F | 1 | 675.0 | | | | Costa Rica 287 | 164.0 |
| Procellaria westlandica | M | 1 | 780.0 | | | | Australia 640 | 165.0 |
| Procellaria cinerea | U | 37 | 1131.0 | 133.00 | 900.0–1520.0 | B | Kerguelen Is. 665 | 166.0 |
| Calonectris diomedea | U | 37 | 535.0 | 60.00 | 400.0–650.0 | B | Crete 507 | 167.0 |
| Puffinus pacificus | U | 124 | 388.0 | 33.40 | 320.0–510.0 | B | Hawaiian Is. 242 | 169.0 |
| Puffinus bulleri | F | 2 | 380.0 | | 342.0–418.0 | | 445 | 170.0 |
| Puffinus carneipes | U | | 568.0 | | | B | 656 | 171.0 |
| Puffinus creatopus | U | 6 | 721.0 | | 665.0–791.0 665.0–791.0 | PB | California, USA 594 | 172.0 |
| Puffinus gravis | U | 58 | 849.0 | 26.70 | | PB | Nova Scotia, Can. 83 | 173.0 |
| Puffinus griseus | U | 100 | 787.0 | | 666.0–978.0 | B | New Zealand 445 | 174.0 |
| Puffinus tenuirostris | B | 25 | 543.0 | | 473.0–614.0 | B | Australia 445 | 175.0 |
| Puffinus nativitatis | U | 99 | 356.0 | 29.85 | 280.0–415.0 | B | Hawaiian Is. 242 | 176.0 |
| Puffinus puffinus | U | 187 | 453.0 | | | B | Wales 70 | 177.0 |
| Puffinus auricularis | B | 11 | 323.0 | | 290.0–358.0 | | Mexico 48, 286 | 179.0 |
| Puffinus opisthomelas | U | 1 | 276.0 | | | | California, USA 591 | 180.0 |

## Body Masses of World Birds (continued)

| Species | Sex | N | Mean | Std dev | Range | Sn | Location | Number |
|---|---|---|---|---|---|---|---|---|
| Puffinus huttoni | U | 17 | 364.0 | | | B | New Zealand 363 | 182.0 |
| Puffinus lherminieri | U | 78 | 168.0 | 4.12 | 128.0–211.0 | B | Galapagos Is. 237 | 183.0 |
| Puffinus assimilis | U | 91 | 226.0 | | 170.0–275.0 | B | S. Atlantic 614 | 184.0 |

### ORDER: PROCELLARIIFORMES                FAMILY: HYDROBATIDAE

| Species | Sex | N | Mean | Std dev | Range | Sn | Location | Number |
|---|---|---|---|---|---|---|---|---|
| Oceanites oceanicus | U | 31 | 32.0 | 3.00 | 27.0–39.0 | B | Crozet Is. 310 | 188.0 |
| Oceanites gracilis | F | 1 | 17.0 | | | | Galapagos Is. 238 | 189.0 |
| Garrodia nereis | U | 38 | 38.2 | 3.50 | 31.0–44.0 | B | Kerguelen Is. 665 | 190.0 |
| Pelagodroma marina | U | 100 | 47.2 | 4.00 | 40.0–62.0 | | New Zealand 115 | 191.0 |
| Fregetta tropica | U | 38 | 52.0 | 3.00 | 43.0–59.0 | B | Crozet Is. 310 | 192.0 |
| Fregetta grallaria | U | | 50.0 | | | | 3 | 193.0 |
| Hydrobates pelagicus | U | 1770 | 25.2 | | 20.3–31.1 | B | Britain 258 | 195.0 |
| Oceanodroma microsoma | U | 11 | 20.5 | 1.11 | 18.2–21.7 | | E. Pacific O. 122 | 196.0 |
| Oceanodroma tethys | U | 154 | 23.5 | 1.83 | 19.0–35.0 | B | Galapagos Is. 238 | 197.0 |
| Oceanodroma castro | U | 389 | 41.8 | 4.37 | 28.5–56.5 | Y | Galapagos Is. 238 | 198.0 |
| Oceanodroma leucorhoa leucorhoa | U | 124 | 39.8 | 5.04 | 29.0–53.0 | | E. Pacific O. 122 | 199.0 |
| Oceanodroma leucorhoa socorroensis | U | 152 | 31.7 | 4.16 | 22.4–44.0 | | Guadelupe Is. 122 | 199.0 |
| Oceanodroma leucorhoa chapmani | U | 78 | 34.8 | 2.68 | 29.0–42.0 | B | Mexico 122 | 199.0 |
| Oceanodroma monorhis | U | 4 | 35.8 | | 23.0–40.0 | | 115 | 200.0 |
| Oceanodroma tristrami | U | 61 | 84.0 | 7.81 | 66.0–105.0 | B | Hawaiian Is. 242 | 202.0 |
| Oceanodroma melania | U | 44 | 59.0 | 3.69 | 50.0–67.0 | | E. Pacific O. 122 | 205.0 |
| Oceanodroma homochroa | U | 20 | 36.9 | 2.27 | 33.3–42.4 | | E. Pacific O. 122 | 206.0 |
| Oceanodroma furcata | U | 55 | 55.3 | 3.70 | 47.5–62.8 | B | California, USA 241 | 208.0 |

## Body Masses of World Birds (continued)

| Species | Sex | N | Mean | Std dev | Range | Sn | Location | Number |
|---|---|---|---|---|---|---|---|---|
| **ORDER: PROCELLARIIFORMES** | | | | | **FAMILY: PELECANOIDIDAE** | | | |
| Pelecanoides georgicus | U | 71 | 121.0 | 13.00 | 90.0–150.0 | B | Crozet Is. 310 | 211.0 |
| Pelecanoides urinatrix | U | 52 | 141.0 | 13.00 | 105.0–165.0 | B | Crozet Is. 310 | 212.0 |
| **ORDER: PELECANIFORMES** | | | | | **FAMILY: PHAETHONTIDAE** | | | |
| Phaethon aethereus | U | | 750.0 | | | | 487a | 213.0 |
| Phaethon rubricauda | U | 84 | 624.0 | 45.80 | 540.0–750.0 | B | Hawaiian Is. 242 | 214.0 |
| Phaethon lepturus | U | 59 | 334.0 | | | | Aldabra Atoll 150 | 215.0 |
| **ORDER: PELECANIFORMES** | | | | | **FAMILY: FREGATIDAE** | | | |
| Fregata magnificens | M | 16 | 1281.0 | 34.80 | | | Panama | 216.0 |
| | F | 6 | 1667.0 | | | | 244 | |
| Fregata aquila | U | | 1620.0 | | | | 78 | 217.0 |
| Fregata minor | M | 316 | 927.0 | 111.00 | 640.0–1350.0 | B | Christmas Is. | 218.0 |
| | F | 312 | 1183.0 | 115.00 | 850.0–1550.0 | | 533 | |
| Fregata ariel | M | 29 | 754.0 | | 625.0–875.0 | | Aldabra Atoll | 219.0 |
| | F | 45 | 858.0 | | 760.0–955.0 | | 151 | |
| **ORDER: PELECANIFORMES** | | | | | **FAMILY: SULIDAE** | | | |
| Morus bassanus | M | 27 | 2932.0 | | 2470.0–3470.0 | | England | 221.0 |
| | F | 27 | 3067.0 | | 2570.0–3610.0 | | 413 | |
| Morus capensis | U | | 2700.0 | | | | 173 | 222.0 |
| Morus serrator | U | 50 | 2350.0 | | | | New Zealand 682 | 223.0 |
| Sula abotti | B | 10 | 1456.0 | 153.00 | 1300.0–1700.0 | | 413 | 224.0 |
| Sula nebouxii | M | 23 | 1283.0 | | 1100.0–1580.0 | | Galapagos Is. | 225.0 |
| | F | 28 | 1801.0 | | 1450.0–2230.0 | | 413 | |
| Sula variegata | U | | 1300.0 | | | | 171 | 226.0 |
| Sula dactylatra personata | M | 26 | 1880.0 | | 1503.0–2211.0 | B | Hawaii | 227.0 |
| | F | 27 | 2095.0 | | 1616.0–2353.0 | | 316 | |
| Sula dactylatra dactylatra | M | 2 | 1450.0 | | 1400.0–1500.0 | | St. Thomas | 227.1 |
| | F | 4 | 1584.0 | | 1450.0–1660.0 | | 177 | |
| Sula sula | M | 20 | 938.0 | | 850.0–1160.0 | | Galapagos Is. | 228.0 |
| | F | 20 | 1068.0 | | 850.0–1210.0 | | 413 | |

## Body Masses of World Birds (continued)

| Species | Sex | N | Mean | Std dev | Range | Sn | Location | Number |
|---|---|---|---|---|---|---|---|---|
| Sula leucogaster | M | 64 | 1093.0 | 230.00 | | B | Sonora, Mexico | 229.0 |
| | F | 69 | 1382.0 | 104.00 | | | 623a | |

### ORDER: PELECANIFORMES  FAMILY: PHALACROCORACIDAE

| Species | Sex | N | Mean | Std dev | Range | Sn | Location | Number |
|---|---|---|---|---|---|---|---|---|
| Phalacrocorax africanus | M | 1 | 685.0 | | | | | 230.0 |
| | F | 1 | 550.0 | | | | 115 | |
| Phalacrocorax coronatus | U | | 800.0 | | | | | 231.0 |
| | | | | | | | 173 | |
| Phalacrocorax pygmaeus | M | 3 | 743.0 | | 650.0–870.0 | | | 232.0 |
| | F | 3 | 615.0 | | 565.0–640.0 | | 115 | |
| Phalacrocorax melanoleucos | B | 8 | 684.0 | | 570.0–920.0 | | | 234.0 |
| | | | | | | | 506 | |
| Phalacrocorax penicillatus | U | 5 | 2103.0 | | | W | California, USA | 236.0 |
| | | | | | | | 16 | |
| Phalacrocorax harrisi | M | 3 | 3993.0 | | 3800.0–4090.0 | | Galapagos Is. | 237.0 |
| | F | 3 | 2787.0 | | 2730.0–2900.0 | | 572 | |
| Phalacrocorax neglectus | U | | 2000.0 | | | | | 238.0 |
| | | | | | | | 173 | |
| Phalacrocorax fuscescens | U | | | | 1360.0–2040.0 | | | 239.0 |
| | | | | | | | 550 | |
| Phalacrocorax brasilianus | M | 10 | 1260.0 | 12.60 | | | Panama | 240.0 |
| | F | 3 | 1070.0 | | | | 244 | |
| Phalacrocorax auritus | M | 33 | 1808.0 | 224.00 | | | Florida, USA | 241.0 |
| | F | 32 | 1540.0 | 215.00 | | | 244 | |
| Phalacrocorax varius | M | | 2200.0 | | 1815.0–2470.0 | | | 243.0 |
| | F | | 1730.0 | | 1360.0–2070.0 | | 78 | |
| Phalacrocorax sulcirostris | M | | 960.0 | | 795.0–1160.0 | | | 244.0 |
| | F | | 765.0 | | 650.0–1080.0 | | 78 | |
| Phalacrocorax carbo | M | 36 | 2283.0 | | 1975.0–2687.0 | B | Germany | 245.0 |
| | F | 17 | 1936.0 | | 1673.0–2174.0 | | 115 | |
| Phalacrocorax capensis | U | | 1200.0 | | | | | 249.0 |
| | | | | | | | 173 | |
| Phalacrocorax bougainvillii | U | | 1800.0 | | | | | 250.0 |
| | | | | | | | 171 | |
| Phalacrocorax verrucosus | U | 27 | 2630.0 | 373.00 | 2100.0–3300.0 | | | 251.0 |
| | | | | | | | 665 | |
| Phalacrocorax atriceps | U | | 2000.0 | | | | | 252.0 |
| | | | | | | | 78 | |
| Phalacrocorax campbelli | F | 1 | 1860.0 | | | B | New Zealand | 254.0 |
| | | | | | | | 671 | |
| Phalacrocorax urile | M | 2 | 2526.0 | | 2497.0–2554.0 | B | Alaska, USA | 261.0 |
| | F | 2 | 1788.0 | | 1646.0–1930.0 | | 445 | |
| Phalacrocorax pelagicus | M | 9 | 2034.0 | | 1816.0–2440.0 | PB | Alaska, USA | 262.0 |
| | F | 5 | 1702.0 | | 1475.0–2043.0 | | 445 | |

## Body Masses of World Birds (continued)

| Species | Sex | N | Mean | Std dev | Range | Sn | Location | Number |
|---|---|---|---|---|---|---|---|---|
| Phalacrocorax aristotelis | M | 4 | 1940.0 | | | B | Scotland | 263.0 |
| | F | 4 | 1598.0 | | | | 654 | |
| Phalacrocorax gaimardi | U | | 1300.0 | | | | | 264.0 |
| | | | | | | | 171 | |
| Phalacrocorax punctatus | U | | 1800.0 | | | | | 265.0 |
| | | | | | | | 11 | |

### ORDER: PELECANIFORMES    FAMILY: ANHINGIDAE

| Species | Sex | N | Mean | Std dev | Range | Sn | Location | Number |
|---|---|---|---|---|---|---|---|---|
| Anhinga anhinga | B | 26 | 1235.0 | | | | Florida, USA | 267.0 |
| | | | | | | | 244 | |
| Anhinga rufa | M | 7 | 1292.0 | | 948.0–1815.0 | | | 268.0 |
| | F | 2 | 1444.0 | | 1358.0–1530.0 | 82 | | |
| Anhinga melanogaster | M | 2 | 1436.0 | | 1058.0–1815.0 | | | 269.0 |
| | | | | | | 115 | | |

### ORDER: PELECANIFORMES    FAMILY: PELECANIDAE

| Species | Sex | N | Mean | Std dev | Range | Sn | Location | Number |
|---|---|---|---|---|---|---|---|---|
| Pelecanus onocrotalus | M | | | | –11000.0 | | | 271.0 |
| | F | | | | | | 149 | |
| Pelecanus rufescens | U | 83 | 5200.0 | | | | | 272.0 |
| | | | | | | | 115 | |
| Pelecanus crispus | U | | 9000.0 | | –13000.0 | | | 273.0 |
| | | | | | | | 149 | |
| Pelecanus conspicillatus | U | | 5000.0 | | 4000.0–7700.0 | | | 275.0 |
| | | | | | | | 550 | |
| Pelecanus erythrorhynchos | U | | 7000.0 | | 4500.0–13600.0 | | | 276.0 |
| | | | | | | | 78 | |
| Pelecanus occidentalis | M | 56 | 3702.0 | 389.00 | | | Florida, USA | 277.0 |
| | F | 47 | 3174.0 | 329.00 | | | 244 | |
| Pelecanus thagus | U | | 6000.0 | | | | | 278.0 |
| | | | | | | | 171 | |

### ORDER: ANSERIFORMES    FAMILY: ANHIMIDAE

| Species | Sex | N | Mean | Std dev | Range | Sn | Location | Number |
|---|---|---|---|---|---|---|---|---|
| Anhima cornuta | U | | 3150.0 | | | | Brazil | 279.0 |
| | | | | | | | 564 | |
| Chauna torquata | F | 1 | 4800.0 | | | | Brazil | 281.0 |
| | | | | | | | 37 | |

### ORDER: ANSERIFORMES    FAMILY: ANATIDAE

| Species | Sex | N | Mean | Std dev | Range | Sn | Location | Number |
|---|---|---|---|---|---|---|---|---|
| Anseranas semipalmata | F | | 2070.0 | | | | | 282.0 |
| | | | | | | | 504 | |
| Dendrocygna guttata | F | | 800.0 | | | | | 283.0 |
| | | | | | | | 504 | |
| Dendrocygna eytoni | F | | 792.0 | | | | | 284.0 |
| | | | | | | | 504 | |

## Body Masses of World Birds (continued)

| Species | Sex | N | Mean | Std dev | Range | Sn | Location | Number |
|---|---|---|---|---|---|---|---|---|
| Dendrocygna bicolor | B | 5 | 710.0 | | 590.0–771.0 | | southern USA 446 | 285.0 |
| Dendrocygna arcuata | F | | 732.0 | | | | 504 | 286.0 |
| Dendrocygna javanica | U | | | | 450.0–600.0 | | 5 | 287.0 |
| Dendrocygna viduata | U | 10 | 690.0 | 35.90 | 635.0–760.0 | | South Africa 255 | 288.0 |
| Dendrocygna arborea | F | | 1150.0 | | | | 504 | 289.0 |
| Dendrocygna autumnalis | M | 44 | 813.0 | | 680.0–952.0 | B | Texas, USA 446 | 290.0 |
| | F | 45 | 849.0 | | 652.0–1020.0 | | | |
| Thalassornis leuconotus | F | | 680.0 | | | | 504 | 291.0 |
| Oxyura dominica | U | 7 | 363.0 | | | | Panama 244 | 292.0 |
| Oxyura jamaicensis | M | 12 | 590.0 | | –816.0 | | | 293.0 |
| | F | 17 | 499.0 | | –635.0 | | 446 | |
| Oxyura leucocephala | M | 3 | 737.0 | | 553.0–865.0 | | Pakistan 115 | 295.0 |
| | F | 3 | 593.0 | | 539.0–631.0 | | | |
| Oxyura maccoa | F | | 677.0 | | | | 504 | 296.0 |
| Oxyura vittata | F | | 560.0 | | | | 504 | 297.0 |
| Oxyura australis | F | | 852.0 | | | | 504 | 298.0 |
| Biziura lobata | F | | 1551.0 | | | | 504 | 299.0 |
| Stictonetta naevosa | F | | 744.0 | | | | 504 | 300.0 |
| Cygnus olor | M | 59 | 11800.0 | 890.00 | 9200.0–14300.0 | W | England 485 | 301.0 |
| | F | 35 | 9670.0 | 640.00 | 7600.0–10600.0 | | | |
| Cygnus atratus | M | 270 | 6200.0 | | 4600.0–8700.0 | | | 302.0 |
| | F | 243 | 5100.0 | | 3700.0–7200.0 | | 541 | |
| Cygnus melanocoryphus | M | 8 | 5400.0 | | 4500.0–6700.0 | | | 303.0 |
| | F | 7 | 4000.0 | | 3500.0–4400.0 | | 541 | |
| Cygnus cygnus | B | 12 | 9350.0 | | 7400.0–14000.0 | | northern Europe 115 | 304.0 |
| Cygnus buccinator | M | 27 | 11400.0 | 727.00 | | | Montana, USA 20 | 305.0 |
| | F | 47 | 10300.0 | 1230.00 | | | | |
| Cygnus columbianus | M | 76 | 7100.0 | | 4700.0–9600.0 | | | 306.0 |
| | F | 86 | 6200.0 | | 4300.0–8200.0 | | 541 | |
| Cygnus columbianus bewickii | M | 96 | 6400.0 | | 4900.0–7800.0 | W | Britain 258 | 306.0 |
| | F | 95 | 5700.0 | | 3400.0–7200.0 | | | |

## Body Masses of World Birds (continued)

| Species | Sex | N | Mean | Std dev | Range | Sn | Location | Number |
|---|---|---|---|---|---|---|---|---|
| Coscoroba coscoroba | M | 2 | 4600.0 | | 3800.0–5400.0 | | | 307.0 |
| | F | 3 | 3800.0 | | 3200.0–4500.0 | | 541 | |
| Anser cygnoides | F | | 3150.0 | | | | | 308.0 |
| | | | | | | | 504 | |
| Anser brachyrhynchus | M | 750 | 2770.0 | 310.00 | 1900.0–3350.0 | F | Britain | 309.0 |
| | F | 796 | 2520.0 | 270.00 | 1810.0–3150.0 | | 115 | |
| Anser fabalis rossicus | M | 126 | 2668.0 | 233.00 | 1970.0–3390.0 | W | Netherlands | 310.0 |
| | F | 117 | 2374.0 | 203.00 | 2000.0–2800.0 | | 115 | |
| Anser fabalis fabalis | M | 68 | 3198.0 | 302.00 | 2690.0–4060.0 | W | Netherlands | 310.0 |
| | F | 58 | 2843.0 | 274.00 | 2220.0–3490.0 | | 115 | |
| Anser albifrons | M | 89 | 2703.0 | | 2330.0–3220.0 | F | Saskatchewan | 311.0 |
| | F | 79 | 2456.0 | | 1920.0–2830.0 | | 446 | |
| Anser erythropus | B | 10 | 1964.0 | 324.00 | 1400.0–2500.0 | | | 312.0 |
| | | | | | | | 115 | |
| Anser anser | M | 94 | 3509.0 | 321.00 | 2600.0–4560.0 | | Scotland | 313.0 |
| | F | 75 | 3108.0 | 274.00 | 2160.0–3800.0 | | 115 | |
| Anser indicus | B | | | | 2000.0–3200.0 | | | 314.0 |
| | | | | | | | 149 | |
| Anser caerulescens caerulescens | M | 467 | 2744.0 | 235.00 | | F | James Bay, Can. | 315.0 |
| | F | 422 | 2517.0 | 217.00 | | | 113 | |
| Anser caerulescens atlantica | M | 10 | 3450.0 | | | F | Quebec, Canada | 315.1 |
| | F | 18 | 3087.0 | | | | 446 | |
| Anser rossii | M | 31 | 1679.0 | | 1320.0–1880.0 | F | Saskatchewan | 316.0 |
| | F | 32 | 1500.0 | | 1270.0–1660.0 | | 446 | |
| Anser canagica | B | 14 | 2743.0 | | 2327.0–3122.0 | | Alaska, USA | 317.0 |
| | | | | | | | 446 | |
| Branta sandvicensis | M | | 2010.0 | 152.00 | | B | captive | 318.0 |
| | F | 27 | 1930.0 | 166.00 | | | 667 | |
| Branta canadensis canadensis | M | 232 | 3814.0 | | –6265.0 | | | 319.0 |
| | F | 159 | 3314.0 | | –5902.0 | | 412 | |
| Branta canadensis interior | M | 128 | 4181.0 | | 3799.0–4727.0 | W | Illinois, USA | 319.1 |
| | F | 121 | 3514.0 | | 3062.0–3912.0 | | 478 | |
| Branta canadensis moffitti | M | 99 | 4741.0 | | | S | Minnesota, USA | 319.2 |
| | F | 104 | 4044.0 | | | | 373 | |
| Branta canadensis parvipes | M | 113 | 2679.0 | | –4767.0 | | | 319.3 |
| | F | 129 | 2542.0 | | –3859.0 | | 412 | |
| Branta canadensis occidentalis | M | 175 | 3690.0 | | | | Alaska, USA | 319.4 |
| | F | 134 | 3043.0 | | | | 477 | |
| Branta canadensis leucopareia | B | 2 | 1940.0 | | 1927.0–1954.0 | B | Aleutian Is., USA | 319.5 |
| | | | | | | | 446 | |
| Branta canadensis hutchinsii | M | 31 | 2043.0 | | –2724.0 | | | 319.6 |
| | F | 37 | 1861.0 | | –2361.0 | | 412 | |
| Branta canadensis minima | M | 52 | 1480.0 | 29.80 | 1240.0–1700.0 | | California, USA | 319.7 |
| | F | 58 | 1264.0 | 29.40 | 940.0–1490.0 | | 479 | |

## Body Masses of World Birds (continued)

| Species | Sex | N | Mean | Std dev | Range | Sn | Location | Number |
|---|---|---|---|---|---|---|---|---|
| Branta leucopsis | M | 366 | 1788.0 | 163.00 | 1350.0–2230.0 | PB | Spitzbergen | 320.0 |
|  | F | 253 | 1586.0 | 122.00 | 1210.0–1950.0 |  | 444 |  |
| Branta bernicla | M | 430 | 1370.0 |  | 1080.0–1790.0 | PB | NW Terr., Can. | 321.0 |
|  | F | 361 | 1230.0 |  | 880.0–1590.0 |  | 62 |  |
| Branta ruficollis | M | 5 | 1375.0 |  | 1200.0–1625.0 |  |  | 322.0 |
|  | F | 2 | 1094.0 |  | 1058.0–1130.0 |  | 115 |  |
| Cereopsis novaehollandiae | F |  | 3560.0 |  |  |  |  | 323.0 |
|  |  |  |  |  |  |  | 504 |  |
| Cyanochen cyanopterus | F |  | 1520.0 |  |  |  |  | 324.0 |
|  |  |  |  |  |  |  | 504 |  |
| Chloephaga melanoptera | F |  | 2900.0 |  |  |  |  | 325.0 |
|  |  |  |  |  |  |  | 504 |  |
| Chloephaga picta | B | 2 | 2781.0 |  | 2724.0–2838.0 |  | Argentina | 326.0 |
|  |  |  |  |  |  |  | 493 |  |
| Chloephaga hybrida | M | 1 | 2611.0 |  |  |  | Argentina | 327.0 |
|  | F | 1 | 2043.0 |  |  |  | 493 |  |
| Chloephaga poliocephala | F |  | 2200.0 |  |  |  |  | 328.0 |
|  |  |  |  |  |  |  | 504 |  |
| Chloephaga rubidiceps | F |  | 2000.0 |  |  |  |  | 329.0 |
|  |  |  |  |  |  |  | 504 |  |
| Neochen jubata | F |  | 1250.0 |  |  |  |  | 330.0 |
|  |  |  |  |  |  |  | 504 |  |
| Alopochen aegyptiacus | M |  |  |  | 1900.0–2250.0 |  |  | 331.0 |
|  | F |  |  |  | 1500.0–1800.0 |  | 115 |  |
| Tadorna ferruginea | M | 15 | 1360.0 |  | 1200.0–1600.0 | F |  | 332.0 |
|  | F | 1 | 1100.0 |  |  |  | 149 |  |
| Tadorna cana | B | 8 | 1182.0 |  | 1012.0–1295.0 |  | South Africa | 333.0 |
|  |  |  |  |  |  |  | 255 |  |
| Tadorna tadornoides | F |  | 1290.0 |  |  |  |  | 334.0 |
|  |  |  |  |  |  |  | 504 |  |
| Tadorna variegata | F |  | 1300.0 |  |  |  |  | 335.0 |
|  |  |  |  |  |  |  | 504 |  |
| Tadorna tadorna | M | 11 | 1261.0 | 110.00 | 1100.0–1450.0 | S |  | 337.0 |
|  | F | 5 | 1043.0 |  | 926.0–1250.0 |  | 115 |  |
| Tadorna radjah | F |  | 839.0 |  |  |  |  | 338.0 |
|  |  |  |  |  |  |  | 504 |  |
| Tachyeres pteneres | F |  | 4228.0 |  |  |  |  | 339.0 |
|  |  |  |  |  |  |  | 504 |  |
| Tachyeres leucocephalus | F |  | 3013.0 |  |  |  |  | 340.0 |
|  |  |  |  |  |  |  | 504 |  |
| Tachyeres brachypterus | F |  | 3450.0 |  |  |  |  | 341.0 |
|  |  |  |  |  |  |  | 504 |  |
| Tachyeres patachonicus | F |  | 2346.0 |  |  |  |  | 342.0 |
|  |  |  |  |  |  |  | 504 |  |

## Body Masses of World Birds (continued)

| Species | Sex | N | Mean | Std dev | Range | Sn | Location | Number |
|---|---|---|---|---|---|---|---|---|
| Plectropterus gambensis | F | | 3560.0 | | | | | 343.0 |
| | | | | | | | 504 | |
| Cairina moschata | M | 3 | 2915.0 | | | | domestic | 344.0 |
| | F | 3 | 2022.0 | | | | 244 | |
| Cairina scutulata | M | | | | 2945.0–3855.0 | | | 345.0 |
| | F | | | | 2150.0–3050.0 | | 5 | |
| Pteronetta hartlaubii | F | | 790.0 | | | | | 346.0 |
| | | | | | | | 504 | |
| Sarkidiornis melanotos | M | | 2610.0 | | | | | 347.0 |
| | F | | | | 1925.0–2325.0 | | 5 | |
| Nettapus pulchellus | F | | 304.0 | | | | | 348.0 |
| | | | | | | | 504 | |
| Nettapus coromandelianus | M | | | | 255.0–312.0 | | | 349.0 |
| | F | | | | 185.0–255.0 | | 5 | |
| Nettapus auritus | U | 5 | 257.0 | | 220.0–290.0 | | | 350.0 |
| | | | | | | | 78 | |
| Callonetta leucophrys | M | 2 | 423.0 | | 370.0–450.0 | | | 351.0 |
| | F | 3 | 321.0 | | 285.0–340.0 | | 37, 111 | |
| Aix sponsa | M | 248 | 681.0 | | –907.0 | F | | 352.0 |
| | F | 163 | 635.0 | | –908.0 | | 446 | |
| Aix galericulata | M | | 628.0 | | 571.0–693.0 | | | 353.0 |
| | F | | 512.0 | | 428.0–608.0 | | 149 | |
| Chenonetta jubata | U | | 870.0 | | | | | 354.0 |
| | | | | | | | 194 | |
| Amazonetta brasiliensis | U | | 500.0 | | | | Brazil | 355.0 |
| | | | | | | | 564 | |
| Merganetta armata | F | | 330.0 | | | | | 356.0 |
| | | | | | | | 504 | |
| Hymenolaimus malacorhynchus | F | | 810.0 | | | | | 357.0 |
| | | | | | | | 504 | |
| Salvadorina waigiuensis | F | | 469.0 | | | | | 358.0 |
| | | | | | | | 504 | |
| Anas penelope | M | 42 | 819.0 | | 610.0–1073.0 | | Denmark | 359.0 |
| | F | 24 | 724.0 | | 552.0–962.0 | | 446 | |
| Anas americana | M | 65 | 792.0 | 79.10 | 635.0–1036.0 | Y | western Canada | 360.0 |
| | F | 68 | 719.0 | 80.60 | 512.0–872.0 | | 684 | |
| Anas sibilatrix | F | | 828.0 | | | | | 361.0 |
| | | | | | | | 504 | |
| Anas falcata | M | 4 | 713.0 | | 590.0–770.0 | | China | 362.0 |
| | F | 5 | 585.0 | | 422.0–700.0 | | 115 | |
| Anas strepera | M | 16 | 990.0 | | | F | Illinois, USA | 363.0 |
| | F | 14 | 849.0 | | | | 446 | |
| Anas formosa | U | | | | 500.0–600.0 | | | 364.0 |
| | | | | | | | 149 | |

## Body Masses of World Birds (continued)

| Species | Sex | N | Mean | Std dev | Range | Sn | Location | Number |
|---|---|---|---|---|---|---|---|---|
| Anas crecca | M | 194 | 364.0 | | −454.0 | F | | 365.0 |
| | F | 81 | 318.0 | | −409.0 | | 446 | |
| Anas flavirostris | F | | 395.0 | | | | | 366.0 |
| | | | | | | | 504 | |
| Anas capensis | U | 63 | 402.0 | | 342.0−590.5 | | South Africa 115 | 367.0 |
| Anas gibberifrons | M | 218 | 508.0 | 38.00 | | | | 369.0 |
| | F | 153 | 469.0 | 36.00 | | | 346 | |
| Anas castanea | M | 89 | 660.0 | 65.00 | | | | 371.0 |
| | F | 58 | 590.0 | 138.00 | | | 346 | |
| Anas nesiotis | M | 1 | 426.0 | | | | Campbell Is. 346 | 372.1 |
| Anas aucklandica aucklandica | M | 10 | 521.0 | 49.00 | | | Auckland Is. 346 | 372.2 |
| | F | 6 | 410.0 | | | | | |
| Anas chlorotis | B | 42 | 582.0 | | | | New Zealand 346 | 372.0 |
| Anas platyrhynchos | U | 5847 | 1082.0 | 129.00 | 720.0−1580.0 | Y | England 443 | 373.0 |
| Anas platyrhynchos diazi | M | | | | 960.0−1060.0 | | | 373.1 |
| | F | | | | 815.0−990.0 | | 446 | |
| Anas platyrhynchos laysanensis | M | 5 | 463.0 | | | | | 374.0 |
| | F | 5 | 427.0 | | | | 667 | |
| Anas platyrhynchos wyvilliana | M | 28 | 644.0 | | | | Hawaiian Is. 667 | 375.0 |
| | F | 19 | 585.0 | | | | | |
| Anas fulvigula | M | 30 | 1030.0 | 107.00 | −1280.0 | | Florida, USA 446 | 376.0 |
| | F | 11 | 968.0 | 76.00 | −1132.0 | | | |
| Anas rubripes | M | 376 | 1400.0 | | 900.0−1800.0 | F | Massachusetts 446 | 377.0 |
| | F | 176 | 1100.0 | | 900.0−1500.0 | | | |
| Anas undulata | U | 42 | 1008.0 | 121.00 | 660.0−1220.0 | | South Africa 255 | 378.0 |
| Anas poecilorhyncha | M | | | | 1230.0−1500.0 | | | 380.0 |
| | F | | | | 790.0−1360.0 | | 5 | |
| Anas luzonica | M | 7 | 906.0 | | 852.0−977.0 | | Philippines 475 | 381.0 |
| | F | 9 | 779.0 | | 752.0−818.0 | | | |
| Anas superciliosa pelewensis | F | 2 | 465.0 | | | | New Britain 219 | 382.0 |
| Anas superciliosa | M | | 1114.0 | | | | Australia 693 | 382.0 |
| | F | | 1025.0 | | | | | |
| Anas sparsa | F | | 909.0 | | | | 504 | 383.0 |
| Anas specularoides | F | | 900.0 | | | | 504 | 384.0 |
| Anas specularis | F | | 975.0 | | | | 504 | 385.0 |

## Body Masses of World Birds (continued)

| Species | Sex | N | Mean | Std dev | Range | Sn | Location | Number |
|---|---|---|---|---|---|---|---|---|
| Anas acuta | M | 232 | 1035.0 | | | F | Illinois, USA | 386.0 |
| | F | 60 | 986.0 | | | | 446 | |
| Anas eatoni | U | 38 | 571.0 | 68.00 | 455.0–695.0 | | 665 | 386.1 |
| Anas georgica spinicauda | F | | 706.0 | | | | 504 | 388.0 |
| Anas georgica georgica | M | 3 | 632.0 | | 610.0–660.0 | B | | 388.0 |
| | F | 2 | 535.0 | | 460.0–610.0 | | 666 | |
| Anas bahamensis galapagoensis | B | 2 | 438.0 | | 425.0–450.0 | | 667 | 389.0 |
| Anas bahamensis | B | 5 | 535.0 | | 475.0–633.0 | | Surinam 245 | 389.0 |
| Anas erythrorhyncha | F | | 523.0 | | | | 504 | 390.0 |
| Anas versicolor | F | | 373.0 | | | | 504 | 392.0 |
| Anas hottentota | F | | 240.0 | | | | 504 | 393.0 |
| Anas querquedula | B | 200 | 326.0 | | | | France 115 | 394.0 |
| Anas discors | M | 105 | 409.0 | | –590.0 | F | | 395.0 |
| | F | 101 | 363.0 | | –545.0 | | 446 | |
| Anas cyanoptera | M | 26 | 408.0 | | –549.0 | F | | 396.0 |
| | F | 19 | 363.0 | | –499.0 | | 446 | |
| Anas platalea | F | | 523.0 | | | | 504 | 397.0 |
| Anas smithii | F | | 598.0 | | | | 504 | 398.0 |
| Anas rhynchotis | F | | 665.0 | | | | 504 | 399.0 |
| Anas clypeata | M | 90 | 636.0 | | –908.0 | F | | 400.0 |
| | F | 71 | 590.0 | | –726.0 | | 446 | |
| Malacorhynchus membranaceus | F | | 344.0 | | | | 504 | 401.0 |
| Marmaronetta angustirostris | U | | 477.0 | | 420.0–500.0 | W | USSR 149 | 402.0 |
| Rhodonessa caryophyllacea | U | | | | 793.0–1360.0 | | 5 | 403.0 |
| Netta rufina | B | 29 | 1118.0 | | 990.0–1300.0 | S | USSR 115 | 404.0 |
| Netta peposaca | U | | 1000.0 | | | | Brazil 564 | 405.0 |
| Netta erythrophthalma | F | | 822.0 | | | | 504 | 406.0 |

## Body Masses of World Birds (continued)

| Species | Sex | N | Mean | Std dev | Range | Sn | Location | Number |
|---|---|---|---|---|---|---|---|---|
| Aythya ferina | B | 321 | 823.0 | | 467.0–1240.0 | W | France 115 | 407.0 |
| Aythya valisineria | M | 743 | 1248.0 | | | W | Maryland | 408.0 |
| | F | 304 | 1190.0 | | | | 419 | |
| Aythya americana | M | 1157 | 1100.0 | | | S | | 409.0 |
| | F | 485 | 990.0 | | | | 446 | |
| Aythya collaris | M | 285 | 730.0 | | –1090.0 | | | 410.0 |
| | F | 151 | 680.0 | | –1180.0 | | 446 | |
| Aythya nyroca | B | 13 | 574.0 | | 470.0–740.0 | S | USSR; France 115 | 411.0 |
| Aythya baeri | F | | 708.0 | | | | 504 | 413.0 |
| Aythya australis | B | | | | 838.0–902.0 | | 693 | 414.0 |
| Aythya fuligula | B | 1043 | 694.0 | | 400.0–950.0 | W | France 115 | 415.0 |
| Aythya novaeseelandiae | F | | 610.0 | | | | 504 | 416.0 |
| Aythya marila | M | 17 | 932.0 | | 844.0–1046.0 | B | Alaska, USA | 417.0 |
| | F | 9 | 957.0 | | 856.0–1117.0 | | 278 | |
| Aythya affinis | M | 112 | 850.0 | | 620.0–1050.0 | | | 418.0 |
| | F | 118 | 790.0 | | 540.0–960.0 | | 446 | |
| Somateria mollissima | M | 22 | 2218.0 | | 1384.0–2800.0 | B | N. Atlantic | 419.0 |
| | F | 32 | 1915.0 | | 1192.0–2895.0 | | 36 | |
| Somateria spectabilis | M | 41 | 1668.0 | | | F | Alaska | 420.0 |
| | F | 141 | 1567.0 | | | | 446 | |
| Somateria fischeri | M | 18 | 1432.0 | 117.00 | 1231.0–1700.0 | B | Alaska | 421.0 |
| | F | 14 | 1304.0 | 170.00 | 1075.0–1675.0 | | 177 | |
| Polysticta stelleri | M | 48 | 773.0 | | 628.0–900.0 | B | Russia | 422.0 |
| | F | 42 | 842.0 | | 651.0–1000.0 | | 446 | |
| Histrionicus histrionicus | M | 9 | 687.0 | | 636.0–770.0 | B | Alaska | 424.0 |
| | F | 8 | 558.0 | | 520.0–594.0 | | 446 | |
| Clangula hyemalis | M | 661 | 932.0 | | | W | Michigan | 425.0 |
| | F | 636 | 814.0 | | | | 446 | |
| Melanitta nigra | M | 8 | 1100.0 | | –1270.0 | | | 426.0 |
| | F | 4 | 800.0 | | –1100.0 | | 446 | |
| Melanitta perspicillata | M | 12 | 1000.0 | | –1100.0 | | | 427.0 |
| | F | 10 | 900.0 | | –1100.0 | | 446 | |
| Melanitta fusca | B | 12 | 1757.0 | | 1519.0–1980.0 | | 115 | 428.0 |
| Melanitta fusca deglandi | M | 13 | 1500.0 | | –1800.0 | | | 428.0 |
| | F | 19 | 1200.0 | | –1500.0 | | 446 | |
| Bucephala clangula | M | 58 | 1000.0 | | –1400.0 | | | 429.0 |
| | F | 53 | 800.0 | | –1100.0 | | 446 | |

## Body Masses of World Birds (continued)

| Species | Sex | N | Mean | Std dev | Range | Sn | Location | Number |
|---|---|---|---|---|---|---|---|---|
| Bucephala islandica | M | 3 | 1090.0 | | −1300.0 | | | 430.0 |
| | F | 7 | 730.0 | | −860.0 | | 446 | |
| Bucephala albeola | M | 29 | 473.0 | 32.80 | 424.0−551.0 | | Oregon | 431.0 |
| | F | 16 | 334.0 | 23.20 | 297.0−374.0 | | 252 | |
| Mergellus albellus | M | | 652.0 | | 540.0−825.0 | F | USSR | 432.0 |
| | F | | 568.0 | | 515.0−630.0 | | 115 | |
| Lophodytes cucullatus | M | 24 | 680.0 | | −910.0 | | | 433.0 |
| | F | 20 | 540.0 | | −680.0 | | 446 | |
| Mergus octosetaceus | U | | 983.0 | | | | estimated 345 | 434.0 |
| Mergus serrator | M | 18 | 1135.0 | | −1317.0 | | | 435.0 |
| | F | 17 | 908.0 | | −1271.0 | | 446 | |
| Mergus squamatus | U | | 1234.0 | | | | estimated 345 | 436.0 |
| Mergus merganser | M | 13 | 1709.0 | | 1528.0−2054.0 | F | | 437.0 |
| | F | 11 | 1232.0 | | 1050.0−1362.0 | | 446 | |
| Mergus australis | U | | 898.0 | | | | estimated 345 | 438.0 |
| Heteronetta atricapilla | M | 1 | 460.0 | | | | Brazil | 439.0 |
| | F | 1 | 605.0 | | | | 37 | |

### ORDER: PHOENICOPTERIFORMES    FAMILY: PHOENICOPTERIDAE

| Species | Sex | N | Mean | Std dev | Range | Sn | Location | Number |
|---|---|---|---|---|---|---|---|---|
| Phoenicopterus ruber roseus | M | 13 | 3540.0 | | 3100.0−4100.0 | B | | 440.0 |
| | F | 12 | 2530.0 | | 2100.0−3300.0 | | 23 | |
| Phoenicopterus minor | U | | 1900.0 | | | | | 442.0 |
| | | | | | | | 115 | |

### ORDER: CICONIIFORMES    FAMILY: ARDEIDAE

| Species | Sex | N | Mean | Std dev | Range | Sn | Location | Number |
|---|---|---|---|---|---|---|---|---|
| Syrigma sibilatrix | F | 1 | 370.0 | | | | Brazil 37 | 445.0 |
| Egretta rufescens | U | | 450.0 | | | | | 446.0 |
| | | | | | | | 570a | |
| Egretta vinaceigula | U | 5 | 288.0 | | 250.0−340.0 | | Zambia 167 | 447.0 |
| Egretta ardesiaca | U | 7 | 324.0 | | 270.0−390.0 | | Zambia 167 | 448.0 |
| Egretta tricolor | M | 35 | 415.0 | 45.50 | | | | 449.0 |
| | F | 5 | 334.0 | | | | 244 | |
| Egretta novaehollandiae | U | | 500.0 | | | | | 450.0 |
| | | | | | | | 78 | |
| Egretta caerulea | M | 11 | 364.0 | 47.10 | | | | 451.0 |
| | F | 8 | 315.0 | | | | 243 | |
| Egretta garzetta | U | | 500.0 | | | | | 452.0 |
| | | | | | | | 149 | |

## Body Masses of World Birds (continued)

| Species | Sex | N | Mean | Std dev | Range | Sn | Location | Number |
|---|---|---|---|---|---|---|---|---|
| Egretta gularis | M | 1 | 400.0 | | | | 115 | 453.0 |
| Egretta thula | U | 17 | 371.0 | 25.00 | | | 244 | 455.0 |
| Egretta eulophotes | U | | 500.0 | | | | 78 | 456.0 |
| Egretta sacra | B | 2 | 356.0 | | 341.0–370.0 | | New Britain 219 | 457.0 |
| Egretta intermedia | U | | 500.0 | | | | 78 | 458.0 |
| Pilherodius pileatus | M F | 2 | 554.0 | | 518.0–591.0 | | 245, 246 | 459.0 |
| Ardea cinerea | B | 30 | 1443.0 | | 1020.0–2073.0 | | Netherlands 115 | 460.0 |
| Ardea herodias | M F | 17 15 | 2576.0 2204.0 | 299.00 337.00 | | | 244 | 461.0 |
| Ardea cocoi | U | | 3200.0 | | | | Brazil 564 | 462.0 |
| Ardea pacifica | U | | 650.0 | | | | 78 | 463.0 |
| Ardea melanocephala | U | 6 | 1060.0 | | 710.0–1650.0 | | 78 | 464.0 |
| Ardea goliath | U | 3 | 4468.0 | | 4310.0–4750.0 | | 78 | 466.0 |
| Ardea sumatrana | U | | 2600.0 | | | | 78 | 468.0 |
| Ardea purpurea | M F | | | | 617.0–1218.0 525.0–1135.0 | | Netherlands 115 | 469.0 |
| Ardea picata | M F | 2 2 | 264.0 234.0 | | 247.0–280.0 225.0–242.0 | | New Guinea 217 | 470.0 |
| Casmerodius albus | M F | 12 9 | 935.0 812.0 | 134.00 | | | 244 | 471.0 |
| Bubulcus ibis | U | 9 | 338.0 | | | | 244 | 472.0 |
| Ardeola ralloides | B | 23 | 287.0 | | 230.0–370.0 | | 115 | 473.0 |
| Ardeola grayii | U | 2 | 253.0 | | 230.0–276.0 | | 115, 361 | 474.0 |
| Butorides striatus | U | 34 | 212.0 | 5.92 | | | Florida, USA 244 | 479.0 |
| Agamia agami | M F | 3 1 | 535.0 475.0 | | | | Panama 244 | 482.0 |
| Nyctanassa violaceus | M F | 8 7 | 716.0 649.0 | | | | Florida, USA 243 | 483.0 |

## Body Masses of World Birds (continued)

| Species | Sex | N | Mean | Std dev | Range | Sn | Location | Number |
|---|---|---|---|---|---|---|---|---|
| Nycticorax nycticorax | B | 5 | 883.0 | | 727.0–1014.0 | B | NE USA 445 | 484.0 |
| Nycticorax caledonicus | M | 2 | 912.0 | | 810.0–1014.0 | | 475, 506 | 485.0 |
| Gorsachius goisagi | U | 1 | 527.0 | | | | Philippines 475 | 488.0 |
| Cochlearius cochlearius | M | 9 | 712.0 | | 680.0–770.0 | | | 490.0 |
| | F | 6 | 602.0 | | 503.0–726.0 | | 162, 593 | |
| Tigrisoma mexicanum | M | 3 | 1274.0 | | | | Panama | 491.0 |
| | F | 4 | 1046.0 | | | | 244 | |
| Tigrisoma fasciatum | U | | 850.0 | | | | 603a | 492.0 |
| Tigrisoma lineatum | M | 5 | 897.0 | | | | Panama | 493.0 |
| | F | 6 | 823.0 | | | | 244 | |
| Zebrilus undulatus | F | 1 | 123.0 | | | | Peru 589 | 496.0 |
| Ixobrychus involucris | U | 2 | 88.7 | | 73.4–104.0 | | Brazil; Peru 37, 193 | 497.0 |
| Ixobrychus minutus | B | 18 | 148.0 | | 140.0–150.0 | B | 115 | 498.0 |
| Ixobrychus sinensis | M | 1 | 104.0 | | | | | 499.0 |
| | F | 1 | 92.0 | | | | 290 | |
| Ixobrychus exilis | U | 20 | 86.3 | 4.26 | | | 244 | 500.0 |
| Ixobrychus cinnamomeus | M | 1 | 106.0 | | | | Philippines 475 | 502.0 |
| Ixobrychus sturmii | U | | 140.0 | | | | 78 | 503.0 |
| Ixobrychus flavicollis | B | 7 | 318.0 | | 275.0–358.0 | | 219, 393, 475 | 504.0 |
| Botaurus lentiginosus | U | 16 | 706.0 | 183.00 | 520.0–1072.0 | B | Ontario 590 | 505.0 |
| Botaurus pinnatus | F | 1 | 584.0 | | | | Mexico 584 | 506.0 |
| Botaurus stellaris | M | | 966.0 | | | W | USSR | 507.0 |
| | F | | 867.0 | | | | 149 | |

**ORDER: CICONIIFORMES    FAMILY: BALAENICIPIDIDAE**

| Species | Sex | N | Mean | Std dev | Range | Sn | Location | Number |
|---|---|---|---|---|---|---|---|---|
| Balaeniceps rex | M | 1 | 6700.0 | | | | | 509.0 |
| | F | 4 | 5268.0 | | 4360.0–5900.0 | | 313 | |

**ORDER: CICONIIFORMES    FAMILY: SCOPIDAE**

| Species | Sex | N | Mean | Std dev | Range | Sn | Location | Number |
|---|---|---|---|---|---|---|---|---|
| Scopus umbretta | U | | | | 415.0–430.0 | | 82 | 510.0 |

## Body Masses of World Birds (continued)

| Species | Sex | N | Mean | Std dev | Range | Sn | Location | Number |
|---------|-----|---|------|---------|-------|----|----------|--------|
| ORDER: CICONIIFORMES | | | | | FAMILY: THRESKIORNITHIDAE | | | |
| Eudocimus albus | M | 12 | 1036.0 | 105.00 | 873.0–1261.0 | B | Florida, USA | 511.0 |
| | F | 16 | 764.0 | 68.40 | 593.0–864.0 | | 328 | |
| Eudocimus ruber | M | 13 | 588.0 | | 505.0–640.0 | | Trinidad | 512.0 |
| | F | 8 | 741.0 | | 710.0–770.0 | | 188 | |
| Phimosus infuscatus | B | 3 | 575.0 | | 550.0–600.0 | | Brazil | 513.0 |
| | | | | | | | 37 | |
| Plegadis falcinellus | M | | | | 557.0–768.0 | S | USSR | 514.0 |
| | F | | | | 530.0–680.0 | | 115 | |
| Plegadis chihi | M | 32 | 697.0 | 58.90 | 563.0–807.0 | B | Utah, USA | 515.0 |
| | F | 35 | 546.0 | 45.30 | 433.0–677.0 | | 177 | |
| Plegadis ridgwayi | M | 1 | 608.0 | | | | Bolivia; Peru | 516.0 |
| | F | 2 | 527.0 | | 500.0–554.0 | | 584 | |
| Theristicus caerulescens | M | 1 | 1500.0 | | | | Bolivia | 518.0 |
| | | | | | | | 584 | |
| Theristicus caudatus | F | 1 | 1550.0 | | | | Brazil | 519.0 |
| | | | | | | | 37 | |
| Mesembrinibis cayennensis | U | 1 | 670.0 | | | | Peru | 522.0 |
| | | | | | | | 193 | |
| Bostrychia hagedash | B | 5 | 1168.0 | | 798.0–1512.0 | | | 523.0 |
| | | | | | | | 588 | |
| Geronticus eremita | B | 11 | 1200.0 | 90.7 | 1000.0–1350.0 | | captive | 527.0 |
| | | | | | | | 163, 353a | |
| Threskiornis aethiopicus | M | 40 | 1618.0 | 140.00 | 1268.0–1963.0 | | South Africa | 530.0 |
| | F | 54 | 1378.0 | 126.00 | 1131.0–1718.0 | | 351 | |
| Threskiornis molucca | U | | 1800.0 | | | | | 533.0 |
| | | | | | | | 78 | |
| Threskiornis spinicollis | U | | 1800.0 | | | | | 534.0 |
| | | | | | | | 78 | |
| Platalea leucorodia | B | 7 | 1892.0 | | 1656.0–2080.0 | | | 539.0 |
| | | | | | | | 313 | |
| Platalea regia | M | 6 | | | 1650.0–2070.0 | | | 540.0 |
| | F | 4 | | | 1400.0–1800.0 | | 313 | |
| Platalea alba | B | 5 | 1521.0 | | 1020.0–1900.0 | | | 541.0 |
| | | | | | | | 163, 255 | |
| Platalea flavipes | M | 4 | 1895.0 | | 1750.0–2000.0 | | | 543.0 |
| | F | 2 | 1600.0 | | 1500.0–1700.0 | | 313 | |
| Ajaia ajaja | M | | | | 1240.0–1750.0 | | Surinam | 544.0 |
| | F | | | | 1400.0–1700.0 | | 247 | |
| ORDER: CICONIIFORMES | | | | | FAMILY: CICONIIDAE | | | |
| Mycteria americana | M | 9 | 2702.0 | | | | | 545.0 |
| | F | 1 | 2050.0 | | | | 244 | |

## Body Masses of World Birds (continued)

| Species | Sex | N | Mean | Std dev | Range | Sn | Location | Number |
|---|---|---|---|---|---|---|---|---|
| Mycteria ibis | M | 1 | 2384.0 | | | | | 547.0 |
| | F | 6 | 1949.0 | | 1190.0–2330.0 | | 313, 553 | |
| Mycteria leucocephala | B | 10 | 3180.0 | | 3030.0–3370.0 | | 313 | 548.0 |
| Anastomus lamelligerus | U | 9 | 1120.0 | | 628.0–1400.0 | | 78 | 550.0 |
| Ciconia nigra | U | | 3000.0 | | | | 115 | 551.0 |
| Ciconia abdimii | B | 10 | 1373.0 | 156.00 | 1071.0–1570.0 | | captive 163 | 552.0 |
| Ciconia episcopus | M | 1 | 2185.0 | | | | India 476 | 553.0 |
| | F | | | | | | | |
| Ciconia maguari | M | 9 | 4200.0 | | | | | 555.0 |
| | F | 5 | 3800.0 | | | | 313 | |
| Ciconia ciconia | B | 68 | 3473.0 | | 2275.0–4400.0 | PB | Germany 115 | 556.0 |
| Ciconia boyciana | M | 7 | 5014.0 | | 4200.0–5900.0 | | captive | 557.0 |
| | F | 8 | 4687.0 | | 3400.0–5200.0 | | 404 | |
| Ephippiorhynchus asiaticus | U | | 4100.0 | | | | 78 | 558.0 |
| Ephippiorhynchus senegalensis | M | 5 | 6378.0 | | 5085.0–7524.0 | | | 559.0 |
| | F | 3 | 5947.0 | | 5000.0–6840.0 | | 313, 588 | |
| Jabiru mycteria | M | 8 | 6892.0 | | 5902.0–8100.0 | | | 560.0 |
| | F | 4 | 5217.0 | | 4300.0–6356.0 | | 313, 553 | |
| Leptoptilos javanicus | B | 1 | 4994.0 | | | | captive | 561.0 |
| | | 1 | 4540.0 | | | | 553 | |
| Leptoptilos crumeniferus | M | 37 | | | 5600.0–8900.0 | | | 562.0 |
| | F | 22 | | | 4000.0–6800.0 | | 313 | |

### ORDER: FALCONIFORMES      FAMILY: CATHARTIDAE

| Species | Sex | N | Mean | Std dev | Range | Sn | Location | Number |
|---|---|---|---|---|---|---|---|---|
| Coragyps atratus | M | 6 | 2172.0 | | | | Florida, USA | 564.0 |
| | F | 6 | 1989.0 | | | | 243 | |
| Cathartes aura | U | 20 | 1467.0 | 132.00 | | | Florida, USA 244 | 565.0 |
| Cathartes burrovianus | B | 5 | 953.0 | | 820.0–1272.0 | | 48, 111, 622 | 566.0 |
| Cathartes melambrotus | U | | 1200.0 | | | | Peru 623 | 567.0 |
| Gymnogyps californianus | U | 9 | 10104.0 | | 8535.0–14074.0 | | California, USA 326 | 568.0 |
| Vultur gryphus | M | | 12500.0 | | 10900.0–13600.0 | | | 569.0 |
| | F | | 10100.0 | | 9600.0–11400.0 | | 650 | |
| Sarcoramphus papa | B | | 3400.0 | | 3100.0–3700.0 | | 650 | 570.0 |

## Body Masses of World Birds (continued)

| Species | Sex | N | Mean | Std dev | Range | Sn | Location | Number |
|---------|-----|---|------|---------|-------|----|----------|--------|
| **ORDER: FALCONIFORMES** | | | | | **FAMILY: PANDIONIDAE** | | | |
| Pandion haliaetus | M | 10 | 1403.0 | | 1220.0–1600.0 | | | 571.0 |
| | F | 14 | 1568.0 | | 1250.0–1900.0 | 81 | | |
| **ORDER: FALCONIFORMES** | | | | | **FAMILY: ACCIPITRIDAE** | | | |
| Aviceda cuculoides | U | 1 | 296.0 | | | | Zimbabwe 285 | 572.0 |
| Aviceda jerdoni | M | 1 | 353.0 | | | | Philippines 475 | 574.0 |
| Aviceda subcristata | B | 11 | 294.0 | | | | 30, 35, 219, 267 | 575.0 |
| Leptodon cayanensis | B | 5 | 484.0 | | 435.0–540.0 | | 188, 244, 245, 311, 510 | 576.0 |
| Chondrohierax uncinatus | B | 7 | 278.0 | | | | Panama 244 | 578.0 |
| Henicopernis longicauda | M | 1 | 447.0 | | | | | 579.0 |
| | F | 1 | 730.0 | | | 81 | | |
| Pernis apivorus | M | 10 | 684.0 | | 510.0–800.0 | | | 581.0 |
| | F | 8 | 832.0 | | 625.0–1050.0 | 81 | | |
| Pernis ptilorhynchus | U | 9 | 1066.0 | | 750.0–1490.0 | | 149 | 582.0 |
| Pernis celebensis | B | 2 | 724.0 | | 696.0–752.0 | | Philippines 475 | 583.0 |
| Elanoides forficatus | B | 14 | 442.0 | | 372.0–510.0 | | 447 | 584.0 |
| Machaeramphus alcinus | U | 1 | 650.0 | | | | 82 | 585.0 |
| Gampsonyx swainsonii | U | 3 | 92.5 | | 80.5–104.0 | | 111, 188, 625 | 586.0 |
| Elanus caeruleus | M | 2 | 316.0 | | 311.0–322.0 | | | 587.0 |
| | F | 5 | 350.0 | | 332.0–375.0 | | 583, 584, 586 | |
| Elanus notatus | U | | 250.0 | | | | 194 | 588.0 |
| Elanus scriptus | U | | 241.0 | | | | 550 | 590.0 |
| Rostrhamus sociabilis | U | 7 | 378.0 | | | | Surinam; Mexico 48, 642 | 592.0 |
| Rostrhamus hamatus | B | 5 | 421.0 | | 377.0–450.0 | | Surinam 245, 246 | 593.0 |
| Harpagus bidentatus | M | 4 | 182.0 | | 165.0–210.0 | | | 594.0 |
| | F | 3 | 239.0 | | 260.0–305.0 | | 193, 244, 510 | |
| Ictinia mississippiensis | M | 11 | 245.0 | | 216.0–269.0 | | Oklahoma, USA | 596.0 |
| | F | 5 | 311.0 | | 278.0–339.0 | | 447 | |

## Body Masses of World Birds (continued)

| Species | Sex | N | Mean | Std dev | Range | Sn | Location | Number |
|---|---|---|---|---|---|---|---|---|
| Ictinia plumbea | B | 10 | 247.0 | | 190.0–272.0 | | Trinidad 188 | 597.0 |
| Lophoictinia isura | M | 1 | 501.0 | | | | Australia 262 | 598.0 |
| Milvus milvus | M | 10 | 947.0 | | 802.0–1052.0 | | | 600.0 |
| | F | 4 | 1213.0 | | 1140.0–1298.0 | | 116 | |
| Milvus migrans | M | 30 | 827.0 | | 630.0–941.0 | | 116 | 601.0 |
| Haliastur sphenurus | U | | 800.0 | | | | 194 | 603.0 |
| Haliastur indus | M | 1 | 587.0 | | | | | 604.0 |
| | F | 2 | 450.0 | | 434.0–466.0 | | 218, 475 | |
| Haliaeetus leucogaster | F | 2 | 2638.0 | | 2475.0–2800.0 | | 78 | 605.0 |
| Haliaeetus sanfordi | F | 3 | | | 2300.0–2500.0 | | 81 | 606.0 |
| Haliaeetus vocifer | M | 3 | | | 1928.0–2497.0 | | | 607.0 |
| | F | 2 | 3400.0 | | 3170.0–3630.0 | | 116 | |
| Haliaeetus leucoryphus | M | 8 | 2527.0 | | 2040.0–3278.0 | | | 609.0 |
| | F | 7 | 3088.0 | | 2100.0–3700.0 | | 116 | |
| Haliaeetus albicilla | M | 12 | 4014.0 | 755.00 | 3075.0–5430.0 | | | 610.0 |
| | F | 18 | 5572.0 | 980.00 | 4080.0–6920.0 | | 116 | |
| Haliaeetus leucocephalus | M | 35 | 4130.0 | | 3637.0–4819.0 | | | 611.0 |
| | F | 37 | 5350.0 | | 3631.0–6400.0 | | 447 | |
| Haliaeetus pelagicus | B | 3 | 7757.0 | | 6800.0–8970.0 | | 149 | 612.0 |
| Ichthyophaga ichthyaetus | M | 1 | 1590.0 | | | | | 614.0 |
| | F | | | | 2270.0–2700.0 | | 81 | |
| Gypohierax angolensis | U | 6 | 1600.0 | | 1361.0–1712.0 | | 78 | 615.0 |
| Gypaetus barbatus | U | 10 | 5680.0 | | 5000.0–6750.0 | | Europe 149 | 616.0 |
| Neophron percnopterus | B | 9 | 2120.0 | | 1829.0–2400.0 | | 116, 149 | 617.0 |
| Necrosyrtes monachus | U | | | | 1524.0–2102.0 | | 116 | 618.0 |
| Gyps africanus | U | | 5300.0 | | | | 269 | 619.0 |
| Gyps rueppellii | U | | 7400.0 | | | | 269 | 623.0 |
| Gyps himalayensis | U | | | | 8000.0–12000.0 | | 81 | 624.0 |
| Gyps fulvus | B | 15 | 7436.0 | | 6200.0–8500.0 | | SE Europe 116 | 625.0 |

## Body Masses of World Birds (continued)

| Species | Sex | N | Mean | Std dev | Range | Sn | Location | Number |
|---|---|---|---|---|---|---|---|---|
| Gyps coprotheres | U | 3 | 8177.0 | | 7587.0–8575.0 | | South Africa 327 | 626.0 |
| Aegypius monachus | M | 20 | | | 7000.0–11500.0 | | Romania | 627.0 |
| | F | 21 | | | 7500.0–12500.0 | | 149 | |
| Torgos tracheliotus | U | | 7500.0 | | | | 650 | 628.0 |
| Trigonoceps occipitalis | U | | 5900.0 | | | | 650 | 629.0 |
| Circaetus gallicus | B | 22 | 1703.0 | | 1180.0–2324.0 | | 116 | 631.0 |
| Circaetus cinereus | U | 26 | 2048.0 | | 1540.0–2465.0 | | 82 | 633.0 |
| Circaetus fasciolatus | M | 2 | 934.0 | | 908.0–960.0 | | | 634.0 |
| | F | 1 | 1100.0 | | | | 82 | |
| Terathopius ecaudatus | U | | | | 1927.0–2950.0 | | 116 | 636.0 |
| Spilornis cheela elgini | M | | | | 790.0–1024.0 | | | 637.0 |
| | F | | | | 1024.0–1450.0 | | 5 | |
| Spilornis holospilus | B | 4 | 684.0 | | 603.0–762.0 | | Philippines 475 | 641.0 |
| Circus aeruginosus | M | 19 | 492.0 | | 320.0–667.0 | Y | Britain | 645.0 |
| | F | 25 | 763.0 | | 540.0–1269.0 | | 258 | |
| Circus ranivorus | U | 21 | 507.0 | | 382.0–606.0 | | 78 | 646.0 |
| Circus buffoni | M | 2 | 410.0 | | 391.0–430.0 | | Surinam | 650.0 |
| | F | 2 | 612.0 | | 580.0–645.0 | | 188 | |
| Circus assimilis | B | | 420.0 | | | | 78 | 651.0 |
| Circus cyaneus | M | 186 | 358.0 | 39.90 | 301.0–472.0 | F | New Jersey, USA | 653.0 |
| | F | 174 | 513.0 | 54.60 | 375.0–661.0 | | 92 | |
| Circus macrourus | M | 4 | 332.0 | | 311.0–374.0 | | | 655.0 |
| | F | 17 | 445.0 | | 402.0–550.0 | | 149 | |
| Circus melanoleucos | M | 1 | 270.0 | | | | | 656.0 |
| | F | 1 | 455.0 | | | | 149 | |
| Circus pygargus | M | 13 | 261.0 | | 227.0–305.0 | Y | Britain | 657.0 |
| | F | 6 | 370.0 | | 319.0–445.0 | | 258 | |
| Polyboroides typus | U | | 570.0 | | | | 81 | 658.0 |
| Kaupifalco monogrammicus | U | | | | 268.0–355.0 | | 81 | 660.0 |
| Melierax metabates | F | 3 | 598.0 | | 580.0–629.0 | | Sudan 680 | 661.0 |
| Melierax canorus | U | 2 | 684.0 | | 684.0–685.0 | | South Africa 255 | 663.0 |

## Body Masses of World Birds (continued)

| Species | Sex | N | Mean | Std dev | Range | Sn | Location | Number |
|---|---|---|---|---|---|---|---|---|
| Melierax gabar | M | | | | 90.0–123.0 | | | 664.0 |
| | F | | | | 167.0–240.0 | | 82 | |
| Accipiter tachiro | M | 4 | 202.0 | | 192.0–218.0 | | Zimbabwe | 669.0 |
| | F | 1 | 381.0 | | | | 280, 281, 284 | |
| Accipiter castanilius | M | 5 | | | 135.0–150.0 | | | 670.0 |
| | F | 7 | | | 152.0–200.0 | | 82 | |
| Accipiter badius | B | 3 | 196.0 | | 129.0–266.0 | | | 671.0 |
| | | | | | | | 149, 680 | |
| Accipiter brevipes | M | 7 | 175.0 | | 150.0–223.0 | | | 673.0 |
| | F | 2 | 254.0 | | 232.0–275.0 | | 116 | |
| Accipiter soloensis | M | 1 | 140.0 | | | | | 674.0 |
| | F | 1 | 204.0 | | | | 82 | |
| Accipiter francesii | M | 6 | 110.0 | | 102.0–116.0 | | Madagascar | 675.0 |
| | F | | | | | | 39 | |
| Accipiter novaehollandiae | M | 3 | 183.0 | | 175.0–197.0 | | New Britain | 677.0 |
| | F | 2 | 334.0 | | 304.0–365.0 | | 219 | |
| Accipiter fasciatus | F | 8 | 510.0 | | | | | 678.0 |
| | | | | | | | 425 | |
| Accipiter melanochlamys | F | 1 | 294.0 | | | | | 679.0 |
| | | | | | | | 81 | |
| Accipiter albogularis | M | 1 | 250.0 | | | | | 680.0 |
| | F | 1 | 425.0 | | | | 81 | |
| Accipiter haplochrous | M | 3 | 155.0 | | 152.0–162.0 | | New Caledonia | 682.0 |
| | F | 3 | 254.0 | | 227.0–268.0 | | 506 | |
| Accipiter luteoschistaceus | M | 2 | 218.0 | | 215.0–222.0 | | New Britain | 684.0 |
| | F | | | | | | 219 | |
| Accipiter imitator | F | 2 | 238.0 | | 225.0–250.0 | | | 685.0 |
| | | | | | | | 81 | |
| Accipiter poliocephalus | B | 3 | 212.0 | | 200.0–225.0 | | | 686.0 |
| | | | | | | | 81 | |
| Accipiter superciliosus | M | 2 | 80.1 | | 74.0–86.2 | | | 688.0 |
| | F | 1 | 116.0 | | | | 193, 601 | |
| Accipiter erythropus | M | 3 | 84.7 | | 78.0–94.0 | | | 690.0 |
| | F | 3 | 146.0 | | 132.0–170.0 | | 82 | |
| Accipiter minullus | M | 2 | 75.7 | | 75.3–76.1 | | Zimbabwe | 691.0 |
| | F | 1 | 101.0 | | | | 280, 282 | |
| Accipiter gularis | M | 5 | 103.0 | | 92.0–142.0 | | | 692.0 |
| | F | 15 | 140.0 | | 111.0–193.0 | | 81 | |
| Accipiter virgatus | M | 10 | 100.0 | | 83.3–142.0 | | | 693.0 |
| | F | 3 | 143.0 | | 103.0–192.0 | | 149, 371, 475 | |
| Accipiter erythrauchen | M | 1 | 156.0 | | | | | 695.0 |
| | F | | | | | | 81 | |
| Accipiter cirrhocephalus | M | 1 | 125.0 | | | | | 696.0 |
| | F | 1 | 235.0 | | | | 78 | |

## Body Masses of World Birds (continued)

| Species | Sex | N | Mean | Std dev | Range | Sn | Location | Number |
|---|---|---|---|---|---|---|---|---|
| Accipiter ovampensis | U | 3 | 195.0 | | 119.0–305.0 | | 82 | 700.0 |
| Accipiter nisus | M | 70 | 150.0 | 8.90 | | B | Scotland | 701.0 |
| | F | 246 | 325.0 | 26.30 | | | 418 | |
| Accipiter rufiventris | U | 2 | 198.0 | | 185.0–210.0 | | 82 | 702.0 |
| Accipiter striatus | M | 435 | 103.0 | 6.40 | 82.0–125.0 | F | Wisconsin, USA | 703.0 |
| | F | 487 | 174.0 | 10.40 | 144.0–208.0 | | 402 | |
| Accipiter striatus venator | M | 13 | 94.9 | | | | Puerto Rico | 703.1 |
| | F | 11 | 171.0 | | | | 148 | |
| Accipiter cooperii | M | 51 | 349.0 | 19.60 | 297.0–380.0 | F | Wisconsin, USA | 707.0 |
| | F | 57 | 529.0 | 36.10 | 460.0–588.0 | | 403 | |
| Accipiter bicolor | M | 3 | 245.0 | | 235.0–250.0 | | | 709.0 |
| | F | 3 | 436.0 | | 425.0–454.0 | | 244, 385, 457, 497 | |
| Accipiter melanoleucos | B | 19 | 695.0 | | 476.0–980.0 | | 82 | 710.0 |
| Accipiter gentilis | M | 77 | 912.0 | 14.90 | 735.0–1099.0 | F | Wisconsin, USA | 712.0 |
| | F | 103 | 1137.0 | 18.60 | 845.0–1364.0 | | 401 | |
| Accipiter radiatus | M | 1 | 640.0 | | | | | 715.0 |
| | F | | 1000.0 | | | | 147 | |
| Urotriorchis macrourus | M | | 492.0 | | | | | 717.0 |
| | F | | | | | | 82 | |
| Butastur rufipennis | F | 2 | 356.0 | | 305.0–408.0 | | 81 | 718.0 |
| Butastur indicus | M | 4 | 397.0 | | 375.0–433.0 | | | 721.0 |
| | | | | | | | | 81 |
| Geranospiza caerulescens | B | 7 | 338.0 | | 230.0–430.0 | | 244, 245, 246, 385 | 722.0 |
| Leucopternis plumbea | U | 1 | 482.0 | | | | Panama | 723.0 |
| | | | | | | | 314 | |
| Leucopternis schistacea | F | | 1000.0 | | | | 81 | 724.0 |
| Leucopternis princeps | F | 1 | 1000.0 | | | | NW Ecuador | 725.0 |
| | | | | | | | 589 | |
| Leucopternis melanops | M | 2 | 307.0 | | 297.0–317.0 | | | 726.0 |
| | F | | | | | | 81 | |
| Leucopternis semiplumbea | M | 1 | 250.0 | | | | Panama | 729.0 |
| | F | 1 | 325.0 | | | | 497 | |
| Leucopternis albicollis | M | 1 | 600.0 | | | | Surinam; Panama | 730.0 |
| | F | 1 | 650.0 | | | | 245, 497 | |
| Leucopternis occidentalis | F | 1 | 660.0 | | | | 589 | 731.0 |
| Buteogallus aequinoctialis | B | 4 | 715.0 | | 634.0–760.0 | | Surinam | 733.0 |
| | | | | | | | 245 | |

## Body Masses of World Birds (continued)

| Species | Sex | N | Mean | Std dev | Range | Sn | Location | Number |
|---|---|---|---|---|---|---|---|---|
| Buteogallus anthracinus | M | 6 | 793.0 | | | | Panama | 734.0 |
| | F | 4 | 1199.0 | | | | 244 | |
| Buteogallus urubitinga | M | 2 | 925.0 | | 853.0–996.0 | | | 736.0 |
| | F | 4 | 1068.0 | | 900.0–1250.0 | | 243, 244, 457, 510 | |
| Parabuteo unicinctus | M | 220 | 690.0 | | 550.0–829.0 | W | Texas, USA | 738.0 |
| | F | 177 | 998.0 | | 825.0–1173.0 | | 447 | |
| Parabuteo unicinctus | M | 37 | 725.0 | | 634.0–877.0 | | Arizona, USA | 738.0 |
| | F | 14 | 1047.0 | | 918.0–1203.0 | | 447 | |
| Busarellus nigricollis | M | 2 | 614.0 | | 502.0–725.0 | | | 739.0 |
| | F | 6 | 995.0 | | 758.0–1195.0 | | 244, 245, 246, 510 | |
| Geranoaetus melanoleucus | U | 2 | 2252.0 | | 2123.0–2380.0 | | | 740.0 |
| | | | | | | | 112, 227 | |
| Harphyaliaetus solitarius | U | | 3000.0 | | | | | 741.0 |
| | | | | | | | 603a | |
| Harphyaliaetus coronatus | U | | 2950.0 | | | | Brazil | 742.0 |
| | | | | | | | 564 | |
| Heterospizias meridionalis | B | 8 | 808.0 | | | | Surinam; Panama | 742.5 |
| | | | | | | | 244, 245 | |
| Asturina plagiata | M | 5 | 416.0 | | | | | 743.0 |
| | F | 4 | 637.0 | | | | 577 | |
| Buteo magnirostris | B | 16 | 269.0 | | | | Panama | 745.0 |
| | | | | | | | 244 | |
| Buteo lineatus | M | 10 | 475.0 | 81.00 | | | Florida, USA | 746.0 |
| | F | 14 | 643.0 | 96.20 | | | 244 | |
| Buteo platypterus | M | 14 | 420.0 | | | | | 748.0 |
| | F | 13 | 490.0 | | | | 577 | |
| Buteo leucorrhous | M | 1 | 290.0 | | | | Peru | 749.0 |
| | | | | | | | 584 | |
| Buteo brachyurus | M | | 460.0 | | 450.0–470.0 | | Surinam | 750.0 |
| | F | | 530.0 | | | | 247 | |
| Buteo swainsoni | M | 5 | 908.0 | | | | | 752.0 |
| | F | 7 | 1069.0 | | | | 577 | |
| Buteo albicaudatus | U | 4 | 884.0 | | | | Surinam | 753.0 |
| | | | | | | | 642 | |
| Buteo albonotatus | M | 3 | 628.0 | | 607.0–667.0 | | Mexico | 757.0 |
| | F | 4 | 886.0 | | 845.0–937.0 | | 587 | |
| Buteo jamaicensis | M | 108 | 1028.0 | | | | | 759.0 |
| | F | 100 | 1224.0 | | | | 577 | |
| Buteo ventralis | F | 1 | 1135.0 | | | | | 760.0 |
| | | | | | | | 81 | |
| Buteo buteo | M | 214 | 781.0 | | 427.0–1183.0 | Y | Britain | 761.0 |
| | F | 261 | 969.0 | | 486.0–1364.0 | | 258 | |
| Buteo oreophilus | U | 1 | 700.0 | | | | | 762.0 |
| | | | | | | | 82 | |

## Body Masses of World Birds (continued)

| Species | Sex | N | Mean | Std dev | Range | Sn | Location | Number |
|---|---|---|---|---|---|---|---|---|
| Buteo rufinus | M | 8 | 1035.0 | | 590.0–1281.0 | | | 764.0 |
| | F | 11 | 1314.0 | | 945.0–1760.0 | | 78 | |
| Buteo hemilasius | M | 2 | 1180.0 | | | | | 765.0 |
| | F | 3 | 1510.0 | | | | 78 | |
| Buteo regalis | M | 15 | 1059.0 | | | | | 766.0 |
| | F | 4 | 1231.0 | | | | 577 | |
| Buteo lagopus | M | 152 | 847.0 | | 600.0–1128.0 | Y | Britain | 767.0 |
| | F | 119 | 1065.0 | | 783.0–1660.0 | | 258 | |
| Buteo auguralis | U | 9 | 670.0 | | 560.0–890.0 | | | 768.0 |
| | | | | | | | 82 | |
| Buteo rufofuscus | M | | | | 880.0–1160.0 | | | 771.0 |
| | F | | | | 1087.0–1530.0 | | 81 | |
| Morphnus guianensis | U | | 1750.0 | | | | Peru | 772.0 |
| | | | | | | | 623 | |
| Harpia harpyja | M | | 4800.0 | | | | Brazil | 773.0 |
| | F | | 7600.0 | | | | 564 | |
| Pithecophaga jefferyi | M | 1 | 4041.0 | | | | Philippines | 775.0 |
| | | | | | | | 475 | |
| Aquila pomarina | M | 16 | 1200.0 | | 1053.0–1509.0 | | | 777.0 |
| | F | 21 | 1540.0 | | 1195.0–2160.0 | | 78 | |
| Aquila clanga | M | 3 | 1733.0 | | 1600.0–2000.0 | | | 778.0 |
| | F | 4 | 2678.0 | | 2150.0–32.0 | | 149 | |
| Aquila rapax rapax | U | | | | 2000.0–2500.0 | | | 779.0 |
| | | | | | | | 80 | |
| Aquila rapax orientalis | U | | | | 2300.0–4800.0 | | | 779.1 |
| | | | | | | | 80 | |
| Aquila heliaca | F | 2 | 3395.0 | | 3160.0–3630.0 | | | 783.0 |
| | | | | | | | 5, 149 | |
| Aquila wahlbergi | M | | 640.0 | | 437.0–845.0 | | | 784.0 |
| | | | | | | | 78 | |
| Aquila chrysaetos | M | 31 | 3477.0 | 562.00 | | | Idaho, USA | 786.0 |
| | F | 18 | 4913.0 | 695.00 | | | 181 | |
| Aquila audax | U | 176 | 3500.0 | | 2500.0–5770.0 | | | 787.0 |
| | | | | | | | 78 | |
| Aquila verreauxii | M | 1 | 3600.0 | | | | | 788.0 |
| | F | 4 | 4600.0 | | 3100.0–5779.0 | | 78 | |
| Hieraaetus fasciatus | M | | 1500.0 | | | | India | 789.0 |
| | F | | 2500.0 | | | | 149 | |
| Hieraaetus pennatus | M | 12 | 701.0 | | 510.0–770.0 | | | 791.0 |
| | F | 11 | 968.0 | | 840.0–1250.0 | | 78 | |
| Hieraaetus morphnoides | M | 5 | 608.0 | 29.00 | 578.0–655.0 | | Australia | 792.0 |
| | F | 5 | 1070.0 | 152.00 | 880.0–1250.0 | | 144 | |
| Hieraaetus dubius | M | 1 | 714.0 | | | | | 793.0 |
| | F | 2 | 910.0 | | 879.0–940.0 | | 81 | |

## Body Masses of World Birds (continued)

| Species | Sex | N | Mean | Std dev | Range | Sn | Location | Number |
|---|---|---|---|---|---|---|---|---|
| Hieraaetus kienerii | M | 1 | 732.0 | | | | Philippines 475 | 794.0 |
| Polemaetus bellicosus | U | 20 | 4230.0 | | 3012.0–6200.0 | | 78 | 795.0 |
| Spizastur melanoleucus | U | | 850.0 | | | | Peru 623 | 796.0 |
| Lophaetus occipitalis | M | | 1140.0 | | | | | 797.0 |
| | F | | 1445.0 | | | | 78 | |
| Spizaetus africanus | U | 3 | 1047.0 | | 938.0–1153.0 | | 39 | 798.0 |
| Spizaetus cirrhatus | F | | | | 1360.0–1810.0 | | 81 | 799.0 |
| Spizaetus nipalensis | M | | 2500.0 | | | | | 800.0 |
| | F | | 3500.0 | | | | 81 | |
| Spizaetus philippensis | F | 1 | 1168.0 | | | | Philippines 475 | 804.0 |
| Spizaetus tyrannus | U | | 1025.0 | | | | Peru 623 | 806.0 |
| Spizaetus ornatus | M | 3 | 1069.0 | | 841.0–1215.0 | | | 807.0 |
| | F | 4 | 1421.0 | | 950.0–1760.0 | | 244,274,277,457, 497,510 | |
| Stephanoaetus coronatus | U | 4 | 3640.0 | | 3175.0–4120.0 | | 78 | 808.0 |

**ORDER: FALCONIFORMES**      **FAMILY: SAGITTARIIDAE**

| Species | Sex | N | Mean | Std dev | Range | Sn | Location | Number |
|---|---|---|---|---|---|---|---|---|
| Sagittarius serpentarius | M | 1 | 3809.0 | | | | | 810.0 |
| | F | 1 | 3405.0 | | | | 81 | |

**ORDER: FALCONIFORMES**      **FAMILY: FALCONIDAE**

| Species | Sex | N | Mean | Std dev | Range | Sn | Location | Number |
|---|---|---|---|---|---|---|---|---|
| Daptrius ater | M | 2 | 342.0 | | 330.0–354.0 | | Surinam 245, 246 | 811.0 |
| Daptrius americanus | F | 1 | 586.0 | | | | Surinam 245 | 812.0 |
| Phalcoboenus megalopterus | M | 1 | 795.0 | | | | 81 | 814.0 |
| Phalcoboenus australis | U | 1 | 1187.0 | | | | 81 | 816.0 |
| Polyborus plancus | M | 14 | 834.0 | 133.00 | | | Panama | 818.0 |
| | F | 10 | 953.0 | 63.20 | | | 244 | |
| Milvago chimachima | M | 4 | 297.0 | | 256.0–325.0 | | | 819.0 |
| | F | 2 | 368.0 | | 326.0–410.0 | | 37, 244, 245, 246 | |
| Milvago chimango | B | 29 | 296.0 | 30.90 | | | 700 | 820.0 |

## Body Masses of World Birds (continued)

| Species | Sex | N | Mean | Std dev | Range | Sn | Location | Number |
|---|---|---|---|---|---|---|---|---|
| Herpetotheres cachinnans | M | 5 | 620.0 | | 567.0–686.0 | | | 821.0 |
| | F | 4 | 715.0 | | 626.0–800.0 | | 244, 245, 272, 510 | |
| Micrastur ruficollis | M | 6 | 161.0 | | | | Panama | 822.0 |
| | F | 6 | 196.0 | | | | 244 | |
| Micrastur gilvicollis | U | 5 | 204.0 | | 172.0–223.0 | | Peru; Brazil | 824.0 |
| | | | | | | | 193, 610a | |
| Micrastur mirandollei | F | 1 | 556.0 | | | | Surinam | 825.0 |
| | | | | | | | 245 | |
| Micrastur semitorquatus | M | 2 | 562.0 | | 479.0–646.0 | | Belize; Panama | 826.0 |
| | F | 1 | 900.0 | | | | 244, 510 | |
| Spiziapteryx circumcinctus | M | 5 | 152.0 | | 137.0–164.0 | | Argentina | 828.0 |
| | F | 4 | 196.0 | | 176.0–228.0 | | 112 | |
| Polihierax semitorquatus | U | 11 | 57.0 | | 44.0–72.0 | | Somalia; Kenya | 829.0 |
| | | | | | | | 81 | |
| Microhierax latifrons | M | 1 | 41.2 | | | | Borneo | 833.0 |
| | | | | | | | 627 | |
| Microhierax erythrogonys | M | 2 | 43.5 | | 40.0–47.0 | | | 834.0 |
| | | | | | | | 81 | |
| Falco berigora | M | | 474.0 | | 417.0–510.0 | | Australia | 836.0 |
| | F | | 625.0 | | 560.0–730.0 | | 131 | |
| Falco naumanni | M | 34 | 141.0 | | 90.0–176.0 | | | 837.0 |
| | F | 25 | 164.0 | | 109.0–208.0 | | 78 | |
| Falco tinnunculus | M | 40 | 186.0 | | 117.0–259.0 | Y | Britain | 838.0 |
| | F | 57 | 217.0 | | 137.0–299.0 | | 258 | |
| Falco newtoni | M | 4 | 105.0 | | 90.0–117.0 | | | 839.0 |
| | F | 6 | 145.0 | | 131.0–159.0 | | 152 | |
| Falco punctatus | M | 6 | 143.0 | | 123.0–146.0 | | | 840.0 |
| | F | 6 | 196.0 | | 173.0–240.0 | | 152 | |
| Falco araea | M | 14 | 72.4 | 16.28 | | | | 841.0 |
| | F | 32 | 87.9 | 26.76 | | | 152 | |
| Falco cenchroides | M | 179 | 168.0 | 13.40 | | | | 843.0 |
| | F | 133 | 186.0 | 23.10 | | | 427 | |
| Falco sparverius | M | 69 | 111.0 | 9.30 | | Y | California | 844.0 |
| | F | 111 | 120.0 | 9.20 | | | 53 | |
| Falco rupicoloides | U | 371 | 260.0 | | 165.0–334.0 | | | 845.0 |
| | | | | | | | 82 | |
| Falco ardosiaceus | U | 12 | 238.0 | | 195.0–300.0 | | | 847.0 |
| | | | | | | | 82 | |
| Falco dickinsoni | U | 42 | 209.0 | | 167.0–246.0 | | | 848.0 |
| | | | | | | | 82 | |
| Falco zoniventris | F | 1 | 174.0 | | | | Madagascar | 849.0 |
| | | | | | | | 39 | |
| Falco chicquera | U | 10 | 198.0 | | 139.0–305.0 | | | 850.0 |
| | | | | | | | 82 | |

## Body Masses of World Birds (continued)

| Species | Sex | N | Mean | Std dev | Range | Sn | Location | Number |
|---|---|---|---|---|---|---|---|---|
| Falco vespertinus | M | 5 | 149.0 | | 130.0–164.0 | | | 851.0 |
| | F | 5 | 182.0 | | 162.0–197.0 | | 149 | |
| Falco eleonorae | U | 20 | 390.0 | | 340.0–450.0 | | 78 | 853.0 |
| Falco concolor | F | 1 | 250.0 | | | | Madagascar 39 | 854.0 |
| Falco femoralis | M | 7 | 260.0 | | 208.0–305.0 | | Mexico | 855.0 |
| | F | 6 | 407.0 | | 310.0–500.0 | | 447 | |
| Falco columbarius | M | 145 | 163.0 | 14.10 | 134.0–223.0 | F | New Jersey | 856.0 |
| | F | 189 | 218.0 | 17.20 | 134.0–281.0 | | 92 | |
| Falco rufigularis | M | 11 | 129.0 | | 108.0–148.0 | | Surinam | 857.0 |
| | F | 6 | 202.0 | | 177.0–242.0 | | 188 | |
| Falco subbuteo | M | 3 | 204.0 | | 200.0–208.0 | | | 858.0 |
| | F | 4 | 276.0 | | 245.0–325.0 | | 149 | |
| Falco cuvieri | U | | 183.0 | | 150.0–224.0 | | 82 | 859.0 |
| Falco severus | M | 1 | 183.0 | | | | | 860.0 |
| | F | 3 | 211.0 | | 192.0–249.0 | | 35, 219, 475 | |
| Falco longipennis | M | 8 | 213.0 | | 177.0–250.0 | | | 861.0 |
| | F | 14 | 293.0 | | 201.0–340.0 | | 133 | |
| Falco hypoleucos | M | 1 | 335.0 | | | | Australia | 863.0 |
| | F | 1 | 624.0 | | | | 132 | |
| Falco subniger | M | 5 | 664.0 | | 620.0–710.0 | | Australia | 864.0 |
| | F | 3 | 907.0 | | 843.0–1000.0 | | 131 | |
| Falco mexicanus | M | 15 | 554.0 | | 500.0–635.0 | PB | Colorado | 865.0 |
| | F | 31 | 863.0 | | 760.0–975.0 | | 447 | |
| Falco biarmicus | U | 58 | 593.0 | | 402.0–910.0 | | 78 | 866.0 |
| Falco jugger | F | 1 | 755.0 | | | | 81 | 867.0 |
| Falco cherrug | M | 4 | | | 820.0–890.0 | | | 868.0 |
| | F | 3 | | | 970.0–1130.0 | | 149 | |
| Falco rusticolus | M | 7 | 1170.0 | | 960.0–1304.0 | | | 870.0 |
| | F | 12 | 1752.0 | | 1396.0–2000.0 | | 81 | |
| Falco peregrinus | M | 12 | 611.0 | | | | | 871.0 |
| | F | 19 | 952.0 | | | | 577 | |
| Falco deiroleucus | F | 1 | 654.0 | | | | Guatemala 188 | 873.0 |
| Falco fasciinucha | U | 2 | 259.0 | | 212.0–306.0 | | 82 | 874.0 |

### ORDER: GALLIFORMES       FAMILY: CRACIDAE

| Species | Sex | N | Mean | Std dev | Range | Sn | Location | Number |
|---|---|---|---|---|---|---|---|---|
| Ortalis vetula | M | 106 | 584.0 | 58.00 | 468.0–794.0 | Y | Texas | 875.0 |
| | F | 102 | 542.0 | 52.00 | 439.0–709.0 | | 364 | |

## Body Masses of World Birds (continued)

| Species | Sex | N | Mean | Std dev | Range | Sn | Location | Number |
|---|---|---|---|---|---|---|---|---|
| Ortalis cinereiceps | U | 2 | 493.0 | | 490.0–495.0 | | Panama 86, 497 | 876.0 |
| Ortalis columbiana | M | 1 | 600.0 | | | | Columbia | 876.0 |
| | F | 1 | 500.0 | | | | 385 | |
| Ortalis garrula | B | 10 | 534.0 | | | | Panama 244, 314 | 877.0 |
| Ortalis ruficauda | U | | 625.0 | | 455.0–800.0 | | 78 | 878.0 |
| Ortalis erythroptera | B | 2 | 632.0 | | 620.0–645.0 | | Peru 672 | 879.0 |
| Ortalis poliocephala | M | 1 | 760.0 | | | | Oaxaca, Mexico 584 | 881.0 |
| Ortalis canicollis | M | 1 | 590.0 | | | | Argentina 112 | 882.0 |
| Ortalis leucogastra | M | 1 | 439.0 | | | | Guatemala | 883.0 |
| | F | 1 | 560.0 | | | | 584 | |
| Ortalis motmot | B | 6 | 500.0 | | 385.0–620.0 | | 37, 161, 245 | 885.0 |
| Penelope montagnii | M | 1 | 460.0 | | | | Peru 668 | 890.0 |
| Penelope marail | B | 3 | 910.0 | | 800.0–1000.0 | | French Guiana 161 | 891.0 |
| Penelope superciliaris | U | | 850.0 | | | | Brazil 564 | 892.0 |
| Penelope purpurascens | B | 4 | 2060.0 | | 2000.0–2150.0 | | Panama 270, 498 | 894.0 |
| Penelope jacquacu | B | 3 | 1282.0 | | 1180.0–1410.0 | | Peru 193 | 897.0 |
| Penelope obscura | M | 1 | 960.0 | | | | Paraguay 610 | 898.0 |
| Penelope pileata | M | 1 | 1100.0 | | | | Brazil | 899.0 |
| | F | 1 | 1420.0 | | | | 226 | |
| Pipile jacutinga | U | | | | 1100.0–1400.0 | | Brazil 564 | 905.0 |
| Aburria aburri | B | 5 | 1423.0 | | 1195.0–1550.0 | | Peru 193 | 906.0 |
| Chamaepetes unicolor | B | 12 | 1135.0 | | | | Panama 244 | 907.0 |
| Chamaepetes goudotii | M | 2 | 645.0 | | 550.0–740.0 | | | 908.0 |
| | F | 1 | 778.0 | | | | 94, 303 | |
| Penelopina nigra | B | 2 | 890.0 | | 866.0–914.0 | | Oaxaca, Mexico 584 | 909.0 |
| Mitu mitu | U | | 3060.0 | | | | Peru 923 | 914.0 |

## Body Masses of World Birds (continued)

| Species | Sex | N | Mean | Std dev | Range | Sn | Location | Number |
|---|---|---|---|---|---|---|---|---|
| Crax rubra | B | 3 | 4133.0 | | 4050.0–4225.0 | | Belize 510 | 917.0 |
| Crax alector | U | | | | 3200.0–3600.0 | | Brazil 564 | 920.0 |
| Crax fasciolata | B | 2 | 2515.0 | | 2280.0–2750.0 | | Brazil 226, 564 | 922.0 |
| Crax blumenbachii | U | | 3500.0 | | | | Brazil 564 | 923.0 |

**ORDER: GALLIFORMES**      **FAMILY: MEGAPODIIDAE**

| Species | Sex | N | Mean | Std dev | Range | Sn | Location | Number |
|---|---|---|---|---|---|---|---|---|
| Megapodius freycinet | M | | | | 595.0–964.0 | | | 928.0 |
| | F | | | | 850.0–1021.0 | 5 | | |
| Leipoa ocellata | U | | | | 1816.0–1930.0 | | 550 | 935.0 |
| Alectura lathami | U | | 2330.0 | | 2210.0–2450.0 | | Australia 147 | 936.0 |
| Talegalla fuscirostris | U | 2 | 1000.0 | | | | estimated 35 | 938.0 |
| Talegalla jobiensis | M | 2 | 1559.0 | | 1531.0–1588.0 | | New Guinea 217 | 939.0 |

**ORDER: GALLIFORMES**      **FAMILY: NUMIDIDAE**

| Species | Sex | N | Mean | Std dev | Range | Sn | Location | Number |
|---|---|---|---|---|---|---|---|---|
| Numida meleagris | B | 10 | 1299.0 | | | | 244 | 945.0 |
| Guttera pucherani | M | | 1149.0 | | 721.0–1573.0 | | 636 | 947.0 |
| Acryllium vulturinum | M | 1 | 1645.0 | | | | | 948.0 |
| | F | 1 | 1135.0 | | | 66 | | |

**ORDER: GALLIFORMES**      **FAMILY: PHASIANIDAE**

| Species | Sex | N | Mean | Std dev | Range | Sn | Location | Number |
|---|---|---|---|---|---|---|---|---|
| Lerwa lerwa | U | | | | 454.0–709.0 | | 5 | 949.0 |
| Ammoperdix griseogularis | U | 6 | 200.0 | | 182.0–205.0 | | 149 | 950.0 |
| Ammoperdix heyi | U | | 181.0 | | | | 298 | 951.0 |
| Tetraogallus caucasicus | B | 2 | 1834.0 | | 1734.0–1933.0 | | 116 | 952.0 |
| Tetraogallus caspius | M | 2 | 2592.0 | | 2500.0–2684.0 | | | 953.0 |
| | F | 2 | 2072.0 | | 1800.0–2344.0 | 116 | | |
| Tetraogallus tibetanus | M | | | | 1500.0–1750.0 | | | 954.0 |
| | F | | | | 1170.0–1600.0 | 298 | | |
| Tetraogallus altaicus | M | | 3000.0 | | | | estimated 298 | 955.0 |
| | F | | 2540.0 | | | | | |

## Body Masses of World Birds (continued)

| Species | Sex | N | Mean | Std dev | Range | Sn | Location | Number |
|---|---|---|---|---|---|---|---|---|
| Tetraogallus himalayensis | B | 14 | 2428.0 | | 2000.0–3100.0 | W | 149 | 956.0 |
| Tetraophasis obscurus | M | 1 | 938.0 | | | | | 957.0 |
| | F | 2 | 780.0 | | 720.0–840.0 | | 313, 553 | |
| Tetraophasis szechenyii | M | 2 | 1260.0 | | 1020.0–1500.0 | | | 958.0 |
| | F | 1 | 880.0 | | | | 298 | |
| Alectoris graeca | M | | | | 550.0–850.0 | | | 959.0 |
| | F | | | | 410.0–650.0 | | 298 | |
| Alectoris chukar | M | 22 | 619.0 | | | | New Mexico | 960.0 |
| | F | 24 | 537.0 | | | | 54 | |
| Alectoris philbyi | U | | 441.0 | | | | estimated 298 | 961.0 |
| Alectoris magna | M | | | | 445.0–710.0 | | | 962.0 |
| | F | | | | 442.0–615.0 | | 298 | |
| Alectoris barbara | M | | 461.0 | | | | estimated | 963.0 |
| | F | | 376.0 | | | | 298 | |
| Alectoris rufa rufa | M | | | | 480.0–547.0 | | | 964.0 |
| | F | | | | 391.0–514.0 | | 298 | |
| Alectoris melanocephala | M | | 724.0 | | | | estimated | 965.0 |
| | F | | 522.0 | | | | 298 | |
| Francolinus pondicerianus | M | 114 | 274.0 | | | | | 966.0 |
| | F | 91 | 228.0 | | | | 298 | |
| Francolinus pintadeanus | M | | | | 347.0–388.0 | | | 967.0 |
| | F | | 310.0 | | | | 298 | |
| Francolinus francolinus asiae | M | 19 | 482.0 | | | | | 968.0 |
| | F | 18 | 424.0 | | | | 5 | |
| Francolinus pictus | U | | | | 242.0–340.0 | | 5 | 969.0 |
| Francolinus gularis | M | | 510.0 | | | | 5 | 970.0 |
| Francolinus lathami | B | | 269.0 | | 254.0–284.0 | | estimated 298 | 971.0 |
| Francolinus coqui | M | 4 | 262.0 | | 227.0–284.0 | | | 972.0 |
| | F | | | | 218.0–259.0 | | 298 | |
| Francolinus albogularis | U | 8 | 276.0 | | | | 298 | 973.0 |
| Francolinus schlegelii | M | | 251.0 | | | | estimated | 974.0 |
| | F | | 223.0 | | | | 298 | |
| Francolinus streptophorus | M | 2 | 385.0 | | 364.0–406.0 | | 298 | 975.0 |
| Francolinus finschi | B | | 560.0 | | | | estimated 298 | 976.0 |
| Francolinus africanus | M | 13 | 423.0 | | 345.0–539.0 | | | 977.0 |
| | F | 3 | 359.0 | | 354.0–369.0 | | 298 | |

## Body Masses of World Birds (continued)

| Species | Sex | N | Mean | Std dev | Range | Sn | Location | Number |
|---|---|---|---|---|---|---|---|---|
| Francolinus levaillantii | M | 3 | 463.0 | | 359.0–567.0 | | | 978.0 |
| | F | 4 | 401.0 | | 354.0–454.0 | | 298 | |
| Francolinus levaillantoides | M | | | | 370.0–528.0 | | | 979.0 |
| | F | | | | 379.0–450.0 | | 298 | |
| Francolinus psilolaemus | M | 2 | 520.0 | | 510.0–530.0 | | | 980.0 |
| | F | 2 | 440.0 | | 370.0–510.0 | | 298 | |
| Francolinus shelleyi | B | 17 | 488.0 | | 397.0–600.0 | | | 981.0 |
| | | | | | | | 298 | |
| Francolinus sephaena | U | 4 | 245.0 | | 205.0–310.0 | | | 982.0 |
| | | | | | | | 78 | |
| Francolinus ahantensis | M | | 608.0 | | | | estimated | 983.0 |
| | F | | 487.0 | | | | 298 | |
| Francolinus squamatus | M | | | | 372.0–565.0 | | | 984.0 |
| | F | | | | 377.0–515.0 | | 298 | |
| Francolinus griseostriatus | B | | 410.0 | | 390.0–430.0 | | estimated | 985.0 |
| | | | | | | | 298 | |
| Francolinus nahani | M | 2 | 310.0 | | 308.0–312.0 | | | 986.0 |
| | F | 3 | | | 234.0–260.0 | | 298 | |
| Francolinus hartlaubi | M | | | | 245.0–290.0 | | | 987.0 |
| | F | | | | 210.0–240.0 | | 298 | |
| Francolinus hildebrandti | M | | | | 600.0–645.0 | | | 988.0 |
| | F | | | | 430.0–480.0 | | 298 | |
| Francolinus natalensis | M | 10 | 606.0 | | 485.0–723.0 | | | 989.0 |
| | F | 5 | 426.0 | | 370.0–482.0 | | 298 | |
| Francolinus bicalcaratus | M | 5 | 507.0 | | | | | 990.0 |
| | F | 3 | 381.0 | | | | 298 | |
| Francolinus clappertoni | M | 12 | 604.0 | | | | | 991.0 |
| | F | 10 | 463.0 | | | | 298 | |
| Francolinus icterorhynchus | M | 7 | 571.0 | | 504.0–588.0 | | | 992.0 |
| | F | | | | 420.0–462.0 | | 298 | |
| Francolinus harwoodi | M | | 545.0 | | | | | 993.0 |
| | F | | | | 413.0–446.0 | | 298 | |
| Francolinus capensis | M | | | | 600.0–915.0 | | | 994.0 |
| | F | | | | 435.0–659.0 | | 298 | |
| Francolinus adspersus | M | 12 | 465.0 | | 340.0–635.0 | | | 995.0 |
| | F | 24 | 394.0 | | 340.0–549.0 | | 298 | |
| Francolinus camerunensis | M | | 593.0 | | | | estimated | 996.0 |
| | F | | 509.0 | | | | 298 | |
| Francolinus swierstrai | M | | 600.0 | | | | estimated | 997.0 |
| | F | | 560.0 | | | | 298 | |
| Francolinus erckelii | U | | 1350.0 | | 1100.0–1600.0 | | | 998.0 |
| | | | | | | | 78 | |
| Francolinus ochropectus | U | | 940.0 | | | | estimated | 999.0 |
| | | | | | | | 298 | |

**Body Masses of World Birds (continued)**

| Species | Sex | N | Mean | Std dev | Range | Sn | Location | Number |
|---|---|---|---|---|---|---|---|---|
| Francolinus castaneicollis | M | | | | 915.0–1200.0 | | | 1000.0 |
| | F | | | | 550.0–650.0 | 298 | | |
| Francolinus nobilis | M | | | | 862.0–895.0 | | | 1001.0 |
| | F | | | | 600.0–670.0 | 298 | | |
| Francolinus jacksoni | M | | | | 1130.0–1164.0 | | | 1002.0 |
| | | | | | | 298 | | |
| Francolinus leucoscepus | M | 173 | 753.0 | | 615.0–896.0 | | | 1003.0 |
| | F | 223 | 545.0 | | 400.0–615.0 | 298 | | |
| Francolinus rufopictus | M | | 848.0 | | 779.0–964.0 | | | 1004.0 |
| | F | | 588.0 | | 439.0–666.0 | 298 | | |
| Francolinus afer | M | | | | 480.0–907.0 | | | 1005.0 |
| | F | | | | 370.0–652.0 | 298 | | |
| Francolinus swainsonii | M | 90 | 706.0 | | 400.0–875.0 | | | 1006.0 |
| | F | 100 | 505.0 | | 340.0–750.0 | 298 | | |
| Perdix perdix | M | 87 | 398.0 | | –454.0 | | | 1007.0 |
| | F | 57 | 381.0 | | –434.0 | 412 | | |
| Perdix dauuricae | M | | | | 294.0–300.0 | | | 1008.0 |
| | F | | | | 200.0–340.0 | 298 | | |
| Perdix hodgsoniae | U | | 450.0 | | | | | 1009.0 |
| | | | | | | 5 | | |
| Rhizothera longirostris | M | | 800.0 | | | | estimated | 1010.0 |
| | F | | 697.0 | | | | 298 | |
| Margaroperdix madagascarensis | B | | 220.0 | | | | estimated 298 | 1011.0 |
| Melanoperdix nigra | B | | 260.0 | | | | estimated 298 | 1012.0 |
| Coturnix coturnix | M | 144 | 90.0 | | 76.0–111.0 | | Africa | 1013.0 |
| | F | 90 | 103.0 | | 81.0–122.0 | 298 | | |
| Coturnix japonica | B | | 90.0 | | | | estimated 298 | 1014.0 |
| Coturnix pectoralis | M | | 114.0 | | | | | 1015.0 |
| | F | | 95.0 | | | 298 | | |
| Coturnix novaezelandiae | U | | 104.0 | | | | | 1016.0 |
| | | | | | | 134 | | |
| Coturnix coromandelica | B | | | | 64.0–85.0 | | | 1017.0 |
| | | | | | | 5 | | |
| Coturnix delegorguei | B | 27 | 76.0 | | 65.0–94.0 | | | 1018.0 |
| | | | | | | 298 | | |
| Coturnix ypsilophora | M | 2 | 79.3 | | 74.5–84.0 | | | 1020.0 |
| | F | 4 | 90.2 | | 88.0–92.0 | 298 | | |
| Coturnix adansonii | B | 3 | 44.5 | | 43.0–46.6 | | | 1021.0 |
| | | | | | | 285, 298 | | |
| Coturnix chinensis | M | 2 | 35.5 | | 34.5–36.5 | | | 1022.0 |
| | F | 6 | 27.2 | | 20.1–33.5 | 216, 475 | | |

## Body Masses of World Birds (continued)

| Species | Sex | N | Mean | Std dev | Range | Sn | Location | Number |
|---|---|---|---|---|---|---|---|---|
| Anurophasis monorthonyx | B | | 401.0 | | | | estimated 298 | 1023.0 |
| Perdicula asiatica | U | | | | 57.0–82.0 | 5 | | 1024.0 |
| Perdicula argoondah | M<br>F | | 62.0<br>59.0 | | | | estimated 298 | 1025.0 |
| Perdicula erythrorhyncha | U | | | | 70.0–85.0 | 5 | | 1026.0 |
| Perdicula erythrorhyncha blewitti | U | | | | 50.0–70.0 | 5 | | 1026.0 |
| Perdicula manipurensis | U | | | | 64.0–78.0 | 5 | | 1027.0 |
| Arborophila torqueola | M<br>F | | | | 325.0–430.0<br>261.0–386.0 | | 298 | 1028.0 |
| Arborophila rufogularis rufogularis | U | | | | 200.0–300.0 | 5 | | 1029.0 |
| Arborophila atrogularis | M<br>F | | 256.0<br>220.0 | | | | 298 | 1030.0 |
| Arborophila crudigularis | M<br>F | 1 | 311.0<br>212.0 | | | | estimated 298 | 1031.0 |
| Arborophila mandellii | B | | 268.0 | | | | estimated 298 | 1032.0 |
| Arborophila brunneopectus | M<br>F | | 317.0<br>268.0 | | | | estimated 298 | 1033.0 |
| Arborophila rufipectus | M<br>F | | | | 410.0–470.0<br>350.0–380.0 | | 298 | 1034.0 |
| Arborophila orientalis | B | | 268.0 | | 263.0–274.0 | | estimated 298 | 1035.0 |
| Arborophila javanica | B | | 272.0 | | 257.0–286.0 | | estimated 298 | 1036.0 |
| Arborophila gingica | U | 1 | 253.0 | | | | 298 | 1037.0 |
| Arborophila davidi | U | | 241.0 | | | | estimated 298 | 1038.0 |
| Arborophila cambodiana | M<br>F | | 318.0<br>257.0 | | | | estimated 298 | 1039.0 |
| Arborophila rubrirostris | M<br>F | | 243.0<br>209.0 | | | | estimated 298 | 1040.0 |
| Arborophila hyperythra | U | | 270.0 | | | | estimated 298 | 1041.0 |
| Arborophila ardens | M<br>F | 1<br>1 | 300.0<br>237.0 | | | | 298 | 1042.0 |
| Arborophila charltonii chloropus | M<br>F | | 290.0<br>250.0 | | | | 298 | 1045.0 |

## Body Masses of World Birds (continued)

| Species | Sex | N | Mean | Std dev | Range | Sn | Location | Number |
|---|---|---|---|---|---|---|---|---|
| Caloperdix oculea | M | 1 | 190.0 | | | | 298 | 1046.0 |
| Haematortyx sanguiniceps | M | 1 | 300.0 | | | | 298 | 1047.0 |
| Rollulus rouloul | M | 7 | 232.0 | | | | | 1048.0 |
| | F | 6 | 202.0 | | | | 298 | |
| Ptilopachus petrosus | M | 2 | 190.0 | | | | 298 | 1049.0 |
| Bambusicola fytchii | M | | | | 285.0–400.0 | | | 1050.0 |
| | F | | 340.0 | | | 5 | | |
| Bambusicola thoracica | M | | | | 242.0–297.0 | | | 1051.0 |
| | F | | | | 200.0–342.0 | 5 | | |
| Galloperdix spadicea | U | | | | 284.0–454.0 | | 5 | 1052.0 |
| Galloperdix lunulata | M | | | | 255.0–285.0 | | | 1053.0 |
| | F | | | | 226.0–255.0 | 5 | | |
| Galloperdix bicalcarata | M | | | | 312.0–368.0 | | | 1054.0 |
| | F | | | | 200.0–312.0 | 5 | | |
| Ophrysia superciliosa | B | | 705.0 | | | | estimated 298 | 1055.0 |
| Ithaginis cruentus berezowskii | M | 5 | | | 520.0–600.0 | | | 1056.0 |
| | F | 3 | | | 410.0–620.0 | 298 | | |
| Tragopan melanocephalus | M | | | | 1800.0–2150.0 | | | 1057.0 |
| | F | | | | 1250.0–1400.0 | 5 | | |
| Tragopan satyra | M | | | | 1600.0–2100.0 | | | 1058.0 |
| | F | | | | 1000.0–1200.0 | 5 | | |
| Tragopan blythii | M | | 1930.0 | | | | | 1059.0 |
| | F | | | | 1000.0–1500.0 | 297 | | |
| Tragopan temminckii | M | 2 | 1404.0 | | 1362.0–1447.0 | | captive | 1060.0 |
| | F | 2 | 964.0 | | 907.0–1021.0 | 297 | | |
| Tragopan caboti | M | 1 | 1400.0 | | | | | 1061.0 |
| | F | | 900.0 | | | | 297 | |
| Pucrasia macrolopha | M | 10 | 1184.0 | | | | | 1062.0 |
| | F | 10 | 932.0 | | | | 297 | |
| Lophorus impejanus | M | | | | 1980.0–2380.0 | | | 1063.0 |
| | F | | | | 1800.0–2150.0 | 5 | | |
| Lophorus sclateri | M | 2 | 2500.0 | | | | | 1064.0 |
| | F | | | | 2126.0–2267.0 | 297 | | |
| Lophorus lhuysii | M | 2 | 3008.0 | | 2837.0–3178.0 | | | 1065.0 |
| | | | | | | 297 | | |
| Gallus gallus jabovillei | M | 10 | 844.0 | | 672.0–1020.0 | | | 1066.0 |
| | F | 1 | 500.0 | | | | 297 | |
| Gallus sonneratii | M | | | | 790.0–1136.0 | | | 1067.0 |
| | F | | | | 705.0–790.0 | 297 | | |

## Body Masses of World Birds (continued)

| Species | Sex | N | Mean | Std dev | Range | Sn | Location | Number |
|---|---|---|---|---|---|---|---|---|
| Gallus lafayetii | M | | | | 790.0–1140.0 | | | 1068.0 |
| | F | | | | 510.0–625.0 | 5 | | |
| Gallus varius | U | | 620.0 | | 454.0–795.0 | | | 1069.0 |
| | | | | | | 78 | | |
| Lophura leucomelanos hamiltonii | M | | | | 910.0–1080.0 | | | 1070.0 |
| | F | | | | 564.0–1024.0 | 5 | | |
| Lophura nycthemera beaulieui | M | 9 | | | 1500.0–2000.0 | | | 1071.0 |
| | F | 2 | 1230.0 | | 1160.0–1300.0 | 297 | | |
| Lophura edwardsi | B | 2 | 1082.0 | | 1050.0–1115.0 | | | 1073.0 |
| | | | | | | 297 | | |
| Lophura swinhoii | U | 2 | 1100.0 | | | | | 1074.0 |
| | | | | | | 297 | | |
| Lophura erythropthalma | M | 2 | 1043.0 | | | | captive | 1077.0 |
| | F | 1 | 837.0 | | | 297 | | |
| Lophura ignita | M | 5 | 2175.0 | | 1812.0–2605.0 | | | 1078.0 |
| | F | | 1600.0 | | | 297 | | |
| Lophura diardi | M | 1 | 1420.0 | | | | captive | 1079.0 |
| | F | 3 | 835.0 | | 680.0–1025.0 | 297 | | |
| Lophura bulweri | M | 4 | 1615.0 | | 1470.0–1800.0 | | | 1080.0 |
| | F | 2 | 960.0 | | 916.0–1004.0 | 297 | | |
| Crossoptilon crossoptilon drouyni | M | 3 | | | 2350.0–2750.0 | | | 1082.0 |
| | F | 4 | | | 1400.0–2050.0 | 297 | | |
| Crossoptilon mantchuricum | M | | | | 1650.0–2475.0 | | | 1083.0 |
| | F | | | | 1450.0–2025.0 | 297 | | |
| Crossoptilon auritum | M | | | | 1700.0–2110.0 | | | 1084.0 |
| | F | | | | 1450.0–1880.0 | 297 | | |
| Catreus wallichi | M | | | | 1475.0–1700.0 | | | 1085.0 |
| | F | | | | 1250.0–1360.0 | 5 | | |
| Syrmaticus ellioti | M | 17 | 1156.0 | | 1044.0–1317.0 | | | 1086.0 |
| | F | 35 | 878.0 | | 726.0–1090.0 | 297 | | |
| Syrmaticus humiae | M | | 1022.0 | | 975.0–1070.0 | | | 1087.0 |
| | F | 1 | 650.0 | | | 297 | | |
| Syrmaticus mikado | M | 2 | 1300.0 | | | | captive | 1088.0 |
| | F | 2 | 1015.0 | | | 297 | | |
| Syrmaticus soemmerringii | U | 1 | 907.0 | | | | | 1089.0 |
| | | | | | | 297 | | |
| Syrmaticus reevesii | M | 24 | 1529.0 | | | | | 1090.0 |
| | F | 30 | 949.0 | | | 297 | | |
| Phasianus colchicus | M | 6378 | 1317.0 | | –1861.0 | | | 1091.0 |
| | F | 759 | 953.0 | | –1453.0 | 412 | | |
| Chrysolophus pictus | M | 5 | | | 575.0–710.0 | | | 1092.0 |
| | F | 5 | | | 550.0–665.0 | 297 | | |
| Chrysolophus amherstiae | M | 5 | | | 675.0–850.0 | | | 1093.0 |
| | F | 5 | | | 624.0–804.0 | 297 | | |

## Body Masses of World Birds (continued)

| Species | Sex | N | Mean | Std dev | Range | Sn | Location | Number |
|---|---|---|---|---|---|---|---|---|
| Polyplectron chalcurum | M | 2 | 508.0 | | 425.0–590.0 | | | 1094.0 |
| | F | | 251.0 | | 238.0–269.0 | | 297 | |
| Polyplectron germaini | M | 1 | 510.0 | | | | | 1096.0 |
| | F | 1 | 397.0 | | | | 297 | |
| Polyplectron bicalcaratum | M | 4 | | | 660.0–710.0 | | | 1097.0 |
| | F | 2 | 480.0 | | 460.0–500.0 | | 297 | |
| Polyplectron malacense | U | 2 | 633.0 | | 586.0–680.0 | | | 1098.0 |
| | | | | | | | 297, 690 | |
| Polyplectron emphanum | M | 2 | 436.0 | | | | captive | 1100.0 |
| | F | 2 | 322.0 | | | | 297 | |
| Argusianus argus | M | 6 | 2361.0 | | 2040.0–2725.0 | | | 1102.0 |
| | F | 3 | 1627.0 | | 1590.0–1700.0 | | 297 | |
| Afropavo congensis | M | 2 | 1418.0 | | 1361.0–1475.0 | | | 1103.0 |
| | F | 2 | 1144.0 | | 1135.0–1154.0 | | 297 | |
| Pavo cristatus | M | | | | 4000.0–6000.0 | | | 1104.0 |
| | F | 3 | | | 2750.0–4000.0 | | 5 | |
| Pavo muticus | M | | | | 3850.0–5000.0 | | | 1105.0 |
| | F | 3 | | | 1060.0–1160.0 | | 297 | |
| Dendragapus falcipennis | U | | 600.0 | | | | estimated | 1106.0 |
| | | | | | | | 149 | |
| Dendragapus canadensis | M | 62 | 492.0 | | 400.0–489.0 | PB | Washington, USA | 1107.0 |
| | F | 84 | 456.0 | | 370.0–513.0 | | 703 | |
| Dendragapus obscurus | M | 359 | 1188.0 | | | PB | Washington, USA | 1108.0 |
| | F | 410 | 891.0 | | | | 702 | |
| Lagopus lagopus | M | 498 | 601.0 | 26.50 | | W | Alaska, USA | 1109.0 |
| | F | 326 | 516.0 | 18.20 | | | 670 | |
| Lagopus mutus | B | 139 | 422.0 | | 359.0–482.0 | W | Alaska, USA | 1110.0 |
| | | | | | | | 664 | |
| Lagopus leucurus | M | 25 | 359.0 | | 298.0–416.0 | | | 1111.0 |
| | F | 30 | 351.0 | | 279.0–381.0 | | 664 | |
| Tetrao tetrix | M | 26 | 1255.0 | | 1000.0–1400.0 | W | | 1112.0 |
| | F | 35 | 910.0 | | 765.0–1050.0 | | 149 | |
| Tetrao urogallus | M | 75 | 4100.0 | | 3600.0–5050.0 | | | 1114.0 |
| | F | 10 | 1800.0 | | 1700.0–1920.0 | | 149 | |
| Tetrao parvirostris kamschaticus | M | 4 | 3640.0 | | 3445.0–3800.0 | | 149 | 1115.0 |
| Bonasa bonasia | B | 56 | 429.0 | | 340.0–500.0 | W | | 1116.0 |
| | | | | | | | 149 | |
| Bonasa umbellus | M | 180 | 621.0 | | | Y | New York, USA | 1118.0 |
| | F | 214 | 532.0 | | | | 85 | |
| Centrocercus urophasianus | M | 465 | 3190.0 | 183.00 | | B | Colorado, USA | 1119.0 |
| | F | 221 | 1745.0 | 151.00 | | | 26 | |
| Tympanuchus phasianellus | M | 236 | 953.0 | | −1090.0 | | | 1120.0 |
| | F | 247 | 817.0 | | −999.0 | | 412 | |

## Body Masses of World Birds (continued)

| Species | Sex | N | Mean | Std dev | Range | Sn | Location | Number |
|---|---|---|---|---|---|---|---|---|
| Tympanuchus cupido | M | 22 | 999.0 | | −1362.0 | | | 1121.0 |
| | F | 16 | 772.0 | | −908.0 | | 412 | |
| Tympanuchus cupido attwateri | M | 8 | 1014.0 | | 760.0−1135.0 | | Texas, USA | 1121.0 |
| | F | 6 | 730.0 | | 708.0−786.0 | | 340 | |
| Tympanuchus pallidicinctus | M | 20 | 784.0 | | 667.0−895.0 | | Oklahoma, USA | 1122.0 |
| | F | 5 | 727.0 | | 676.0−781.0 | | 340 | |
| Meleagris gallopavo | M | 54 | 7400.0 | | −10800.0 | | | 1123.0 |
| | F | 55 | 4222.0 | | −5584.0 | | 412 | |
| Agriocharis ocellata | U | 1 | 5525.0 | | | | Belize 510 | 1124.0 |
| Dendrortyx barbatus | B | 2 | 432.0 | | 405.0−459.0 | | estimated 298 | 1125.0 |
| Dendrortyx macroura | B | 4 | 431.0 | | 374.0−455.0 | | Morelos, Mexico 659 | 1126.0 |
| Dendrortyx leucophrys | U | | 350.0 | | | | 603a | 1127.0 |
| Oreortyx pictus | U | 56 | 233.0 | | −293.0 | | 412 | 1128.0 |
| Callipepla squamata | M | 143 | 191.0 | | −234.0 | | | 1129.0 |
| | F | 132 | 177.0 | | −218.0 | | 293a | |
| Callipepla douglasii | M | | 175.0 | | | | | 1130.0 |
| | F | | 169.0 | | | | 298 | |
| Callipepla californica | M | 418 | 176.0 | | −207.0 | | | 1131.0 |
| | F | 272 | 170.0 | | −207.0 | | 412 | |
| Callipepla gambelii | M | 145 | 170.0 | | −207.0 | | | 1132.0 |
| | F | 103 | 162.0 | | −193.0 | | 412 | |
| Philortyx fasciatus | B | 4 | 130.0 | | 125.0−136.0 | | Mexico 458 | 1133.0 |
| Colinus virginianus | U | 847 | 178.0 | | | W | Illinois, USA 630 | 1134.0 |
| Colinus nigrogularis | M | 9 | 127.0 | | 109.0−152.0 | | | 1135.0 |
| | F | 6 | 131.0 | | 120.0−146.0 | | 272, 323, 457, 510 | |
| Colinus cristatus | M | 7 | 139.0 | | 117.0−153.0 | | | 1136.0 |
| | F | 2 | 129.0 | | 127.0−131.0 | | 245, 246, 384 | |
| Odontophorus gujanensis | U | | 300.0 | | | | 603a | 1137.0 |
| Odontophorus capueira | M | | 457.0 | | | | estimated | 1138.0 |
| | F | | 396.0 | | | | 298 | |
| Odontophorus erythrops | U | | 280.0 | | | | 603a | 1140.0 |
| Odontophorus atrifrons | B | | 304.0 | | 298.0−311.0 | | estimated 298 | 1141.0 |
| Odontophorus hyperythrus | M | | 392.0 | | | | estimated | 1142.0 |
| | F | | 352.0 | | | | 298 | |

## Body Masses of World Birds (continued)

| Species | Sex | N | Mean | Std dev | Range | Sn | Location | Number |
|---|---|---|---|---|---|---|---|---|
| Odontophorus melanonotus | U | | 322.0 | | | | estimated 298 | 1143.0 |
| Odontophorus speciosus | B | | 317.0 | | 302.0–332.0 | | estimated 298 | 1144.0 |
| Odontophorus strophium | U | | 302.0 | | | | estimated 298 | 1146.0 |
| Odontophorus columbianus | B | | 340.0 | | 336.0–343.0 | | estimated 298 | 1147.0 |
| Odontophorus leucolaemus | U | | 275.0 | | | | 603a | 1148.0 |
| Odontophorus stellatus | U | | 310.0 | | | | Peru 623 | 1150.0 |
| Odontophorus guttatus | M | 18 | 314.0 | 21.60 | | | Panama 244 | 1151.0 |
| | F | 4 | 294.0 | | | | | |
| Dactylortyx thoracicus | M | | | | 180.0–266.0 | | | 1152.0 |
| | F | | | | 168.0–206.0 | | 298 | |
| Cyrtonyx montezumae | M | 45 | 195.0 | | –224.0 | | Mexico | 1153.0 |
| | F | 22 | 176.0 | 15.90 | –200.0 | | 341 | |
| Cyrtonyx ocellatus | M | | 218.0 | | | | cstimatcd | 1154.0 |
| | F | | 182.0 | | | | 298 | |
| Rhynchortyx cinctus | U | | 150.0 | | | | 603a | 1155.0 |

### ORDER: GRUIFORMES    FAMILY: TURNICIDAE

| Species | Sex | N | Mean | Std dev | Range | Sn | Location | Number |
|---|---|---|---|---|---|---|---|---|
| Turnix sylvatica | U | | | | 36.0–43.0 | | 5 | 1156.0 |
| Turnix maculosa | M | 1 | 26.0 | | | | New Guinea 154 | 1157.0 |
| | F | | | | | | | |
| Turnix hottentotta | M | 2 | 40.1 | | 40.0–40.2 | | | 1159.0 |
| | F | 2 | 60.0 | | 57.5–62.4 | | 636 | |
| Turnix tanki | U | | | | 36.0–43.0 | | 5 | 1160.0 |
| Turnix suscitator | M | | | | 43.0–57.0 | | | 1162.0 |
| | F | | | | 43.0–72.0 | | 5 | |
| Turnix varia | U | | 88.0 | | | | 194 | 1165.0 |
| Turnix pyrrhothorax | M | 4 | 32.5 | | 27.0–37.0 | | Australia | 1170.0 |
| | F | 9 | 50.7 | | 31.0–83.0 | | 394 | |
| Turnix velox | U | | 41.0 | | | | 134 | 1171.0 |
| Ortyxelos meiffrenii | M | 2 | 17.6 | | 15.7–19.5 | | 636 | 1172.0 |

### ORDER: GRUIFORMES    FAMILY: RALLIDAE

| Species | Sex | N | Mean | Std dev | Range | Sn | Location | Number |
|---|---|---|---|---|---|---|---|---|
| Sarothrura pulchra | M | 9 | 45.3 | | 39.0–49.0 | | | 1173.0 |
| | F | 2 | 42.0 | | 41.0–43.0 | | 636 | |

## Body Masses of World Birds (continued)

| Species | Sex | N | Mean | Std dev | Range | Sn | Location | Number |
|---|---|---|---|---|---|---|---|---|
| Sarothrura elegans | M | 3 | 40.3 | | 32.0–45.0 | | | 1174.0 |
| | F | 3 | 50.2 | | 49.5–51.0 | | 636 | |
| Sarothrura rufa | M | 4 | 33.8 | | 30.0–42.0 | | | 1175.0 |
| | | | | | | | 636 | |
| Sarothrura boehmi | F | 1 | 21.4 | | | | Zimbabwe | 1177.0 |
| | | | | | | | 285 | |
| Sarothrura affinis | M | 1 | 28.8 | | | | | 1178.0 |
| | | | | | | | 636 | |
| Sarothrura insularis | F | 1 | 30.0 | | | | Madagascar | 1179.0 |
| | | | | | | | 39 | |
| Sarothrura ayresi | F | 1 | 14.0 | | | | | 1180.0 |
| | | | | | | | 636 | |
| Himantornis haematopus | F | 1 | 390.0 | | | | | 1182.0 |
| | | | | | | | 636 | |
| Canirallus kioloides | U | 4 | 269.0 | | 258.0–280.0 | | Madagascar | 1184.0 |
| | | | | | | | 39 | |
| Coturnicops noveboracensis | U | 26 | 51.6 | | | S | Michigan, USA | 1186.0 |
| | | | | | | | 645 | |
| Rallina rubra | U | 2 | 73.5 | | 71.0–76.0 | | New Guinea | 1189.0 |
| | | | | | | | 216 | |
| Rallina forbesi | B | 2 | 87.5 | | 87.0–88.0 | | New Guinea | 1191.0 |
| | | | | | | | 216 | |
| Rallina tricolor | U | 4 | 209.0 | | | | | 1193.0 |
| | | | | | | | 35 | |
| Rallina eurizonoides | B | 2 | 111.0 | | 110.0–112.0 | | Philippines | 1196.0 |
| | | | | | | | 475, 584 | |
| Anurolimnas viridis | U | 2 | 65.9 | | 58.8–73.0 | | | 1198.0 |
| | | | | | | | 246, 437 | |
| Laterallus melanophaius | U | 6 | 51.4 | | | | Peru | 1200.0 |
| | | | | | | | 193 | |
| Laterallus ruber | B | 8 | 45.3 | | 41.6–48.9 | | | 1202.0 |
| | | | | | | | 272, 510 | |
| Laterallus albigularis | M | 13 | 49.7 | 5.77 | | | Panama | 1203.0 |
| | F | 12 | 45.0 | 6.34 | | | 244 | |
| Latcrallus exilis | U | | 33.0 | | | | | 1204.0 |
| | | | | | | | 603a | |
| Laterallus jamaicensis | U | 4 | 33.9 | | 24.6–39.9 | | | 1205.0 |
| | | | | | | | 510, 585, 586 | |
| Laterallus leucopyrrhus | F | 1 | 46.5 | | | | Brazil | 1207.0 |
| | | | | | | | 37 | |
| Laterallus xenopterus | U | 1 | 53.0 | | | | Brazil | 1208.0 |
| | | | | | | | 563 | |
| Gallirallus australis | M | | 820.0 | | 532.0–1117.0 | | | 1211.0 |
| | F | | 700.0 | | 382.0–1010.0 | | 78 | |

## Body Masses of World Birds (continued)

| Species | Sex | N | Mean | Std dev | Range | Sn | Location | Number |
|---|---|---|---|---|---|---|---|---|
| Gallirallus torquatus | F | 1 | 241.0 | | | | Philippines 475 | 1216.0 |
| Gallirallus philippensis | U | 16 | 180.0 | | 142.0–218.0 | | Tonga 491 | 1218.0 |
| Gallirallus owstoni | M | 2 | 230.0 | | 222.0–239.0 | | captive 598 | 1219.0 |
| Gallirallus striatus | U | | | | 100.0–142.0 | | 5 | 1225.0 |
| Rallus pectoralis | B | 10 | 77.6 | 8.90 | 65.0–92.0 | | 154, 216 | 1227.0 |
| Rallus longirostris | M | 13 | 323.0 | 20.70 | 300.0–350.0 | | S. Carolina,USA 376 | 1229.0 |
| | F | 7 | 271.0 | | 250.0–275.0 | | | |
| Rallus elegans | M | 9 | 415.0 | | 340.0–490.0 | | eastern USA 376 | 1230.0 |
| | F | 9 | 306.0 | | 253.0–325.0 | | | |
| Rallus limicola | M | 9 | 89.0 | | 64.0–120.0 | | Ontario, Canada 590 | 1232.0 |
| | F | 3 | 74.9 | | 67.0–79.6 | | | |
| Rallus aquaticus | U | 50 | 120.0 | | 92.0–164.0 | S | Britain 258 | 1235.0 |
| Rallus caerulescens | U | 2 | 112.0 | | 80.0–144.0 | | 78 | 1236.0 |
| Rallus madagascariensis | M | 1 | 148.0 | | | | Madagascar 39 | 1237.0 |
| Crecopsis egregia | U | 6 | 119.0 | | 110.0–137.0 | | 285, 636 | 1238.0 |
| Crex crex | M | 28 | 169.0 | 18.70 | 135.0–202.0 | B | Netherlands 116 | 1239.0 |
| | F | 3 | 142.0 | | 140.0–145.0 | | | |
| Dryolimnas cuvieri | M | 32 | 189.0 | | 145.0 | | Madagascar | 1242.0 |
| | F | 21 | 176.0 | | 138.0–223.0 | | | 39 |
| Atlantisia rogersi | M | 6 | 41.8 | | 38.0–49.0 | | Tristan de Cund 514 | 1243.0 |
| | F | 7 | 36.9 | | 34.0–42.0 | | | |
| Aramides mangle | F | 1 | 164.0 | | | | Brazil 621 | 1244.0 |
| Aramides axillaris | M | 2 | 268.0 | | 262.0–275.0 | | Honduras; Mexico 584 | 1245.0 |
| Aramides cajanea | B | 18 | 397.0 | | | | Panama 244 | 1246.0 |
| Aramides ypecaha | M | 1 | 860.0 | | | | Paraguay 610 | 1248.0 |
| | F | 1 | 765.0 | | | | | |
| Aramides saracura | M | 1 | 540.0 | | | | Brazil 37 | 1249.0 |
| Amaurolimnas concolor | M | 1 | 133.0 | | | | Brazil 622 | 1251.0 |
| Gymnocrex plumbeiventris | U | | 300.0 | | | | estimated 35 | 1253.0 |

## Body Masses of World Birds (continued)

| Species | Sex | N | Mean | Std dev | Range | Sn | Location | Number |
|---|---|---|---|---|---|---|---|---|
| Amaurornis akool | M | | | | 114.0–170.0 | | | 1254.0 |
| | F | | | | 110.0–140.0 | | 5 | |
| Amaurornis olivaceus | B | 14 | 209.0 | | | | | 1256.0 |
| | | | | | | | 35, 154, 219, 475 | |
| Amaurornis phoenicurus | F | 2 | 173.0 | | 166.0–180.0 | | Philippines | 1257.0 |
| | | | | | | | 475 | |
| Porzana parva | U | | | | 42.0–56.0 | | | 1261.0 |
| | | | | | | | 5 | |
| Porzana pusilla | M | 2 | 32.5 | | 30.0–35.0 | | Australia | 1262.0 |
| | | | | | | | 369 | |
| Porzana porzana | U | 16 | 77.9 | | 65.0–96.0 | Y | Britain | 1264.0 |
| | | | | | 65.0–96.0 | | 258 | |
| Porzana carolina | U | 11 | 74.6 | | | | | 1266.0 |
| | | | | | | | 592 | |
| Porzana albicollis | M | 2 | 112.0 | | 110.0–114.0 | | Peru | 1268.0 |
| | | | | | | | 584 | |
| Porzana fusca | U | | 60.0 | | | | | 1270.0 |
| | | | | | | | 78 | |
| Porzana tabuensis | B | 2 | 45.5 | | 44.0–47.0 | | New Guinea | 1272.0 |
| | | | | | | | 154 | |
| Porzana flaviventer | U | | 25.0 | | | | | 1275.0 |
| | | | | | | | 603a | |
| Aenigmatolimnas marginalis | F | 1 | 61.0 | | | | Zaire | 1277.0 |
| | | | | | | | 116 | |
| Neocrex erythrops | U | 3 | 62.1 | | 55.3–70.0 | | Columbia | 1280.0 |
| | | | | | | | 387 | |
| Pardirallus maculatus | B | 16 | 171.0 | 18.20 | 140.0–198.0 | | | 1281.0 |
| | | | | | | | 37, 46, 112, 496, 589 | |
| Pardirallus nigricans | M | 1 | 217.0 | | | | Brazil | 1282.0 |
| | | | | | | | 37 | |
| Pardirallus sanguinolentus | M | 3 | 197.0 | | 170.0–213.0 | | Brazil | 1283.0 |
| | | | | | | | 37 | |
| Gallicrex cinerea | M | 7 | 546.0 | | 476.0–650.0 | | Philippines | 1287.0 |
| | F | 3 | 356.0 | | 298.0–434.0 | | 475 | |
| Porphyrio porphyrio | M | 2 | 840.0 | | 840.0–840.0 | B | New Caledonia | 1288.0 |
| poliocephalus | F | 3 | 733.0 | | 690.0–820.0 | | 506 | |
| Porphyrio alleni | U | | 140.0 | | | | Ghana | 1291.0 |
| | | | | | | | 228 | |
| Porphyrio martinicus | M | 20 | 257.0 | 27.10 | | | | 1292.0 |
| | F | 8 | 215.0 | | | | 244 | |
| Porphyrio flavirostris | M | 2 | 98.5 | | 93.0–104.0 | | Argentina | 1293.0 |
| | | | | | | | 111 | |
| Gallinula nesiotis | U | | 400.0 | | | | | 1296.0 |
| | | | | | | | 78 | |

## Body Masses of World Birds (continued)

| Species | Sex | N | Mean | Std dev | Range | Sn | Location | Number |
|---------|-----|---|------|---------|-------|----|----------|--------|
| Gallinula chloropus | M | 103 | 340.0 | | 186.0–493.0 | | | 1297.0 |
| | F | 110 | 265.0 | | 146.0–375.0 | | 78 | |
| Gallinula tenebrosa | U | | 547.0 | | | | | 1298.0 |
| | | | | | | | 134 | |
| Gallinula angulata | U | | 150.0 | | | | | 1299.0 |
| | | | | | | | 78 | |
| Gallinula melanops | U | 1 | 154.0 | | | | Brazil | 1300.0 |
| | | | | | | | 37 | |
| Gallinula ventralis | U | | 383.0 | | | | W. Australia | 1301.0 |
| | | | | | | | 550 | |
| Gallinula mortierii | U | | 400.0 | | | | | 1302.0 |
| | | | | | | | 134 | |
| Fulica cristata | U | 37 | 826.0 | 96.80 | 585.0–1080.0 | | South Africa | 1303.0 |
| | | | | | | | 255 | |
| Fulica atra | U | 102 | | | 555.0–1150.0 | W | Caspian Sea | 1304.0 |
| | | | | | | | 149 | |
| Fulica americana | M | 27 | 724.0 | | 576.0–848.0 | F | | 1306.0 |
| | F | 20 | 560.0 | | 427.0–628.0 | | 517 | |
| Fulica leucoptera | F | 4 | | | 400.0–500.0 | | Brazil | 1308.0 |
| | | | | | | | 37 | |
| Fulica rufifrons | M | 1 | 685.0 | | | | Brazil | 1311.0 |
| | F | 1 | 550.0 | | | | 37 | |

**ORDER: GRUIFORMES**    **FAMILY: HELIORNITHIDAE**

| Species | Sex | N | Mean | Std dev | Range | Sn | Location | Number |
|---------|-----|---|------|---------|-------|----|----------|--------|
| Podica senegalensis | U | 4 | 599.0 | | 338.0–879.0 | | | 1314.0 |
| | | | | | | | 636 | |
| Heliornis fulica | M | 4 | 140.0 | | | | Panama | 1316.0 |
| | F | 3 | 130.0 | | | | 244 | |

**ORDER: GRUIFORMES**    **FAMILY: RHYNOCHETIDAE**

| Species | Sex | N | Mean | Std dev | Range | Sn | Location | Number |
|---------|-----|---|------|---------|-------|----|----------|--------|
| Rhynochetos jubatus | M | 1 | 860.0 | | | | captive | 1317.0 |
| | | | | | | | 223 | |

**ORDER: GRUIFORMES**    **FAMILY: EURYPYGIDAE**

| Species | Sex | N | Mean | Std dev | Range | Sn | Location | Number |
|---------|-----|---|------|---------|-------|----|----------|--------|
| Eurypyga helias | B | 4 | 222.0 | | 178.0–295.0 | | | 1318.0 |
| | | | | | | | 193, 245 | |

**ORDER: GRUIFORMES**    **FAMILY: GRUIDAE**

| Species | Sex | N | Mean | Std dev | Range | Sn | Location | Number |
|---------|-----|---|------|---------|-------|----|----------|--------|
| Grus grus | U | 16 | 5500.0 | 129.00 | | W | Spain | 1322.0 |
| | | | | | | | 472 | |
| Grus nigricollis | F | 1 | 6000.0 | | | | | 1323.0 |
| | | | | | | | 296 | |
| Grus monacha | M | 7 | 3930.0 | | | | captive | 1324.0 |
| | F | 4 | 3540.0 | | | | 296 | |

## Body Masses of World Birds (continued)

| Species | Sex | N | Mean | Std dev | Range | Sn | Location | Number |
|---|---|---|---|---|---|---|---|---|
| Grus canadensis canadensis | M | 33 | 3350.0 | 22.30 | 2700.0–3700.0 | W | New Mexico, USA | 1325.0 |
|  | F | 31 | 2982.0 | 19.00 | 2450.0–3300.0 |  | 177 |  |
| Grus canadensis tabida | M | 61 | 5797.0 | 31.50 | 5040.0–6700.0 | W | New Mexico, USA | 1325.1 |
|  | F | 28 | 5345.0 | 29.80 | 4900.0–6030.0 |  | 177 |  |
| Grus canadensis pratensis | U | 6 | 5089.0 |  | 4426.0–5788.0 |  | captive | 1325.2 |
|  |  |  |  |  |  |  | 177 |  |
| Grus japonensis | U | 14 | 8786.0 | 1200.00 |  |  |  | 1326.0 |
|  |  |  |  |  |  |  | 489 |  |
| Grus americana | U | 3 | 5826.0 |  | 5448.0–6356.0 |  | captive | 1327.0 |
|  |  |  |  |  |  |  | 177 |  |
| Grus vipio | U | 3 | 4663.0 |  |  |  |  | 1328.0 |
|  |  |  |  |  |  |  | 489 |  |
| Grus leucogeranus | M | 7 | 6387.0 |  | 5100.0–7400.0 | B |  | 1329.0 |
|  | F | 4 | 5475.0 |  | 4900.0–6000.0 |  | 296 |  |
| Grus antigone | U | 4 | 8863.0 |  |  |  |  | 1330.0 |
|  |  |  |  |  |  |  | 489 |  |
| Grus rubicunda | M | 321 | 6383.0 |  | 4761.0–8729.0 |  |  | 1331.0 |
|  | F | 217 | 5663.0 |  | 3628.0–7255.0 |  | 296 |  |
| Grus carunculatus | B | 3 | 8159.0 |  | 7225.0–8966.0 |  |  | 1332.0 |
|  |  |  |  |  |  |  | 78, 296 |  |
| Anthropoides virgo | B | 4 | 2308.0 |  | 2100.0–2500.0 | B |  | 1333.0 |
|  |  |  |  |  |  |  | 296 |  |
| Anthropoides paradisea | M | 1 | 5675.0 |  |  |  | captive | 1334.0 |
|  | F | 1 | 3633.0 |  |  |  | 296 |  |
| Balearica pavonina | U | 21 | 3590.0 |  | 2725.0–4100.0 |  |  | 1335.0 |
|  |  |  |  |  |  |  | 78 |  |
| Balearica regulorum | F | 2 | 3772.0 |  | 3575.0–3970.0 |  |  | 1336.0 |
|  |  |  |  |  |  |  | 636 |  |

**ORDER: GRUIFORMES**  **FAMILY: ARAMIDAE**

| Species | Sex | N | Mean | Std dev | Range | Sn | Location | Number |
|---|---|---|---|---|---|---|---|---|
| Aramus guarauna | U | 31 | 1080.0 | 110.00 | 900.0–1270.0 | PB | Florida, USA | 1337.0 |
|  |  |  |  |  |  |  | 415 |  |

**ORDER: GRUIFORMES**  **FAMILY: PSOPHIIDAE**

| Species | Sex | N | Mean | Std dev | Range | Sn | Location | Number |
|---|---|---|---|---|---|---|---|---|
| Psophia crepitans | M | 1 | 1050.0 |  |  |  | Fr. Guiana | 1338.0 |
|  |  |  |  |  |  |  | 161 |  |
| Psophia leucoptera | U |  | 990.0 |  |  |  | Peru | 1339.0 |
|  |  |  |  |  |  |  | 623 |  |
| Psophia viridis | U |  | 1000.0 |  |  |  | Brazil | 1340.0 |
|  |  |  |  |  |  |  | 564 |  |

**ORDER: GRUIFORMES**  **FAMILY: CARIAMIDAE**

| Species | Sex | N | Mean | Std dev | Range | Sn | Location | Number |
|---|---|---|---|---|---|---|---|---|
| Cariama cristata | U |  | 1400.0 |  |  |  | Brazil | 1341.0 |
|  |  |  |  |  |  |  | 564 |  |

## Body Masses of World Birds (continued)

| Species | Sex | N | Mean | Std dev | Range | Sn | Location | Number |
|---------|-----|---|------|---------|-------|----|----------|--------|
| Chunga burmeisteri | B | 4 | 1298.0 | | 1224.0–1395.0 | | Bolivia 112, 523 | 1342.0 |

**ORDER: GRUIFORMES**          **FAMILY: OTIDIDAE**

| Species | Sex | N | Mean | Std dev | Range | Sn | Location | Number |
|---------|-----|---|------|---------|-------|----|----------|--------|
| Tetrax tetrax | U | | | | 600.0–900.0 | | 5 | 1343.0 |
| Otis tarda | M | | | | 7200.0–11200.0 | S | USSR | 1344.0 |
| | F | | | | 4000.0–8000.0 | | 149 | |
| Neotis denhami | M | 1 | 4120.0 | | | | Sudan | 1345.0 |
| | F | | | | | | 680 | |
| Neotis ludwigii | U | | | | 3100.0–7300.0 | | 636 | 1346.0 |
| Neotis nuba | M | | 5440.0 | | | | 116 | 1347.0 |
| Neotis heuglinii | M | 2 | 6000.0 | | 4000.0–8000.0 | | Kenya | 1348.0 |
| | F | 2 | 2800.0 | | 2600.0–3000.0 | | 636 | |
| Ardeotis arabs | M | 2 | 7850.0 | | 5700.0–10000.0 | | | 1349.0 |
| | F | 1 | 4500.0 | | | | 636 | |
| Ardeotis kori kori | U | | | | 13500.0–19000.0 | | 636 | 1350.0 |
| Ardeotis kori struthiuniculus | M | 1 | 10900.0 | | | | | 1350.0 |
| | F | 2 | 5900.0 | | | | 636 | |
| Ardeotis nigriceps | M | | | | 8000.0–14500.0 | | | 1351.0 |
| | F | | | | 3500.0–6750.0 | | 5 | |
| Ardeotis australis | M | | 7200.0 | | 6350.0–12760.0 | | | 1352.0 |
| | F | | 5400.0 | | 4500.0–6350.0 | | 78 | |
| Chlaymdotis undulata | M | 4 | 1758.0 | | 1150.0–2380.0 | | Mongolia; USSR | 1353.0 |
| | F | 2 | 1100.0 | | 1100.0–1100.0 | | 149 | |
| Eupodotis ruficrista | M | 3 | 680.0 | | 550.0–770.0 | | 636 | 1356.0 |
| Eupodotis afra | U | 102 | 690.0 | | 500.0–878.0 | | 78 | 1358.0 |
| Eupodotis senegalensis | U | | 1400.0 | | | | 636 | 1362.0 |
| Eupodotis caerulescens | U | 2 | 1366.0 | | 1120.0–1612.0 | | South Africa 255 | 1363.0 |
| Eupodotis melanogaster | M | | 1020.0 | | | | 66 | 1364.0 |
| Eupodotis hartlaubii | F | | 1190.0 | | | | 66 | 1365.0 |
| Eupodotis bengalensis | B | | | | 1800.0–2250.0 | | 5 | 1366.0 |
| Eupodotis indica | B | | | | 510.0–740.0 | | 5 | 1367.0 |

## Body Masses of World Birds (continued)

| Species | Sex | N | Mean | Std dev | Range | Sn | Location | Number |
|---|---|---|---|---|---|---|---|---|
| **ORDER: CHARADRIIFORMES** | | | | | **FAMILY: JACANIDAE** | | | |
| Actophilornis africanus | M | 8 | 143.0 | | | | | 1368.0 |
| | F | 5 | 261.0 | | | | 289 | |
| Actophilornis albinucha | F | 1 | 239.0 | | | | | 1369.0 |
| | | | | | | | 294 | |
| Microparra capensis | F | 1 | 41.3 | | | | | 1370.0 |
| | | | | | | | 294 | |
| Irediparra gallinacea | M | 6 | 75.1 | | 68.0–84.0 | | | 1371.0 |
| | F | 7 | 130.0 | | 120.0–149.0 | | 294 | |
| Hydrophasianus chirurgus | M | 5 | 126.0 | | 113.0–135.0 | | | 1372.0 |
| | F | 3 | 231.0 | | 205.0–260.0 | | 294 | |
| Metopidius indicus | B | 10 | 155.0 | | 94.0–210.0 | | | 1373.0 |
| | | | | | | | 294 | |
| Jacana spinosa | M | 20 | 78.9 | 8.50 | | | Panama | 1374.0 |
| | F | 16 | 112.0 | 10.80 | | | 244 | |
| Jacana jacana | M | 16 | 108.0 | | | | | 1375.0 |
| | F | 15 | 143.0 | | | | 289 | |
| **ORDER: CHARADRIIFORMES** | | | | | **FAMILY: ROSTRATULIDAE** | | | |
| Rostratula benghalensis | B | 34 | 121.0 | | 90.0–164.0 | | | 1376.0 |
| | | | | | | | 294 | |
| Rostratula semicollaris | B | 5 | 76.6 | | 68.0–86.0 | | | 1377.0 |
| | | | | | | | 294 | |
| **ORDER: CHARADRIIFORMES** | | | | | **FAMILY: SCOLOPACIDAE** | | | |
| Scolopax rusticola | M | 250 | 306.0 | 26.10 | 250.0–410.0 | | Ireland | 1378.0 |
| | F | 234 | 313.0 | 25.50 | 205.0–420.0 | | 116 | |
| Scolopax saturata rosenbergii | F | 1 | 220.0 | | | | | 1380.0 |
| | | | | | | | 294 | |
| Scolopax minor | M | 390 | 176.0 | | –222.0 | | | 1383.0 |
| | F | 313 | 219.0 | | –278.0 | | 412 | |
| Gallinago solitaria | M | | | | 130.0–148.0 | | | 1384.0 |
| | F | | | | 126.0–159.0 | | 294 | |
| Gallinago hardwickii | B | 499 | 156.0 | | | | | 1385.0 |
| | | | | | | | 289 | |
| Gallinago nemoricola | U | | | | 148.0–198.0 | | | 1386.0 |
| | | | | | | | 5 | |
| Gallinago stenura | B | 472 | 113.0 | | 85.0–134.0 | | | 1387.0 |
| | | | | | | | 5 | |
| Gallinago megala | U | 7 | 140.0 | | 112.0–164.0 | | | 1388.0 |
| | | | | | | | 294, 475 | |
| Gallinago media | M | 143 | 157.0 | 6.40 | | B | Norway | 1389.0 |
| | F | 67 | 184.0 | 14.50 | | | 264 | |

## Body Masses of World Birds (continued)

| Species | Sex | N | Mean | Std dev | Range | Sn | Location | Number |
|---------|-----|---|------|---------|-------|-----|----------|--------|
| Gallinago gallinago | M | 15 | 128.0 | | −156.0 | | | 1390.0 |
| | F | 14 | 116.0 | | −156.0 | | 412 | |
| Gallinago gallinago raddei | M | 20 | 97.0 | | | | China | 1390.0 |
| | F | 16 | 113.0 | | | | 289 | |
| Gallinago nigripennis | U | 28 | 112.0 | | 90.0−164.0 | | | 1391.0 |
| | | | | | | | 78 | |
| Gallinago macrodactyla | F | 1 | 216.0 | | | | | 1392.0 |
| | | | | | | | 294 | |
| Gallinago paraguaiae | M | 3 | 77.6 | | 65.0−90.0 | | | 1393.0 |
| | F | 4 | 97.0 | | 90.0−105.0 | | 294 | |
| Gallinago nobilis | M | 1 | 188.0 | | | | | 1395.0 |
| | F | 1 | 197.0 | | | | 294 | |
| Gallinago undulata | M | 5 | 294.0 | | 270.0−320.0 | | | 1396.0 |
| | F | 3 | 332.0 | | 282.0−363.0 | | 294 | |
| Gallinago jamesoni | U | 8 | 166.0 | | 140.0−224.0 | | | 1397.0 |
| | | | | | | | 294 | |
| Lymnocryptes minimus | M | 17 | 53.7 | 6.90 | 41.0−63.0 | W | Netherlands | 1400.0 |
| | F | 8 | 46.7 | | 33.0−73.0 | | 117 | |
| Coenocorypha pusilla | F | 24 | 85.4 | | | | | 1401.0 |
| | | | | | | | 388 | |
| Coenocorypha aucklandica | M | 29 | 101.0 | 4.60 | 92.5−112.0 | | New Zealand | 1402.0 |
| | F | 21 | 111.0 | 5.80 | 103.0−120.0 | | 388 | |
| Limosa limosa | M | 11 | 252.0 | | 235.0−367.0 | | | 1403.0 |
| | F | 11 | 330.0 | | 297.0−362.0 | | 149 | |
| Limosa haemastica | M | 6 | 222.0 | | | | | 1404.0 |
| | F | 6 | 289.0 | | | | 289 | |
| Limosa lapponica | M | 69 | 309.0 | 25.20 | 233.0−360.0 | W | Britain | 1405.0 |
| | F | 20 | 376.0 | 25.10 | 348.0−455.0 | | 117 | |
| Limosa fedoa | M | 10 | 320.0 | | 281.0−362.0 | | | 1406.0 |
| | F | 9 | 421.0 | | 240.0−510.0 | | 294 | |
| Numenius minutus | B | 8 | 351.0 | | 300.0−440.0 | | | 1408.0 |
| | | | | | | | 294 | |
| Numenius phaeopus | M | 29 | 355.0 | 22.10 | 310.0−403.0 | B | Manitoba, Canada | 1409.0 |
| | F | 36 | 404.0 | 29.10 | 345.0−459.0 | | 567 | |
| Numenius tahitiensis | M | 10 | 378.0 | | | | | 1410.0 |
| | F | 10 | 489.0 | | | | 289 | |
| Numenius tenuirostris | F | 2 | 308.0 | | 255.0−360.0 | | | 1411.0 |
| | | | | | | | 294 | |
| Numenius arquata | M | 124 | 742.0 | 68.70 | 540.0−900.0 | F | Netherlands | 1412.0 |
| | F | 97 | 869.0 | 68.20 | 720.0−1050.0 | | 117 | |
| Numenius americanus | M | 12 | 531.0 | 32.70 | 493.0−597.0 | B | Idaho, USA | 1413.0 |
| | F | 24 | 642.0 | 31.50 | 570.0−689.0 | | 177 | |
| Numenius madagascariensis | U | 12 | 792.0 | | 565.0−1150.0 | W | | 1414.0 |
| | | | | | | | 294 | |

## Body Masses of World Birds (continued)

| Species | Sex | N | Mean | Std dev | Range | Sn | Location | Number |
|---|---|---|---|---|---|---|---|---|
| Bartramia longicauda | M | 8 | 137.0 | | | | | 1415.0 |
| | F | 6 | 164.0 | | | | 289 | |
| Tringa erythropus | U | 65 | 158.0 | 13.10 | 140.0–210.0 | F | Britain 117 | 1416.0 |
| Tringa totanus | B | 200 | 129.0 | | 107.0–152.0 | B | 294 | 1417.0 |
| Tringa stagnatilis | B | 61 | 77.5 | | 55.0–120.0 | M | China 117 | 1418.0 |
| Tringa nebularia | B | 51 | 174.0 | | 129.0–245.0 | Y | 117 | 1419.0 |
| Tringa guttifer | M | 3 | | | 136.0–141.0 | | | 1420.0 |
| | F | 1 | 158.0 | | | | 294 | |
| Tringa melanoleuca | U | 15 | 171.0 | 15.40 | 124.0–224.0 | W | Surinam 117 | 1421.0 |
| Tringa flavipes | B | 20 | 81.0 | | 69.0–94.0 | B | Alaska, USA 278 | 1422.0 |
| Tringa solitaria | U | 104 | 48.4 | 8.60 | 31.1–65.1 | S | Venezuela 626 | 1423.0 |
| Tringa ochropus | U | 19 | 71.4 | 10.40 | 53.0–93.0 | M | north Africa 117 | 1424.0 |
| Tringa glareola | M | 16 | 62.0 | | 52.0–81.0 | B | USSR | 1425.0 |
| | F | 11 | 73.0 | | 51.0–89.0 | | 117 | |
| Tringa cinerea | B | 33 | 72.0 | | 58.0–108.0 | B | Finland; USSR 117 | 1426.0 |
| Tringa hypoleucos | B | 38 | 51.7 | | 38.0–70.0 | F | 117 | 1427.0 |
| Tringa macularia | U | 56 | 40.4 | 6.15 | 29.4–59.8 177 | S | Pennsylvania, USA | 1428.0 |
| Tringa brevipes | B | 6 | 107.0 | | 79.0–148.0 | | Alaska, USA 88, 213 | 1429.0 |
| Tringa incana | M | 13 | 101.0 | | 87.0–114.0 | B | Alaska, USA | 1430.0 |
| | F | 16 | 116.0 | | 98.0–130.0 | | 278 | |
| Catoptrophorus semipalmatus | U | 41 | 215.0 | | | B | Virginia, USA 271 | 1431.0 |
| Prosobonia cancellata | B | 7 | 36.0 | | 32.0–44.0 | | 294 | 1432.0 |
| Arenaria interpres | M | 16 | 110.0 | | | | | 1434.0 |
| | F | 7 | 120.0 | | | | 289 | |
| Arenaria melanocephala | M | 12 | 114.0 | | | | | 1435.0 |
| | F | 9 | 124.0 | | | | 289 | |
| Limnodromus griseus griseus | M | 12 | 110.0 | | 73.0–152.0 | M | New Jersey, USA | 1436.0 |
| | F | 30 | 116.0 | | 82.5–154.0 | | 23 | |
| Limnodromus griseus hendersoni | M | 7 | 100.0 | | 75.0–129.0 | F | | 1436.0 |
| | F | 3 | 117.0 | | 114.0–122.0 | | 117 | |

## Body Masses of World Birds (continued)

| Species | Sex | N | Mean | Std dev | Range | Sn | Location | Number |
|---|---|---|---|---|---|---|---|---|
| Limnodromus scolopaceus | M | 28 | 100.0 | | 90.0–114.0 | B | Alaska, USA | 1437.0 |
| | F | 11 | 109.0 | | 93.0–119.0 | | 278 | |
| Limnodromus semipalmatus | B | 8 | 187.0 | | 165.0–245.0 | | | 1438.0 |
| | | | | | | | 294 | |
| Aphriza virgata | M | 27 | 183.0 | 16.10 | 156.0–222.0 | S | Alaska, USA | 1439.0 |
| | F | 6 | 205.0 | | 182.0–226.0 | | 177 | |
| Calidris tenuirostris | B | 25 | 167.0 | | 135.0–207.0 | M | China | 1440.0 |
| | | | | | | | 117 | |
| Calidris canutus | M | 13 | 126.0 | | 112.0–136.0 | B | northern Canada | 1441.0 |
| | F | 9 | 148.0 | | 135.0–169.0 | | 452 | |
| Calidris alba | B | 45 | 57.0 | | 47.0–72.5 | B | Greenland | 1442.0 |
| | | | | | | | 258 | |
| Calidris pusilla | U | 1364 | 31.3 | 5.54 | | F | Maine, USA | 1443.0 |
| | | | | | | | 174 | |
| Calidris mauri | U | 42 | 23.3 | 2.69 | 18.0–30.0 | F | California, USA | 1444.0 |
| | | | | | | | 177 | |
| Calidris minuta | U | 261 | 23.0 | 3.20 | | F | Morocco | 1445.0 |
| | | | | | | | 117 | |
| Calidris ruficollis | M | 62 | 32.0 | | 21.0–47.0 | | | 1446.0 |
| | F | 29 | 36.0 | | 27.0–51.0 | | 294 | |
| Calidris temminckii | B | 111 | 23.0 | 3.00 | 14.7–36.0 | F | France | 1447.0 |
| | | | | | | | 117 | |
| Calidris subminuta | B | 20 | 30.2 | | 23.0–37.0 | | | 1448.0 |
| | | | | | | | 294 | |
| Calidris minutilla | U | 276 | 23.2 | 3.60 | 16.0–34.0 | S | Venezuela | 1449.0 |
| | | | | | | | 626 | |
| Calidris fuscicollis | U | 367 | 34.7 | 4.60 | 27.5–45.7 | S | Venezuela | 1450.0 |
| | | | | | | | 626 | |
| Calidris bairdii | M | 46 | 38.6 | 4.00 | | B | Alaska, USA | 1451.0 |
| | F | 16 | 43.5 | 4.90 | | | 355 | |
| Calidris melanotos | M | 74 | 97.8 | | | | | 1452.0 |
| | F | 38 | 65.0 | | | | 355 | |
| Calidris acuminata | M | 10 | 70.3 | | | | | 1453.0 |
| | F | 10 | 63.5 | | | | 289 | |
| Calidris maritima | M | 72 | 76.8 | | 57.0–90.0 | W | | 1454.0 |
| | F | 92 | 86.2 | | 63.0–102.0 | | 149 | |
| Calidris ptilocnemis | M | 91 | 76.3 | 6.00 | 61.1–86.0 | F | Alaska, USA | 1455.0 |
| | F | 51 | 83.0 | 8.58 | 65.8–106.0 | | 177 | |
| Calidris alpina sakhalina | M | 267 | 55.4 | | | B | Alaska, USA | 1456.0 |
| | F | 177 | 59.7 | | | | 356 | |
| Calidris alpina schintzii | M | 92 | 44.2 | | | | | 1456.1 |
| | F | 92 | 49.6 | | | | 289 | |
| Calidris ferruginea | U | 265 | 67.8 | 9.85 | | F | France | 1457.0 |
| | | | | | | | 117 | |

## Body Masses of World Birds (continued)

| Species | Sex | N | Mean | Std dev | Range | Sn | Location | Number |
|---|---|---|---|---|---|---|---|---|
| Micropalama himantopus | M | 24 | 53.8 | | | | | 1458.0 |
| | F | 15 | 60.9 | | | | 289 | |
| Tryngites subruficollis | M | 4 | 71.0 | | 64.2–80.5 | B | Alaska, USA | 1459.0 |
| | F | 6 | 53.0 | | 50.0–58.0 | | 278 | |
| Eurynorhynchus pygmeus | M | | 29.5 | | | | | 1460.0 |
| | F | 1 | 34.0 | | | | 294 | |
| Limicola falcinellus sibirica | U | 35 | 40.0 | | 32.0–56.0 | W | | 1461.0 |
| | | | | | | | 294 | |
| Philomachus pugnax | M | 36 | 171.0 | | | F | Africa | 1462.0 |
| | F | 225 | 104.0 | | | | 289 | |
| Steganopus tricolor | M | 155 | 51.8 | 4.10 | | | Canada | 1463.0 |
| | F | 48 | 68.1 | 7.10 | | | 109 | |
| Phalaropus lobatus | M | 43 | 32.7 | | | | | 1464.0 |
| | F | 14 | 34.9 | | | | 289 | |
| Phalaropus fulicaria | M | 132 | 50.2 | | | | | 1465.0 |
| | F | 78 | 61.1 | | | | 312 | |

### ORDER: CHARADRIIFORMES    FAMILY: DROMADIDAE

| Species | Sex | N | Mean | Std dev | Range | Sn | Location | Number |
|---|---|---|---|---|---|---|---|---|
| Dromas ardeola | F | 1 | 325.0 | | | | | 1466.0 |
| | | | | | | | 66 | |

### ORDER: CHARADRIIFORMES    FAMILY: CHIONIDIDAE

| Species | Sex | N | Mean | Std dev | Range | Sn | Location | Number |
|---|---|---|---|---|---|---|---|---|
| Chionis alba | U | 1 | 400.0 | | | | Brazil | 1467.0 |
| | | | | | | | 37 | |
| Chionis minor | U | 78 | 579.0 | 52.80 | 480.0–730.0 | B | Kerguelen Is. | 1468.0 |
| | | | | | | | 665 | |

### ORDER: CHARADRIIFORMES    FAMILY: PLUVIANELLIDAE

| Species | Sex | N | Mean | Std dev | Range | Sn | Location | Number |
|---|---|---|---|---|---|---|---|---|
| Pluvianellus socialis | M | | 89.0 | | 79.0–102.0 | | | 1469.0 |
| | F | | 79.5 | | 69.5–87.0 | | 294 | |

### ORDER: CHARADRIIFORMES    FAMILY: PEDIONOMIDAE

| Species | Sex | N | Mean | Std dev | Range | Sn | Location | Number |
|---|---|---|---|---|---|---|---|---|
| Pedionomus torquatus | M | 1 | 43.0 | | | | Australia | 1470.0 |
| | F | 1 | 60.0 | | | | 120 | |

### ORDER: CHARADRIIFORMES    FAMILY: THINOCORIDAE

| Species | Sex | N | Mean | Std dev | Range | Sn | Location | Number |
|---|---|---|---|---|---|---|---|---|
| Attagis gayi | B | 4 | 311.0 | | 283.0–310.0 | | Peru; Bolivia | 1471.0 |
| | | | | | | | 584 | |
| Thinocorus orbignyianus | B | 16 | 115.0 | 13.00 | 96.0–140.0 | | Peru; Bolivia | 1473.0 |
| | | | | | | | 584 | |
| Thinocorus rumicovorus | B | 4 | 44.0 | | 42.0–45.0 | | Peru | 1474.0 |
| | | | | | | | 584 | |

## Body Masses of World Birds (continued)

| Species | Sex | N | Mean | Std dev | Range | Sn | Location | Number |
|---------|-----|---|------|---------|-------|----|----------|--------|
| ORDER: CHARADRIIFORMES | | | | | FAMILY: BURHINIDAE | | | |
| Burhinus oedicnemus | B | 13 | 459.0 | | 290.0–535.0 | | France 117 | 1475.0 |
| Burhinus vermiculatus | F | 1 | 320.0 | | | | 66 | 1477.0 |
| Burhinus capensis | U | 5 | 423.0 | | 400.0–450.0 | | 78 | 1478.0 |
| Burhinus bistriatus | M | 1 | 787.0 | | | | Oaxaca, Mexico 584 | 1479.0 |
| Burhinus recurvirostris | U | | 790.0 | | | | 5 | 1482.0 |
| Burhinus giganteus | U | | 1025.0 | | | | 5 | 1483.0 |
| ORDER: CHARADRIIFORMES | | | | | FAMILY: HAEMATOPODIDAE | | | |
| Haematopus ostralegus ostralegus | U | 282 | 526.0 | 45.50 | 430.0–675.0 | B | Netherlands 117 | 1484. |
| Haematopus moquini | B | 109 | 693.0 | | | | 289 | 1486.0 |
| Haematopus bachmani | M | 5 | 607.0 | | 555.0–648.0 | B | Alaska, USA 177 | 1488.0 |
| | F | 5 | 689.0 | | 618.0–750.0 | | | |
| Haematopus palliatus | U | 20 | 632.0 | 38.00 | 575.0–730.0 | S | S. Carolina, USA 590 | 1489.0 |
| Haematopus unicolor | B | 144 | 702.0 | | | | 289 | 1491.0 |
| Haematopus fuliginosus | B | 15 | 833.0 | | | | 289 | 1492.0 |
| Haematopus ater | B | 5 | 678.0 | | 585.0–708.0 | | 294 | 1493.0 |
| Haematopus leucopodus | B | 5 | 618.0 | | 585.0–700.0 | | 294 | 1494.0 |
| ORDER: CHARADRIIFORMES | | | | | FAMILY: IBIDORHYNCHIDAE | | | |
| Ibidorhyncha struthersii | B | 7 | 294.0 | | 270.0–320.0 | | 294 | 1495.0 |
| ORDER: CHARADRIIFORMES | | | | | FAMILY: RECURVIROSTRIDAE | | | |
| Himantopus himantopus | B | 20 | 161.0 | | 138.0–208.0 | | northern Australia 165 | 1496.0 |
| Himantopus leucocephalus | B | 18 | 176.0 | | | | Australia 464 | 1497.0 |
| Himantopus leucocephalus | B | 29 | 193.0 | | | | New Zealand 464 | 1497.0 |

## Body Masses of World Birds (continued)

| Species | Sex | N | Mean | Std dev | Range | Sn | Location | Number |
|---|---|---|---|---|---|---|---|---|
| Himantopus novaezelandiae | M | 2 | 219.0 | | | | New Zealand | 1498.0 |
| | F | 2 | 227.0 | | | | 464 | |
| Himantopus mexicanus | U | 18 | 166.0 | | | | California, USA | 1499.0 |
| | | | | | | | 233 | |
| Cladorhynchus leucocephala | M | 7 | 232.0 | | 213.0–250.0 | | | 1501.0 |
| | F | 3 | 197.0 | | 110.0–247.0 | | 294 | |
| Recurvirostra avosetta | B | 19 | 306.0 | 57.00 | 228.0–435.0 | | Netherlands | 1502.0 |
| | | | | | | | 117 | |
| Recurvirostra americana | U | 33 | 316.0 | | | | California, USA | 1503.0 |
| | | | | | | | 233 | |
| Recurvirostra novaehollandiae | B | 7 | 307.0 | | 278.0–360.0 | | | 1504.0 |
| | | | | | | | 294 | |
| Recurvirostra andina | B | 5 | 361.0 | | 315.0–410.0 | | | 1505.0 |
| | | | | | | | 294 | |

### ORDER: CHARADRIIFORMES          FAMILY: GLAREOLIDAE

| Species | Sex | N | Mean | Std dev | Range | Sn | Location | Number |
|---|---|---|---|---|---|---|---|---|
| Pluvianus aegyptius | U | 10 | 82.0 | | 73.4–90.5 | B | Ethiopia | 1506.0 |
| | | | | | | | 117 | |
| Rhinoptilus africanus | U | 2 | 84.1 | | 80.0–88.2 | | South Africa | 1507.0 |
| | | | | | | | 255 | |
| Rhinoptilus chalcopterus | M | 6 | 151.0 | | 117.0–172.0 | | Botswana | 1508.0 |
| | F | 3 | 155.0 | | 148.0–160.0 | | 636 | |
| Rhinoptilus cinctus | M | | 125.0 | | | | | 1509.0 |
| | | | | | | | 78 | |
| Cursorius cursor | B | 16 | 138.0 | 19.40 | 115.0–198.0 | B | India | 1511.0 |
| | | | | | | | 117 | |
| Cursorius rufus | U | 1 | 75.0 | | | | Namibia | 1512.0 |
| | | | | | | | 636 | |
| Cursorius temminckii | U | 5 | 69.0 | | 64.0–73.2 | | | 1513.0 |
| | | | | | | | 78, 255 | |
| Glareola pratincola | B | 12 | 80.1 | 7.88 | 68.0–90.0 | | | 1515.0 |
| | | | | | | | 117 | |
| Glareola maldivarum | B | 17 | 86.8 | | | | | 1516.0 |
| | | | | | | | 289 | |
| Glareola nordmanni | B | 10 | 97.2 | | 87.0–105.0 | | | 1517.0 |
| | | | | | | | 149 | |
| Glareola ocularis | M | 1 | 103.0 | | | | Madagascar | 1518.0 |
| | F | 1 | 82.0 | | | | 39 | |
| Glareola nuchalis | U | | | | 43.0–52.0 | | | 1519.0 |
| | | | | | | | 636 | |
| Glareola lactea | U | 2 | 37.5 | | 37.0–38.0 | | India | 1521.0 |
| | | | | | | | 5 | |
| Stiltia isabella | F | 1 | 48.0 | | | | New Guinea | 1522.0 |
| | | | | | | | 154 | |

## Body Masses of World Birds (continued)

| Species | Sex | N | Mean | Std dev | Range | Sn | Location | Number |
|---|---|---|---|---|---|---|---|---|
| **ORDER: CHARADRIIFORMES** | | | | | **FAMILY: CHARADRIIDAE** | | | |
| Pluvialis apricaria | U | 402 | 214.0 | 15.50 | 165.0–260.0 | S | Netherlands 117 | 1523.0 |
| Pluvialis fulva | B | 27 | 153.0 | | 122.0–192.0 | S | Wake Is. 117 | 1524.0 |
| Pluvialis dominica dominica | U | 60 | 145.0 | | 126.0–169.0 | B | Alaska, USA 278 | 1525.0 |
| Pluvialis squatarola | U | 31 | 220.0 | 24.40 | 181.0–263.0 | B | 177 | 1526.0 |
| Charadrius obscurus | M | 1 | 160.0 | | | | New Zealand 294 | 1527.0 |
| | F | 5 | 110.0 | | | | | |
| Charadrius hiaticula | U | 75 | 64.0 | | 55.4–74.5 | B | Netherlands 23 | 1528.0 |
| Charadrius semipalmatus | M | 26 | 47.4 | 5.90 | 37.6–57.4 | S | eastern USA | 1529.0 |
| | F | 24 | 46.1 | 4.70 | 39.2–56.5 | | 177 | |
| Charadrius placidus | B | 19 | 62.7 | | 41.0–70.0 | | 294 | 1530.0 |
| Charadrius dubius | B | 461 | 38.7 | | 33.0–48.3 | B | Netherlands 117 | 1531.0 |
| Charadrius wilsonia | B | 39 | 55.1 | 7.10 | | W | Panama 611 | 1532.0 |
| Charadrius vociferus | M | 10 | 92.1 | 10.40 | 83.9–109.0 | B | Great Plains, USA | 1533.0 |
| | F | 6 | 101.0 | | 87.7–121.0 | | 177 | |
| Charadrius pecuarius | U | 886 | 34.0 | | 19.0–49.0 | | 78 | 1536.0 |
| Charadrius tricollaris | U | 11 | 31.2 | | 35.0–38.0 | | 294 | 1537.0 |
| Charadrius melodus | U | 87 | 55.2 | | 46.4–63.7 | B | New York, USA 675 | 1539.0 |
| Charadrius venustus pallidus | U | 17 | 34.8 | | | | 294 | 1540.0 |
| Charadrius alexandrinus | U | 38 | 41.4 | 2.54 | 37.0–49.0 | B | California, USA 177 | 1541.0 |
| Charadrius marginatus | U | 262 | 48.3 | | | | 294 | 1542.0 |
| Charadrius ruficapillus | B | 140 | 35.2 | | 30.0–47.0 | | 294 | 1543.0 |
| Charadrius peronii | U | | 42.0 | | | | 78 | 1544.0 |
| Charadrius collaris | B | 18 | 28.3 | 1.68 | 25.8–30.9 | | Panama 611 | 1546.0 |
| Charadrius bicinctus | B | 102 | 57.0 | | 46.0–75.0 | | 294 | 1547.0 |
| Charadrius falklandicus | B | 6 | 65.0 | | 62.0–72.0 | | 294 | 1548.0 |

## Body Masses of World Birds (continued)

| Species | Sex | N | Mean | Std dev | Range | Sn | Location | Number |
|---|---|---|---|---|---|---|---|---|
| Charadrius alticola | B | 5 | 45.0 | | 41.0–49.0 | | 294 | 1549.0 |
| Charadrius mongolus | B | 13 | 57.7 | | 52.0–68.0 | | 149 | 1550.0 |
| Charadrius leschenaultii | B | 16 | 91.4 | | 78.0–103.0 | B | 117 | 1551.0 |
| Charadrius asiaticus | B | 21 | 77.1 | 7.91 | 63.0–91.0 | | USSR 117 | 1552.0 |
| Charadrius veredus | B | 7 | 79.7 | | 77.0–88.0 | | 294 | 1553.0 |
| Charadrius montanus | M | 14 | 102.0 | 10.20 | 88.3–119.0 | B | Colorado, USA | 1554.0 |
| | F | 8 | 114.0 | | 89.9–144.0 | | 177 | |
| Charadrius modestus | B | 5 | 78.0 | | 71.0–89.0 | | 294 | 1555.0 |
| Charadrius rubricollis | U | | | | 45.0–50.0 | | 294 | 1556.0 |
| Erythrogonys cinctus | B | 26 | 54.0 | | 46.0–64.0 | | 294 | 1558.0 |
| Eudromias morinellus | M | 10 | 100.0 | | 88.0–116.0 | B | USSR | 1559.0 |
| | F | 5 | 117.0 | | 98.9–130.0 | | 23 | |
| Oreopholus ruficollis ruficollis | U | 6 | 133.0 | | 120.0–145.0 | | 294 | 1560.0 |
| Anarhynchus frontalis | B | 85 | 59.0 | | 47.0–70.5 | | 294 | 1561.0 |
| Phegornis mitchellii | B | 5 | 35.5 | | 28.0–46.0 | | 289 | 1562.0 |
| Peltohyas australis | M | 1 | 100.0 | | | | | 1563.0 |
| | F | 7 | 84.9 | | | | 294 | |
| Elsyornis melanops | M | | | | 28.0–29.0 | | | 1564.0 |
| | F | | | | 30.0–33.0 | | 294 | |
| Vanellus vanellus | M | 32 | 211.0 | 17.90 | 140.0–242.0 | B | | 1565.0 |
| | F | 40 | 226.0 | 21.80 | 180.0–317.0 | | 117 | |
| Vanellus crassirostris | B | 2 | 170.0 | | 170.0–170.0 | | 66 | 1566.0 |
| Vanellus malabaricus | U | 3 | 140.0 | | 108.0–203.0 | | 294 | 1567.0 |
| Vanellus macropterus | U | | 325.0 | | | | estimated 294 | 1568.0 |
| Vanellus tricolor | U | 9 | 184.0 | | 140.0–216.0 | | Australia 140 | 1569.0 |
| Vanellus miles | U | 10 | 379.0 | | 290.0–440.0 | | Australia 140 | 1570.0 |
| Vanellus armatus | B | 264 | 156.0 | | 114.0–211.0 | | 294 | 1571.0 |

## Body Masses of World Birds (continued)

| Species | Sex | N | Mean | Std dev | Range | Sn | Location | Number |
|---|---|---|---|---|---|---|---|---|
| Vanellus spinosus | U | 8 | 152.0 | | 127.0–170.0 | | 227, 294 | 1572.0 |
| Vanellus tectus | F | | 100.0 | | | | 294 | 1574.0 |
| Vanellus melanocephalus | U | 2 | 214.0 | | 199.0–228.0 | | 294 | 1575.0 |
| Vanellus cinereus | B | 7 | 270.0 | | 236.0–296.0 | | 294 | 1576.0 |
| Vanellus indicus | U | 21 | 181.0 | | 110.0–230.0 | | 294 | 1577.0 |
| Vanellus albiceps | U | 1 | 201.0 | | | | 294 | 1578.0 |
| Vanellus senegallus | U | | | | 197.0–277.0 | | 294 | 1579.0 |
| Vanellus lugubris | U | | 121.0 | | 107.0–130.0 | | 655 | 1580.0 |
| Vanellus melanopterus | U | 3 | 168.0 | | 163.0–170.0 | | 294 | 1581.0 |
| Vanellus coronatus | B | 6 | 167.0 | | 148.0–200.0 | | 294 | 1582.0 |
| Vanellus superciliosus | U | | 150.0 | | | | estimated 294 | 1583.0 |
| Vanellus gregarius | M | 2 | 252.0 | | 245.0–260.0 | | | 1584.0 |
| | F | 7 | 200.0 | | 180.0–252.0 | | 294 | |
| Vanellus leucurus | B | 6 | 137.0 | | 114.0–198.0 | | 117 | 1585.0 |
| Vanellus cayanus | M | 4 | 72.2 | | 55.0–80.0 | | | 1586.0 |
| | F | 2 | 83.0 | | 82.0–84.0 | | 294 | |
| Vanellus chilensis | B | 3 | 327.0 | | 277.0–425.0 | | 37, 86, 294 | 1587.0 |
| Vanellus resplendens | B | 12 | 214.0 | | 193.0–230.0 | | 294 | 1588.0 |

### ORDER: CHARADRIIFORMES      FAMILY: LARIDAE

| Species | Sex | N | Mean | Std dev | Range | Sn | Location | Number |
|---|---|---|---|---|---|---|---|---|
| Larus pacificus | U | | | | 900.0–1135.0 | | W. Australia 550 | 1590.0 |
| Larus belcheri | U | | 600.0 | | | | 171 | 1591.0 |
| Larus crassirostris | M | 7 | 573.0 | | 492.0–640.0 | | | 1593.0 |
| | F | 4 | 494.0 | | 436.0–535.0 | | 149 | |
| Larus modestus | U | | 360.0 | | | | 171 | 1594.0 |
| Larus heermanni | U | 10 | 500.0 | 67.90 | 371.0–643.0 | | Sonora, Mexico 594 | 1595.0 |

## Body Masses of World Birds (continued)

| Species | Sex | N | Mean | Std dev | Range | Sn | Location | Number |
|---|---|---|---|---|---|---|---|---|
| Larus leucopthalmus | M | 10 | 357.0 | | 325.0–415.0 | | Egypt | 1596.0 |
| | F | 9 | 303.0 | | 275.0–355.0 | | 636 | |
| Larus hemprichii | M | 1 | 510.0 | | | | | 1597.0 |
| | F | 1 | 400.0 | | | | 66 | |
| Larus canus | M | 96 | 432.0 | | 340.0–552.0 | B | N. Atlantic | 1598.0 |
| | F | 72 | 375.0 | | 290.0–530.0 | | 36 | |
| Larus canus kamtschatschensis | F | 2 | 490.0 | | 394.0–586.0 | | 149 | 1598.0 |
| Larus audouinii | U | 1 | 770.0 | | | | 66 | 1599.0 |
| Larus delawarensis | M | 48 | 566.0 | 42.00 | | B | Ontario, Canada | 1600.0 |
| | F | 51 | 471.0 | 46.00 | | | 516 | |
| Larus californicus californicus | M | 64 | 657.0 | 85.30 | 490.0–885.0 | | | 1601.0 |
| | F | 84 | 556.0 | 53.40 | 432.0–695.0 | | 288 | |
| Larus californicus albertaensis | M | 32 | 841.0 | 103.00 | 653.0–1045.0 | | | 1601.1 |
| | F | 19 | 710.0 | 91.20 | 568.0–903.0 | | 288 | |
| Larus marinus | M | 116 | 1829.0 | | 1380.0–2272.0 | B | N. Atlantic | 1602.0 |
| | F | 93 | 1488.0 | | 1033.0–2085.0 | | 36 | |
| Larus dominicanus | U | | 900.0 | | | | 171 | 1603.0 |
| Larus glaucescens | U | 110 | 1010.0 | 136.00 | 730.0–1400.0 | | western Canada | 1604.0 |
| | | | | | | | 540 | |
| Larus occidentalis | U | 48 | 1011.0 | 25.10 | 800.0–1190.0 | | western USA | 1605.0 |
| | | | | | | | 540 | |
| Larus livens | U | 2 | 1322.0 | | 1173.0–1470.0 | | Sonora, Mexico | 1606.0 |
| | | | | | | | 179 | |
| Larus hyperboreus | M | 39 | 1576.0 | 121.00 | 1280.0–1820.0 | | Iceland | 1607.0 |
| | F | 26 | 1249.0 | 107.00 | 1070.0–1430.0 | | 117 | |
| Larus glaucoides kumlieni | U | 1 | 557.0 | | | PB | Alaska, USA | 1608.0 |
| | | | | | | | 230 | |
| Larus glaucoides | M | 1 | 863.0 | | | F | Britain | 1608.0 |
| | F | | | | | | 258 | |
| Larus thayeri | M | 3 | 1093.0 | | 1028.0–1152.0 | | Ellesmere Is. | 1609.0 |
| | F | 4 | 899.0 | | 846.0–980.0 | | 452 | |
| Larus argentatus | M | 220 | 1226.0 | | 755.0–1495.0 | B | N. Atlantic | 1610.0 |
| | F | 139 | 1044.0 | | 717.0–1385.0 | | 36 | |
| Larus cachinnans michahellis | M | 80 | 1275.0 | 98.00 | 1040.0–1500.0 | | France | 1611.0 |
| | F | 80 | 1033.0 | 92.00 | 800.0–1400.0 | | 117 | |
| Larus schistisagus | B | 6 | 1327.0 | | 1100.0–1694.0 | | Alaska; Japan 213, 598a, 660 | 1613.0 |
| Larus fuscus graellsii | M | 22 | 880.0 | 61.00 | 770.0–1000.0 | | Britain | 1614.0 |
| | F | 31 | 755.0 | 58.00 | 620.0–908.0 | | 117 | |
| Larus fuscus fuscus | M | 52 | 768.0 | 59.90 | 670.0–945.0 | | Norway | 1614.0 |
| | F | 64 | 662.0 | 64.40 | 545.0–840.0 | | 117 | |

## Body Masses of World Birds (continued)

| Species | Sex | N | Mean | Std dev | Range | Sn | Location | Number |
|---|---|---|---|---|---|---|---|---|
| Larus ichthyaetus | M | 8 | 1599.0 | | 1130.0–2000.0 | | USSR; Pakistan | 1615.0 |
| | F | 8 | 1215.0 | | 960.0–1500.0 | | 117 | |
| Larus cirrocephalus | B | 4 | 309.0 | | 255.0–335.0 | | Kenya; S.Africa | 1617.0 |
| | | | | | | | 117 | |
| Larus hartlaubii | B | 18 | 292.0 | | 235.0–340.0 | | South Africa | 1618.0 |
| | | | | | | | 636 | |
| Larus novaehollandiae | M | 3 | 323.0 | | 320.0–328.0 | | Campbell Is. | 1619.0 |
| | F | | | | | | 671 | |
| Larus maculipennis | B | 3 | 320.0 | | 290.0–361.0 | | Brazil | 1622.0 |
| | | | | | | | 37 | |
| Larus ridibundus | U | 324 | 284.0 | 24.60 | 195.0–327.0 | B | Netherlands | 1623.0 |
| | | | | | | | 117 | |
| Larus genei | B | 18 | 281.0 | | 223.0–350.0 | | | 1624.0 |
| | | | | | | | 117 | |
| Larus philadelphia | U | 12 | 212.0 | 28.20 | 162.0–270.0 | W | Oklahoma, USA | 1625.0 |
| | | | | | | | 590a | |
| Larus melanocephalus | U | 2 | 256.0 | | 232.0–280.0 | | | 1628.0 |
| | | | | | | | 227 | |
| Larus relictus | M | 5 | 519.0 | | | | | 1629.0 |
| | F | 6 | 463.0 | | | | 322 | |
| Larus atricilla | U | 39 | 325.0 | 15.90 | 270.0–400.0 | B | Florida, USA | 1631.0 |
| | | | | | | | 534 | |
| Larus pipixcan | U | 40 | 280.0 | 9.69 | 220.0–335.0 | B | Minnesota, USA | 1632.0 |
| | | | | | | | 117 | |
| Larus minutus | U | 87 | 118.0 | | 88.0–162.0 | S | Britain | 1633.0 |
| | | | | | | | 258 | |
| Pagophila eburnea | U | 8 | 616.0 | | 519.0–701.0 | | Bering Sea | 1634.0 |
| | | | | | | | 177 | |
| Rhodostethia rosea | U | 19 | 187.0 | 9.50 | 170.0–210.0 | PB | Alaska | 1635.0 |
| | | | | | | | 598a | |
| Xema sabini | M | 4 | 205.0 | | 190.0–214.0 | | Alaska, USA | 1636.0 |
| | F | 4 | 177.0 | | 158.0–190.0 | | 117 | |
| Creagrus furcatus | B | 17 | 687.0 | | 610.0–780.0 | B | Galapagos Is. | 1637.0 |
| | | | | | | | 239 | |
| Rissa tridactyla | M | 223 | 421.0 | | 305.0–512.0 | B | N. Atlantic | 1638.0 |
| | F | 183 | 393.0 | | 305.0–525.0 | | 36 | |
| Rissa brevirostris | M | 94 | 400.0 | 32.40 | 325.0–450.0 | | Pribilof Is. | 1639.0 |
| | F | 85 | 382.0 | 29.60 | 340.0–510.0 | | 177 | |
| Chlidonias hybridus | B | 11 | 88.2 | | 79.0–94.0 | | China | 1640.0 |
| | | | | | | | 118 | |
| Chlidonias leucopterus | U | 110 | 54.2 | 4.90 | 42.0–66.0 | | | 1641.0 |
| | | | | | | | 118 | |
| Chlidonias niger | U | 36 | 65.3 | 4.65 | 60.3–74.1 | B | Ontario, Canada | 1642.0 |
| | | | | | | | 177 | |

## Body Masses of World Birds (continued)

| Species | Sex | N | Mean | Std dev | Range | Sn | Location | Number |
|---|---|---|---|---|---|---|---|---|
| Phaetusa simplex | B | 5 | 232.0 | | 208.0–247.0 | | 37, 112, 245 | 1643.0 |
| Sterna nilotica nilotica | B | 12 | 233.0 | 27.30 | 189.0–292.0 | PB | Netherlands 118 | 1644.0 |
| Sterna nilotica aranea | B | 6 | 170.0 | | 160.0–184.0 | B | Virginia, USA 177 | 1644.0 |
| Sterna caspia | U | 84 | 655.0 | | 574.0–782.0 | B | Texas 474 | 1645.0 |
| Sterna maxima maxima | U | 28 | 470.0 | 19.70 | | | 244 | 1646.0 |
| Sterna maxima albididorsalis | U | 22 | 367.0 | 14.60 | 350.0–395.0 | | Mauritania 118 | 1646.1 |
| Sterna bergii thalassina | M | 4 | 342.0 | | 325.0–350.0 | | 118 | 1647.0 |
| Sterna bengalensis | M | 2 | 188.0 | | 185.0–190.0 | | Kenya | 1649.0 |
|  | F | 2 | 220.0 | | 205.0–235.0 | | 118 | |
| Sterna sandvicensis | U | 10 | 208.0 | 12.70 | 193.0–238.0 | B | Virginia, USA 177 | 1650.0 |
| Sterna elegans | U | 14 | 257.0 | 22.30 | 217.0–300.0 | | 591 | 1651.0 |
| Sterna dougallii | U | 299 | 110.0 | | | | New York, USA 414 | 1652.0 |
| Sterna sumatrana | U | | 100.0 | | | | 331 | 1654.0 |
| Sterna hirundinacea | F | 4 | | | 172.0–196.0 | | Brazil 37 | 1655.0 |
| Sterna hirundo | U | 265 | 120.0 | | 103.0–145.0 | | New York, USA 339 | 1656.0 |
| Sterna paradisaea | U | 261 | 110.0 | 8.30 | 86.0–127.0 | B | NE USA & Canada 177 | 1657.0 |
| Sterna vittata | U | | 140.0 | | | | 3 | 1658.0 |
| Sterna forsteri | U | 24 | 158.0 | 16.80 | 127.0–193.0 | | Oklahoma, USA 590a | 1661.0 |
| Sterna trudeaui | B | 2 | 153.0 | | 146.0–160.0 | | Brazil 37 | 1662.0 |
| Sterna repressa | U | 12 | 90.0 | | 78.0–105.0 | | Egypt 636 | 1663.0 |
| Sterna balaenarum | U | | 60.0 | | | | 173 | 1664.0 |
| Sterna aleutica | U | 16 | 120.0 | 12.40 | 83.4–140.0 | | Alaska, USA 598a | 1665.0 |
| Sterna lunata | U | 83 | 146.0 | 9.11 | 115.0–177.0 | | Hawaiian Is. 242 | 1666.0 |

## Body Masses of World Birds (continued)

| Species | Sex | N | Mean | Std dev | Range | Sn | Location | Number |
|---|---|---|---|---|---|---|---|---|
| Sterna anaethetus anarctica | U | 69 | 95.6 | 6.58 | | | Seychelles 118 | 1667.0 |
| Sterna fuscata | U | 95 | 180.0 | | 147.0–220.0 | | Trinidad 188 | 1668.0 |
| Sterna albifrons | B | 30 | 57.0 | | 50.0–63.0 | | Morocco 118 | 1671.0 |
| Sterna saundersi | F | 3 | 42.3 | | 40.0–45.0 | | 636 | 1672.0 |
| Sterna antillarum | U | 18 | 43.1 | 2.12 | 39.0–47.6 | B | Kansas, USA 177 | 1673.0 |
| Sterna superciliaris | B | 29 | 46.4 | | 40.0–57.0 | | Surinam 188 | 1674.0 |
| Sterna nereis | U | | 57.0 | | | | 550 | 1676.0 |
| Larosterna inca | U | | 180.0 | | | | 171 | 1677.0 |
| Procelsterna cerulea | U | 52 | 53.0 | 7.21 | 46.0–65.0 | B | Hawaiian Is. 242 | 1678.0 |
| Anous stolidus | U | 12 | 198.0 | 29.90 | 156.0–272.0 | | Hawaiian Is. 177 | 1680.0 |
| Anous minutus | U | 102 | 119.0 | 4.83 | 98.0–144.0 | Y | Ascension Is. 10 | 1681.0 |
| Anous tenuirostris | B | 4 | 111.0 | | 97.4–120.0 | | 636 | 1682.0 |
| Gygis alba | U | 109 | 111.0 | 10.40 | 92.0–139.0 | | Hawaiian Is. 242 | 1683.0 |

### ORDER: CHARADRIIFORMES          FAMILY: STERCORARIIDAE

| Species | Sex | N | Mean | Std dev | Range | Sn | Location | Number |
|---|---|---|---|---|---|---|---|---|
| Catharacta skua | M | 219 | 413.0 | | 306.0–523.0 | B | N. Atlantic 36 | 1685.0 |
| | F | 189 | 478.0 | | 306.0–604.0 | | | |
| Catharacta antarctica | U | | 1400.0 | | | | 173 | 1686.0 |
| Catharacta lonnbergi | U | 15 | 1922.0 | 163.00 | 1670.0–2130.0 | B | Kerguelen Is. 665 | 1687.0 |
| Catharacta maccormicki | U | 80 | 1156.0 | 98.00 | | B | Antarctica 370 | 1689.0 |
| Stercorarius pomarinus | M | 73 | 648.0 | 6.25 | 542.0–797.0 | | Alaska, USA 359 | 1690.0 |
| | F | 52 | 740.0 | 11.70 | 576.0–917.0 | | | |
| Stercorarius parasiticus | M | 20 | 421.0 | 11.60 | 301.0–540.0 | | Alaska, USA 359 | 1691.0 |
| | F | 11 | 508.0 | 24.40 | 346.0–644.0 | | | |
| Stercorarius longicaudus | M | 26 | 280.0 | 6.10 | 236.0–343.0 | | Alaska, USA 359 | 1692.0 |
| | F | 18 | 313.0 | 7.40 | 258.0–358.0 | | | |

### ORDER: CHARADRIIFORMES          FAMILY: RYNCHOPIDAE

| Species | Sex | N | Mean | Std dev | Range | Sn | Location | Number |
|---|---|---|---|---|---|---|---|---|
| Rynchops niger | M | 56 | 349.0 | 25.74 | 260.0–392.0 | B | Texas, USA 474 | 1693.0 |
| | F | 73 | 254.0 | 17.94 | 212.0–292.0 | | | |

## Body Masses of World Birds (continued)

| Species | Sex | N | Mean | Std dev | Range | Sn | Location | Number |
|---|---|---|---|---|---|---|---|---|
| Rynchops flavirostris | B | 10 | 164.0 | | 111.0–204.0 | | 636 | 1694.0 |

**ORDER: CHARADRIIFORMES**      **FAMILY: ALCIDAE**

| Species | Sex | N | Mean | Std dev | Range | Sn | Location | Number |
|---|---|---|---|---|---|---|---|---|
| Alle alle | B | 86 | 163.0 | 12.05 | 134.0–192.0 | B | Spitzbergen 118 | 1696.0 |
| Uria aalge | M | 121 | 1006.0 | 80.00 | 775.0–1202.0 | B | Newfoundland | 1697.0 |
| | F | 117 | 979.0 | 76.00 | 815.0–1187.0 | | 628 | |
| Uria lomvia | U | 139 | 964.0 | | | | 599 | 1698.0 |
| Alca torda | U | 1442 | 719.0 | | | | 599 | 1699.0 |
| Cepphus grylle | U | 258 | 405.0 | 31.50 | | B | Canada 89 | 1701.0 |
| Cepphus columba | U | 5 | 487.0 | | 433.0–543.0 | B | California, USA 594 | 1702.0 |
| Cepphus carbo | B | 3 | 490.0 | | 415.0–545.0 | | 149 | 1703.0 |
| Brachyramphus marmoratus | U | 76 | 222.0 | | | | 542 | 1704.0 |
| Brachyramphus brevirostris | U | 14 | 224.0 | | | | 542 | 1705.0 |
| Synthliboramphus hypoleucus | B | 375 | 167.0 | 13.00 | 136.0–215.0 | B | California, USA 408 | 1706.0 |
| Synthliboramphus craveri | U | 8 | 151.0 | | | | 29 | 1707.0 |
| Synthliboramphus antiquus | U | 154 | 206.0 | | 177.0–249.0 | B | western Canada 543 | 1708.0 |
| Ptychoramphus aleuticus | U | 25 | 188.0 | 16.00 | | B | western Canada 639 | 1710.0 |
| Cyclorrhynchus psittacula | U | 42 | 258.0 | 19.00 | 215.0–292.0 | | 143 | 1711.0 |
| Aethia cristatella | U | 192 | 264.0 | 19.00 | 195.0–330.0 | | 143 | 1712.0 |
| Aethia pygmaea | U | 60 | 121.0 | 7.00 | 102.0–138.0 | | 143 | 1713.0 |
| Aethia pusilla | U | 457 | 84.0 | 7.00 | 72.0–98.0 | | 143 | 1714.0 |
| Cerorhinca monocerata | U | 48 | 520.0 | 36.70 | | B | western Canada 639 | 1715.0 |
| Fratercula arctica | U | 165 | 381.0 | 29.90 | | B | England 240 | 1716.0 |
| Fratercula corniculata | U | 36 | 619.0 | | | | 599 | 1717.0 |

## Body Masses of World Birds (continued)

| Species | Sex | N | Mean | Std dev | Range | Sn | Location | Number |
|---|---|---|---|---|---|---|---|---|
| Fratercula cirrhata | U | 16 | 779.0 | | | | 599 | 1718.0 |

### ORDER: GAVIIFORMES             FAMILY: GAVIIDAE

| Species | Sex | N | Mean | Std dev | Range | Sn | Location | Number |
|---|---|---|---|---|---|---|---|---|
| Gavia stellata | U | 12 | 1551.0 | 258.00 | 1150.0–1980.0 | | Ontario, Canada 590 | 1719.0 |
| Gavia arctica | M | 2 | 3355.0 | | 3310.0–3400.0 | | | 1720.0 |
| | F | 3 | | | 2037.0–2471.0 | | 115 | |
| Gavia pacifica | U | 17 | 1659.0 | 397.00 | 990.0–2450.0 | | Alaska, USA 598a | 1721.0 |
| Gavia immer | U | 5 | 4134.0 | | 3600.0–4480.0 | W | 115 | 1722.0 |
| Gavia adamsii | M | 7 | 5500.0 | | 4400.0–6400.0 | B | Alaska 445 | 1723.0 |

### ORDER: PTEROCLIDIFORMES       FAMILY: PTEROCLIDIDAE

| Species | Sex | N | Mean | Std dev | Range | Sn | Location | Number |
|---|---|---|---|---|---|---|---|---|
| Syrrhaptes paradoxus | B | 13 | 266.0 | | 235.0–300.0 | B | Mongolia 118 | 1725.0 |
| Pterocles alchata | M | | 250.0 | | 230.0–290.0 | | | 1726.0 |
| | F | | 225.0 | | | | 149 | |
| Pterocles namaqua | U | 32 | 177.0 | | 143.0–193.0 | | 78 | 1727.0 |
| Pterocles exustus | M | | | | 184.0–234.0 | | India | 1728.0 |
| | F | | | | 170.0–213.0 | | 118 | |
| Pterocles senegallus | M | 2 | 272.0 | | | | Morocco; Iraq | 1729.0 |
| | F | 1 | 255.0 | | | | 118 | |
| Pterocles gutturalis | U | 9 | 338.0 | | 285.0–400.0 | | 66 | 1730.0 |
| Pterocles orientalis | M | 9 | 428.0 | | 400.0–460.0 | F | USSR | 1731.0 |
| | F | 11 | 383.0 | | 300.0–420.0 | | 118 | |
| Pterocles coronatus | M | 1 | 300.0 | | | | Morocco | 1732.0 |
| | F | | | | | | 118 | |
| Pterocles decoratus | U | 4 | 188.0 | | 167.0–210.0 | | 66 | 1734.0 |
| Pterocles bicinctus | B | 28 | 237.0 | | 210.0–280.0 | | 636 | 1735.0 |
| Pterocles indicus | U | | | | 170.0–227.0 | | 5 | 1737.0 |
| Pterocles lichtensteinii | M | 2 | 248.0 | | 240.0–255.0 | | Morocco 118 | 1738.0 |
| Pterocles burchelli | U | 1 | 193.0 | | | | 66 | 1739.0 |

### ORDER: COLUMBIFORMES         FAMILY: COLUMBIDAE

| Species | Sex | N | Mean | Std dev | Range | Sn | Location | Number |
|---|---|---|---|---|---|---|---|---|
| Columba livia | M | 41 | 369.0 | 38.60 | | | Kansas, USA | 1743.0 |
| | F | 37 | 340.0 | 34.70 | | | 307 | |

## Body Masses of World Birds (continued)

| Species | Sex | N | Mean | Std dev | Range | Sn | Location | Number |
|---|---|---|---|---|---|---|---|---|
| Columba leuconota leuconota | M | 3 | | | 277.0–300.0 | | | 1745.0 |
| | F | 3 | | | 255.0–307.0 | | 5 | |
| Columba guinea | B | 14 | 352.0 | | 307.0–371.0 | B | Ethiopia 681 | 1746.0 |
| Columba albitorques | M | 2 | 277.0 | | 262.0–292.0 | | 636 | 1747.0 |
| Columba oenas | M | 12 | 302.0 | 18.80 | 280.0–334.0 | B | | 1748.0 |
| | F | 7 | 280.0 | 16.20 | 254.0–302.0 | | 118 | |
| Columba eversmanni | B | 8 | 201.0 | | 183.0–234.0 | B | Afghanistan 118 | 1750.0 |
| Columba palumbus | B | 518 | 490.0 | | 284.0–614.0 | Y | 118 | 1751.0 |
| Columba unicincta | M | 5 | 423.0 | | 357.0–490.0 | | | 1755.0 |
| | F | 3 | 358.0 | | 356.0–360.0 | | 636 | |
| Columba thomensis | M | 2 | 525.0 | | 520.0–530.0 | | | 1757.0 |
| | F | 1 | 350.0 | | | | 636 | |
| Columba arquatrix | B | 4 | 381.0 | | | | Zambia 166 | 1758.0 |
| Columba albinucha | U | 1 | 325.0 | | | | 66 | 1761.0 |
| Columba elphinstonii | U | 1 | 379.0 | | | | India 5 | 1763.0 |
| Columba punicea | U | | | | 370.0–510.0 | | 5 | 1765.0 |
| Columba palumboides | M | 1 | 520.0 | | | | India 5 | 1767.0 |
| Columba vitiensis | M | 2 | 420.0 | | 410.0–430.0 | | | 1769.0 |
| | F | 2 | 510.0 | | 510.0–511.0 | | 30, 475, 506 | |
| Columba pallidiceps | M | 1 | 459.0 | | | | New Britain 219 | 1773.0 |
| Columba leucocephala | M | 17 | 263.0 | 27.00 | 210.0–309.0 | | Florida, USA 611a | 1774.0 |
| | F | 22 | 231.0 | 14.00 | 200.0–267.0 | | | |
| Columba speciosa | M | 13 | 262.0 | 11.54 | | | Panama 244 | 1775.0 |
| | F | 6 | 225.0 | | | | | |
| Columba squamosa | U | | 250.0 | | | | 185 | 1776.0 |
| Columba picazuro | M | 1 | 402.0 | | | | Brazil 37 | 1778.0 |
| Columba maculosa | M | 4 | 347.0 | | 308.0–345.0 | | 37, 411 | 1779.0 |
| Columba fasciata fasciata | M | 5888 | 353.0 | | 270.0–460.0 | | Colorado, USA 63 | 1780.0 |
| | F | 5291 | 332.0 | | 226.0–424.0 | | | |
| Columba fasciata monilis | M | 1880 | 398.0 | | 300.0–515.0 | | California, USA 63 | 1780.1 |
| | F | 942 | 386.0 | | 300.0–470.0 | | | |

## Body Masses of World Birds (continued)

| Species | Sex | N | Mean | Std dev | Range | Sn | Location | Number |
|---------|-----|---|------|---------|-------|----|---------|--------|
| Columba caribaea | U | | 250.0 | | | | 185 | 1782.0 |
| Columba cayennensis | U | 9 | 240.0 | | 167.0–262.0 | | 37,188, 245, 246, 385, 510 | 1783.0 |
| Columba flavirostris | U | 6 | 324.0 | | 268.0–424.0 | | Mexico 594 | 1784.0 |
| Columba inornata | U | | 250.0 | | | | 185 | 1786.0 |
| Columba plumbea | U | 5 | 207.0 | | 172.0–221.0 | | Peru 193 | 1787.0 |
| Columba subvinacea | M | 6 | 180.0 | | | | | 1788.0 |
| | F | 4 | 164.0 | | | | 193, 244, 589 | |
| Columba nigrirostris | B | 5 | 176.0 | | 128.0–236.0 | | 86, 244, 618 | 1789.0 |
| Columba iriditorques | M | 15 | 130.0 | 8.00 | | | | 1791.0 |
| | F | 18 | 122.0 | 10.00 | | | 636 | |
| Columba delegorguei | U | 2 | 138.0 | | 133.0–142.0 | | 66 | 1793.0 |
| Columba picturata | M | 11 | 184.0 | | 160.0–210.0 | | Aldabra Is. | 1794.0 |
| | F | 7 | 165.0 | | 150.0–185.0 | | 40 | |
| Columba larvata | U | 4 | 117.0 | | 81.7–140.0 | | 66 | 1795.0 |
| Columba mayeri | M | 33 | 315.0 | 43.08 | 240.0–410.0 | | | 1796.0 |
| | F | 29 | 291.0 | 38.23 | 213.0–369.0 | | 152 | |
| Streptopelia turtur | U | 246 | 132.0 | | 85.0–170.0 | F | Portugal 118 | 1797.0 |
| Rhipidura aureola | U | 20 | 155.0 | 10.85 | 136.0–169.0 | | Ethiopia 633 | 1799.0 |
| Streptopelia orientalis | M | 9 | 202.0 | | 165.0–236.0 | | | 1800.0 |
| | F | 5 | 227.0 | | 204.0–238.0 | | 118 | |
| Streptopelia senegalensis | U | 1157 | 101.0 | 9.20 | 72.0–139.0 | Y | South Africa 118 | 1801.0 |
| Streptopelia chinensis | U | 343 | 159.0 | 11.00 | 128.0–194.0 | | California, USA 177 | 1802.0 |
| Streptopelia decipiens | B | 5 | 134.0 | | 125.0–140.0 | | 66 | 1803.0 |
| Streptopelia vinacea | B | 6 | 108.0 | | 85.0–120.0 | | 66 | 1804.0 |
| Streptopelia capicola | U | 50 | 142.0 | | 92.0–188.0 | | 78 | 1805.0 |
| Streptopelia tranquebarica | M | 1 | 104.0 | | | | India 5 | 1806.0 |
| Streptopelia semitorquata | U | 20 | 176.0 | | 145.0–275.0 | | 78 | 1807.0 |

## Body Masses of World Birds (continued)

| Species | Sex | N | Mean | Std dev | Range | Sn | Location | Number |
|---|---|---|---|---|---|---|---|---|
| Streptopelia decaocto | M | 87 | 152.0 | 11.60 | 115.0–184.0 | Y | India | 1808.0 |
| | F | 80 | 146.0 | 10.30 | 113.0–176.0 | | 118 | |
| Streptopelia roseogrisea | B | 4 | 155.0 | | 135.0–172.0 | | Chad; Niger | 1809.0 |
| | | | | | | | 118 | |
| Streptopelia reichenowi | U | 14 | 119.0 | | 98.0–135.0 | | | 1810.0 |
| | | | | | | | 636 | |
| Streptopelia bitorquata | B | 5 | 158.0 | | 146.0–174.0 | | Philippines | 1811.0 |
| | | | | | | | 475 | |
| Macropygia unchall | U | 2 | 162.0 | | 153.0–172.0 | | Malaysia | 1812.0 |
| | | | | | | | 371 | |
| Macropygia rufipennis | U | | | | 230.0–285.0 | | | 1813.0 |
| | | | | | | | 5 | |
| Macropygia amboinensis | B | 17 | 143.0 | | 107.0–179.0 | | New Guinea | 1816.0 |
| | | | | | | | 154 | |
| Macropygia phasianella | B | 7 | 180.0 | | 168.0–199.0 | | Philippines | 1818.0 |
| | | | | | | | 475 | |
| Macropygia ruficeps | U | 3 | 80.0 | | 74.0–88.0 | | Malaysia | 1819.0 |
| | | | | | | | 371 | |
| Macropygia nigrirostris | B | | 90.1 | 19.00 | 66.0–104.0 | | New Guinea | 1820.0 |
| | | | | | | | 154 | |
| Reinwardtoena reinwardtsi | U | 2 | 209.0 | | | | New Guinea | 1822.0 |
| | | | | | | | 35 | |
| Reinwardtoena browni | B | 4 | 304.0 | | 279.0–325.0 | | | 1823.0 |
| | | | | | | | 30, 219 | |
| Turtur abyssinicus | U | 12 | 58.5 | | 55.0–66.5 | | Ghana | 1827.0 |
| | | | | | | | 228 | |
| Turtur chalcospilos | B | 11 | 60.6 | 5.84 | 50.0–67.3 | | Zimbabwe | 1828.0 |
| | | | | | | | 280, 282 | |
| Turtur afer | U | 10 | 65.6 | | 57.5–74.0 | | Ghana | 1829.0 |
| | | | | | | | 228 | |
| Turtur tympanistria | B | 15 | 67.7 | | 54.0–79.0 | | Somalia | 1830.0 |
| | | | | | | | 691 | |
| Turtur brehmeri | U | 1 | 116.0 | | | | Liberia | 1831.0 |
| | | | | | | | 314a | |
| Oena capensis | B | 138 | 40.6 | | 32.0–54.0 | Y | South Africa | 1832.0 |
| | | | | | | | 118 | |
| Chalcophaps indica | B | 14 | 124.0 | | 108.0–134.0 | | Philippines | 1833.0 |
| | | | | | | | 475 | |
| Chalcophaps stephani | U | 15 | 118.0 | | | | New Guinea | 1834.0 |
| | | | | | | | 35 | |
| Henicophaps albifrons | F | 1 | 247.0 | | | | New Guinea | 1835.0 |
| | | | | | | | 35 | |
| Phaps chalcoptera | U | | 310.0 | | | | | 1837.0 |
| | | | | | | | 550 | |

## Body Masses of World Birds (continued)

| Species | Sex | N | Mean | Std dev | Range | Sn | Location | Number |
|---|---|---|---|---|---|---|---|---|
| Phaps elegans | U | | 200.0 | | | | 550 | 1838.0 |
| Phaps histrionica | U | | 227.0 | | | | 550 | 1839.0 |
| Geophaps lophotes | U | | | | 185.0–225.0 | | 550 | 1840.0 |
| Geophaps plumifera | U | | 110.0 | | | | 550 | 1841.0 |
| Geophaps smithii | U | | 194.0 | | 171.0–225.0 | | 147 | 1843.0 |
| Geopelia cuneata | U | | | | 28.0–43.0 | | 550 | 1846.0 |
| Geopelia striata placida | U | | 56.0 | | | | 194 | 1847.0 |
| Geopelia humeralis | U | | 114.0 | | | | 550 | 1850.0 |
| Zenaida macroura | M | 140 | 123.0 | 1.85 | | PB | Illinois, USA | 1853.0 |
|  | F | 95 | 115.0 | 1.76 | | | 234 | |
| Zenaida graysoni | B | 6 | 192.0 | | 165.0–215.0 | | captive 18 | 1854.0 |
| Zenaida auriculata | B | 16 | 114.0 | | 95.0–149.0 | | 37, 188, 384 | 1855.0 |
| Zenaida aurita | B | 13 | 159.0 | | 136.0–190.0 | | 186, 323, 432, 457 | 1856.0 |
| Zenaida asiatica | U | 30 | 153.0 | 13.20 | 125.0–187.0 | B | Arizona, USA | 1857.0 |
|  | | | | | 125.0–187.0 | | 177 | |
| Columbina inca | U | 125 | 47.5 | 4.41 | 33.0–57.0 | | Arizona, USA 177 | 1859.0 |
| Columbina squammata | U | 18 | 54.2 | 3.84 | 48.0–60.0 | | Venezuela 625 | 1860.0 |
| Columbina passerina | U | 284 | 30.1 | 0.35 | 22.4–41.2 | W | Puerto Rico 186 | 1861.0 |
| Columbina minuta | B | 21 | 33.2 | 4.53 | 26.0–42.2 | | 48, 245, 625 | 1862.0 |
| Columbina talpacoti | B | 74 | 46.5 | | 35.5–56.5 | | Trinidad 576 | 1863.0 |
| Columbina picui | B | 6 | 50.0 | | 45.0–59.0 | | 37, 111 | 1865.0 |
| Claravis pretiosa | B | 15 | 67.3 | | 52.0–77.0 | | Trinidad 188 | 1868.0 |
| Claravis mondetoura | M | 5 | 89.7 | | | | Panama; Peru 244, 668 | 1870.0 |
| Metriopelia melanoptera | U | | 125.0 | | | | Chile 292 | 1873.0 |

## Body Masses of World Birds (continued)

| Species | Sex | N | Mean | Std dev | Range | Sn | Location | Number |
|---|---|---|---|---|---|---|---|---|
| Leptotila verreauxi | U | 34 | 153.0 | 7.11 | | | Panama 244 | 1876.0 |
| Leptotila plumbeiceps | B | 10 | 170.0 | 19.60 | 139.0–205.0 | | Belize 510 | 1878.0 |
| Leptotila wellsi | U | 1 | 200.0 | | | | Grenada 220 | 1880.0 |
| Leptotila rufaxilla | B | 16 | 157.0 | 19.60 | 131.0–183.0 | | 111, 161, 188, 245, 246 | 1881.0 |
| Leptotila jamaicensis | B | 10 | 160.0 | 22.70 | 117.0–190.0 | | 323, 432, 457 | 1882.0 |
| Leptotila cassini | U | 12 | 159.0 | 14.26 | 132.0–179.0 | | 86, 510, 611 | 1884.0 |
| Geotrygon lawrencii | U | | 220.0 | | | | 603a | 1887.0 |
| Geotrygon costaricensis | M | 2 | 320.0 | | 310.0–330.0 | | Panama 244 | 1888.0 |
| | F | 2 | 254.0 | | 225.0–283.0 | | | |
| Geotrygon caniceps | U | | 210.0 | | | | 185 | 1890.0 |
| Geotrygon versicolor | U | | 225.0 | | | | 185 | 1891.0 |
| Geotrygon veraguensis | U | 1 | 155.0 | | | | Panama 315 | 1892.0 |
| Geotrygon albifacies | F | 2 | 316.0 | | 294.0–339.0 | | Honduras 584 | 1893.0 |
| Geotrygon chiriquensis | B | 19 | 308.0 | | | | Panama 244 | 1894.0 |
| Geotrygon goldmani | U | 1 | 258.0 | | | | Panama 50 | 1895.0 |
| Geotrygon linearis | B | 4 | 245.0 | | 230.0–284.0 | | Trinidad 311 | 1896.0 |
| Geotrygon frenata | B | 10 | 311.0 | | | | Columbia; Peru 387, 668 | 1897.0 |
| Geotrygon chrysia | U | 4 | 171.0 | 21.20 | 148.0–199.0 | W | Puerto Rico 186 | 1898.0 |
| Geotrygon mystacea | U | | 230.0 | | | | 185 | 1899.0 |
| Geotrygon violacea | U | 2 | 97.8 | | 93.5–102.0 | | 105, 315 | 1900.0 |
| Geotrygon montana | U | 62 | 115.0 | | | | Peru 193 | 1901.0 |
| Starnoenas cyanocephala | U | 1 | 242.0 | | | | Cuba 648 | 1902.0 |
| Caloenas nicobarica | M | 2 | 492.0 | | 460.0–525.0 | | India 5 | 1903.0 |
| | F | 4 | | | 490.0–600.0 | | | |

## Body Masses of World Birds (continued)

| Species | Sex | N | Mean | Std dev | Range | Sn | Location | Number |
|---|---|---|---|---|---|---|---|---|
| Gallicolumba luzonica | B | 6 | 196.0 | | 181.0–204.0 | | Philippines 475 | 1904.0 |
| Gallicolumba rufigula | U | 2 | 130.0 | | | | New Guinea 35 | 1909.0 |
| Gallicolumba jobiensis | U | 11 | 135.0 | | | | New Guinea 35 | 1911.0 |
| Gallicolumba xanthonura | M | 2 | 128.0 | | 117.0–140.0 | | Guam 290 | 1914.0 |
| | F | 4 | 95.6 | | 58.5–119.0 | | | |
| Gallicolumba beccarii | M | 4 | 93.5 | | 84.0–104.0 | | | 1920.0 |
| | F | 1 | 59.0 | | | | 154, 219 | |
| Trugon terrestris | M | 1 | 400.0 | | | | New Guinea 35 | 1923.0 |
| Otidiphaps nobilis | U | 1 | 500.0 | | | | New Guinea 35 | 1925.0 |
| Phapitreron leucotis | B | 54 | 108.0 | | 89.3–136.0 | | Philippines 475 | 1926.0 |
| Phapitreron amethystina | B | 16 | 136.0 | | 112.0–149.0 | | Philippines 475 | 1927.0 |
| Phapitreron cinereiceps | B | 3 | 146.0 | | 131.0–159.0 | | Philippines 475 | 1928.0 |
| Treron fulvicollis | U | | 167.0 | | | | 330 | 1929.0 |
| Treron olax | U | | 77.0 | | | | 330 | 1930.0 |
| Treron bicincta | U | 2 | 174.0 | | 155.0–194.0 | | India 466 | 1932.0 |
| Treron pompadora | B | 10 | 234.0 | | 218.0–257.0 | | Philippines 475 | 1933.0 |
| Treron curvirostra | B | 9 | 148.0 | | 112.0–186.0 | | Borneo 627 | 1934.0 |
| Treron capellei | U | | 411.0 | | | | 330 | 1939.0 |
| Treron phoenicoptera | B | 6 | 235.0 | | 226.0–248.0 | | India 361 | 1940.0 |
| Treron waalia | U | 2 | 260.0 | | 251.0–268.0 | | 228, 680 | 1941.0 |
| Treron calva uellensis | M | 17 | 218.0 | | 160.0–250.0 | | | 1942.0 |
| | F | 6 | 179.0 | | 130.0–225.0 | | 636 | |
| Treron australis | B | | 226.0 | | 180.0–240.0 | | 66 | 1944.0 |
| Treron apicauda | M | | | | 185.0–255.0 | | 5 | 1945.0 |
| Treron sphenura | U | 3 | 210.0 | | 205.0–214.0 | | 5, 371 | 1948.0 |

**Body Masses of World Birds (continued)**

| Species | Sex | N | Mean | Std dev | Range | Sn | Location | Number |
|---------|-----|---|------|---------|-------|----|----------|--------|
| Ptilinopus occipitalis | B | 23 | 238.0 | | 185.0–278.0 | | Philippines 475 | 1957.0 |
| Ptilinopus jambu | U | | 135.0 | | | | 330 | 1959.0 |
| Ptilinopus leclancheri | B | 3 | 162.0 | | 153.0–174.0 | | Philippines 475 | 1960.0 |
| Ptilinopus magnificus | U | 15 | 189.0 | | | | New Guinea 35 | 1963.0 |
| Ptilinopus perlatus | U | 7 | 210.0 | | | | New Guinea 35 | 1964.0 |
| Ptilinopus ornatus | B | 10 | 163.0 | 12.60 | 142.0–185.0 | | New Guinea 154 | 1965.0 |
| Ptilinopus aurantiifrons | B | 3 | 159.0 | | 148.0–173.0 | | New Guinea 217 | 1967.0 |
| Ptilinopus superbus | U | 11 | 122.0 | | | | New Guinea 35 | 1969.0 |
| Ptilinopus perousii | B | 2 | 93.5 | | 85.0–102.0 | | Tonga 491 | 1970.0 |
| Ptilinopus coronulatus | U | 9 | 77.0 | | | | New Guinea 35 | 1972.0 |
| Ptilinopus pulchellus | U | 15 | 70.0 | | | | New Guinea 35 | 1973.0 |
| Ptilinopus porphyraceus | U | 11 | 105.0 | | 85.4–121.0 | | Tonga 491 | 1978.0 |
| Ptilinopus rivoli | B | 16 | 149.0 | 15.20 | 132.0–188.0 | | New Guinea 154, 216, 219 | 1988.0 |
| Ptilinopus viridis | U | | 125.0 | | | | New Guinea 154 | 1990.0 |
| Ptilinopus iozonus | U | 8 | 115.0 | | | | New Guinea 35 | 1994.0 |
| Ptilinopus insolitus | M | 1 | 162.0 | | | | New Guinea 219 | 1995.0 |
| | F | 1 | 115.0 | | | | | |
| Ptilinopus nanus | U | 3 | 47.0 | | | | New Guinea 35 | 1996.0 |
| Drepanoptila holosericea | U | 2 | 215.0 | | 210.0–220.0 | | New Caledonia 506 | 2002.0 |
| Alectroenas sganzini | M | 2 | 146.0 | | 134.0–158.0 | | Aldabra Is. 40 | 2004.0 |
| | F | 1 | 171.0 | | | | | |
| Alectroenas pulcherrima | B | 2 | 165.0 | | 161.0–169.0 | | Seychelles 211 | 2006.0 |
| Ducula poliocephala | M | 2 | 537.0 | | 510.0–564.0 | | Philippines 475 | 2007.0 |
| Ducula aenae | B | 8 | 560.0 | | 454.0–644.0 | | Philippines 475 | 2012.0 |

## Body Masses of World Birds (continued)

| Species | Sex | N | Mean | Std dev | Range | Sn | Location | Number |
|---|---|---|---|---|---|---|---|---|
| Ducula rubricera | U | 1 | 715.0 | | | | New Ireland 30 | 2019.0 |
| Ducula rufigaster | U | 2 | 436.0 | | | | New Guinea 35 | 2026.0 |
| Ducula finschii | M | | 384.0 | | | | New Britain 219 | 2027.0 |
| | F | | 308.0 | | | | | |
| Ducula goliath | F | 1 | 600.0 | | | | New Caledonia 506 | 2032.0 |
| Ducula pinon | U | 1 | 802.0 | | | | New Guinea 35 | 2033.0 |
| Ducula melanochroa | M | 1 | 665.0 | | | | New Ireland 30 | 2034.0 |
| Ducula mullerii | U | | 800.0 | | | | estimated 35 | 2035.0 |
| Ducula zoeae | U | 2 | 590.0 | | | | New Guinea 35 | 2036.0 |
| Ducula badia | M | 2 | 622.0 | | 580.0–665.0 | 5 | | 2037.0 |
| Ducula bicolor | B | 3 | 483.0 | | 465.0–510.0 | 5 | | 2040.0 |
| Lopholaimus antarcticus | U | | 518.0 | | | 147 | | 2042.0 |
| Hemiphaga novaeseelandiae | U | 2 | 600.0 | | 582.0–619.0 | | New Zealand 500 | 2043.0 |
| Gymnophaps albertisii | B | 6 | 259.0 | | | | New Guinea 35, 216, 218 | 2045.0 |
| Goura cristata | U | | 2000.0 | | | | 520 | 2048.0 |
| Goura sheepmakeri | U | | 2000.0 | | | | estimated 35 | 2050.0 |

### ORDER: PSITTACIFORMES                    FAMILY: PSITTACIDAE

| Species | Sex | N | Mean | Std dev | Range | Sn | Location | Number |
|---|---|---|---|---|---|---|---|---|
| Nestor notabilis | M | 91 | 956.0 | 172.00 | | | New Zealand 56 | 2052.0 |
| | F | 38 | 779.0 | 136.00 | | | | |
| Nestor meridionalis | M | 16 | 487.0 | 33.90 | 440.0–560.0 | | Kapiti Is., NZ 390 | 2053.0 |
| | F | 8 | 418.0 | | 390.0–450.0 | | | |
| Micropsitta keiensis | B | 3 | 11.7 | | 11.0–12.0 | | New Guinea 267 | 2055.0 |
| Micropsitta pusio | U | 6 | 11.0 | | | | New Guinea 35 | 2057.0 |
| Micropsitta bruijnii | M | 3 | 14.8 | | 13.5–15.5 | | New Britain 219 | 2060.0 |
| | F | 2 | 12.2 | | 11.5–13.0 | | | |
| Opopsitta gulielmitertii | B | 3 | 31.0 | | 27.0–33.0 | | New Guinea 154 | 2061.0 |

## Body Masses of World Birds (continued)

| Species | Sex | N | Mean | Std dev | Range | Sn | Location | Number |
|---|---|---|---|---|---|---|---|---|
| Opopsitta diophthalma | F | 1 | 56.0 | | | | New Guinea 154 | 2062.0 |
| Psittaculirostris desmarestii | B | 3 | 117.0 | | 108.0−126.0 | | New Guinea 154 | 2063.0 |
| Bolbopsittacus lunulatus | M | 8 | 71.6 | | 62.5−77.1 | | Philippines 475 | 2066.0 |
| Psittacella brehmii | M | 2 | 94.5 | | 90.0−99.0 | | New Guinea 154, 216 | 2068.0 |
| Psittacella madaraszi | M | 2 | 42.3 | | 41.5−43.0 | | New Guinea 154 | 2071.0 |
|  | F | 2 | 35.7 | | 35.3−36.0 | | | |
| Geoffroyus geoffroyi | B | 9 | 148.0 | | | | New Guinea 35 | 2072.0 |
| Geoffroyus simplex | U | 1 | 161.0 | | | | New Guinea 35 | 2073.0 |
| Geoffroyus heteroclitus | B | 4 | 170.0 | | 154.0−180.0 | | New Britain 219 | 2074.0 |
| Prioniturus montanus | M | 9 | 126.0 | | 116.0−142.0 | | Philippines 475 | 2075.0 |
|  | F | 3 | 111.0 | | 102.0−120.0 | | | |
| Prioniturus discurus | M | 5 | 137.0 | | 127.0−149.0 | | Philippines 475 | 2079.0 |
|  | F | 3 | 160.0 | | 136.0−176.0 | | | |
| Tanygnathus lucionensis | M | 4 | 215.0 | | 199.0−231.0 | | Philippines 475 | 2085.0 |
| Tanygnathus sumatranus | M | 2 | 248.0 | | 230.0−267.0 | | Philippines 475 | 2086.0 |
|  | F | 2 | 293.0 | | 252.0−334.0 | | | |
| Eclectus roratus | B | 12 | 428.0 | | | | 35, 219 | 2088.0 |
| Prosopeia tabuensis | F | | 280.0 | | | | Tonga 492 | 2090.0 |
| Alisterus scapularis | U | | 235.0 | | | | 520 | 2092.0 |
| Alisterus amboinensis | B | 3 | 153.0 | | 145.0−163.0 | | New Guinea 267 | 2093.0 |
| Alisterus chloropterus | U | 10 | 156.0 | | | | New Guinea 35 | 2094.0 |
| Aprosmictus erythropterus | U | | 156.0 | | | | 520 | 2096.0 |
| Polytelis swainsonii | U | | 149.0 | | | | 520 | 2097.0 |
| Polytelis anthopeplus | U | | 114.0 | | | | 550 | 2098.0 |
| Purpureicephalus spurius | U | | 128.0 | | | | W. Australia 550 | 2100.0 |
| Barnardius zonarius | U | | | | −200.0 | | 550 | 2101.0 |

## Body Masses of World Birds (continued)

| Species | Sex | N | Mean | Std dev | Range | Sn | Location | Number |
|---|---|---|---|---|---|---|---|---|
| Barnardius barnardi | U | | 118.0 | | | | | 2102.0 |
| | | | | | | | 520 | |
| Platycercus caledonicus | U | | 110.0 | | | | | 2103.0 |
| | | | | | | | 134 | |
| Platycercus elegans | U | | 116.0 | | | | | 2104.0 |
| | | | | | | | 480a | |
| Platycercus venustus | U | | 85.0 | | | | | 2106.0 |
| | | | | | | | 520 | |
| Platycercus adscitus | U | | 121.0 | | | | | 2107.0 |
| | | | | | | | 134 | |
| Platycercus eximius | U | | 110.0 | | | | | 2108.0 |
| | | | | | | | 194 | |
| Platycercus icterotis | M | 63 | 65.4 | 6.70 | 42.0–80.0 | | W. Australia | 2109.0 |
| | F | 52 | 61.2 | 6.60 | 28.0–71.0 | | 347 | |
| Northiella haematogaster | U | | 80.0 | | | | | 2110.0 |
| | | | | | | | 520 | |
| Psephotus haematonotus | U | | 65.0 | | | | | 2111.0 |
| | | | | | | | 194 | |
| Psephotus varius | U | | 60.0 | | | | | 2112.0 |
| | | | | | | | 520 | |
| Psephotus chrysopterygius | U | | 54.0 | | | | | 2114.0 |
| | | | | | | | 520 | |
| Cyanoramphus novaezelandiae | U | 1 | 72.0 | | | | New Caledonia | 2117.0 |
| | | | | | | | 506 | |
| Neophema bourkii | U | | 39.0 | | | | | 2122.0 |
| | | | | | | | 358 | |
| Neophema chrysostoma | U | | 45.0 | | | | | 2123.0 |
| | | | | | | | 520 | |
| Neophema elegans | U | | 43.0 | | | | | 2124.0 |
| | | | | | | | 520 | |
| Neophema petrophila | U | | 52.0 | | | | | 2125.0 |
| | | | | | | | 520 | |
| Neophema chrysogaster | U | | 42.0 | | | | | 2126.0 |
| | | | | | | | 520 | |
| Neophema pulchella | U | | 38.0 | | | | | 2127.0 |
| | | | | | | | 520 | |
| Neophema splendida | U | | 37.0 | | | | | 2128.0 |
| | | | | | | | 520 | |
| Lathamus discolor | U | | 65.0 | | | | | 2129.0 |
| | | | | | | | 134 | |
| Melopsittacus undulatus | M | 173 | 29.2 | 1.92 | | | | 2130.0 |
| | F | 164 | 28.8 | 2.09 | | | 699 | |
| Pezoporus wallicus | U | | 130.0 | | | | | 2131.0 |
| | | | | | | | 520 | |

## Body Masses of World Birds (continued)

| Species | Sex | N | Mean | Std dev | Range | Sn | Location | Number |
|---------|-----|---|------|---------|-------|----|----------|--------|
| Strigops habroptilus | M | 39 | 2060.0 | | 1500.0–3000.0 | | | 2133.0 |
| | F | 18 | 1280.0 | | 950.0–1640.0 | | 381 | |
| Coracopsis nigra | U | 5 | 141.0 | | 132.0–153.0 | | Seychelles 211, 381 | 2136.0 |
| Psittacus erithacus | U | 2 | 250.0 | | 200.0–300.0 | | 78 | 2137.0 |
| Poicephalus robustus | M | 3 | 370.0 | | 310.0–401.0 | | | 2138.0 |
| | F | 3 | 318.0 | | 280.0–364.0 | | 207 | |
| Poicephalus gulielmi | M | 1 | 290.0 | | | | | 2139.0 |
| | F | 1 | 260.0 | | | | 207 | |
| Poicephalus senegalus | U | 5 | 147.0 | | 120.0–161.0 | | 207 | 2140.0 |
| Poicephalus meyeri | M | 18 | 120.0 | | 102.0–134.0 | | | 2142.0 |
| | F | 10 | 115.0 | | 100.0–123.0 | | 207 | |
| Poicephalus flavifrons | B | 21 | 174.0 | | 140.0–205.0 | | 207 | 2143.0 |
| Poicephalus rufiventris | M | 1 | 120.0 | | | | 66 | 2144.0 |
| Poicephalus cryptoxanthus | B | 13 | 140.0 | | 121.0–156.0 | | 207 | 2145.0 |
| Poicephalus rueppellii | M | 4 | 115.0 | | 105.0–132.0 | | | 2146.0 |
| | F | 1 | 120.0 | | | | 207 | |
| Agapornis canus | U | | 30.0 | | 28.0–31.0 | | 78 | 2147.0 |
| Agapornis pullaria | U | 3 | 37.5 | | 29.5–43.0 | | 207 | 2148.0 |
| Agapornis taranta | M | | 59.0 | | 53.0–66.0 | | | 2149.0 |
| | F | | 56.0 | | 49.0–64.0 | | 207 | |
| Agapornis swinderniana | M | 4 | | | 39.0–41.0 | | 207 | 2150.0 |
| Agapornis roseicollis | U | 29 | | | 46.0–63.0 | | 207 | 2151.0 |
| Agapornis fischeri | U | 4 | 48.3 | | 43.0–53.0 | | 78 | 2152.0 |
| Agapornis personatus | M | 8 | 49.0 | | | | | 2153.0 |
| | F | 9 | 56.0 | | | | 207 | |
| Agapornis lilianae | M | 11 | 38.0 | | | | | 2154.0 |
| | F | 12 | 43.0 | | | | 207 | |
| Loriculus philippensis | B | 19 | 34.9 | | 32.3–39.5 | | 475 | 2158.0 |
| Loriculus tener | U | 1 | 12.0 | | | | New Britain 219 | 2164.0 |
| Psittacula eupatria | F | 3 | 214.0 | | 198.0–225.0 | | India 5, 361 | 2168.0 |

## Body Masses of World Birds (continued)

| Species | Sex | N | Mean | Std dev | Range | Sn | Location | Number |
|---------|-----|---|------|---------|-------|-----|----------|--------|
| Psittacula krameri | M | 2 | 126.0 | | 119.0–134.0 | | | 2170.0 |
| | F | 4 | 107.0 | | 95.8–120.0 | | 152, 361, 680 | |
| Psittacula echo | F | 1 | 163.0 | | | | 152 | 2171.0 |
| Psittacula cyanocephala | B | 12 | 66.0 | | 56.0–71.5 | | 5 | 2176.0 |
| Psittacula alexandri | U | | 150.0 | | 133.0–168.0 | | 78 | 2181.0 |
| Anodorhynchus hyacinthinus | U | | 1500.0 | | | | Brazil 564 | 2184.0 |
| Anodorhynchus leari | U | | 940.0 | | | | Brazil 564 | 2185.0 |
| Ara ararauna | U | | 1125.0 | | | | Peru 623 | 2188.0 |
| Ara militaris | M | 1 | 1134.0 | | | | Mexico 593 | 2190.0 |
| Ara ambigua | U | | 1300.0 | | | | 603a | 2191.0 |
| Ara macao | U | | 1015.0 | | | | Peru 623 | 2192.0 |
| Ara chloroptera | U | | 1250.0 | | | | Peru 623 | 2193.0 |
| Ara severa | U | | 430.0 | | | | Peru 623 | 2197.0 |
| Ara manilata | U | | 370.0 | | | | Peru 623 | 2198.0 |
| Ara couloni | B | 2 | 250.0 | | 207.0–294.0 | | Peru 193 | 2199.0 |
| Ara maracana | B | 2 | 256.0 | | 246.0–266.0 | | Paraguay 610 | 2200.0 |
| Ara nobilis | F | 2 | 136.0 | | 129.0–144.0 | | Surinam 245, 246 | 2202.0 |
| Aratinga wagleri | U | 6 | 196.0 | | 162.0–217.0 | | Columbia 105, 384, 385 | 2209.0 |
| Aratinga finschi | U | | 150.0 | | | | 603a | 2212.0 |
| Aratinga leucopthalmus | B | 8 | 157.0 | | 100.0–218.0 | | Surinam; Peru 161, 193, 245 | 2213.0 |
| Aratinga weddellii | B | 6 | 108.0 | | 96.0–129.0 | | Peru 193, 610a | 2219.0 |
| Aratinga nana | B | 5 | 76.9 | | 72.4–84.0 | | 125a, 323, 510 | 2220.0 |
| Aratinga canicularis | M | 2 | 70.8 | | 68.0–73.5 | | Mexico | 2221.0 |
| | F | 1 | 79.6 | | | | 458, 593 | |

## Body Masses of World Birds (continued)

| Species | Sex | N | Mean | Std dev | Range | Sn | Location | Number |
|---|---|---|---|---|---|---|---|---|
| Aratinga aurea | U | | 84.0 | | | | Brazil 564 | 2222.0 |
| Aratinga pertinax | B | 4 | 86.2 | | 76.8–100.0 | | 245, 625 | 2223.0 |
| Nandayus nenday | B | 4 | 128.0 | | 120.0–141.0 | | Paraguay 610 | 2225.0 |
| Rhynchopsitta terrisi | B | 3 | 442.0 | | 392.0–468.0 | | Coahuila, Mexico 634 | 2229.0 |
| Cyanoliseus patagonus | B | 4 | 273.0 | | 254.0–303.0 | | Argentina 411 | 2231.0 |
| Pyrrhura frontalis | M | 3 | | | 82.5–94.0 | | Brazil | 2234.0 |
| | F | 2 | 72.0 | | 72.0–72.0 | | 37 | |
| Pyrrhura perlata | M | 2 | 75.0 | | 70.0–80.0 | | Brazil 226 | 2235.0 |
| Pyrrhura rhodogaster | B | 3 | 97.3 | | 88.0–102.0 | | Bolivia; Brazil 22, 610a | 2236.0 |
| Pyrrhura picta | U | 9 | 64.7 | | 60.0–69.0 | | 161, 245, 610a, 668 | 2238.0 |
| Pyrrhura rupicola | U | | 75.0 | | | | Peru 623 | 2244.0 |
| Pyrrhura hoematotis | B | 3 | 69.7 | | 64.0–74.2 | | Venezuela 105 | 2247.0 |
| Pyrrhura hoffmanni | B | 34 | 82.2 | | | | Panama 244 | 2249.0 |
| Myiopsitta monachus | B | 9 | 101.0 | | 90.1–114.0 | | Argentina 411 | 2252.0 |
| Bolborhynchus lineola | B | | 53.6 | | 45.2–59.4 | | 244, 387, 668 | 2255.0 |
| Bolborhynchus orbygnesius | B | 2 | 49.0 | | 48.0–50.0 | | Peru 434 | 2256.0 |
| Forpus cyanopygius | U | 8 | 33.2 | | 30.0–36.6 | | Mexico 593, 594 | 2258.0 |
| Forpus passerinus | B | 13 | 25.2 | | 21.5–28.3 | | Venezuela 625 | 2259.0 |
| Forpus xanthopterygius | U | | 26.0 | | | | Brazil 564 | 2260.0 |
| Forpus conspicillatus | U | 7 | 26.4 | | 24.0–28.2 | | Columbia 384, 387 | 2261.0 |
| Forpus sclateri | U | | 25.0 | | | | Peru 623 | 2262.0 |
| Forpus coelestis | M | 2 | 27.9 | | | | Peru | 2263.0 |
| | F | 2 | 24.5 | | | | 672 | |
| Brotogeris tirica | F | 1 | 63.0 | | | | Brazil 622 | 2265.0 |

## Body Masses of World Birds (continued)

| Species | Sex | N | Mean | Std dev | Range | Sn | Location | Number |
|---|---|---|---|---|---|---|---|---|
| Brotogeris versicolurus | M | 1 | 52.5 | | | | Bolivia | 2266.0 |
| | F | 1 | 68.3 | | | | 22 | |
| Brotogeris pyrrhopterus | M | 1 | 68.0 | | | | Peru | 2268.0 |
| | F | 2 | 60.0 | | | | 672 | |
| Brotogeris jugularis | M | 14 | 61.0 | 4.38 | | | Panama | 2269.0 |
| | F | 9 | 65.5 | | | | 244 | |
| Brotogeris cyanoptera | U | | 67.0 | | | | Peru | 2270.0 |
| | | | | | | | 623 | |
| Brotogeris chrysopterus | B | 6 | 59.3 | | 49.0–69.0 | | Surinam | 2271.0 |
| | | | | | | | 245, 246 | |
| Brotogeris sanctithomae | U | 16 | 59.0 | 3.50 | | | Peru | 2272.0 |
| | | | | | | | 193 | |
| Nannopsittaca panychlora | M | 1 | 42.0 | | | | | 2273.0 |
| | | | | | | | 434 | |
| Nannopsittaca dachilleae | B | 18 | 41.5 | | 37.5–46.0 | | | 2273.1 |
| | | | | | | | 434 | |
| Touit batavica | U | 7 | 55.6 | | 52.0–59.5 | | Trinidad | 2274.0 |
| | | | | | | | 576 | |
| Touit huetii | B | 2 | 60.0 | | 58.0–62.0 | | | 2275.0 |
| | | | | | | | 434 | |
| Touit dilectissima | M | 2 | 65.0 | | 59.0–71.0 | | Columbia | 2277.0 |
| | | | | | | | 260 | |
| Touit purpurata | U | 6 | 59.7 | | 52.6–62.6 | | Brazil | 2278.0 |
| | | | | | | | 437 | |
| Touit stictoptera | M | 3 | 78.8 | | 70.5–84.0 | | Peru | 2281.0 |
| | | | | | | | 138 | |
| Pionites melanocephala | U | 3 | 148.0 | | 130.0–164.0 | | | 2282.0 |
| | | | | | | | 245, 610a | |
| Pionites leucogaster | U | | 155.0 | | | | Peru | 2283.0 |
| | | | | | | | 623 | |
| Pionopsitta haematotis | B | 12 | 149.0 | | | | Mexico; Panama | 2285.0 |
| | | | | | | | 244, 457, 497 | |
| Pionopsitta barrabandi | U | | 140.0 | | | | Peru | 2287.0 |
| | | | | | | | 623 | |
| Gypopsitta vulturina | M | 3 | 162.0 | | 160.0–165.0 | | Brazil | 2290.0 |
| | F | 1 | 145.0 | | | | 136 | |
| Hapalopsittaca amazonina | M | 2 | 106.0 | | 97.0–115.0 | | | 2292.0 |
| | | | | | | | 224, 451 | |
| Pionus menstruus | B | 7 | 247.0 | | 213.0–263.0 | | | 2294.0 |
| | | | | | | | 161, 245 | |
| Pionus sordidus | M | 1 | 272.0 | | | | Venezuela | 2295.0 |
| | | | | | | | 105 | |
| Pionus maximiliani | M | 1 | 293.0 | | | | Brazil | 2296.0 |
| | | | | | | | 37 | |

## Body Masses of World Birds (continued)

| Species | Sex | N | Mean | Std dev | Range | Sn | Location | Number |
|---|---|---|---|---|---|---|---|---|
| Pionus senilis | B | 11 | 212.0 | | | | Panama 244 | 2298.0 |
| Pionus chalcopterus | M | 1 | 210.0 | | | | Peru 672 | 2299.0 |
| Pionus fuscus | B | 3 | 202.0 | | 198.0–210.0 | | 161, 245, 246 | 2300.0 |
| Amazona leucocephala | B | 2 | 227.0 | | 222.0–232.0 | | Cuba 590 | 2301.0 |
| Amazona albifrons | B | 15 | 206.0 | 17.90 | 176.0–242.0 | | Mexico; Belize 323, 457, 510 | 2304.0 |
| Amazona xantholora | B | 10 | 217.0 | 12.48 | 200.0–232.0 | | Yucatan, Mexico 323, 457 | 2305.0 |
| Amazona vittata | U | | | | 250.0–300.0 | | 578 | 2307.0 |
| Amazona viridigenalis | U | 1 | 294.0 | | | | Mexico 368 | 2310.0 |
| Amazona finschi | B | 3 | 302.0 | | 282.0–312.0 | | W. Mexico 593, 594 | 2311.0 |
| Amazona autumnalis | B | 16 | 416.0 | | | | Panama 244 | 2312.0 |
| Amazona aestiva | U | | 400.0 | | | | Brazil 564 | 2319.0 |
| Amazona ochrocephala | U | | 510.0 | | | | Peru 623 | 2322.0 |
| Amazona amazonica | U | 3 | 338.0 | | 298.0–403.0 | | Surinam 245, 246 | 2323.0 |
| Amazona mercenaria | F | 1 | 340.0 | | | | Peru 668 | 2324.0 |
| Amazona farinosa | B | 8 | 610.0 | | 535.0–644.0 | | Panama; Mexico 269a, 457, 497 | 2325.0 |
| Deroptyus accipitrinus | U | 4 | 231.0 | | 190.0–270.0 | | Surinam 245, 246 | 2331.0 |
| Triclaria malachitacea | M | 1 | 152.0 | | | | Brazil 37 | 2332.0 |

### ORDER: PSITTACIFORMES — FAMILY: CACATUIDAE

| Species | Sex | N | Mean | Std dev | Range | Sn | Location | Number |
|---|---|---|---|---|---|---|---|---|
| Probosciger aterrimus | U | | 760.0 | | | | 520 | 2333.0 |
| Calyptorhynchus baudinii | B | 75 | 620.0 | | 540.0–770.0 | | 519 | 2334.0 |
| Calyptorhynchus latirostris | B | 186 | 612.0 | | 480.0–750.0 | | 519 | 2335.0 |
| Calyptorhynchus funereus funereus | U | | 801.0 | | | | 520 | 2336.0 |

## Body Masses of World Birds (continued)

| Species | Sex | N | Mean | Std dev | Range | Sn | Location | Number |
|---|---|---|---|---|---|---|---|---|
| Calyptorhynchus funereus xanthanotus | U | | 719.0 | | | | 520 | 2336.0 |
| Calyptorhynchus banksii magnificus | U | | 625.0 | | | | 520 | 2337.0 |
| Calyptorhynchus lathami | U | | 430.0 | | | | 520 | 2338.0 |
| Callocephalon fimbriatum | U | | 219.0 | | | | 480a | 2339.0 |
| Eolophus roseicapillus | U | | 320.0 | | | | 145 | 2340.0 |
| Cacatua leadbeateri | U | | 310.0 | | | | 550 | 2341.0 |
| Cacatua galerita | U | | 892.0 | | | | 520 | 2343.0 |
| Cacatua haematuropygia | B | 5 | 288.0 | | 256.0–302.0 | | Philippines 475 | 2347.0 |
| Cacatua sanguinea | M | 22 | 562.0 | | 466.0–626.0 | | W. Australia | 2350.0 |
| | F | 17 | 488.0 | | 370.0–540.0 | | 518 | |
| Cacatua pastinator pastinator | U | | 611.0 | | | | 520 | 2351.0 |
| Cacatua tenuirostris | U | | 740.0 | | | | 550 | 2352.0 |
| Leptolophus hollandicus | U | | | | 85.0–114.0 | | 550 | 2353.0 |

### ORDER: PSITTACIFORMES      FAMILY: LORIIDAE

| Species | Sex | N | Mean | Std dev | Range | Sn | Location | Number |
|---|---|---|---|---|---|---|---|---|
| Chalcopsitta sintillata | U | 2 | 200.0 | | | | New Guinea | 2356.0 |
| Pseudeos fuscata | U | 10 | 149.0 | | | | New Guinea 35 | 2364.0 |
| Trichoglossus haematodus | U | 16 | 122.0 | | | | New Guinea 35 | 2366.0 |
| Trichoglossus johnstoniae | B | 11 | 55.2 | | 48.0–62.1 | | Philippines 475 | 2370.0 |
| Trichoglossus chlorolepidotus | U | | 78.0 | | 71.0–83.0 | | 133 | 2372.0 |
| Psitteuteles goldiei | M | 2 | 61.0 | | | | New Guinea 35 | 2375.0 |
| Lorius lory | M | 1 | 240.0 | | | | New Guinea 267 | 2378.0 |
| Lorius hypoinochrous | B | 5 | 207.0 | | 187.0–225.0 | | 30, 219 | 2379.0 |
| Lorius albidinuchus | F | 4 | 132.0 | | 120.0–146.0 | | New Ireland 30 | 2380.0 |

## Body Masses of World Birds (continued)

| Species | Sex | N | Mean | Std dev | Range | Sn | Location | Number |
|---|---|---|---|---|---|---|---|---|
| Glossopsitta concinna | U | | 75.8 | | | | 454 | 2388.0 |
| Glossopsitta pusilla | U | | 45.0 | | | | 194a | 2389.0 |
| Glossopsitta porphyrocephala | U | | 43.8 | | | | 454 | 2390.0 |
| Charmosyna rubrigularis | M | | | | 33.0–37.0 | | | 2392.0 |
| | F | | | | 31.5–34.0 | | 30, 219 | |
| Charmosyna placentis | B | 17 | 33.7 | 4.20 | 26.5–42.0 | | New Guinea 154 | 2398.0 |
| Charmosyna pulchella | B | 6 | 33.6 | | | | New Guinea 35, 216 | 2402.0 |
| Charmosyna papou | M | 6 | 87.2 | | 74.0–95.0 | | New Guinea | 2404.0 |
| | F | 4 | 78.8 | | 77.0–84.0 | | 216 | |
| Oreopsittacus arfaki | B | 13 | 20.2 | 1.92 | 17.8–22.5 | | New Guinea 154, 216 | 2405.0 |
| Neopsittacus musschenbroekii | B | 13 | 50.9 | | 49.0–62.0 | | New Guinea 154 | 2406.0 |
| Neopsittacus pullicauda | B | 7 | 32.3 | | 28.0–35.5 | | New Guinea 216 | 2407.0 |

### ORDER: COLIIFORMES     FAMILY: COLIIDAE

| Species | Sex | N | Mean | Std dev | Range | Sn | Location | Number |
|---|---|---|---|---|---|---|---|---|
| Colius striatus | U | 102 | 51.1 | | 36.8–64.0 | | Kenya 67 | 2408.0 |
| Colius leucocephalus | M | 8 | 37.8 | | 31.0–42.0 | | | 2409.0 |
| | F | 4 | 34.0 | | 28.0–39.0 | | 207 | |
| Colius castanotus | B | 5 | 62.2 | | 50.0–72.0 | | 207 | 2410.0 |
| Colius colius | U | 15 | 41.4 | | 29.0–54.0 | | South Africa 255 | 2411.0 |
| Urocolius macrourus | U | 4 | 48.4 | | 40.0–52.5 | | 66 | 2412.0 |
| Urocolius indicus | M | 5 | 61.8 | | 58.4–66.8 | | | 2413.0 |
| | F | 5 | 51.0 | | 48.4–55.5 | | 282, 255 | |

### ORDER: MUSOPHAGIFORMES     FAMILY: MUSOPHAGIDAE

| Species | Sex | N | Mean | Std dev | Range | Sn | Location | Number |
|---|---|---|---|---|---|---|---|---|
| Tauraco persa | B | 5 | 297.0 | | 280.0–308.0 | | 207 | 2414.0 |
| Tauraco schuettii | B | 16 | 235.0 | | 199.0–272.0 | | 207 | 2415.0 |
| Tauraco fischeri | B | 11 | 250.0 | | 227.0–283.0 | | 207 | 2417.0 |
| Tauraco livingstonii | F | 1 | 249.0 | | | | 66 | 2418.0 |

## Body Masses of World Birds (continued)

| Species | Sex | N | Mean | Std dev | Range | Sn | Location | Number |
|---|---|---|---|---|---|---|---|---|
| Tauraco bannermani | U | 4 | 224.0 | | 200.0–250.0 | | 207 | 2420.0 |
| Tauraco macrorhynchus | M | 3 | 268.0 | | 261.0–272.0 | | | 2422.0 |
| | F | 8 | 225.0 | | | | 207 | |
| Tauraco leucotis | U | 15 | 265.0 | | 200.0–315.0 | | 207 | 2423.0 |
| Tauraco ruspolii | U | 5 | 263.0 | | 200.0–290.0 | | 207 | 2424.0 |
| Tauraco hartlaubi | B | 7 | 224.0 | | 200.0–255.0 | | 66 | 2425.0 |
| Tauraco leucolophus | B | 5 | 215.0 | | 198.0–226.0 | | 207 | 2426.0 |
| Musophaga johnstoni | B | 3 | 240.0 | | 232.0–247.0 | | 207 | 2427.0 |
| Musophaga porphyreolopha | U | 1 | 280.0 | | | | 66 | 2428.0 |
| Musophaga violacea | U | 1 | 360.0 | | | | 207 | 2429.0 |
| Musophaga rossae | M | 4 | 422.0 | | 390.0–444.0 | | | 2430.0 |
| | F | 2 | 396.0 | | 395.0–398.0 | | 207 | |
| Corythaixoides concolor | B | 24 | 268.0 | | 202.0–305.0 | | 207 | 2431.0 |
| Corythaixoides personata | U | 6 | 250.0 | | 210.0–300.0 | | 207 | 2432.0 |
| Corythaixoides leucogaster | M | 2 | 198.0 | | 170.0–225.0 | | | 2433.0 |
| | F | 3 | 236.0 | | 225.0–250.0 | | 207 | |
| Crinifer zonurus | U | 8 | 527.0 | | 392.0–737.0 | | | 2435.0 |
| | | | | | | | 66, 207, 677 | |
| Corythaeola cristata | B | 10 | 965.0 | | 857.0–1231.0 | | 207 | 2436.0 |

**ORDER: CUCULIFORMES**          **FAMILY: CUCULIDAE**

| Species | Sex | N | Mean | Std dev | Range | Sn | Location | Number |
|---|---|---|---|---|---|---|---|---|
| Oxylophus jacobinus | M | 5 | 64.9 | | 56.0–71.0 | | | 2437.0 |
| | F | 5 | 79.8 | | 76.0–84.0 | | 118, 255 | |
| Oxylophus levaillantii | B | 13 | 122.0 | | 102.0–141.0 | | 207 | 2438.0 |
| Clamator glandarius | M | 6 | 169.0 | | 153.0–192.0 | B | Spain | 2440.0 |
| | F | 1 | 138.0 | | | | 118 | |
| Pachycoccyx audeberti | M | 1 | 92.0 | | | | | 2441.0 |
| | F | 4 | 115.0 | | 100.0–120.0 | | 207 | |
| Cuculus sparverioides | M | 2 | 124.0 | | 116.0–131.0 | | | 2443.0 |
| | | | | | | | 5 | |
| Cuculus varius | B | 3 | 104.0 | | 100.0–108.0 | | India | 2444.0 |
| | | | | | | | 5, 361 | |

## Body Masses of World Birds (continued)

| Species | Sex | N | Mean | Std dev | Range | Sn | Location | Number |
|---|---|---|---|---|---|---|---|---|
| Cuculus fugax | B | 6 | 83.6 | | 78.5–89.2 | | | 2446.0 |
| | | | | | | | 475, 627 | |
| Cuculus solitarius | B | 3 | 71.7 | | 66.5–74.5 | | | 2447.0 |
| | | | | | | | 255, 281, 354 | |
| Cuculus clamosus | U | 3 | 84.9 | | 78.0–91.5 | | | 2448.0 |
| | | | | | | | 78 | |
| Cuculus micropterus | U | 2 | 47.5 | | 47.0–48.0 | | India | 2449.0 |
| | | | | | | | 466 | |
| Cuculus canorus | B | 136 | 113.0 | | | B | Britain | 2450.0 |
| | | | | | | | 118 | |
| Cuculus gularis | M | 6 | 104.0 | | 95.0–113.0 | | | 2451.0 |
| | F | 2 | 97.5 | | 96.0–99.0 | | 207 | |
| Cuculus saturatus | M | 11 | 116.0 | | 105.0–128.0 | B | Siberia | 2452.0 |
| | F | 6 | 77.0 | | 75.0–85.0 | | 118 | |
| Cuculus poliocephalus | U | | 41.0 | | | | | 2453.0 |
| | | | | | | | 28 | |
| Cuculus rochii | M | 2 | 64.5 | | 64.0–65.0 | | | 2454.0 |
| | | | | | | 207 | | |
| Cercococcyx mechowi | F | 1 | 50.0 | | | | | 2456.0 |
| | | | | | | | 66 | |
| Cercococcyx olivinus | M | 2 | 65.0 | | 64.0–66.0 | | | 2457.0 |
| | | | | | | | 207 | |
| Cercococcyx montanus | U | 1 | 56.0 | | | | | 2458.0 |
| | | | | | | | 51 | |
| Cacomantis sonneratii | U | | 32.0 | | | | | 2459.0 |
| | | | | | | | 28 | |
| Cacomantis merulinus | U | | 26.0 | | | | | 2461.0 |
| | | | | | | | 28 | |
| Cacomantis sepulcralis | B | 14 | 33.4 | | 24.4–39.5 | | Philippines | 2462.0 |
| | | | | | | | 475 | |
| Cacomantis variolosus | M | 1 | 30.8 | | | | Borneo | 2463.0 |
| | | | | | | | 627 | |
| Cacomantis castaneiventris | B | 9 | 34.9 | | 25.0–38.0 | | New Guinea | 2464.0 |
| | | | | | | | 154, 216 | |
| Cacomantis flabelliformis | F | 1 | 50.0 | | | | New Guinea | 2466.0 |
| | | | | | | | 154 | |
| Chrysococcyx minutillus | F | 1 | 20.0 | | | | New Guinea | 2468.0 |
| | | | | | | | 35 | |
| Chrysococcyx lucidus | U | 19 | 24.8 | 1.62 | 21.9–27.5 | | New Zealand | 2472.0 |
| | | | | | | | 500 | |
| Chrysococcyx basalis | U | | 22.8 | | | | | 2473.0 |
| | | | | | | | 194 | |
| Chrysococcyx meyeri | B | 4 | 19.0 | | | | New Guinea | 2475.0 |
| | | | | | | | 35 | |

**Body Masses of World Birds (continued)**

| Species | Sex | N | Mean | Std dev | Range | Sn | Location | Number |
|---|---|---|---|---|---|---|---|---|
| Chrysococcyx flavigularis | B | 8 | | | 27.5–31.0 | | 207 | 2479.0 |
| Chrysococcyx klaas | U | 4 | 24.0 | | 19.9–23.0 | | 66, 228 | 2480.0 |
| Chrysococcyx cupreus | B | 32 | 37.7 | | 30.0–46.0 | | 207 | 2481.0 |
| Chrysococcyx caprius | M | 24 | 29.0 | | 24.0–36.0 | | | 2482.0 |
| | F | 14 | 35.0 | | 29.0–44.0 | | 207 | |
| Caliechthrus leucolophus | M | 3 | 116.0 | | 110.0–125.0 | | New Guinea 154, 267 | 2483.0 |
| Surniculus lugubris | B | 10 | 35.7 | | 32.6–39.0 | | Philippines 475 | 2484.0 |
| Eudynamys scolopacea | B | 11 | 238.0 | 41.50 | 192.0–327.0 | | 5, 219, 361, 475 | 2486.0 |
| Eudynamys taitensis | U | 4 | 126.0 | | 111.0–140.0 | | New Zealand 500 | 2489.0 |
| Ceuthmochares aereus | B | 35 | 63.8 | | 52.0–80.0 | | 207 | 2491.0 |
| Phaenicophaeus diardi | F | 2 | 57.0 | | 55.8–58.2 | | Borneo 627 | 2492.0 |
| Phaenicophaeus tristis | M | 1 | 124.0 | | | | India 361 | 2494.0 |
| Phaenicophaeus javanicus | B | 2 | 97.5 | | 97.0–98.0 | | Borneo | 2498.0 |
| Phaenicophaeus curvirostris | B | 3 | 126.0 | | 111.0–144.0 | | Borneo | 2500.0 |
| Coua ruficeps | B | 4 | 196.0 | | 186.0–204.0 | | Madagascar 39 | 2512.0 |
| Coua cristata | B | 2 | 115.0 | | 110.0–120.0 | | Madagascar 39 | 2513.0 |

**ORDER: CUCULIFORMES**     **FAMILY: CENTROPODIDAE**

| Species | Sex | N | Mean | Std dev | Range | Sn | Location | Number |
|---|---|---|---|---|---|---|---|---|
| Centropus violaceus | F | 1 | 500.0 | | | | New Britain 219 | 2518.0 |
| Centropus menbeki | F | 4 | 510.0 | | | | New Guinea 35 | 2519.0 |
| Centropus ateralbus | M | 1 | 330.0 | | | | New Britain 219 | 2520.0 |
| Centropus phasianinus | U | | 510.0 | | | | W. Australia 550 | 2524.0 |
| Centropus sinensis | U | 3 | 277.0 | | 230.0–362.0 | | India 5, 361 | 2527.0 |
| Centropus viridis | B | 16 | 142.0 | | 106.0–172.0 | | Philippines 475 | 2530.0 |

## Body Masses of World Birds (continued)

| Species | Sex | N | Mean | Std dev | Range | Sn | Location | Number |
|---|---|---|---|---|---|---|---|---|
| Centropus grillii | M | 6 | 100.0 | | 94.0–108.0 | | | 2532.0 |
| | F | 1 | 151.0 | | | | 207 | |
| Centropus bengalensis | M | 4 | 88.0 | | 82.0–94.0 | | Philippines | 2533.0 |
| | F | 1 | 152.0 | | | | 475 | |
| Centropus leucogaster | M | 1 | 293.0 | | | | | 2535.0 |
| | F | 2 | 336.0 | | 327.0–346.0 | | 475 | |
| Centropus monachus | U | 2 | 232.0 | | 225.0–240.0 | | | 2538.0 |
| | | | | | | | 66 | |
| Centropus cupreicaudus | M | 6 | 272.0 | | 250.0–293.0 | | | 2539.0 |
| | F | 5 | 299.0 | | 245.0–342.0 | | 207 | |
| Centropus senegalensis | U | 2 | 156.0 | | 135.0–178.0 | | Botswana | 2540.0 |
| | | | | | | | 118, 680 | |
| Centropus superciliosus | B | 5 | 152.0 | | 135.0–170.0 | | | 2541.0 |
| | | | | | | | 66 | |
| Centropus melanops | B | 27 | 224.0 | | 133.0–265.0 | | Philippines | 2543.0 |
| | | | | | | | 475 | |

**ORDER: CUCULIFORMES**    **FAMILY: COCCYZIDAE**

| Species | Sex | N | Mean | Std dev | Range | Sn | Location | Number |
|---|---|---|---|---|---|---|---|---|
| Coccyzus pumilus | U | 1 | 33.2 | | | | Venezuela | 2546.0 |
| | | | | | | | 626 | |
| Coccyzus cinereus | F | 4 | | | 43.0–57.5 | | Brazil | 2547.0 |
| | | | | | | | 37 | |
| Coccyzus erythropthalmus | U | 104 | 51.1 | 6.81 | 39.6–65.0 | | Pennsylvania, USA | 2548.0 |
| | | | | | | | 177 | |
| Coccyzus americanus | U | 103 | 64.0 | 9.07 | 50.0–84.6 | | Pennsylvania, USA | 2549.0 |
| | | | | | | | 177 | |
| Coccyzus euleri | F | 1 | 61.0 | | | | Brazil | 2550.0 |
| | | | | | | | 610a | |
| Coccyzus minor | U | 18 | 63.9 | 10.80 | 50.0–85.5 | W | Puerto Rico | 2551.0 |
| | | | | | | | 186 | |
| Coccyzus melacoryphus | B | 15 | 48.0 | 4.86 | 44.0–63.0 | | | 2553.0 |
| | | | | | | | 37, 188, 384 | |
| Hyetornis pluvialis | U | | 163.0 | | | | | 2556.0 |
| | | | | | | | 185 | |
| Piaya cayana | B | 20 | 108.0 | | | | Panama | 2557.0 |
| | | | | | | | 244 | |
| Piaya melanogaster | M | 1 | 98.0 | | | | Brazil | 2558.0 |
| | | | | | | | 610a | |
| Piaya minuta | B | 15 | 40.2 | | | | | 2559.0 |
| | | | | | | | 111, 188, 244, 245, 611, 668 | |
| Saurothera merlini | M | 1 | 145.0 | | | | Cuba | 2560.0 |
| | | | | | | | 429 | |
| Saurothera longirostris | U | | 110.0 | | | | | 2562.0 |
| | | | | | | | 185 | |

## Body Masses of World Birds (continued)

| Species | Sex | N | Mean | Std dev | Range | Sn | Location | Number |
|---|---|---|---|---|---|---|---|---|
| Saurothera vieilloti | B | 9 | 80.5 | | 69.2–96.9 | | Puerto Rico 186, 431 | 2563.0 |

### ORDER: CUCULIFORMES  FAMILY: OPISTHOCOMIDAE

| Species | Sex | N | Mean | Std dev | Range | Sn | Location | Number |
|---|---|---|---|---|---|---|---|---|
| Opisthocomus hoazin | U | | 855.0 | | | | Peru 623 | 2564.0 |

### ORDER: CUCULIFORMES  FAMILY: CROTOPHAGIDAE

| Species | Sex | N | Mean | Std dev | Range | Sn | Location | Number |
|---|---|---|---|---|---|---|---|---|
| Crotophaga major | M | 16 | 157.0 | 12.64 | | | Panama | 2565.0 |
| | F | 9 | 140.0 | | | | 244 | |
| Crotophaga ani | M | 10 | 119.0 | 7.10 | 108.0–133.0 | Y | Panama | 2566.0 |
| | F | 12 | 91.0 | 6.68 | 81.8–106.0 | | 611 | |
| Crotophaga sulcirostris | M | 16 | 87.3 | 4.02 | 80.6–93.0 | | Mexico | 2567.0 |
| | F | 19 | 77.1 | 9.64 | 58.7–96.1 | | 585 | |
| Guira guira | M | 3 | 136.0 | | 124.0–157.0 | | | 2568.0 |
| | F | 2 | 168.0 | | 153.0–182.0 | | 37, 111 | |

### ORDER: CUCULIFORMES  FAMILY: NEOMORPHIDAE

| Species | Sex | N | Mean | Std dev | Range | Sn | Location | Number |
|---|---|---|---|---|---|---|---|---|
| Tapera naevia | B | 10 | 52.1 | | | | Panama 244 | 2569.0 |
| Morococcyx erythropygus | B | 3 | 65.1 | | 60.5–70.4 | | Chiapas, Mexico 584 | 2570.0 |
| Dromococcyx phasianellus | U | 4 | 84.5 | | | | 37, 244, 668 | 2571.0 |
| Dromococcyx pavoninus | U | 2 | 43.2 | | 40.5–45.9 | | Brazil 437 | 2572.0 |
| Geococcyx californianus | U | 23 | 376.0 | 90.00 | 221.0–538.0 | Y | Oklahoma, USA 590a | 2573.0 |
| Geococcyx velox | M | 1 | 203.0 | | | | Mexico 548 | 2574.0 |
| Neomorphus geoffroyi | U | | 340.0 | | | | Peru 623 | 2575.0 |
| Neomorphus squamiger | M | 1 | 340.0 | | | | Brazil 226 | 2576.0 |

### ORDER: STRIGIFORMES  FAMILY: TYTONIDAE

| Species | Sex | N | Mean | Std dev | Range | Sn | Location | Number |
|---|---|---|---|---|---|---|---|---|
| Tyto multipunctata | F | 1 | 540.0 | | | | 529 | 2580.0 |
| Tyto tenebricosa | M | | 600.0 | | | | | 2581.0 |
| | F | | 875.0 | | | | 315a | |
| Tyto novaehollandiae | M | | 545.0 | | | | | 2584.0 |
| | F | | 673.0 | | | | 315a | |
| Tyto alba pratincola | M | 33 | 479.0 | 34.80 | | | Utah, USA | 2590.0 |
| | F | 41 | 568.0 | 58.40 | | | 367 | |

## Body Masses of World Birds (continued)

| Species | Sex | N | Mean | Std dev | Range | Sn | Location | Number |
|---|---|---|---|---|---|---|---|---|
| Tyto alba guttata | B | 23 | 294.0 | | 250.0–400.0 | F | Netherlands 118 | 2590.0 |
| Tyto capensis | B | 8 | 419.0 | | | | 315a | 2592.0 |

ORDER: STRIGIFORMES      FAMILY: STRIGIDAE

| Species | Sex | N | Mean | Std dev | Range | Sn | Location | Number |
|---|---|---|---|---|---|---|---|---|
| Otus saggitatus | U | 1 | 121.0 | | | | 366 | 2596.0 |
| Otus rufescens | U | 1 | 77.0 | | | | Malaysia 690 | 2597.0 |
| Otus icterorhynchus | B | 7 | 73.3 | | | | 315a | 2598.0 |
| Otus ireneae | U | 3 | 50.3 | | | | 366 | 2599.0 |
| Otus spilocephalus | U | 16 | 67.5 | | 60.0–77.0 | | Malaysia 371 | 2601.0 |
| Otus flammeolus | M | 56 | 53.9 | | 45.0–63.0 | | | 2604.0 |
| | F | 9 | 57.2 | | 51.0–63.0 | | 180 | |
| Otus scops scops | U | 169 | 92.0 | 14.30 | 64.0–135.0 | F | France 118 | 2605.0 |
| Otus senegalensis | U | 3 | 57.0 | | 49.2–61.6 | | Zimbabwe 285 | 2606.0 |
| Otus sunia | U | 5 | 83.5 | | 75.0–95.0 | | 149 | 2607.0 |
| Otus manadensis manadensis | U | 4 | 88.3 | | | | 366 | 2608.0 |
| Otus umbra | U | 1 | 95.0 | | | | 366 | 2613.0 |
| Otus hartlaubi | U | | 79.0 | | | | estimated 315a | 2614.0 |
| Otus mantananensis sibutuensis | U | 11 | 106.0 | 7.80 | | | 366 | 2616.0 |
| Otus magicus | U | 4 | 165.0 | | | | 366 | 2618.0 |
| Otus rutilus | M | 1 | 107.0 | | | | Madagascar 39 | 2620.0 |
| Otus bakkamoena lettia | M | 3 | 108.0 | | | | | 2622.0 |
| | F | 4 | 142.0 | | | | 366 | |
| Otus bakkamoena everetti | M | 4 | 125.0 | | | | | 2622.0 |
| | F | 3 | 152.0 | | | | 366 | |
| Otus megalotis | U | 2 | 245.0 | | 180.0–310.0 | | 221, 366 | 2625.0 |
| Otus leucotis | B | 16 | 204.0 | | | | 315a | 2628.0 |

## Body Masses of World Birds (continued)

| Species | Sex | N | Mean | Std dev | Range | Sn | Location | Number |
|---|---|---|---|---|---|---|---|---|
| Otus kennicottii kennicottii | M | 14 | 152.0 | | 130.0–178.0 | | | 2629.0 |
| | F | 11 | 186.0 | | 152.0–215.0 | | 180 | |
| Otus kennicottii cineraceus | M | 35 | 111.0 | | 88.0–137.0 | | | 2629.1 |
| | F | 18 | 123.0 | | 92.0–160.0 | | 180 | |
| Otus kennicottii quercinus | M | 26 | 134.0 | | 108.0–170.0 | | | 2629.2 |
| | F | 10 | 152.0 | | 130.0–164.0 | | 180a | |
| Otus asio naevius | M | 31 | 167.0 | 16.30 | 140.0–210.0 | Y | Ohio, USA | 2630.0 |
| | F | 66 | 194.0 | 16.30 | 150.0–235.0 | | 253 | |
| Otus asio vinaceus | M | 1 | 100.0 | | | | | 2630.0 |
| | | | | | | | 584 | |
| Otus seductus | M | 2 | 160.0 | | 158.0–161.0 | | Colima, Mexico | 2631.0 |
| | | | | | | | 584 | |
| Otus cooperi | M | 3 | 149.0 | | 145.0–153.0 | | Mexico | 2632.0 |
| | | | | | | | 584 | |
| Otus trichopsis | M | 23 | 84.5 | | 70.0–104.0 | | | 2633.0 |
| | F | 8 | 92.2 | | 79.0–121.0 | | 180 | |
| Otus choliba | B | 8 | 135.0 | | 114.0–143.0 | | | 2634.0 |
| | | | | | | | 244, 245, 385, 439, 576 | |
| Otus roboratus roboratus | M | 7 | 144.0 | | | | | 2636.0 |
| | F | 3 | 162.0 | | | | 301 | |
| Otus roboratus pacificus | B | 17 | 87.2 | | | | | 2636.0 |
| | | | | | | | 301 | |
| Otus clarkii | M | 3 | 135.0 | | 123.0–150.0 | | | 2638.0 |
| | F | 1 | 186.0 | | | | 497, 669 | |
| Otus barbarus | F | 1 | 69.0 | | | | | 2639.0 |
| | | | | | | | 669 | |
| Otus ingens | M | 3 | 151.0 | | 134.0–168.0 | | Peru | 2640.0 |
| | F | 2 | 196.0 | | 181.0–210.0 | | 193, 260, 668 | |
| Otus watsonii | M | 4 | 117.0 | | 115.0–122.0 | | Brazil | 2644.0 |
| | F | 2 | 134.0 | | 127.0–141.0 | | 226 | |
| Otus atricapillus | M | 7 | 170.0 | | 155.0–194.0 | | Brazil | 2646.0 |
| | F | 8 | 190.0 | | 174.0–211.0 | | 37 | |
| Otus guatemalae | B | 9 | 107.0 | | 91.0–123.0 | | | 2647.0 |
| | | | | | | | 669 | |
| Otus nudipes | B | 5 | 130.0 | | 103.0–154.0 | | Puerto Rico | 2648.0 |
| | | | | | | | 431 | |
| Otus lawrencii | F | 1 | 80.0 | | | | Cuba | 2649.0 |
| | | | | | | | 429 | |
| Otus albogularis | F | 1 | 185.0 | | | | Peru | 2651.0 |
| | | | | | | | 668 | |
| Bubo virginianus virginianus | M | 22 | 1318.0 | | 985.0–1588.0 | | | 2653.0 |
| | F | 29 | 1768.0 | | 1417.0–2503.0 | | 180 | |
| Bubo virginianus occidentalis | M | 18 | 1154.0 | | 865.0–1460.0 | | | 2653.1 |
| | F | 18 | 1555.0 | | 1112.0–2046.0 | | 180 | |

## Body Masses of World Birds (continued)

| Species | Sex | N | Mean | Std dev | Range | Sn | Location | Number |
|---|---|---|---|---|---|---|---|---|
| Bubo virginianus pallescens | M | 18 | 914.0 | | 724.0–1257.0 | | | 2653.2 |
| | F | 12 | 1142.0 | | 801.0–1550.0 | | 180 | |
| Bubo bubo | M | 14 | 2380.0 | | 1835.0–2810.0 | W | Norway | 2654.0 |
| | F | 12 | 2992.0 | | 2280.0–4200.0 | | 118 | |
| Bubo capensis | M | 4 | 929.0 | | | | | 2657.0 |
| | F | 3 | 1347.0 | | | | 315a | |
| Bubo africanus | M | | 585.0 | | | | | 2658.0 |
| | F | | 685.0 | | | | 315a | |
| Bubo poensis | M | 1 | 575.0 | | | | | 2659.0 |
| | F | 4 | 746.0 | | | | 315a | |
| Bubo shelleyi | F | 1 | 1257.0 | | | | | 2662.0 |
| | | | | | | | 315a | |
| Bubo lacteus | M | 4 | 1704.0 | | | | | 2663.0 |
| | F | 6 | 2625.0 | | | | 315a | |
| Bubo leucostictus | M | 2 | 511.0 | | | | | 2665.0 |
| | F | 3 | 555.0 | | | | 315a | |
| Ketupa zeylonensis | F | 1 | 1105.0 | | | | India | 2668.0 |
| | | | | | | | 5 | |
| Scotopelia peli | F | 4 | 2188.0 | | | | | 2671.0 |
| | | | | | | | 315a | |
| Scotopelia ussheri | M | 1 | 743.0 | | | | | 2672.0 |
| | F | | 834.0 | | | | 315a | |
| Scotopelia bouvieri | F | | 637.0 | | | | estimated | 2673.0 |
| | | | | | | | 315a | |
| Nyctea scandiaca | M | 23 | 1806.0 | 144.00 | 1606.0–2043.0 | W | Alberta, Canada | 2674.0 |
| | F | 21 | 2279.0 | 261.00 | 1838.0–2951.0 | | 318 | |
| Strix aluco | M | 11 | 426.0 | 34.60 | 385.0–500.0 | W | Italy | 2678.0 |
| | F | 7 | 524.0 | | 415.0–620.0 | | 118 | |
| Strix butleri | F | 5 | 219.0 | | | | Negev, Sinai | 2679.0 |
| | | | | | | | 315a, 379 | |
| Strix occidentalis | M | 10 | 582.0 | | 518.0–694.0 | | | 2680.0 |
| | F | 10 | 637.0 | | 548.0–760.0 | | 180 | |
| Strix varia | M | 20 | 632.0 | | 468.0–774.0 | | | 2681.0 |
| | F | 24 | 801.0 | | 610.0–1051.0 | | 180 | |
| Strix hylophila | M | 2 | 302.0 | | 285.0–320.0 | | Brazil | 2683.0 |
| | F | 1 | 395.0 | | | | 37 | |
| Strix uralensis macroura | M | 40 | 706.0 | 112.00 | 503.0–950.0 | Y | Romania | 2685.0 |
| | F | 57 | 863.0 | 137.00 | 569.0–1307.0 | | 118 | |
| Strix nebulosa | M | 17 | 789.0 | 175.00 | 568.0–1100.0 | W | Finland; Sweden | 2687.0 |
| | F | 21 | 1159.0 | 306.00 | 680.0–1900.0 | | 118 | |
| Strix virgata | B | 8 | 250.0 | | 187.0–333.0 | | | 2688.0 |
| | | | | | | | 188, 244, 315, | |
| | | | | | | | 368, 387, 457 | |
| Strix nigrolineata | B | 5 | 446.0 | | 403.0–500.0 | | | 2689.0 |
| | | | | | | | 244, 510, 618 | |

## Body Masses of World Birds (continued)

| Species | Sex | N | Mean | Std dev | Range | Sn | Location | Number |
|---|---|---|---|---|---|---|---|---|
| Strix huhula | U | | 370.0 | | | | Peru 623 | 2690.0 |
| Strix woodfordii | B | 2 | 257.0 | | 243.0–271.0 | | 280, 495 | 2692.0 |
| Jubula lettii | F | 1 | 183.0 | | | | 315a | 2693.0 |
| Lophostrix cristata | M | 2 | 468.0 | | 425.0–510.0 | | Mexico; Peru | 2694.0 |
| | F | 1 | 620.0 | | | | 584 | |
| Pulsatrix perspicillata | B | 13 | 873.0 | 170.00 | 591.0–1250.0 | | 37, 86, 161, 244, 245, 497, 510 | 2695.0 |
| Surnia ulula | M | 16 | 299.0 | | 273.0–326.0 | | | 2698.0 |
| | F | 14 | 345.0 | | 306.0–392.0 | | 180 | |
| Glaucidium passerinum | M | 5 | | | 47.0–62.0 | | Finland | 2699.0 |
| | F | 12 | | | 55.0–70.0 | | 118 | |
| Glaucidium brodiei | M | 2 | 52.5 | | 52.0–53.0 | | | 2700.0 |
| | F | 1 | 63.0 | | | | 5 | |
| Glaucidium perlatum | M | 12 | 69.0 | | | | | 2701.0 |
| | F | 13 | 91.0 | | | | 315a | |
| Glaucidium gnoma | M | 42 | 61.9 | | 54.0–74.0 | | | 2702.0 |
| | F | 10 | 73.0 | | 64.0–87.0 | | 180 | |
| Glaucidium minutissimum | U | 15 | 47.8 | 5.49 | 40.4–59.0 | | Mexico 584 | 2703.0 |
| Glaucidium brasilianum | M | 29 | 61.4 | | 46.0–74.0 | | | 2704.0 |
| | F | 16 | 75.1 | | 62.0–95.0 | | 180 | |
| Glaucidium nanum | M | 8 | 64.1 | | | | Chile; Argentina | 2705.0 |
| | F | 8 | 80.1 | | | | 292 | |
| Glaucidium jardinii | U | 4 | 60.5 | | | | Peru 668 | 2706.0 |
| Glaucidium siju siju | M | 3 | 55.7 | | 55.0–57.0 | | Cuba | 2707.0 |
| | F | 2 | 70.0 | | 66.5–73.5 | | 496 | |
| Glaucidium siju vittatum | M | 2 | 66.5 | | 65.0–68.0 | | Isle of Pines | 2707.0 |
| | F | 6 | 87.2 | | 84.0–92.0 | | 496 | |
| Glaucidium tephronotum | B | 10 | 87.4 | | | | 315a | 2708.0 |
| Glaucidium sjostedti | F | | 139.0 | | | | estimated 315a | 2709.0 |
| Glaucidium cuculoides cuculoides | M | 2 | 164.0 | | 159.0–169.0 | | India 5 | 2710.0 |
| Glaucidium radiatum | B | 7 | 101.0 | | 88.0–114.0 | | India 5, 361 | 2713.0 |
| Glaucidium capense | B | 9 | 120.0 | | | | 315a | 2714.0 |
| Glaucidium albertinum | F | | 73.0 | | | | estimated 315a | 2718.0 |

## Body Masses of World Birds (continued)

| Species | Sex | N | Mean | Std dev | Range | Sn | Location | Number |
|---|---|---|---|---|---|---|---|---|
| Xenoglaux loweryi | B | 3 | 48.0 | | 46.0–51.0 | | Peru 433 | 2719.0 |
| Micrathene whitneyi | U | 20 | 41.0 | | 35.9–44.1 | B | Arizona, USA 653 | 2720.0 |
| Athene noctua | B | 21 | 164.0 | | 146.0–193.0 | PB | Netherlands 118 | 2721.0 |
| Athene brama | B | 2 | 112.0 | | 110.0–114.0 | | India 5, 361 | 2722.0 |
| Athene blewitti | M | 1 | 241.0 | | | | India 5 | 2723.0 |
| Athene cunicularia | M | 15 | 151.0 | | 129.0–185.0 | | | 2724.0 |
| | F | 31 | 159.0 | | 120.0–228.0 | | 180 | |
| Aegolius funereus | M | 74 | 101.0 | | 90.0–113.0 | B | Germany | 2725.0 |
| | F | 96 | 167.0 | | 126.0–194.0 | | 23 | |
| Aegolius acadicus | M | 27 | 74.9 | | 54.0–96.0 | | | 2726.0 |
| | F | 18 | 90.8 | | 65.0–124.0 | | 180 | |
| Aegolius ridgwayi | U | | 80.0 | | | | 603a | 2727.0 |
| Aegolius harrisii | B | 10 | 119.0 | | 104.0–135.0 | | Peru; Bolivia 37, 451, 584 | 2728.0 |
| Ninox rufa | M | | | | 1150.0–1300.0 | | Australia | 2729.0 |
| | F | | | | 700.0–1020.0 | | 184 | |
| Ninox strenua | M | | | | 1130.0–1700.0 | | SE Australia | 2730.0 |
| | F | | | | 1050.0–1600.0 | | 184 | |
| Ninox connivens | B | | 462.0 | | | | 315a | 2731.0 |
| Ninox novaeseelandiae | B | 60 | 174.0 | 14.32 | 140.0–216.0 | | New Zealand 500 | 2734.0 |
| Ninox scutulata | B | 3 | 195.0 | | 172.0–227.0 | | India 5, 361 | 2735.0 |
| Ninox superciliaris | B | 2 | 236.0 | | 235.0–236.0 | | Madagascar 39 | 2737.0 |
| Ninox philippensis | M | 1 | 125.0 | | | | Philippines 475 | 2738.0 |
| Ninox squamipila natalis | B | 7 | 148.0 | | 130.0–190.0 | | Christmas Is. 428 | 2742.0 |
| Ninox odiosa | F | 1 | 209.0 | | | | New Britain 219 | 2747.0 |
| Asio stygius | F | 1 | 675.0 | | | | Columbia 385 | 2752.0 |
| Asio otus | M | 38 | 245.0 | | 178.0–314.0 | | | 2753.0 |
| | F | 28 | 279.0 | | 210.0–342.0 | | 180 | |
| Asio clamator | M | 2 | 341.0 | | 335.0–347.0 | | | 2756.0 |
| | F | 3 | 459.0 | | 400.0–502.0 | | 243, 246 | |

## Body Masses of World Birds (continued)

| Species | Sex | N | Mean | Std dev | Range | Sn | Location | Number |
|---|---|---|---|---|---|---|---|---|
| Asio flammeus | M | 20 | 315.0 | | 206.0–368.0 | | | 2757.0 |
| | F | 27 | 378.0 | | 284.0–475.0 | | 180 | |
| Asio capensis | U | 16 | 310.0 | 38.70 | 227.0–355.0 | | | 2758.0 |
| | | | | | | | 118 | |

### ORDER: CAPRIMULGIFORMES — FAMILY: STEATORNITHIDAE

| Species | Sex | N | Mean | Std dev | Range | Sn | Location | Number |
|---|---|---|---|---|---|---|---|---|
| Steatornis caripensis | U | 8 | 414.0 | | 375.0–480.0 | | Trinidad | 2760.0 |
| | | | | | | | 576 | |

### ORDER: CAPRIMULGIFORMES — FAMILY: PODARGIDAE

| Species | Sex | N | Mean | Std dev | Range | Sn | Location | Number |
|---|---|---|---|---|---|---|---|---|
| Podargus strigoides | U | | 350.0 | | | | | 2761.0 |
| | | | | | | | 194 | |
| Podargus papuensis | M | | | | 350.0–570.0 | | | 2762.0 |
| | F | | | | 300.0–400.0 | | 529 | |
| Podargus ocellatus | M | 2 | 218.0 | | 150.0–286.0 | | | 2763.0 |
| | F | 2 | 140.0 | | 140.0–141.0 | | 154, 218, 383 | |
| Batrachostomus septimus | M | 6 | 94.9 | | | | Philippines | 2766.0 |
| | F | 3 | 79.3 | | 75.2–81.7 | | 366, 475 | |
| Batrachostomus stellatus | U | 3 | 47.0 | | | | | 2767.0 |
| | | | | | | | 366, 690 | |
| Batrachostomus hodgsoni | U | 2 | 51.0 | | | | | 2769.0 |
| | | | | | | | 366 | |
| Batrachostomus affinis | U | 4 | 46.7 | | | | | 2772.0 |
| | | | | | | | 366 | |
| Batrachostomus javensis | U | 7 | 46.0 | | | | | 2773.0 |
| | | | | | | | 366 | |

### ORDER: CAPRIMULGIFORMES — FAMILY: AEGOTHELIDAE

| Species | Sex | N | Mean | Std dev | Range | Sn | Location | Number |
|---|---|---|---|---|---|---|---|---|
| Aegotheles insignis | B | 13 | 70.6 | 6.60 | 61.0–82.0 | | New Guinea | 2776.0 |
| | | | | | | | 157 | |
| Aegotheles cristatus | U | | | | 43.0–57.0 | | | 2777.0 |
| | | | | | | | 550 | |
| Aegotheles wallacii | M | 1 | 48.5 | | | | New Guinea | 2780.0 |
| | | | | | | | 154 | |
| Aegotheles archboldi | F | 1 | 35.0 | | | | New Guinea | 2781.0 |
| | | | | | | | 216 | |
| Aegotheles albertisi | B | 4 | 38.0 | | 36.0–40.0 | | New Guinea | 2782.0 |
| | | | | | | | 154, 157 | |

### ORDER: CAPRIMULGIFORMES — FAMILY: NYCTIBIIDAE

| Species | Sex | N | Mean | Std dev | Range | Sn | Location | Number |
|---|---|---|---|---|---|---|---|---|
| Nyctibius grandis | B | | 575.0 | | | | Peru | 2783.0 |
| | | | | | | | 623 | |
| Nyctibius aethereus | U | | | | 434.0–447.0 | | Brazil | 2784.0 |
| | | | | | | | 564 | |

## Body Masses of World Birds (continued)

| Species | Sex | N | Mean | Std dev | Range | Sn | Location | Number |
|---|---|---|---|---|---|---|---|---|
| Nyctibius griseus | B | 21 | 185.0 | | | | Panama 244 | 2786.0 |
| Nyctibius leucopterus | F | 2 | 165.0 | | 145.0–185.0 | | Bolivia 451, 484 | 2788.0 |
| Nyctibius bracteatus | U | | 125.0 | | | | estimated 623 | 2789.0 |

### ORDER: CAPRIMULGIFORMES    FAMILY: EUROSTOPODIDAE

| Species | Sex | N | Mean | Std dev | Range | Sn | Location | Number |
|---|---|---|---|---|---|---|---|---|
| Eurostopodus argus | U | | 85.0 | | | | W. Australia 550 | 2790.0 |
| Eurostopodus mystacalis | F | 1 | 145.0 | | | | New Guinea 154 | 2791.0 |
| Eurostopodus temminckii | U | 3 | 87.8 | | | | Malaysia 690 | 2795.0 |
| Eurostopodus macrotis | F | 1 | 151.0 | | | | Philippines 475 | 2796.0 |

### ORDER: CAPRIMULGIFORMES    FAMILY: CAPRIMULGIDAE

| Species | Sex | N | Mean | Std dev | Range | Sn | Location | Number |
|---|---|---|---|---|---|---|---|---|
| Lurocalis semitorquatus | U | 1 | 87.0 | | | | Peru 193 | 2797.0 |
| Chordeiles acutipennis | U | 13 | 49.9 | 7.08 | 41.0–64.0 | | Texas, USA 585 | 2800.0 |
| Chordeiles minor | B | 13 | 61.5 | | | | 244 | 2801.0 |
| Nyctiprogne leucopyga | F | 1 | 20.0 | | | | Brazil 136 | 2803.0 |
| Podager nacunda | B | 3 | 173.0 | | 162.0–188.0 | | Brazil; Columbia 37, 384 | 2804.0 |
| Nyctidromus albicollis | B | 37 | 53.2 | | | | Panama 244 | 2805.0 |
| Phalaenoptilus nuttallii | U | 12 | 51.6 | 4.47 | 42.8–58.1 | | western USA 595 | 2806.0 |
| Nyctiphrynus mcleodii | M | 2 | 35.0 | | 33.5–36.4 | | Mexico 584 | 2809.0 |
| Nyctiphrynus yucatanicus | B | 2 | 24.5 | | 21.3–27.7 | | Belize; Mexico 457, 510 | 2810.0 |
| Nyctiphrynus ocellatus | U | 12 | 42.4 | | | | Peru 193 | 2811.0 |
| Caprimulgus carolinensis | U | 12 | 120.0 | | | B | 505 | 2812.0 |
| Caprimulgus rufus | U | 1 | 94.6 | | | | Panama 315 | 2813.0 |
| Caprimulgus cubanensis | U | 8 | 63.5 | | 60.0–80.0 | | Cuba 210 | 2815.0 |

## Body Masses of World Birds (continued)

| Species | Sex | N | Mean | Std dev | Range | Sn | Location | Number |
|---|---|---|---|---|---|---|---|---|
| Caprimulgus salvini | B | 5 | 59.0 | | 51.2–65.5 | | Mexico; Belize 323, 368, 457, 510 | 2816.0 |
| Caprimulgus badius | F | 1 | 64.3 | | | | Belize 584 | 2817.0 |
| Caprimulgus sericocaudatus | M | 1 | 83.0 | | | | Paraguay 610 | 2818.0 |
| Caprimulgus ridgwayi | U | 6 | 48.0 | | 39.8–61.0 | | Mexico 584, 586, 594 | 2819.0 |
| Caprimulgus vociferus | M | 32 | 55.3 | | 43.0–63.7 | | Florida, USA 190 | 2820.0 |
| | F | 39 | 50.6 | | 44.5–61.2 | | | |
| Caprimulgus noctitherus | M | 3 | 35.7 | | 35.0–36.6 | | Puerto Rico 431 | 2821.0 |
| Caprimulgus saturatus | U | | 55.0 | | | | 603a | 2822.0 |
| Caprimulgus longirostris | U | 2 | 43.2 | | 36.4–50.0 | | Brazil; Peru 37, 668 | 2823.0 |
| Caprimulgus cayennensis | U | 7 | 34.9 | | 31.0–38.0 | | Surinam; Tobago 246, 311 | 2824.0 |
| Caprimulgus parvulus | M | 2 | 32.0 | | 25.0–39.0 | | | 2827.0 |
| | F | 3 | 41.8 | | 34.0–46.5 | | 37, 111 | |
| Caprimulgus nigrescens | B | 3 | 37.6 | | 35.0–39.0 | | 161, 246, 610a | 2830.0 |
| Caprimulgus binotatus | U | 1 | 63.0 | | | | 207 | 2833.0 |
| Caprimulgus ruficollis | U | 2 | 68.5 | | 62.0–75.0 | | 118 | 2834.0 |
| Caprimulgus indicus | B | 4 | 87.2 | | 69.0–107.0 | | 5, 149 | 2835.0 |
| Caprimulgus europaeus | U | 36 | 67.0 | 8.10 | 56.0–85.0 | F | France 118 | 2836.0 |
| Caprimulgus rufigena | B | 4 | 52.0 | | 51.3–52.4 | | South Africa 255 | 2838.0 |
| Caprimulgus aegyptius | B | 7 | 77.3 | | 68.0–93.0 | | 78, 118 | 2839.0 |
| Caprimulgus madagascariensis | B | 5 | 40.9 | | 38.0–44.0 | | Madagascar 39, 40 | 2844.0 |
| Caprimulgus macrurus | U | 3 | 78.0 | | | | New Guinea 35, 494 | 2845.0 |
| Caprimulgus atripennis | U | 1 | 55.0 | | | | India 494 | 2846.0 |
| Caprimulgus donaldsoni | U | 14 | 29.0 | | 21.0–36.0 | | 207 | 2848.0 |
| Caprimulgus nigriscapularis | M | 1 | 48.0 | | | | Uganda 207 | 2849.0 |

## Body Masses of World Birds (continued)

| Species | Sex | N | Mean | Std dev | Range | Sn | Location | Number |
|---|---|---|---|---|---|---|---|---|
| Caprimulgus pectoralis | B | 5 | 47.2 | | 41.0–52.2 | | Zimbabwe 282, 283, 284 | 2850.0 |
| Caprimulgus poliocephalus | B | 4 | 41.8 | | 40.0–43.0 | | Zambia 166 | 2851.0 |
| Caprimulgus ruwenzorii | M | 10 | 46.4 | | 44.0–50.0 | | | 2852.0 |
| | F | 9 | 51.6 | | 47.0–57.0 | | 207 | |
| Caprimulgus asiaticus | B | 3 | 42.0 | | 40.0–46.0 | | India 5, 361 | 2853.0 |
| Caprimulgus natalensis | F | 1 | 65.0 | | | | 66 | 2854.0 |
| Caprimulgus inornatus | U | 4 | 55.4 | | 46.0–61.4 | | 66 | 2855.0 |
| Caprimulgus stellatus | U | 1 | 41.0 | | | | 66 | 2856.0 |
| Caprimulgus affinis | B | 2 | 75.0 | | 72.0–78.0 | | India 361 | 2857.0 |
| Caprimulgus tristigma | F | 1 | 87.0 | | | | 66 | 2858.0 |
| Caprimulgus enarratus | M | 1 | 47.5 | | | | Madagascar | 2861.0 |
| | F | 1 | 57.0 | | | | 39 | |
| Caprimulgus batesi | F | 2 | 100.0 | | 89.0–112.0 | | 207 | 2862.0 |
| Caprimulgus climacurus | B | 4 | 39.3 | | 36.3–43.5 | | Ghana 228 | 2863.0 |
| Caprimulgus clarus | B | 19 | 42.5 | | 34.0–53.0 | | 207 | 2864.0 |
| Caprimulgus fossii | B | 7 | 51.4 | | 35.4–68.3 | | 66, 282, 283 | 2865.0 |
| Macrodipteryx longipennis | B | 32 | 48.0 | | 37.0–65.0 | | Chad 207 | 2866.0 |
| Macrodipteryx vexillarius | M | 107 | 67.7 | 4.40 | 59.0–79.0 | B | Zambia | 2867.0 |
| | F | 14 | 77.1 | 5.80 | 67.0–825 | | 65a | |
| Hydropsalis climacocerca | M | 1 | 34.0 | | | | Surinam 246 | 2868.0 |
| Hydropsalis brasiliana | M | 6 | | | 51.0–63.0 | | Brazil | 2869.0 |
| | F | 2 | 58.5 | | 57.0–60.0 | | 37 | |
| Uropsalis segmentata | B | 3 | 42.9 | | | | Peru 668 | 2870.0 |
| Uropsalis lyra | B | 2 | 71.2 | | 68.5–74.0 | | Ecuador 589 | 2871.0 |

**ORDER: APODIFORMES**          **FAMILY: HEMIPROCNIDAE**

| Species | Sex | N | Mean | Std dev | Range | Sn | Location | Number |
|---|---|---|---|---|---|---|---|---|
| Hemiprocne coronata | B | 5 | 22.2 | | 20.2–26.0 | | 475 | 2874.0 |

## Body Masses of World Birds (continued)

| Species | Sex | N | Mean | Std dev | Range | Sn | Location | Number |
|---|---|---|---|---|---|---|---|---|
| Hemiprocne longipennis | F | 2 | 29.0 | | 28.0–30.0 | | India 361 | 2875.0 |
| Hemiprocne mystacea | B | 8 | 74.5 | | 69.0–80.0 | | New Guinea 154, 216, 267 | 2876.0 |
| Hemiprocne comata comata | B | 10 | 18.2 | 1.72 | 15.6–21.8 | | 594a | 2877.0 |
| Hemiprocne comata major | B | 20 | 23.4 | 1.47 | 20.2–25.3 | | 586a | 2874.0 |

### ORDER: APODIFORMES                        FAMILY: APODIDAE

| Species | Sex | N | Mean | Std dev | Range | Sn | Location | Number |
|---|---|---|---|---|---|---|---|---|
| Cypseloides rutilus | B | 43 | 20.2 | | 17.8–24.2 | | Trinidad 188 | 2878.0 |
| Cypseloides phelpsi | U | 3 | 24.0 | | 22.0–26.0 | | Venezuela 586a | 2879.0 |
| Cypseloides niger | U | 12 | 45.6 | 2.06 | 41.5–48.9 | | California, USA 177 | 2880.0 |
| Cypseloides fumigatus | B | | | | 40.0–44.0 | | Brazil 37 | 2883.0 |
| Cypseloides cherriei | U | 5 | 22.1 | | 21.8–22.8 | | Venezuela 105 | 2884.0 |
| Cypseloides cryptus | B | 13 | 35.3 | 1.92 | 32.5–39.4 | | Costa Rica 594a | 2885.0 |
| Cypseloides senex | B | 33 | 99.8 | | 86.1–110.0 | | Argentina 596 | 2886.0 |
| Streptoprocne zonaris | U | 19 | 98.1 | 5.44 | 85.8–107.0 | | Venezuela 105 | 2887.0 |
| Streptoprocne biscutata | U | 1 | 127.0 | | | | Brazil 37 | 2888.0 |
| Streptoprocne semicollaris | U | 2 | 175.0 | | 170.0–180.0 | | 597a | 2889.0 |
| Hydrochrous gigas | U | 3 | 37.0 | | 35.0–39.9 | | 580 | 2890.0 |
| Collocalia esculenta | B | 7 | 7.4 | | 6.3–8.3 | | Philippines 475 | 2891.0 |
| Collocalia marginata | U | 5 | 7.1 | | 6.3–8.3 6.3–8.3 | | 586a | 2892.0 |
| Collocalia troglodytes | U | 21 | 5.4 | 0.48 | 4.5–6.6 | | Philippines 586a | 2894.0 |
| Collocalia francica | B | 19 | 8.9 | | 7.9–11.4 | | 152 | 2896.0 |
| Collocalia unicolor | M | 2 | 11.0 | | 11.0–11.0 | | India 5 | 2897.0 |
| Collocalia hirundinacea | B | 18 | 9.0 | 0.37 | 8.0–10.0 | | New Guinea 154 | 2899.0 |

## Body Masses of World Birds (continued)

| Species | Sex | N | Mean | Std dev | Range | Sn | Location | Number |
|---|---|---|---|---|---|---|---|---|
| Collocalia spodiopygius | B | 55 | 6.8 | | 6.0–8.5 | | New Caledonia 506 | 2900.0 |
| Collocalia spodiopygius | B | 13 | 8.2 | 0.22 | | | Fiji 617 | 2900.0 |
| Collocalia brevirostris | B | 3 | 12.5 | | 12.0–13.0 | | India 5 | 2902.0 |
| Collocalia salangana | U | | 11.3 | | | | 27 | 2907.0 |
| Collocalia vanikorensis | U | 6 | 11.0 | | | | New Guinea 35 | 2908.0 |
| Collocalia fuciphaga | U | 315 | 10.7 | 0.76 | 8.7–14.8 | | 200a | 2916.0 |
| Schoutedenapus myoptilus | B | 5 | 26.0 | | 22.0–30.0 | | Kenya; Uganda 584, 597a | 2920.0 |
| Mearnsia novaeguineae | U | 7 | 32.6 | | 29.0–39.0 | | New Guinea 596 | 2923.0 |
| Zoonavena thomensis | U | | 8.0 | | | | estimated 207 | 2925.0 |
| Telecanthura ussheri stictilaema | B | 8 | 31.2 | | 28.5–34.6 | | 73a | 2927.0 |
| Telecanthura ussheri benguellensis | B | 14 | 34.1 | | 31.0–37.5 | | Zimbabwe 73a | 2927.0 |
| Telecanthura melanopygia | F | 1 | 52.0 | | | | 207 | 2928.0 |
| Rhaphidura leucopygialis | U | 10 | 13.6 | 1.30 | 11.9–15.4 | | Borneo 627 | 2929.0 |
| Rhaphidura sabini | B | 15 | 16.3 | 0.67 | 15.0–18.0 | | Kenya; Uganda 583a, 597a | 2930.0 |
| Neafrapus cassini | M | 1 | 41.5 | | | | | 2931.0 |
| | F | 2 | 38.5 | | | | 207 | |
| Neafrapus boehmi | U | 2 | 21.4 | | 20.5–22.3 | | Zambia 75 | 2932.0 |
| Hirundapus caudacautus | B | 16 | 120.0 | | 101.0–140.0 | | Siberia 416 | 2933.0 |
| Hirundapus cochinchinensis | U | 2 | 80.8 | | 76.1–85.5 | | Thailand 108 | 2934.0 |
| Hirundapus giganteus indicus | U | 4 | 137.0 | | 123.0–167.0 | | Thailand 108 | 2935.0 |
| Hirundapus celebensis | U | 22 | 180.0 | | 170.0–203.0 | | 396 | 2936.0 |
| Chaetura spinicauda | U | 68 | 15.2 | | 13.0–20.0 | | Trinidad 188 | 2937.0 |
| Chaetura martinica | U | 10 | 12.5 | | | | Dominica 471, 597a | 2938.0 |

## Body Masses of World Birds (continued)

| Species | Sex | N | Mean | Std dev | Range | Sn | Location | Number |
|---|---|---|---|---|---|---|---|---|
| Chaetura cinereiventris | U | 128 | 13.9 | | 11.8–18.0 | | 188 | 2939.0 |
| Chaetura egregia | F | 1 | 25.0 | | | | Peru 138 | 2940.0 |
| Chaetura pelagica | U | 1805 | 23.6 | | 17.0–29.8 | S | Illinois, USA 701 | 2941.0 |
| Chaetura vauxi | U | 72 | 17.1 | 1.30 | 15.0–20.9 | | California, USA 177 | 2942.0 |
| Chaetura chapmani | U | 16 | 24.7 | | 22.0–28.0 | | 188 | 2943.0 |
| Chaetura brachyura | U | 240 | 18.3 | 1.15 | 15.5–22.0 | | Trinidad 104 | 2944.0 |
| Chaetura andrei | U | 1 | 19.5 | | | | Brazil 37 | 2945.0 |
| Aeronautes saxatalis | U | 20 | 32.1 | 2.53 | 27.8–36.0 | B | California, USA 177 | 2946.0 |
| Aeronautes montivagus | M | 50 | 20.6 | 0.95 | 18.2–22.8 | | Venezuela | 2947.0 |
| | F | 39 | 19.7 | 1.09 | 17.2–21.8 | | 105 | |
| Tachornis phoenicobia | U | 3 | 9.3 | | 9.0–9.5 | | Dominican Rep. 600 | 2949.0 |
| Tachornis squamata | B | 13 | 10.6 | 0.65 | 9.3–11.5 | | Venezuela 103 | 2951.0 |
| Panyptila sanctihieronymi | B | 3 | 48.1 | | 46.9–50.3 | | Mexico 48 | 2952.0 |
| Panyptila cayennensis | B | 5 | 21.1 | | 15.6–28.0 | | 188, 510 | 2953.0 |
| Cypsiurus parvus | U | 61 | 13.6 | | 10.0–18.0 | | 118 | 2954.0 |
| Cypsiurus batasiensis | B | 3 | 8.1 | | 8.0–8.2 8.0–8.2 | | Sabah 594a | 2955.0 |
| Tachymarptis melba melba | U | 45 | 104.0 | | 76.0–120.0 | B | Switzerland 118 | 2956.0 |
| Tachymarptis melba africanus | U | 12 | 76.0 | | 67.0–87.0 | | Angola 74 | 2956.0 |
| Tachymarptis melba turneti | U | | | | 95.0–110.0 | | 2 | 2956.0 |
| Tachymarptis aequatorialis | B | 26 | 93.5 | | 83.0–105.0 | | Kenya 73 | 2957.0 |
| Apus apus | U | 218 | 37.6 | | 31.0–43.0 | B | Britain 118 | 2959.0 |
| Apus niansae | U | 22 | 34.0 | 2.80 | 25.0–37.5 | | Kenya 584 | 2961.0 |
| Apus pallidus | U | 661 | 41.9 | 2.78 | 34.5–50.0 | | Spain 502 | 2962.0 |

## Body Masses of World Birds (continued)

| Species | Sex | N | Mean | Std dev | Range | Sn | Location | Number |
|---|---|---|---|---|---|---|---|---|
| Apus barbatus | B | 18 | 42.8 | | 39.0–50.0 | | 75 | 2963.0 |
| Apus berliozi | U | 8 | 43.8 | | 38.5–51.0 | | Kenya 597a | 2964.0 |
| Apus bradfieldi | B | 35 | 42.4 | | 33.0–50.0 | | Angola 72 | 2965.0 |
| Apus pacificus | M | 7 | 48.1 | | 38.0–54.0 | B | Mongolia 118 | 2967.0 |
| | F | 2 | 42.5 | | 38.0–47.0 | | | |
| Apus affinis affinis | B | 154 | 17.9 | | 14.5–21.5 | | 409 | 2969.0 |
| Apus affinis theresae | U | 30 | 24.9 | | 19.0–30.0 | | Angola 74a | 2969.0 |
| Apus affinis kuntzi | U | 14 | 24.3 | | 20.0–35.0 | | Taiwan 72 | 2969.0 |
| Apus affinis bannermani | U | 5 | 23.9 | | 23.0–25.0 | | 596a | 2969.0 |
| Apus horus | U | | 28.0 | | 25.4–31.3 | | Kenya 106a | 2971.0 |
| Apus caffer | U | 54 | 22.1 | | 18.0–28.0 | | 118 | 2973.0 |

## ORDER: TROCHILIFORMES                FAMILY: TROCHILIDAE

| Species | Sex | N | Mean | Std dev | Range | Sn | Location | Number |
|---|---|---|---|---|---|---|---|---|
| Glaucis aenea | B | 4 | 4.8 | | 4.0–5.4 | | 584 | 2975.0 |
| Glaucis hirsuta | U | 224 | 7.0 | | 5.0–9.5 | | Trinidad 576 | 2976.0 |
| Threnetes leucurus | U | 42 | 5.6 | 0.10 | | | 79 | 2978.0 |
| Threnetes ruckeri | U | 13 | 6.4 | 0.32 | 6.0–6.8 | | Panama 50 | 2979.0 |
| Phaethornis yaruqui | U | 1 | 6.0 | | | | 79 | 2980.0 |
| Phaethornis guy | U | 86 | 6.3 | | 5.5–8.0 | | Trinidad 576 | 2981.0 |
| Phaethornis syrmatophorus | U | 3 | 5.9 | | | | 79, 387 | 2982.0 |
| Phaethornis superciliosus | U | 277 | 6.0 | 0.27 | | | 79 | 2983.0 |
| Phaethornis malaris | M | 2 | 8.8 | | 7.5–10.0 | | French Guiana 161 | 2984.0 |
| Phaethornis eurynome | U | 16 | 5.4 | | 4.5–7.0 | | Brazil 441 | 2985.0 |
| Phaethornis hispidus | U | 26 | 4.9 | 0.12 | | | 79 | 2986.0 |

## Body Masses of World Birds (continued)

| Species | Sex | N | Mean | Std dev | Range | Sn | Location | Number |
|---|---|---|---|---|---|---|---|---|
| Phaethornis anthophilus | U | 1 | 4.6 | | | | 79 | 2987.0 |
| Phaethornis bourcieri | U | 8 | 4.1 | | | | Surinam 573 | 2988.0 |
| Phaethornis koepckeae | M | 16 | 5.4 | | 4.7–5.8 | | Peru | 2989.0 |
| | F | 10 | 4.7 | | 4.4–4.9 | | 138 | |
| Phaethornis philippii | B | 4 | 4.8 | | 4.2–5.7 | | Brazil 610a | 2990.0 |
| Phaethornis squalidus | U | | 3.7 | | | | 453 | 2991.0 |
| Phaethornis augusti | U | 5 | 5.3 | | 5.0–5.8 | | Venezuela 625 | 2992.0 |
| Phaethornis pretrei | U | 4 | 5.5 | | 4.9–6.0 | | 227, 440, 441 | 2993.0 |
| Phaethornis nattereri | U | 1 | 3.1 | | | | 227 | 2995.0 |
| Phaethornis ruber | U | 8 | 2.3 | | | | 79, 227, 437, 439, 610a | 2997.0 |
| Phaethornis stuarti | U | 2 | 2.9 | | | | Peru 668 | 2998.0 |
| Phaethornis griseogularis | U | 3 | 2.3 | | | | Peru 79, 672 | 2999.0 |
| Phaethornis longuemareus | U | 58 | 3.0 | 0.38 | | | 79 | 3000.0 |
| Phaethornis idaliae | U | 1 | 2.4 | | | | 227 | 3001.0 |
| Eutoxeres aquila | U | 13 | 11.1 | 0.93 | 10.0–12.5 | | Panama 50 | 3002.0 |
| Eutoxeres condamini | U | 24 | 9.4 | 0.27 | | | 79 | 3003.0 |
| Androdon aequatorialis | U | 3 | 6.5 | | 5.5–8.0 | | 79, 497 | 3004.0 |
| Ramphodon naevius | U | 54 | 6.8 | | 5.3–8.5 | | Brazil 441 | 3005.0 |
| Ramphodon dohrnii | U | | 6.5 | | | | 453 | 3006.0 |
| Doryfera johannae | U | 9 | 4.1 | 0.10 | | | 79 | 3007.0 |
| Doryfera ludoviciae | U | 20 | 5.7 | 0.20 | | | 79 | 3008.0 |
| Phaeochroa cuvierii | U | 37 | 8.9 | 0.41 | | | 79 | 3009.0 |
| Campylopterus curvipennis | U | 20 | 6.2 | | | | Belize; Mexico 79, 457, 510 | 3010.0 |

## Body Masses of World Birds (continued)

| Species | Sex | N | Mean | Std dev | Range | Sn | Location | Number |
|---|---|---|---|---|---|---|---|---|
| Campylopterus largipennis | U | 29 | 8.3 | 0.21 | | | 79 | 3012.0 |
| Campylopterus rufus | M | 1 | 9.0 | | | | Mexico | 3013.0 |
| | F | 3 | 6.7 | | | | 46 | |
| Campylopterus hemileucurus | U | 56 | 10.5 | 0.50 | | | 79 | 3016.0 |
| Campylopterus ensipennis | U | 9 | 9.7 | | | | 79, 188 | 3017.0 |
| Campylopterus falcatus | U | 4 | 7.6 | | | | 79, 105 | 3018.0 |
| Campylopterus villaviscensio | M | 8 | 8.4 | | 7.4–9.3 | | Peru | 3020.0 |
| | F | 16 | 6.5 | | 5.2–7.4 | | 138 | |
| Eupetomena macroura | U | 29 | 8.7 | | | | 79 | 3021.0 |
| Florisuga mellivora | U | 28 | 7.4 | 0.63 | | | 79 | 3022.0 |
| Melanotrochilus fuscus | U | 150 | 8.1 | | | | 79 | 3023.0 |
| Colibri delphinae | U | 11 | 6.3 | | 5.5–8.0 | | 105, 311, 497, 510 | 3024.0 |
| Colibri thalassinus | U | 84 | 5.9 | 0.34 | | | 79 | 3025.0 |
| Colibri coruscans | U | 17 | 7.6 | 0.49 | | | 79 | 3026.0 |
| Colibri serrirostris | U | 45 | 6.7 | | | | 79 | 3027.0 |
| Anthracothorax viridigula | U | 25 | 7.5 | 0.49 | | | 79 | 3028.0 |
| Anthracothorax prevostii | B | 14 | 6.4 | 0.44 | 5.7–7.0 | | 510, 511 | 3029.0 |
| Anthracothorax nigricollis | U | 51 | 7.0 | 0.55 | | | 79 | 3030.0 |
| Anthracothorax mango | U | 5 | 8.0 | | | | 79, 600 | 3031.0 |
| Anthracothorax dominicus | U | 30 | 5.4 | 0.80 | 4.0–7.2 | W | Puerto Rico 186 | 3032.0 |
| Anthracothorax viridis | B | 5 | 6.8 | | 6.5–7.1 | | Puerto Rico 431 | 3033.0 |
| Eulampis jugularis | U | 19 | 9.3 | 1.35 | 7.5–12.0 | | Martinique 535 | 3035.0 |
| Eulampis holosericeus | U | 21 | 5.6 | 0.40 | | | 79 | 3036.0 |
| Chrysolampis mosquitus | U | 20 | 3.9 | 0.39 | | | 79 | 3037.0 |

**Body Masses of World Birds (continued)**

| Species | Sex | N | Mean | Std dev | Range | Sn | Location | Number |
|---|---|---|---|---|---|---|---|---|
| Orthorhyncus cristatus | M | 18 | 2.8 | | 2.2–3.2 | | | 3038.0 |
| | F | 11 | 2.4 | | 2.0–3.0 | | 295 | |
| Klais guimeti | B | 12 | 2.6 | | | | | 3039.0 |
| | | | | | | | 79, 244, 315 | |
| Abeillia abeillei | U | 4 | 2.7 | | | | | 3040.0 |
| | | | | | | | 46, 79 | |
| Stephanoxis lalandi | U | 8 | 4.0 | | | | | 3041.0 |
| | | | | | | | 37, 79, 227, 441 | |
| Lophornis ornatus | U | 11 | 2.3 | | | | | 3042.0 |
| | | | | | | | 79, 188 | |
| Lophornis gouldii | F | 1 | 2.4 | | | | Brazil | 3043.0 |
| | | | | | | | 226 | |
| Lophornis magnificus | B | 5 | 2.1 | | | | | 3044.0 |
| | | | | | | | 79, 226 | |
| Lophornis delattrei | U | | 2.8 | | | | | 3046.0 |
| | | | | | | | 603a | |
| Lophornis chalybeus verreauxi | F | 1 | 3.0 | | | | 227 | 3048.0 |
| Lophornis helenae | U | 4 | 2.6 | | | | | 3050.0 |
| | | | | | | | 79 | |
| Lophornis adorabilis | U | 4 | 2.7 | | | | | 3051.0 |
| | | | | | | | 79 | |
| Popelairia popelairii | U | | 2.5 | | | | Peru | 3052.0 |
| | | | | | | | 623 | |
| Popelairia langsdorffi | U | 3 | 3.0 | | | | | 3053.0 |
| | | | | | | | 79 | |
| Popelairia conversii | U | 5 | 3.0 | | | | | 3055.0 |
| | | | | | | | 79 | |
| Discosura longicauda | U | 2 | 3.4 | | | | | 3056.0 |
| | | | | | | | 79, 227 | |
| Chlorestes notatus | B | 30 | 4.2 | | 3.5–5.0 | | Trinidad | 3057.0 |
| | | | | | | | 188 | |
| Chlorostilbon canivetii | B | 19 | 2.5 | | | | Costa Rica | 3058.0 |
| | | | | | | | 187 | |
| Chlorostilbon mellisugus | U | 3 | 2.6 | | 2.3–2.9 | | | 3060.0 |
| | | | | | | | 469, 625, 668 | |
| Chlorostilbon aureoventris | U | 6 | 3.2 | | | | | 3061.0 |
| | | | | | | | 79, 227,440, 441 | |
| Chlorostilbon ricordii | B | 7 | 3.4 | | 3.2–3.9 | | | 3062.0 |
| | | | | | | | 295 | |
| Chlorostilbon bracei | U | 9 | 2.8 | | 2.5–3.1 | | Bahamas | 3063.0 |
| | | | | | | | 600 | |
| Chlorostilbon swainsonii | U | 1 | 4.9 | | | | | 3064.0 |
| | | | | | | | 79 | |

## Body Masses of World Birds (continued)

| Species | Sex | N | Mean | Std dev | Range | Sn | Location | Number |
|---|---|---|---|---|---|---|---|---|
| Chlorostilbon maugaeus | U | 45 | 2.9 | | | | 79 | 3065.0 |
| Chlorostilbon gibsoni | U | 8 | 2.8 | | | | 79, 387 | 3066.0 |
| Cynanthus sordidus | U | 3 | 4.4 | | 4.3–4.7 | | Mexico 46, 47 | 3071.0 |
| Cynanthus latirostris | B | 19 | 3.1 | 0.38 | 2.5–4.0 | | W. Mexico 593 | 3072.0 |
| Cyanophaia bicolor | U | 8 | 4.6 | | | | 79 | 3073.0 |
| Thalurania colombica | B | 114 | 4.5 | 0.53 | | | Costa Rica 50a | 3074.0 |
| Thalurania furcata | U | 72 | 4.4 | 0.24 | | | 79 | 3075.0 |
| Thalurania watertonii | U | 3 | 4.6 | | | | 79, 227 | 3076.0 |
| Thalurania glaucopis | U | 12 | 4.9 | | 4.2–6.0 | | Brazil 441 | 3077.0 |
| Panterpe insignis | U | 47 | 5.7 | 0.39 | | | 79 | 3078.0 |
| Damophila julie | U | 51 | 3.3 | 0.15 | | | 79 | 3079.0 |
| Lepidopyga coeruleogularis | F | 3 | 4.2 | | | | Panama 244 | 3080.0 |
| Lepidopyga goudoti | 3 | 4 | 4.0 | | | | 79 | 3082.0 |
| Hylocharis xantusii | U | 11 | 3.5 | 0.18 | | | 79 | 3083.0 |
| Hylocharis leucotis | M | 158 | 3.6 | 0.30 | | | | 3084.0 |
| | F | 51 | 3.2 | 0.20 | | | 177 | |
| Hylocharis eliciae | U | 13 | 3.6 | | | | 79 | 3085.0 |
| Hylocharis sapphirina | U | 3 | 4.1 | | | | 79, 227 | 3086.0 |
| Hylocharis cyanus | U | 13 | 3.6 | | | | 79, 227, 246 | 3087.0 |
| Hylocharis chrysura | B | 4 | 4.4 | | | | 111, 227 | 3089.0 |
| Hylocharis grayi | U | 2 | 5.9 | | 5.5–6.3 | | Columbia 387 | 3090.0 |
| Chrysuronia oenone | M | 28 | 5.3 | 0.41 | 4.7–6.3 | | Venezuela 105 | 3091.0 |
| | F | 8 | 4.8 | 0.37 | 4.3–5.3 | | | |
| Goldmania violiceps | B | 16 | 4.1 | | 3.2–4.5 | | Panama 50 | 3092.0 |

## Body Masses of World Birds (continued)

| Species | Sex | N | Mean | Std dev | Range | Sn | Location | Number |
|---|---|---|---|---|---|---|---|---|
| Goethalsia bella | B | 8 | | | 3.0–4.0 | | Panama 497 | 3093.0 |
| Trochilus polytmus | M | 5 | 5.0 | | 4.5–5.3 | | Jamaica | 3094.0 |
| | F | 11 | 4.7 | 0.30 | 4.1–5.1 | | 600 | |
| Leucochloris albicollis | U | 26 | 6.5 | | | | 79 | 3095.0 |
| Polytmus guainumbi | U | 5 | 4.6 | | | | 79, 576 | 3096.0 |
| Polytmus milleri | U | 1 | 4.1 | | | | 227 | 3097.0 |
| Polytmus theresiae | U | 6 | 3.6 | | 3.5–3.9 | | Surinam 246 | 3098.0 |
| Leucippus taczanowskii | U | 1 | 6.0 | | | | 79 | 3101.0 |
| Taphrospilus hypostictus | B | 10 | 7.0 | | | | Peru 193 | 3103.0 |
| Amazilia viridicauda | M | 1 | 6.4 | | | | Peru 193 | 3104.0 |
| Amazilia chionogaster | U | 10 | 4.2 | | | | 79, 668 | 3105.0 |
| Amazilia candida | B | 13 | 3.8 | 0.36 | 2.9–4.3 | | Mexico; Belize 323, 457, 510 | 3106.0 |
| Amazilia chionopectus | U | 62 | 4.7 | | | | 79 | 3107.0 |
| Amazilia versicolor | U | 4 | 4.1 | | 4.0–4.5 | | Brazil 37, 79, 441 | 3108.0 |
| Amazilia fimbriata | U | 15 | 5.0 | | 3.5–6.2 | | Venezuela 625 | 3110.0 |
| Amazilia lactea | U | 11 | 4.6 | 0.10 | | | 79 | 3112.0 |
| Amazilia amabilis | U | 18 | 4.3 | 0.55 | | | 79 | 3113.0 |
| Amazilia boucardi | U | | 4.5 | | | | 603a | 3116.0 |
| Amazilia franciae | U | 5 | 5.0 | | | | 79 | 3117.0 |
| Amazilia leucogaster | U | 10 | 4.6 | | | | 79, 395 | 3118.0 |
| Amazilia cyanocephala | U | 13 | 5.8 | 0.63 | | | 79 | 3119.0 |
| Amazilia cyanifrons | U | | 5.0 | | | | 453 | 3120.0 |
| Amazilia beryllina | M | 13 | 4.9 | | 4.4–5.7 | | | 3121.0 |
| | F | 8 | 4.4 | | 4.0–4.8 | | 295, 586 | |

## Body Masses of World Birds (continued)

| Species | Sex | N | Mean | Std dev | Range | Sn | Location | Number |
|---|---|---|---|---|---|---|---|---|
| Amazilia cyanura | M | 1 | 3.9 | | | | Honduras | 3122.0 |
| | F | | | | | | 584 | |
| Amazilia saucerrottei | U | 139 | 4.5 | 0.37 | | | | 3123.0 |
| | | | | | | | 79 | |
| Amazilia tobaci | U | 83 | 4.7 | | 3.5–6.0 | | Trinidad | 3124.0 |
| | | | | | | | 576 | |
| Amazilia viridigaster | U | 5 | 6.7 | | | | | 3125.0 |
| | | | | | | | 79 | |
| Amazilia edward | U | 58 | 4.7 | 0.20 | | | | 3126.0 |
| | | | | | | | 79 | |
| Amazilia rutila | B | 14 | 5.0 | 0.74 | 3.3–5.7 | | Mexico | 3127.0 |
| | | | | | | | 593 | |
| Amazilia yucatanensis | M | 7 | 4.0 | | 3.0–4.7 | | | 3128.0 |
| | F | 7 | 3.7 | | 2.9–4.5 | | 295 | |
| Amazilia tzacatl | U | 105 | 5.0 | 0.28 | | | | 3129.0 |
| | | | | | | | 79 | |
| Amazilia amazilia | U | 18 | 4.3 | 0.55 | | | | 3131.0 |
| | | | | | | | 79 | |
| Amazilia viridifrons | B | 10 | 5.4 | 1.02 | 3.2–6.7 | | Mexico | 3132.0 |
| | | | | | | | 584 | |
| Amazilia violiceps | M | 10 | 5.7 | | 5.5–6.2 | | | 3133.0 |
| | F | 6 | 5.1 | | 5.0–5.5 | | 294, 586 | |
| Eupherusa eximia | U | 126 | 4.3 | 0.30 | | | | 3135.0 |
| | | | | | | | 79 | |
| Eupherusa cyanophrys | M | 5 | 4.9 | | 4.5–5.2 | | Oaxaca, Mexico | 3136.0 |
| | | | | | | | 584 | |
| Eupherusa nigriventris | U | 1 | 3.0 | | | | | 3137.0 |
| | | | | | | | 79 | |
| Elvira chionura | U | 16 | 3.0 | 0.21 | | | | 3138.0 |
| | | | | | | | 79 | |
| Elvira cupreiceps | U | 13 | 3.0 | | | | | 3139.0 |
| | | | | | | | 79 | |
| Microchera albocoronata | U | 8 | 2.7 | 0.20 | | | | 3140.0 |
| | | | | | | | 79 | |
| Chalybura buffonii | B | 16 | 6.4 | | | | | 3141.0 |
| | | | | | | | 314, 611, 625, 672 | |
| Chalybura urochrysia | M | 7 | 6.8 | | 6.4–7.2 | | Panama | 3142.0 |
| | F | 6 | 6.0 | | 5.4–7.2 | | 50 | |
| Aphanotochroa cirrochloris | U | 30 | 7.3 | | | | | 3143.0 |
| | | | | | | | 79 | |
| Lampornis clemenciae | M | 195 | 8.4 | 0.24 | | | | 3144.0 |
| | F | 67 | 6.8 | 0.31 | | | 177 | |
| Lampornis amethystinus | U | 26 | 6.7 | 0.87 | | | | 3146.0 |
| | | | | | | | 79 | |

**Body Masses of World Birds (continued)**

| Species | Sex | N | Mean | Std dev | Range | Sn | Location | Number |
|---|---|---|---|---|---|---|---|---|
| Lampornis viridipallens | U | 11 | 5.4 | | | 79 | | 3147.0 |
| Lampornis hemileucus | U | 1 | 5.6 | | | 79 | | 3149.0 |
| Lampornis castaneoventris | U | 40 | 5.6 | 0.27 | | 79 | | 3150.0 |
| Lamprolaima rhami | B | 3 | 6.1 | | 5.2–7.0 | 584 | | 3151.0 |
| Adelomyia melanogenys | U | 35 | 3.8 | 0.31 | | 79 | | 3152.0 |
| Phlogophilus hemileucurus | U | 9 | 2.6 | | 2.2–3.0 | | Peru 450 | 3154.0 |
| Phlogophilus harterti | B | 26 | 2.7 | | | | Peru 193 | 3155.0 |
| Clytolaema rubricauda | U | 9 | 7.2 | | | 79, 227, 441 | | 3156.0 |
| Polyplancta aurescens | U | 8 | 6.3 | 0.10 | | 79 | | 3157.0 |
| Heliodoxa rubinoides | U | 8 | 7.7 | | | 79 | | 3158.0 |
| Heliodoxa leadbeateri | U | 32 | 7.4 | 0.18 | | 79 | | 3159.0 |
| Heliodoxa jacula | M | 6 | 9.1 | | 8.6–9.5 | | Panama | 3160.0 |
| | F | 7 | 7.2 | | 5.6–8.2 | | 50 | |
| Heliodoxa xanthogonys | M | 1 | 6.4 | | | | Venezuela 626 | 3161.0 |
| Heliodoxa schreibersii | M | 4 | 9.9 | | | | Peru | 3162.0 |
| | F | 14 | 8.5 | 0.40 | | | 193 | |
| Heliodoxa branickii | U | 3 | 5.1 | | | | Peru 668 | 3164.0 |
| Eugenes fulgens | M | 119 | 7.7 | 0.40 | | | Oaxaca, Mexico | 3166.0 |
| | F | 24 | 6.4 | 0.50 | | | 295 | |
| Sternoclyta cyanopectus | M | 10 | 8.8 | 0.34 | 8.4–9.4 | | Venezuela | 3168.0 |
| | F | 6 | 9.5 | 0.45 | 9.0–10.3 | | 105 | |
| Topaza pella | U | 6 | 12.1 | | | 79 | | 3170.0 |
| Oreotrochilus estella | U | 37 | 8.1 | 0.39 | | 79 | | 3172.0 |
| Oreotrochilus melanogaster | U | 2 | 4.4 | | | 79 | | 3174.0 |
| Urochroa bougueri | U | 1 | 8.7 | | | | Peru 450 | 3176.0 |
| Patagona gigas | U | 10 | 20.2 | 1.29 | | 79 | | 3177.0 |

## Body Masses of World Birds (continued)

| Species | Sex | N | Mean | Std dev | Range | Sn | Location | Number |
|---|---|---|---|---|---|---|---|---|
| Aglaeactis cupripennis | U | 23 | 7.6 | 0.59 | | | 79 | 3178.0 |
| Lafresnaya layfesnayi | B | 40 | 5.3 | | 4.5–6.7 | | Ecuador 320 | 3182.0 |
| Pterophanes cyanopterus | B | 3 | 10.3 | | | | Peru 668 | 3183.0 |
| Coeligena coeligena | U | 34 | 6.8 | 0.10 | | | 79 | 3184.0 |
| Coeligena wilsoni | U | 2 | 7.0 | | 7.0–7.1 | | 79, 227 | 3185.0 |
| Coeligena prunellei | U | | 7.4 | | | | 453 | 3186.0 |
| Coeligena torquata | U | 19 | 7.1 | 0.20 | | | 79 | 3187.0 |
| Coeligena helianthea | U | 3 | 8.2 | | | | 79 | 3191.0 |
| Coeligena lutetiae | B | 14 | 6.9 | | 6.2–7.7 | | Peru 451 | 3192.0 |
| Coeligena violifer | U | 7 | 7.4 | 0.30 | | | 79 | 3193.0 |
| Coeligena iris | B | 53 | 6.9 | | 6.0–8.0 | | Ecuador 320 | 3194.0 |
| Ensifera ensifera | U | 4 | 12.0 | | | | 79, 227 | 3195.0 |
| Sephanoides sephaniodes | M | 15 | 5.7 | | | | Chile | 3196.0 |
| | F | 12 | 4.7 | | | | 110 | |
| Sephanoides fernandensis | M | 15 | 10.9 | | | | Chile | 3197.0 |
| | F | 2 | 7.0 | | | | 110 | |
| Sephanoides galeritus | U | | 5.0 | | | | Chile 292 | 3197.1 |
| Boissonneaua flavescens | U | 8 | 8.0 | | | | 79 | 3198.0 |
| Boissonneaua matthewsii | U | 6 | 7.0 | | | | 79, 469, 668 | 3199.0 |
| Boissonneaua jardini | U | 3 | 8.2 | | | | 79, 227 | 3200.0 |
| Heliangelus amethysticollis | U | 28 | 5.3 | 0.10 | | | 79 | 3203.0 |
| Heliangelus exortis | U | 1 | 4.0 | | | | 79 | 3205.0 |
| Heliangelus viola | U | 3 | 5.3 | | | | 79 | 3206.0 |
| Heliangelus regalis | M | 2 | 4.0 | | 3.5–4.5 | | Peru 138 | 3207.0 |

## Body Masses of World Birds (continued)

| Species | Sex | N | Mean | Std dev | Range | Sn | Location | Number |
|---|---|---|---|---|---|---|---|---|
| Eriocnemis vestitus | B | 22 | 4.4 | | 3.9–5.0 | | Peru 451 | 3209.0 |
| Eriocnemis luciani | B | 31 | 6.0 | | 5.4–6.4 | | Ecuador 320 | 3211.0 |
| Eriocnemis cupreoventris | U | | 5.6 | | | | 453 | 3212.0 |
| Eriocnemis mosquera | U | 1 | 5.8 | | | | 79 | 3213.0 |
| Haplophaedia aurelia | B | 10 | 5.2 | | 4.7–6.0 | | 536 | 3218.0 |
| Urosticte benjamini | B | 4 | 4.0 | | 3.9–4.1 | | 138, 469 | 3220.0 |
| Ocreatus underwoodii | U | 19 | 3.0 | 0.14 | | | 79 | 3222.0 |
| Lesbia victoriae | U | 1 | 5.1 | | | | 79 | 3223.0 |
| Lesbia nuna | U | 6 | 3.8 | | | | 79 | 3224.0 |
| Sappho sparganura | U | 4 | 5.9 | | | | 79 | 3225.0 |
| Polyonymus caroli | U | 4 | 5.0 | | | | 79 | 3226.0 |
| Ramphomicron microrhynchum | M | 1 | 3.5 | | | | Ecuador 320 | 3227.0 |
| Metallura williami | U | 12 | 4.5 | | | | 79, 319 | 3229.0 |
| Metallura baroni | B | 9 | 4.3 | | | | Ecuador 319 | 3230.0 |
| Metallura odomae | B | 7 | 5.0 | | 4.7–5.4 | | Peru 451 | 3231.0 |
| Metallura theresiae | U | 2 | 5.0 | | | | 79 | 3232.0 |
| Metallura eupogon | U | 10 | 4.6 | 0.34 | | | Peru 668 | 3233.0 |
| Metallura phoebe | U | 2 | 5.4 | | | | 79 | 3235.0 |
| Metallura tyrianthina | B | 93 | 3.8 | | | | Ecuador 319 | 3236.0 |
| Chalcostigma ruficeps | U | 2 | 3.8 | | | | Peru 668 | 3238.0 |
| Chalcostigma stanleyi | U | 11 | 5.8 | | | | 79 | 3240.0 |
| Chalcostigma herrani | B | 7 | 5.9 | | 5.1–7.0 | | Peru 451 | 3242.0 |

## Body Masses of World Birds (continued)

| Species | Sex | N | Mean | Std dev | Range | Sn | Location | Number |
|---|---|---|---|---|---|---|---|---|
| Oxypogon guerinii | U | 4 | 4.8 | | | | 79, 644 | 3243.0 |
| Aglaiocercus kingi | U | 17 | 4.8 | 0.58 | | | 79 | 3246.0 |
| Oreonympha nobilis | U | 1 | 9.0 | | | | 79 | 3248.0 |
| Augastes lumachellus | U | 2 | 4.0 | | | | 79 | 3249.0 |
| Augastes scutatus | U | 1 | 3.0 | | | | 79 | 3250.0 |
| Augastes geoffroyi | U | 13 | 3.6 | 0.22 | | | 79 | 3251.0 |
| Heliothryx barroti | B | 9 | 5.5 | 0.20 | | | 79 | 3252.0 |
| Heliothryx aurita | U | 7 | 5.4 | | | | 79 | 3253.0 |
| Heliactin cornuta | U | 4 | 7.8 | | | | 79 | 3254.0 |
| Loddigesia mirabilis | U | 1 | 3.0 | | | | 79 | 3255.0 |
| Heliomaster constantii | B | 9 | 7.3 | | 6.5–9.0 | | 295 | 3256.0 |
| Heliomaster longirostris | U | 8 | 6.6 | | | | 79 | 3257.0 |
| Heliomaster squamosus | U | 1 | 5.0 | | | | 79 | 3258.0 |
| Heliomaster furcifer | B | 4 | 5.5 | | 5.0–6.5 | | 37, 79, 227 | 3259.0 |
| Rhodopis vesper | U | 4 | 3.9 | | | | 79 | 3260.0 |
| Thaumastura cora | U | 1 | 2.0 | | | | 79 | 3261.0 |
| Philodice bryantae | U | 21 | 3.3 | | | | 79 | 3262.0 |
| Philodice mitchellii | U | 3 | 3.1 | | | | 79 | 3263.0 |
| Doricha enicura | U | 1 | 2.4 | | | | 79 | 3264.0 |
| Doricha eliza | M | 1 | 2.3 | | | | Yucatan, Mexico | 3265.0 |
| | F | 3 | 2.6 | | 2.5–2.7 | | 457 | |
| Tilmatura dupontii | U | 12 | 2.2 | 0.15 | | | 79 | 3266.0 |
| Calothorax lucifer | B | 9 | 3.3 | | 2.9–3.5 | F | Arizona, USA 509 | 3268.0 |

## Body Masses of World Birds (continued)

| Species | Sex | N | Mean | Std dev | Range | Sn | Location | Number |
|---------|-----|---|------|---------|-------|----|----------|--------|
| Calothorax pulcher | U | 3 | 2.9 | | | | 79 | 3269.0 |
| Archilochus colubris | M | 202 | 3.0 | 0.30 | 2.4−4.1 | | Pennsylvania,USA | 3270.0 |
| | F | 489 | 3.3 | 0.15 | 2.7−4.8 | | 101 | |
| Archilochus alexandri | M | 1207 | 3.2 | 0.35 | 2.3−4.8 | F | Arizona, USA | 3271.0 |
| | F | 733 | 3.6 | 0.35 | 2.5−5.1 | | 509 | |
| Calypte anna | M | 293 | 4.4 | 0.42 | 3.6−6.3 | F | Arizona, USA | 3272.0 |
| | F | 368 | 4.0 | 0.35 | 3.2−5.5 | | 509 | |
| Calypte costae | M | 33 | 3.0 | | 2.5−5.2 | | | 3273.0 |
| | F | 27 | 3.2 | | 2.5−3.4 | | 295 | |
| Calliphlox evelynae | F | 2 | 2.4 | | 2.2−2.6 | | | 3274.0 |
| | | | | | | | 295 | |
| Calliphlox amethystina | U | 24 | 2.4 | | | | | 3275.0 |
| | | | | | | | 79 | |
| Mellisuga helenae | U | | 2.0 | | | | | 3276.0 |
| | | | | | | | 185 | |
| Mellisuga minima | U | 6 | 2.4 | | | | | 3277.0 |
| | | | | | | | 79 | |
| Stellula calliope | M | 46 | 2.5 | | 1.9−3.2 | | | 3278.0 |
| | F | 26 | 2.8 | | 2.2−3.2 | | 295 | |
| Atthis heloisa | B | 15 | 2.2 | | 2.0−2.7 | | | 3279.0 |
| | | | | | | | 295 | |
| Myrtis fanny | U | 1 | 2.3 | | | | | 3281.0 |
| | | | | | | | 79 | |
| Acestrura mulsant | U | 6 | 3.8 | | | | | 3284.0 |
| | | | | | | | 79 | |
| Selasphorus platycercus | M | 97 | 3.5 | 0.30 | | B | Colorado, USA | 3290.0 |
| | F | 109 | 3.6 | 0.40 | | | 91 | |
| Selasphorus rufus | B | 112 | 3.5 | | 2.8−4.5 | F | Arizona, USA | 3291.0 |
| | | | | | | | 509 | |
| Selasphorus sasin sasin | M | 38 | 3.1 | | 2.5−3.8 | | | 3292.0 |
| | F | 18 | 3.2 | | 2.8−3.5 | | 295 | |
| Selasphorus sasin sedentarius | M | 19 | 3.5 | | | | | 3292.0 |
| | F | 26 | 3.7 | | | | 295 | |
| Selasphorus flammula simoni | B | 7 | 2.7 | | 2.4−3.1 | | Costa Rica | 3293.0 |
| | | | | | | | 602 | |
| Selasphorus flammula torridus | B | 30 | 2.6 | | 2.0−3.0 | | Costa Rica | 3293.0 |
| | | | | | 2.0−3.0 | | 602 | |
| Selasphorus scintilla | U | 39 | 2.2 | 0.16 | | | | 3294.0 |
| | | | | | | | 79 | |

### ORDER: TROGONIFORMES       FAMILY: TROGONIDAE

| Species | Sex | N | Mean | Std dev | Range | Sn | Location | Number |
|---------|-----|---|------|---------|-------|----|----------|--------|
| Apaloderma narina | B | 15 | 61.0 | | 51.0−72.0 | | 207 | 3296.0 |

## Body Masses of World Birds (continued)

| Species | Sex | N | Mean | Std dev | Range | Sn | Location | Number |
|---|---|---|---|---|---|---|---|---|
| Apaloderma vittatum | B | 5 | 55.1 | | 50.0–57.0 | | Zambia 166 | 3298.0 |
| Pharomachrus mocinno | B | 17 | 206.0 | | | | Panama 244 | 3299.0 |
| Priotelus temnurus | U | 3 | 58.0 | | 53.5–60.4 | | Cuba 374 | 3305.0 |
| Trogon massena | B | 26 | 141.0 | | | | Panama 244 | 3307.0 |
| Trogon melanurus | U | 5 | 119.0 | | | | Panama 244 | 3308.0 |
| Trogon clathratus | U | | 130.0 | | | | 603a | 3309.0 |
| Trogon comptus | M | 2 | 104.0 | | 100.0–108.0 | | Ecuador 589 | 3310.0 |
| Trogon viridis | U | 5 | 87.6 | | 77.0–99.0 | | 86, 161, 188 | 3311.0 |
| Trogon citreolus | B | 7 | 83.1 | | 72.6–91.0 | | Yucatan, Mexico 457 | 3313.0 |
| Trogon mexicanus | M | 1 | 69.3 | | | | Mexico 139 | 3315.0 |
| Trogon elegans | U | 10 | 67.3 | | 60.0–78.6 | | Mexico 368, 584, 586 | 3316.0 |
| Trogon collaris | M | 29 | 63.4 | 3.60 | | | Panama | 3317.0 |
| | F | 18 | 65.4 | 4.07 | | | 244 | |
| Trogon aurantiiventris | B | 3 | 65.9 | | 61.5–68.5 | | Panama 244, 611 | 3318.0 |
| Trogon personatus | U | 2 | 64.3 | | 60.3–68.3 | | 387, 668 | 3319.0 |
| Trogon rufus | B | 13 | 52.8 | 4.78 | 40.6–58.0 | | 50, 86, 161, 314 436, 497, | 3320.0 |
| Trogon surrucura | B | 3 | 73.3 | | 70.0–78.0 | | Brazil 37 | 3321.0 |
| Trogon curucui | B | 6 | 52.2 | | | | Panama 244 | 3322.0 |
| Trogon violaceus | B | 11 | 51.5 | | 48.0–55.0 | | 188 | 3323.0 |
| Harpactes fasciatus | M | 1 | 62.0 | | | | 5 | 3325.0 |
| Harpactes kasumba | M | 1 | 72.1 | | | | Borneo 627 | 3326.0 |
| Harpactes diardii | U | 5 | 101.0 | | | | Malaysia;Borneo 627, 690 | 3327.0 |
| Harpactes ardens | B | 11 | 96.3 | | 82.6–114.0 | | Philippines 475 | 3328.0 |

## Body Masses of World Birds (continued)

| Species | Sex | N | Mean | Std dev | Range | Sn | Location | Number |
|---|---|---|---|---|---|---|---|---|
| Harpactes orrhophaeus | U | 12 | 52.7 | | | | Malaysia 690 | 3330.0 |
| Harpactes duvaucelii | U | 5 | 35.2 | | | | Malaysia; Borneo 627, 690 | 3331.0 |
| Harpactes erythrocephalus | U | 4 | 80.0 | | 75.0–84.0 | | Malaysia 371 | 3333.0 |
| Harpactes wardi | B | 4 | 119.0 | | 115.0–120.0 | | India 5 | 3334.0 |

### ORDER: CORACIIFORMES          FAMILY: ALCEDINIDAE

| Species | Sex | N | Mean | Std dev | Range | Sn | Location | Number |
|---|---|---|---|---|---|---|---|---|
| Alcedo atthis atthis | U | 15 | 27.0 | 2.35 | 23.0–33.0 | PB | Iran 118 | 3336.0 |
| Alcedo atthis ispida | U | 162 | 35.8 | 2.11 | | B | Britain 118 | 3336.0 |
| Alcedo semitorquata | B | 3 | 38.4 | | 35.0–40.2 | | 207 | 3337.0 |
| Alcedo quadribrachys | U | 9 | 35.9 | | 33.0–39.5 | | Ghana 228 | 3338.0 |
| Alcedo meninting | U | 4 | 18.9 | | | | Malaysia 690 | 3339.0 |
| Alcedo azurea | U | 11 | 36.6 | | | | New Guinea 34, 35 | 3340.0 |
| Alcedo websteri | M | 2 | 55.5 | | 54.0–57.0 | | New Britain 219a | 3341.0 |
| | F | 1 | 67.0 | | | | | |
| Alcedo cristata | B | 11 | 15.5 | | 11.0–21.1 | | 228, 255, 280 | 3344.0 |
| Alcedo leucogaster | U | 5 | 14.5 | | 13.5–14.8 | | Liberia 314a | 3348.0 |
| Alcedo pusilla | U | 8 | 13.2 | | | | New Guinea 217 | 3350.0 |
| Ceyx argentatus | M | 12 | 16.5 | | 13.5–18.2 | | Philippines 475 | 3351.0 |
| | F | 14 | 19.3 | | 17.2–22.4 | | | |
| Ceyx lepidus | U | 10 | 14.0 | | | | New Guinea 35 | 3352.0 |
| Ceyx erithacus | U | 9 | 16.9 | | | | Malaysia 690 | 3353.0 |
| Ceyx rufidorsa | U | 6 | 17.9 | | | | Malaysia 690 | 3354.0 |
| Ceyx melanurus | B | 2 | 23.2 | | 22.5–23.9 | | Philippines 475 | 3355.0 |
| Ispidina madagascariensis | M | 9 | 17.2 | | 14.5–20.5 | | Madagascar 39 | 3357.0 |
| | F | 2 | 19.8 | | 17.0–22.5 | | | |
| Ispidina picta | U | 18 | 11.4 | | 9.6–13.5 | | Ghana 228 | 3358.0 |

## Body Masses of World Birds (continued)

| Species | Sex | N | Mean | Std dev | Range | Sn | Location | Number |
|---------|-----|---|------|---------|-------|-----|----------|--------|
| Ispidina lecontei | U | 5 | 9.5 | | 8.0–10.5 | | 66, 314a | 3359.0 |
| Lacedo pulchella | U | 6 | 48.3 | | | | Malaysia 690 | 3360.0 |
| Dacelo novaguineae | U | | 305.0 | | | | 194 | 3361.0 |
| Dacelo leachii | U | | 311.0 | | 277.0–391.0 | | 147 | 3362.0 |
| Dacelo gaudichaud | U | 22 | 143.0 | | | | New Guinea 35 | 3364.0 |
| Clytoceyx rex | M | 2 | 242.0 | | 238.0–245.0 | | New Guinea 216 | 3365.0 |
| Melidora macrorrhina | U | 7 | 97.0 | | | | New Guinea 35 | 3366.0 |
| Halcyon coromanda | U | 1 | 85.0 | | | | Malaysia 690 | 3371.0 |
| Halcyon badia | B | 13 | 57.9 | | 47.0–64.5 | | Uganda 207 | 3372.0 |
| Halcyon smyrnensis | B | 17 | 91.4 | | 80.5–108.0 | | Philippines 475 | 3373.0 |
| Halcyon pileata | U | 25 | 84.0 | 10.00 | | W | Malaysia 420 | 3374.0 |
| Halcyon leucocephala | B | 22 | 41.4 | | 34.0–47.0 | | 118 | 3376.0 |
| Halcyon senegalensis | B | 6 | 58.6 | | 54.0–67.0 | | 66, 228, 680 | 3377.0 |
| Halcyon senegaloides | B | 5 | 61.8 | | 57.0–66.0 | | Uganda 207 | 3378.0 |
| Halcyon malimbica | U | 12 | 91.8 | | 79.0–105.0 | | Ghana 228 | 3379.0 |
| Halcyon albiventris | B | 9 | 65.1 | | 59.4–69.4 | | South Africa 255 | 3380.0 |
| Halcyon chelicuti | U | 5 | 35.3 | | 31.0–41.5 | | 67, 228 | 3381.0 |
| Todirhamphus winchelli | M | 2 | 61.8 | | 59.5–64.1 | | Philippines 475 | 3383.0 |
|  | F | 2 | 73.5 | | 67.0–80.0 | | | |
| Todirhamphus macleayii | B | 8 | 43.2 | | 41.0–48.0 | | New Guinea 154 | 3386.0 |
| Todirhamphus albonotatus | M | 1 | 145.0 | | | | New Britain 219a | 3387.0 |
|  | F | | | | | | | |
| Todirhamphus pyrrhopygia | U | | 43.0 | | | | W. Australia 550 | 3390.0 |
| Todirhamphus cinnamominus | M | 18 | 59.1 | 5.16 | 50.7–69.5 | | captive 553 | 3392.0 |

## Body Masses of World Birds (continued)

| Species | Sex | N | Mean | Std dev | Range | Sn | Location | Number |
|---|---|---|---|---|---|---|---|---|
| Todirhamphus chloris | B | 17 | 61.5 | | 50.0–88.0 | | | 3394.0 |
| | | | | | | | 219a, 475, 491 | |
| Todirhamphus saurophaga | M | 1 | 114.0 | | | | New Britain | 3396.0 |
| | | | | | | | 219a | |
| Todirhamphus sanctus | U | 45 | 64.2 | 4.61 | 55.0–75.0 | | New Zealand | 3398.0 |
| | | | | | | | 500 | |
| Actenoides concretus | U | 10 | 73.6 | | | | Malaysia | 3406.0 |
| | | | | | | | 690 | |
| Actenoides hombroni | B | 9 | 117.0 | | 106.0–147.0 | | Philippines | 3408.0 |
| | | | | | | | 475 | |
| Syma torotoro | U | 8 | 37.0 | | | | New Guinea | 3411.0 |
| | | | | | | | 35 | |
| Syma megarhyncha | F | 1 | 63.0 | | | | New Guinea | 3412.0 |
| | | | | | | | 154 | |
| Tanysiptera galatea | U | 71 | 50.0 | | | | New Guinea | 3414.0 |
| Tanysiptera nympha | U | 2 | 57.0 | | 57.0–57.0 | | New Guinea | 3418.0 |
| | | | | | | | 218 | |
| Tanysiptera sylvia | U | | 35.0 | | | | New Guinea | 3420.0 |
| | | | | | | | 34 | |
| Megaceryle maxima | U | 5 | 311.0 | | 255.0–375.0 | | | 3421.0 |
| | | | | | | | 228, 255 | |
| Megaceryle lugubris | F | 2 | 255.0 | | 230.0–280.0 | | India | 3422.0 |
| | | | | | | | 5 | |
| Megaceryle alcyon | U | 29 | 148.0 | 20.80 | 125.0–215.0 | | Pennsylvania,USA | 3423.0 |
| | | | | | | | 177 | |
| Megaceryle torquata | B | 15 | 317.0 | | | | | 3424.0 |
| | | | | | | | 86, 111, 244 | |
| Ceryle rudis | M | 189 | 82.4 | 6.03 | 68.0–100.0 | B | Kenya | 3425.0 |
| | F | 96 | 86.4 | 7.40 | 71.0 | | 118 | |
| Chloroceryle amazona | M | 11 | 121.0 | 7.96 | | | Panama | 3426.0 |
| | F | 6 | 132.0 | | | | 244 | |
| Chloroceryle americana | U | 20 | 37.5 | 2.12 | | | Panama | 3427.0 |
| | | | | | | | 244 | |
| Chloroceryle inda | B | 10 | 49.6 | | 38.4–60.0 | | | 3428.0 |
| | | | | | | | 86, 245, 437, 441 | |
| Chloroceryle aenea | M | 6 | 14.0 | | 13.0–15.1 | | Venezuela | 3429.0 |
| | F | 6 | 15.8 | | 14.0–16.9 | | 625 | |

### ORDER: CORACIIFORMES      FAMILY: TODIDAE

| Species | Sex | N | Mean | Std dev | Range | Sn | Location | Number |
|---|---|---|---|---|---|---|---|---|
| Todus multicolor | U | 1 | 5.8 | | | | Cuba | 3430.0 |
| | | | | | | | 429 | |
| Todus angustirostris | B | 3 | 8.3 | | 8.0–9.0 | | Hispaniola | 3431.0 |
| | | | | | | | 600 | |

## Body Masses of World Birds (continued)

| Species | Sex | N | Mean | Std dev | Range | Sn | Location | Number |
|---|---|---|---|---|---|---|---|---|
| Todus mexicanus | U | 26 | 5.4 | 0.40 | 4.8–6.0 | W | Puerto Rico 186 | 3432.0 |
| Todus todus | B | 18 | 6.5 | | 5.5–7.2 | | Jamaica 600 | 3433.0 |
| Todus subulatus | F | 1 | 8.0 | | | | Hispaniola 600 | 3434.0 |

### ORDER: CORACIIFORMES          FAMILY: MOMOTIDAE

| Species | Sex | N | Mean | Std dev | Range | Sn | Location | Number |
|---|---|---|---|---|---|---|---|---|
| Hylomanes momotula | B | 9 | 29.3 | | 20.3–38.0 | | Panama; Belize 497, 510 | 3435.0 |
| Aspatha gularis | B | 6 | 60.9 | | 57.0–65.3 | | Mexico 584 | 3436.0 |
| Electron platyrhynchum | U | 7 | 60.4 | | 53.0–64.0 | | Panama 86, 314, 497, 611 | 3437.0 |
| Electron carinatum | M | 1 | 64.9 | | | | Honduras 584 | 3438.0 |
| Eumomota superciliosa | B | 17 | 62.5 | 7.43 | 43.6–74.0 | | Yucatan, Mexico 323, 457 | 3439.0 |
| Baryphthengus martii | B | 18 | 151.0 | | | | Peru 193 | 3440.0 |
| Baryphthengus ruficapillus | U | 4 | 175.0 | | | | Panama 315 | 3441.0 |
| Momotus mexicanus | B | 6 | 75.7 | | 68.0–91.0 | | Mexico 458, 548, 593 | 3442.0 |
| Momotus momota lessoni | B | 15 | 133.0 | | | | Panama 244 | 3443.0 |
| Momotus momota conexus | M | 11 | 112.0 | 6.63 | | | Panama 244 | 3443.0 |
| | F | 5 | 102.0 | | | | | |
| Momotus momota aequatorialis | B | 9 | 158.0 | | 123.0–176.0 | | 193, 387 | 3443.0 |

### ORDER: CORACIIFORMES          FAMILY: MEROPIDAE

| Species | Sex | N | Mean | Std dev | Range | Sn | Location | Number |
|---|---|---|---|---|---|---|---|---|
| Nyctyornis athertoni | B | 3 | 87.3 | | 84.0–93.0 | | India 5 | 3445.0 |
| Merops gularis | B | 31 | 27.3 | | 22.0–33.0 | | Liberia 207 | 3447.0 |
| Merops muelleri | B | 24 | 22.5 | | 17.4–25.0 | | 207 | 3448.0 |
| Merops bullocki | U | 64 | 23.1 | 1.27 | | | Nigeria 119a | 3449.0 |
| Merops bullockoides | U | 39 | 34.8 | 1.50 | 31.5–38.0 | | South Africa 255 | 3450.0 |
| Merops pusillus | B | 7 | 13.2 | | 11.0–16.0 | | Zimbabwe 282, 283 | 3451.0 |

## Body Masses of World Birds (continued)

| Species | Sex | N | Mean | Std dev | Range | Sn | Location | Number |
|---|---|---|---|---|---|---|---|---|
| Merops variegatus | B | 66 | 22.5 | | 19.5–28.4 | | 207 | 3452.0 |
| Merops oreobates | M | 16 | 26.1 | | 20.0–38.0 | | | 3453.0 |
| | F | 6 | 22.0 | | 17.0–28.0 | | 207 | |
| Merops hirundinaceus | U | 9 | 21.8 | | 19.5–23.6 | | Ghana 228 | 3454.0 |
| Merops breweri | M | 1 | 54.0 | | | | 207 | 3455.0 |
| Merops revoilii | U | 1 | 13.1 | | | | 207 | 3456.0 |
| Merops albicollis | B | 8 | 25.9 | | 24.0–28.0 | | 66 | 3457.0 |
| Merops orientalis | B | 5 | 14.8 | | 14.0–16.0 | | India 361 | 3458.0 |
| Merops boehmi | B | 50 | 16.6 | | 14.4–19.5 | | Malawi 207 | 3459.0 |
| Merops viridis | U | | 32.0 | | | | 78 | 3460.0 |
| Merops persicus | B | 37 | 49.3 | | 41.3–57.0 | F | USSR 207 | 3461.0 |
| Merops superciliosus | B | 6 | 48.3 | | 45.0–53.0 | | India 5 | 3462.0 |
| Merops philippinus | M | 2 | 37.6 | | 37.0–38.2 | | 217, 475 | 3463.0 |
| Merops ornatus | B | 10 | 28.8 | 2.20 | 26.0–33.0 | | New Guinea 154 | 3464.0 |
| Merops apiaster | B | 382 | 56.6 | | | B | France 342 | 3465.0 |
| Merops leschenaulti | F | 2 | 29.5 | | 26.0–33.0 | | India 5 | 3466.0 |
| Merops malimbicus | U | | 45.0 | | | | estimated 207 | 3467.0 |
| Merops nubicus | U | 105 | 54.4 | | | | 69 | 3468.0 |
| Merops nubicoides | U | 129 | 60.0 | 4.10 | 52.0–75.0 | B | Zimbabwe 65 | 3469.0 |

### ORDER: CORACIIFORMES  FAMILY: CORACIIDAE

| Species | Sex | N | Mean | Std dev | Range | Sn | Location | Number |
|---|---|---|---|---|---|---|---|---|
| Coracias garrulus | B | 17 | 146.0 | | 127.0–160.0 | | USSR 118 | 3470.0 |
| Coracias abyssinica | B | 3 | 102.0 | | 99.5–104.0 | | 118, 680 | 3471.0 |
| Coracias caudata | B | 10 | 108.0 | | 87.0–125.0 | | 78, 255 | 3472.0 |

## Body Masses of World Birds (continued)

| Species | Sex | N | Mean | Std dev | Range | Sn | Location | Number |
|---|---|---|---|---|---|---|---|---|
| Coracias spatulata | M | 1 | 111.0 | | | | 207 | 3473.0 |
| Coracias naevia | B | 11 | 168.0 | | 140.0–200.0 | | Botswana 207 | 3474.0 |
| Coracias benghalensis | M | 2 | 143.0 | | 120.0–166.0 | | India | 3475.0 |
| | F | 2 | 171.0 | | 166.0–176.0 | | 5, 361 | |
| Coracias cyanogaster | B | 22 | 142.0 | | 110.0–178.0 | | Ivory Coast 207 | 3477.0 |
| Eurystomus glaucurus | B | 13 | 110.0 | | 98.0–112.0 | | 118 | 3478.0 |
| Eurystomus gularis | B | 40 | 96.3 | | 82.0–110.0 | | Ivory Coast 207 | 3479.0 |
| Eurystomus orientalis | B | 11 | 162.0 | 19.50 | 132.0–192.0 | | New Guinea 154 | 3480.0 |

### ORDER: CORACIIFORMES  FAMILY: BRACHYPTERACIIDAE

| Species | Sex | N | Mean | Std dev | Range | Sn | Location | Number |
|---|---|---|---|---|---|---|---|---|
| Atelornis pittoides | B | 4 | 92.5 | | 83.0–108.0 | | Madagascar 39 | 3484.0 |

### ORDER: CORACIIFORMES  FAMILY: LEPTOSOMATIDAE

| Species | Sex | N | Mean | Std dev | Range | Sn | Location | Number |
|---|---|---|---|---|---|---|---|---|
| Leptosomus discolor | M | 2 | 220.0 | | 219.0–222.0 | | Madagascar 39 | 3487.0 |

### ORDER: CORACIIFORMES  FAMILY: UPUPIDAE

| Species | Sex | N | Mean | Std dev | Range | Sn | Location | Number |
|---|---|---|---|---|---|---|---|---|
| Upupa epops | U | 75 | 61.4 | 8.10 | 41.0–83.0 | S | France 118 | 3488.0 |

### ORDER: CORACIIFORMES  FAMILY: PHOENICULIDAE

| Species | Sex | N | Mean | Std dev | Range | Sn | Location | Number |
|---|---|---|---|---|---|---|---|---|
| Phoeniculus purpureus | M | 5 | 86.6 | | 72.2–99.0 | | | 3490.0 |
| | F | 3 | 61.9 | | 56.0–66.3 | | 228, 680 | |
| Phoeniculus damarensis granti | M | 5 | 85.8 | | 78.0–90.0 | | | 3491.0 |
| | F | 3 | 78.0 | | 72.0–82.0 | | 207 | |
| Phoeniculus bollei | U | 5 | 54.6 | | 47.0–64.0 | | 51, 66 | 3493.0 |
| Phoeniculus castaneiceps | B | 13 | 24.4 | | 22.3–27.0 | | Liberia 207 | 3494.0 |
| Rhinopomastus aterrimus | U | 1 | 22.8 | | | | Ghana 228 | 3495.0 |
| Rhinopomastus cyanomelas | F | 2 | 34.8 | | 33.0–36.6 | | South Africa 255 | 3496.0 |
| Rhinopomastus minor | U | 3 | 21.6 | | 21.0–23.0 | | 66 | 3497.0 |

## Body Masses of World Birds (continued)

| Species | Sex | N | Mean | Std dev | Range | Sn | Location | Number |
|---|---|---|---|---|---|---|---|---|
| ORDER: CORACIIFORMES | | | | | FAMILY: BUCEROTIDAE | | | |
| Tockus albocristatus | B | 6 | 292.0 | | 276.0–315.0 | | 207 | 3498.0 |
| Tockus hartlaubi | M | 7 | 110.0 | | 96.0–135.0 | | | 3499.0 |
| | F | 4 | 88.2 | | | | 207 | |
| Tockus camurus | M | 8 | 111.0 | | 101.0–122.0 | | | 3500.0 |
| | F | 7 | 97.0 | | 84.0–115.0 | | 207 | |
| Tockus erythrorhynchus | M | 75 | 150.0 | | | | | 3502.0 |
| | F | 75 | 128.0 | | | | 207 | |
| Tockus flavirostris | M | 3 | 258.0 | | 225.0–275.0 | | | 3503.0 |
| | F | 3 | 182.0 | | 170.0–191.0 | | 207 | |
| Tockus leucomelas | M | 75 | 211.0 | | | | | 3504.0 |
| | F | 75 | 168.0 | | | | 207 | |
| Tockus alboterminatus | M | 1 | 224.0 | | | | Zambia 166 | 3507.0 |
| Tockus bradfieldi | B | 5 | 194.0 | | 170.0–204.0 | | 207 | 3508.0 |
| Tockus fasciatus | M | 15 | 278.0 | | 250.0–316.0 | | | 3509.0 |
| | F | 7 | 241.0 | | 227.0–260.0 | | 207 | |
| Tockus hemprichii | F | 1 | 297.0 | | | | 207 | 3510.0 |
| Tockus nasutus | M | 1 | 234.0 | | | | Sudan 680 | 3511.0 |
| Tockus pallidirostris | F | 1 | 170.0 | | | | 66 | 3512.0 |
| Ocyceros griseus | M | 1 | 238.0 | | | | India 5 | 3513.0 |
| Anthracoceros coronatus | M | 1 | 300.0 | | | | India 361 | 3516.0 |
| Buceros hydrocorax | B | 33 | 1417.0 | | 1016.0–1662.0 | | Philippines 475 | 3523.0 |
| Penelopides panini | M | 15 | 512.0 | | 453.0–584.0 | | Philippines 475 | 3529.0 |
| | F | 13 | 435.0 | | 335.0–514.0 | | | |
| Aceros nipalensis | B | 2 | 2385.0 | | 2270.0–2500.0 | | India 5 | 3534.0 |
| Aceros leucocephalus | M | 3 | 1086.0 | | 1012.0–1140.0 | | Philippines 475 | 3537.0 |
| Aceros narcondami | M | 4 | | | 700.0–750.0 | | | 3540.0 |
| | F | 3 | | | 600.0–750.0 | | 5 | |
| Aceros plicatus | M | 1 | 1827.0 | | | | New Guinea 35 | 3543.0 |
| | F | | | | | | | |
| Ceratogymna bucinator | M | 7 | 739.0 | | 607.0–941.0 | | | 3544.0 |
| | F | 10 | 567.0 | | 452.0–670.0 | | 207 | |

## Body Masses of World Birds (continued)

| Species | Sex | N | Mean | Std dev | Range | Sn | Location | Number |
|---------|-----|---|------|---------|-------|----|----------|--------|
| Ceratogymna fistulator | M | 3 | 554.0 | | 463.0–710.0 | | | 3545.0 |
| | F | 3 | 466.0 | | 413.0–500.0 | 207 | | |
| Ceratogymna brevis | M | 5 | 1308.0 | | 1265.0–1400.0 | | | 3546.0 |
| | F | 5 | 1162.0 | | 1050.0–1450.0 | 207 | | |
| Ceratogymna subcylindricus | M | 9 | 1311.0 | | 1078.0–1390.0 | | | 3547.0 |
| | F | 4 | 1090.0 | | 1000.0–1200.0 | 207 | | |
| Ceratogymna cylindricus | F | 1 | 921.0 | | | | | 3548.0 |
| | | | | | | 207 | | |
| Ceratogymna atrata | M | 3 | 1264.0 | | 1069.0–1513.0 | | | 3550.0 |
| | F | 3 | 1059.0 | | 907.0–1182.0 | 207 | | |
| Ceratogymna elata | M | 1 | 2100.0 | | | | | 3551.0 |
| | F | 2 | 1750.0 | | | 207 | | |

### ORDER: CORACIIFORMES          FAMILY: BUCORVIDAE

| Species | Sex | N | Mean | Std dev | Range | Sn | Location | Number |
|---------|-----|---|------|---------|-------|----|----------|--------|
| Bucorvus abyssinicus | M | 1 | 4000.0 | | | | estimated 207 | 3552.0 |
| Bucorvus cafer | M | 5 | 4358.0 | | 3500.0–6180.0 | | | 3553.0 |
| | F | 7 | 3324.0 | | 2230.0–4580.0 | 207 | | |

### ORDER: PICIFORMES          FAMILY: GALBULIDAE

| Species | Sex | N | Mean | Std dev | Range | Sn | Location | Number |
|---------|-----|---|------|---------|-------|----|----------|--------|
| Galbalcyrhynchus purusianus | U | 7 | 50.0 | | | | Peru 193 | 3555.0 |
| Brachygalba salmoni | M | 1 | 18.5 | | | | Panama 497 | 3556.0 |
| | F | 1 | 16.0 | | | | | |
| Brachygalba lugubris | F | 3 | 19.3 | | 17.0–23.0 | | Bolivia 22 | 3558.0 |
| Galbula albirostris | B | 11 | 22.1 | 1.38 | 20.0–24.0 | | Brazil 226 | 3561.0 |
| Galbula cyanicollis | F | 2 | 25.8 | | 25.5–26.0 | | Brazil 610a | 3562.0 |
| Galbula ruficauda | B | 14 | 26.5 | | 24.0–33.5 | | Trinidad 188 | 3563.0 |
| Galbula galbula | B | 5 | 20.2 | | 18.0–22.0 | | Surinam 245 | 3564.0 |
| Galbula cyanescens | U | 6 | 25.3 | 1.60 | | | Peru 668 | 3567.0 |
| Galbula leucogastra | U | 3 | 16.7 | | | | Brazil 439 | 3569.0 |
| Galbula dea | B | 11 | 27.4 | 3.61 | 22.0–31.5 | | 161, 245 | 3570.0 |
| Jacamerops aurea | B | 5 | 65.2 | | 58.0–76.0 | | Panama; Surinam 86, 161, 246 | 3571.0 |

## Body Masses of World Birds (continued)

| Species | Sex | N | Mean | Std dev | Range | Sn | Location | Number |
|---|---|---|---|---|---|---|---|---|
| | | | | | ORDER: PICIFORMES | | FAMILY: BUCCONIDAE | |
| Notharchus macrorhynchos | B | 17 | 95.9 | | | | Panama 244 | 3572.0 |
| Notharchus pectoralis | M | 1 | 69.1 | | | | Panama 611 | 3573.0 |
| Notharchus tectus | B | 5 | 26.4 | | 21.0–33.0 | | Surinam; Panama 86, 245 | 3575.0 |
| Bucco macrodactylus | U | | 25.0 | | | | Peru 623 | 3576.0 |
| Bucco tamatia | U | 3 | 33.9 | | 33.0–35.0 | | Surinam; Brazil 245, 439 | 3577.0 |
| Bucco capensis | U | 15 | 54.0 | 3.97 | | | Brazil 45 | 3579.0 |
| Nystalus radiatus | U | 2 | 61.0 | | 59.0–63.0 | | Panama 86, 497 | 3580.0 |
| Nystalus chacuru | M | 2 | 62.5 | | 61.0–64.0 | | Brazil 37 | 3581.0 |
| Nystalus striolatus | U | | 47.0 | | | | Peru 623 | 3582.0 |
| Hypnelus ruficollis | U | 11 | 49.8 | 4.04 | 43.3–57.0 | | Venezuela 625 | 3584.0 |
| Malacoptila fusca | F | 1 | 39.0 | | | | Fr. Guiana 161 | 3585.0 |
| Malacoptila semicincta | U | | 44.0 | | | | Peru 623 | 3586.0 |
| Malacoptila striata | U | 6 | 44.1 | | 41.5–46.5 | | Brazil 441 | 3587.0 |
| Malacoptila fulvogularis | U | 1 | 65.0 | | | | Peru 668 | 3588.0 |
| Malacoptila rufa | U | 3 | 40.3 | | 36.0–42.6 | | Brazil 437 | 3589.0 |
| Malacoptila panamensis | U | 21 | 42.6 | 4.90 | | | 315 | 3590.0 |
| Micromonacha lanceolata | U | | 19.0 | | | | 603a | 3592.0 |
| Nonnula rubecula | U | 2 | 20.6 | | | | Brazil 45 | 3593.0 |
| Nonnula frontalis | U | 7 | 15.7 | | 14.5–19.5 | | Panama 86, 244, 611 | 3596.0 |
| Nonnula ruficapilla | U | | 22.0 | | | | Peru 623 | 3597.0 |
| Hapaloptila castanea | B | 5 | 80.6 | | 75.0–84.0 | | Peru 451 | 3599.0 |

## Body Masses of World Birds (continued)

| Species | Sex | N | Mean | Std dev | Range | Sn | Location | Number |
|---|---|---|---|---|---|---|---|---|
| Monasa atra | B | 7 | 92.8 | | 84.0–104.0 | | 161, 245 | 3600.0 |
| Monasa nigrifrons | U | 15 | 80.7 | 5.00 | | | Peru 668 | 3601.0 |
| Monasa morphoeus | U | 11 | 106.0 | 11.20 | 90.5–122.0 | | Panama 86, 497, 611 | 3602.0 |
| Chelidoptera tenebrosa | B | 6 | 33.7 | | 29.0–39.0 | | Surinam 245, 246 | 3604.0 |

### ORDER: PICIFORMES               FAMILY: MEGALAIMIDAE

| Species | Sex | N | Mean | Std dev | Range | Sn | Location | Number |
|---|---|---|---|---|---|---|---|---|
| Psilopogon pyrolophus | U | 18 | 129.0 | | 115.0–149.0 | | Malaysia 371 | 3605.0 |
| Megalaima virens | M | 3 | | | 250.0–295.0 | | | 3606.0 |
| | F | 3 | | | 192.0–206.0 | | 5 | |
| Megalaima zeylanica | M | 2 | 97.5 | | 87.0–108.0 | | India | 3608.0 |
| | F | 1 | 114.0 | | | | 5, 361 | |
| Megalaima lineata | M | 1 | 139.0 | | | | India | 3609.0 |
| | F | 2 | 160.0 | | 149.0–170.0 | | 5 | |
| Megalaima viridis | U | 16 | 80.5 | | 71.0–90.0 | | India 5 | 3610.0 |
| Megalaima chrysopogon | B | 6 | 167.0 | | 149.0–182.0 | | Borneo 627 | 3613.0 |
| Megalaima mystacophanos | M | 1 | 79.2 | | | | Borneo | 3615.0 |
| | F | 1 | 60.3 | | | | 627 | |
| Megalaima flavifrons | M | 2 | 58.5 | | 57.0–60.0 | | Sri Lanka 5 | 3617.0 |
| Megalaima franklinii | U | 54 | 63.5 | | 50.0–73.0 | | Malaysia 371 | 3618.0 |
| Megalaima asiatica | B | 13 | | | 62.0–100.0 | | India 5 | 3620.0 |
| Megalaima henricii | B | 3 | 76.6 | | 72.8–83.2 | | Borneo 627 | 3623.0 |
| Megalaima australis | B | 5 | | | 32.0–38.0 | | India 5 | 3626.0 |
| Megalaima rubricapilla | B | 3 | 35.7 | | 32.0–39.0 | | Sri Lanka 5 | 3628.0 |
| Megalaima haemacephala | B | 10 | | | 32.0–47.0 | | India 5 | 3630.0 |
| Calorhamphus fuliginosus | U | 4 | 42.2 | | 38.8–45.0 | | Malaysia; Borneo 627, 690 | 3631.0 |

### ORDER: PICIFORMES               FAMILY: LYBIIDAE

| Species | Sex | N | Mean | Std dev | Range | Sn | Location | Number |
|---|---|---|---|---|---|---|---|---|
| Gymnobucco peli | B | 30 | | | 43.0–63.0 | | 207 | 3633.0 |

**Body Masses of World Birds (continued)**

| Species | Sex | N | Mean | Std dev | Range | Sn | Location | Number |
|---|---|---|---|---|---|---|---|---|
| Gymnobucco bonapartei | B | 3 | 73.6 | | 65.0–81.0 | | 66 | 3635.0 |
| Stactolaema leucotis | B | 5 | 56.0 | | 51.5–58.6 | | 66 | 3636.0 |
| Stactolaema anchietae | B | 40 | | | 43.0–57.0 | | Angola 207 | 3637.0 |
| Stactolaema whytii | B | 5 | 49.8 | | 47.0–51.4 | | Zimbabwe 283 | 3638.0 |
| Stactolaema olivacea | U | 12 | 47.1 | | 43.1–50.0 | | 66 | 3639.0 |
| Pogoniulus scolopaceus | M | 25 | | | 13.0–18.0 | | Angola 207 | 3640.0 |
| Pogoniulus coryphaeus | B | 8 | | | 11.0–13.0 | | Angola 207 | 3641.0 |
| Pogoniulus leucomystax | F | 1 | 11.2 | | | | Zambia 166 | 3642.0 |
| Pogoniulus simplex | B | 4 | 8.9 | | 8.3–10.0 | | 66 | 3643.0 |
| Pogoniulus atroflavus | B | 15 | 17.9 | | 14.0–21.5 | | 207 | 3644.0 |
| Pogoniulus subsulphureus | U | 1 | 9.0 | | | | Liberia 314a | 3645.0 |
| Pogoniulus bilineatus | U | 9 | 13.6 | | | | 51 | 3646.0 |
| Pogoniulus chrysoconus | B | 8 | 11.7 | | 9.6–14.2 | | 228, 282, 284 | 3647.0 |
| Pogoniulus pusillus | B | 6 | 10.5 | | 9.0–12.5 | | 66 | 3648.0 |
| Buccanodon duchaillui | U | 1 | 36.5 | | | | Liberia 314a | 3649.0 |
| Tricholaema hirsuta | B | 20 | | | 43.0–55.0 | | 207 | 3650.0 |
| Tricholaema diademata | B | 8 | 29.9 | | 22.0–34.6 | | 207 | 3651.0 |
| Tricholaema frontata | M | 1 | 25.5 | | | | Angola 207 | 3652.0 |
| Tricholaema leucomelaina | B | 171 | 32.2 | | 23.0–39.0 | | Transvaal 207 | 3653.0 |
| Tricholaema lachrymosa | B | 18 | 23.6 | | 21.0–27.0 | | 207 | 3654.0 |
| Tricholaema melanocephala | M | 9 | 22.1 | | 18.0–23.5 | | | 3655.0 |
| | F | 5 | 18.2 | | 17.0–20.0 | | 207 | |
| Lybius undatus | B | 4 | | | 31.8–40.7 | | 207 | 3656.0 |

## Body Masses of World Birds (continued)

| Species | Sex | N | Mean | Std dev | Range | Sn | Location | Number |
|---|---|---|---|---|---|---|---|---|
| Lybius vieilloti | B | 35 | 35.0 | | 28.0–43.0 | | 207 | 3657.0 |
| Lybius leucocephalus | M | 12 | 63.8 | | 55.0–70.0 | | | 3658.0 |
| | F | 11 | 61.8 | | 55.0–65.0 | | 207 | |
| Lybius guifsobalito | M | 1 | 42.0 | | | | | 3661.0 |
| | F | | | | | | 66 | |
| Lybius torquatus | B | 34 | 51.3 | | 44.0–59.3 | | 255, 282, 283, 284 | 3662.0 |
| Lybius melanopterus | M | 1 | 41.0 | | | | | 3663.0 |
| | F | 1 | 53.8 | | | | 66, 207 | |
| Lybius minor | B | 17 | 46.7 | | 40.0–50.0 | | 207 | 3664.0 |
| Lybius bidentatus | U | 2 | 77.9 | | 71.0–84.8 | | 66 | 3665.0 |
| Lybius dubius | U | 18 | 90.7 | | 80.0–105.0 | | Ghana 228 | 3666.0 |
| Trachyphonus purpuratus | U | 4 | 73.8 | | 66.0–85.0 | | 66 | 3668.0 |
| Trachyphonus vaillantii | B | 14 | 67.4 | | 58.5–76.0 | | 78, 255 | 3669.0 |
| Trachyphonus margaritatus | U | 2 | 47.0 | | 46.2–47.9 | | 207 | 3670.0 |
| Trachyphonus erythrocephalus | B | 4 | 53.0 | | 46.9–59.5 | | 207 | 3671.0 |
| Trachyphonus darnaudii | U | 17 | 32.8 | | 26.0–38.0 | | 66 | 3672.0 |

### ORDER: PICIFORMES    FAMILY: CAPITONIDAE

| Species | Sex | N | Mean | Std dev | Range | Sn | Location | Number |
|---|---|---|---|---|---|---|---|---|
| Capito maculicoronatus | B | 4 | 47.8 | | 44.0–52.0 | | Panama 497 | 3675.0 |
| Capito dayi | M | 2 | 64.8 | | 64.5–65.0 | | Brazil 610a | 3678.0 |
| Capito niger | B | 7 | 54.8 | | 44.0–60.0 | | 161, 245, 246 | 3680.0 |
| Eubucco richardsoni | U | | 35.0 | | | | Peru 623 | 3682.0 |
| Eubucco bourcierii | B | 20 | 33.5 | | | | Panama 244 | 3683.0 |
| Eubucco tucinkae | U | | 41.0 | | | | Peru 623 | 3684.0 |
| Eubucco versicolor | M | 3 | 37.8 | | | | Peru | 3685.0 |
| | F | 1 | 33.0 | | | | 668 | |
| Semnornis frantzii | U | 10 | 57.3 | 4.57 | | | Costa Rica 50a | 3686.0 |

## Body Masses of World Birds (continued)

| Species | Sex | N | Mean | Std dev | Range | Sn | Location | Number |
|---|---|---|---|---|---|---|---|---|
| Semnornis ramphastinus | B | 6 | 97.5 | | 91.8–102.0 | | 387, 589 | 3687.0 |
| **ORDER: PICIFORMES** | | | | | | **FAMILY: INDICATORIDAE** | | |
| Indicator maculatus | U | 1 | 46.5 | | | | Liberia 314a | 3688.0 |
| Indicator variegatus | M | 5 | 51.3 | | 47.8–60.5 | | | 3689.0 |
| | F | 5 | 44.9 | | 35.6–53.5 | | 203, 280, 281, 284 | |
| Indicator indicator | B | 5 | 52.2 | | 49.0–58.8 | | 66, 203, 228, 255, 282 | 3690.0 |
| Indicator archipelagicus | U | 2 | 38.5 | | | | Malaysia 690 | 3691.0 |
| Indicator minor | B | 23 | 26.6 | 2.48 | 22.5–32.0 | | 203,228,255,280, 281,282,283 | 3692.0 |
| Indicator conirostris | B | 8 | 31.2 | | 24.5–38.0 | | Kenya 207 | 3693.0 |
| Indicator willcocksi | M | 2 | 20.0 | | 19.5–20.5 | | | 3694.0 |
| | F | 7 | 16.0 | | 13.0–17.7 | | 207 | |
| Indicator exilis | F | 3 | 17.5 | | 17.0–18.0 | | 66 | 3695.0 |
| Indicator pumilio | U | 2 | 16.6 | | 16.2–17.0 | | Kenya 362 | 3696.0 |
| Indicator meliphilus | M | 1 | 15.0 | | | | 66 | 3697.0 |
| Indicator xanthonotus | B | 3 | 28.0 | | 26.0–29.0 | | 5, 203 | 3698.0 |
| Melichneutes robustus | M | 4 | | | 52.3–61.5 | | | 3699.0 |
| | F | 4 | | | 46.9–57.0 | | 207 | |
| Melignomon eisentrauti | B | 13 | 24.2 | | 18.0–29.0 | | 207 | 3700.0 |
| Melignomon zenkeri | F | 2 | 24.5 | | 24.0–25.0 | | 207 | 3701.0 |
| Prodotiscus insignis | B | 4 | 12.8 | | 10.2–16.5 | | 66, 203 | 3702.0 |
| Prodotiscus zambesiae | B | 7 | 10.1 | | 9.5–11.0 | | 207 | 3703.0 |
| Prodotiscus regulus | M | 1 | 17.6 | | | | Zimbabwe 280 | 3704.0 |
| **ORDER: PICIFORMES** | | | | | | **FAMILY: RAMPHASTIDAE** | | |
| Aulacorhynchus prasinus | M | 15 | 160.0 | 13.52 | | | Panama 244 | 3705.0 |
| | F | 16 | 149.0 | 16.20 | | | | |

## Body Masses of World Birds (continued)

| Species | Sex | N | Mean | Std dev | Range | Sn | Location | Number |
|---------|-----|---|------|---------|-------|----|---------|--------|
| Aulacorhynchus sulcatus | M | 1 | 173.0 | | | | Venezuela 105 | 3706.0 |
| Aulacorhynchus derbianus | B | 13 | 220.0 | | | | Peru 193 | 3707.0 |
| Aulacorhynchus haematopygus | B | 8 | 215.0 | | 200.0–232.0 | | Columbia 387 | 3708.0 |
| Aulacorhynchus coeruleicinctis | U | 2 | 208.0 | | | | Peru 668 | 3710.0 |
| Pteroglossus inscriptus | U | 4 | 126.0 | | | | Peru 193 | 3711.0 |
| Pteroglossus viridis | U | 11 | 134.0 | 11.00 | 120.0–156.0 | | 161, 245, 246 | 3712.0 |
| Pteroglossus bitorquatus | B | 9 | 147.0 | | 131.0–171.0 | | Bolivia 22 | 3713.0 |
| Pteroglossus flavirostris | U | | 135.0 | | | | Peru 623 | 3714.0 |
| Pteroglossus mariae | B | 9 | 140.0 | | | | Peru 193 | 3715.0 |
| Pteroglossus castanotis | U | | 310.0 | | | | Peru 623 | 3716.0 |
| Pteroglossus aracari | U | 7 | 232.0 | | 152.0–279.0 | | Surinam 245, 246 | 3717.0 |
| Pteroglossus torquatus | M<br>F | 10<br>8 | 236.0<br>216.0 | 22.14 | | | Panama 244 | 3718.0 |
| Pteroglossus frantzii | B | 26 | 264.0 | | | | Panama 244 | 3719.0 |
| Pteroglossus pluricinctus | M | 1 | 232.0 | | | | Brazil 610a | 3722.0 |
| Pteroglossus beauharnaesii | U | | 203.0 | | | | Peru 623 | 3723.0 |
| Baillonius bailloni | U | 1 | 139.0 | | | | Brazil 441 | 3724.0 |
| Andigena laminirostris | M<br>F | 2 | 335.0 | | 315.0–355.0 | | 589 | 3725.0 |
| Andigena hypoglauca | B | 3 | 310.0 | | | | Peru 668 | 3726.0 |
| Selenidera spectabilis | B | 6 | 224.0 | | 175.0–252.0 | | Panama 50, 86, 497 | 3729.0 |
| Selenidera reinwardtii | M<br>F | 15<br>8 | 117.0<br>160.0 | 14.20 | | | Peru 193 | 3730.0 |
| Selenidera culik | B | 5 | 136.0 | | 129.0–140.0 | | 161, 245 | 3732.0 |
| Selenidera gouldii | B | 3 | 177.0 | | 172.0–188.0 | | Bolivia 22 | 3734.0 |

## Body Masses of World Birds (continued)

| Species | Sex | N | Mean | Std dev | Range | Sn | Location | Number |
|---------|-----|---|------|---------|-------|-----|----------|--------|
| Ramphastos sulfuratus | U | 15 | 339.0 | 34.80 | | | 270 | 3735.0 |
| Ramphastos culminatus | U | 2 | 369.0 | | 350.0–388.0 | | Peru 193 | 3738.0 |
| Ramphastos vitellinus | M | 2 | 390.0 | | 358.0–423.0 | | Trinidad | 3739.0 |
| | F | 4 | 333.0 | | 298.0–387.0 | | 188 | |
| Ramphastos dicolorus | F | 1 | 400.0 | | | | Brazil 37 | 3740.0 |
| Ramphastos swainsonii | M | 9 | 660.0 | | | | | 3741.0 |
| | F | 2 | 584.0 | | | | 269a | |
| Ramphastos tucanus | B | 7 | 530.0 | | 387.0–690.0 | | 161, 245, 246 | 3743.0 |
| Ramphastos cuvieri | U | | 734.0 | | | | Peru 623 | 3744.0 |
| Ramphastos toco | U | | 540.0 | | | | Brazil 564 | 3745.0 |

### ORDER: PICIFORMES　　　　FAMILY: PICIDAE

| Species | Sex | N | Mean | Std dev | Range | Sn | Location | Number |
|---------|-----|---|------|---------|-------|-----|----------|--------|
| Jynx torquilla tschusii | U | 115 | 33.5 | 5.30 | 26.0–50.0 | Γ | Malta 118 | 3746.0 |
| Jynx ruficollis | M | 3 | 51.0 | | 49.0–52.0 | | 66 | 3747.0 |
| Picumnus innominatus | U | 4 | | | 10.5–12.5 | | India 5 | 3748.0 |
| Picumnus aurifrons | M | 5 | 9.2 | | | | Peru | 3749.0 |
| | F | 2 | 8.6 | | 8.4–8.8 | | 193 | |
| Picumnus lafresnayi | U | | | | 9.0–10.0 | | 560 | 3751.0 |
| Picumnus exilis | U | | | | 8.5–9.5 | | 560 | 3752.0 |
| Picumnus sclateri | B | 3 | 10.7 | | | | Peru 672 | 3753.0 |
| Picumnus squamulatus | U | 6 | 10.6 | | 9.9–11.0 | | Venezuela 625 | 3754.0 |
| Picumnus spilogaster | B | 2 | 13.2 | | 13.0–13.5 | | Brazil | 3755.0 |
| | | | | | 13.0–13.5 | | 610a | |
| Picumnus minutissimus | B | 3 | 12.0 | | 11.0–13.0 | | Surinam 245 | 3756.0 |
| Picumnus steindachneri | U | 5 | 9.9 | | 9.2–10.5 | | Peru 450 | 3758.0 |
| Picumnus cirratus | U | | 11.5 | | | | Brazil 564 | 3760.0 |
| Picumnus temminckii | B | 6 | 11.7 | | 11.0–12.5 | | Brazil 37, 441 | 3762.0 |

## Body Masses of World Birds (continued)

| Species | Sex | N | Mean | Std dev | Range | Sn | Location | Number |
|---|---|---|---|---|---|---|---|---|
| Picumnus albosquamatus | M | | | | 9.0–11.0 | | 560 | 3763.0 |
| Picumnus rufiventris | U | | 21.0 | | | | Peru 623 | 3765.0 |
| Picumnus nebulosus | B | 2 | 11.8 | | 11.5–12.0 | | Brazil 37 | 3768.0 |
| Picumnus castelnau | F | 1 | 11.4 | | | | Columbia 483 | 3769.0 |
| Picumnus subtilis | M | 4 | 10.4 | | | | Peru 193 | 3770.0 |
| Picumnus olivaceus | B | 13 | 10.6 | | | | Panama 244 | 3771.0 |
| Picumnus granadensis | M | 1 | 12.2 | | | | Columbia 387 | 3772.0 |
| Sasia africana | U | 6 | 9.4 | | 9.0–10.5 | | Cameroon 207 | 3774.0 |
| Sasia abnormis | U | 12 | 9.2 | | | | Malaysia 690 | 3775.0 |
| Sasia ochracea | U | 1 | 10.0 | | | | India 5 | 3776.0 |
| Nesoctites micromegas | U | | 33.0 | | | | 185 | 3777.0 |
| Melanerpes candidus | M | 1 | 130.0 | | | | Brazil 37 | 3778.0 |
| Melanerpes lewis | B | 5 | 116.0 | | 108.0–138.0 | | California, USA 229 | 3779.0 |
| Melanerpes herminieri | U | | 100.0 | | | | 185 | 3780.0 |
| Melanerpes portoricensis | M | 5 | 60.4 | | 53.9  67.1 | | Puerto Rico 431 | 3781.0 |
| | F | 7 | 52.8 | | 45.7–59.4 | | | |
| Melanerpes erythrocephalus | U | 89 | 71.6 | 7.57 | 56.1–90.5 | S | Ontario, Canada 177 | 3782.0 |
| Melanerpes formicivorus | M | 47 | 82.9 | 4.20 | | | California, USA 325 | 3783.0 |
| | F | 39 | 78.1 | 5.40 | | | | |
| Melanerpes pucherani | M | 8 | 66.0 | | | | 269a | 3784.0 |
| | F | 3 | 59.0 | | | | | |
| Melanerpes chrysauchen | M | 4 | 58.6 | | | | Panama 244 | 3785.0 |
| | F | 5 | 47.0 | | | | | |
| Melanerpes cruentatus | U | | 59.0 | | | | Peru 623 | 3786.0 |
| Melanerpes flavifrons | M | 2 | 63.0 | | 62.0–64.0 | | Brazil 37 | 3787.0 |
| | F | 1 | 53.0 | | | | | |
| Melanerpes cactorum | F | 3 | 33.2 | | 31.0–35.5 | | Paraguay 610 | 3788.0 |

## Body Masses of World Birds (continued)

| Species | Sex | N | Mean | Std dev | Range | Sn | Location | Number |
|---|---|---|---|---|---|---|---|---|
| Melanerpes striatus | U | | 65.0 | | | | 185 | 3789.0 |
| Melanerpes radiolatus | U | 25 | 108.0 | | | | Jamaica 125a | 3790.0 |
| Melanerpes chrysogenys | M | 4 | 72.0 | | 60.0–87.3 | | Mexico | 3791.0 |
| | F | 4 | 63.7 | | 57.8–70.3 | | 458, 593 | |
| Melanerpes hypopolius | M | 1 | 52.4 | | | | Oaxaca, Mexico 584 | 3792.0 |
| Melanerpes pygmaeus | M | 12 | 40.4 | 2.00 | 37.0–42.5 | | Mexico; Belize | 3793.0 |
| | F | 4 | 36.8 | | 35.1–41.6 | | 18a, 323, 457, 510 | |
| Melanerpes rubricapillus | M | 19 | 55.9 | 4.23 | | | Panama | 3794.0 |
| | F | 12 | 49.0 | 3.85 | | | 244 | |
| Melanerpes uropygialis | M | 20 | 69.7 | 6.36 | 54.6–80.6 | Y | Arizona, USA | 3795.0 |
| | F | 24 | 60.0 | 3.13 | 53.8–67.0 | | 177 | |
| Melanerpes carolinus | M | 22 | 67.2 | 5.86 | | | Florida, USA | 3796.0 |
| | F | 9 | 56.2 | | | | 244 | |
| Melanerpes superciliaris | M | 11 | 84.2 | 8.29 | 71.3–99.3 | | Cayman Is. | 3797.0 |
| | F | 5 | 71.0 | | 67.3–73.5 | | 128 | |
| Melanerpes aurifrons | M | 29 | 85.4 | 7.00 | 73.0–99.0 | | Mexico | 3798.0 |
| | F | 14 | 76.4 | 6.36 | 66.0–90.0 | | 546 | |
| Sphyrapicus varius varius | U | 52 | 50.3 | 4.29 | 40.7–62.2 | M | Pennsylvania, USA 177 | 3800.0 |
| Sphyrapicus nuchalis | B | 16 | 45.9 | 4.82 | 36.0–54.9 | W | Arizona, USA 594 | 3801.0 |
| Sphyrapicus ruber daggettii | U | 13 | 48.9 | 4.58 | 40.1–54.7 | Y | California, USA 229 | 3802.0 |
| Sphyrapicus thyroideus | B | 19 | 47.6 | | 44.4–55.3 | B | Nevada, USA 300 | 3803.0 |
| Xiphidiopicus percussus | U | | | | 48.0–97.0 | | 560 | 3804.0 |
| Campethera punctuligera | B | 2 | 69.0 | | 67.0–71.0 | | Ghana 228 | 3805.0 |
| Campethera nubica | U | 11 | 60.9 | | 46.5–66.6 | | 66 | 3806.0 |
| Campethera bennettii | B | 22 | 70.1 | | 61.0–83.5 | | Zimbabwe 207 | 3807.0 |
| Campethera abingoni | B | 3 | 64.7 | | 55.0–74.5 | | 228, 256, 281 | 3809.0 |
| Campethera cailliautii | B | 5 | 42.3 | | 40.0–45.7 | | 66 | 3813.0 |
| Campethera tullbergi | U | 2 | 50.5 | | 49.0–52.0 | | 66 | 3814.0 |
| Campethera nivosa | B | 4 | 38.8 | | 34.5–45.5 | | 314a | 3815.0 |

## Body Masses of World Birds (continued)

| Species | Sex | N | Mean | Std dev | Range | Sn | Location | Number |
|---|---|---|---|---|---|---|---|---|
| Campethera caroli | B | 4 | 59.8 | | 58.0–62.0 | | 66 | 3816.0 |
| Geocolaptes olivaceus | M | 3 | 120.0 | | 109.0–130.0 | | South Africa 256 | 3817.0 |
| Dendropicos elachus | M | 3 | | | 17.0–21.0 | | 207 | 3818.0 |
| Dendropicos poecilolaemus | U | 2 | 28.1 | | 27.0–29.2 | | Kenya 66 | 3819.0 |
| Dendropicos abyssinicus | B | 4 | | | 23.0–26.0 | | 207 | 3820.0 |
| Dendropicos fuscescens | U | 16 | 26.0 | | 21.0–35.7 | | 78 | 3821.0 |
| Dendropicos gabonensis | B | 3 | | | 24.0–28.0 | | 207 | 3823.0 |
| Dendropicos stierlingi | B | 2 | 28.0 | | 25.0–31.0 | | 207 | 3824.0 |
| Dendropicos namaquus | F | 1 | 61.7 | | | | Zimbabwe 284 | 3825.0 |
| Dendropicos pyrrhogaster | B | 6 | 68.5 | | 65.0–74.0 | | 207 | 3826.0 |
| Dendropicos xantholophus | F | 2 | 53.8 | | 51.5–56.0 | | Kenya 66 | 3827.0 |
| Dendropicos elliotii | M | 1 | 40.0 | | | | 207 | 3828.0 |
| Dendropicos goertae | M | 2 | 51.0 | | 50.0–52.0 | | | 3829.0 |
| | F | 2 | 44.2 | | 42.9–45.5 | | 228, 680 | |
| Dendropicos griseocephalus | M | 10 | 44.0 | | 36.0–50.0 | | | 3831.0 |
| | F | 6 | 38.0 | | 33.0–42.0 | | 207 | |
| Dendropicos obsoletus | B | 6 | 21.4 | | 18.6–24.0 | | 207 | 3832.0 |
| Dendrocopos maculatus | M | 5 | 27.1 | | 25.4–30.3 | | Philippines 475 | 3834.0 |
| Dendrocopos nanus | B | 3 | 15.0 | | 12.9–17.0 | | India 5, 361 | 3835.0 |
| Dendrocopos moluccensis | U | | | | 14.5–18.0 | | 560 | 3836.0 |
| Dendrocopos canicapillus mitchellii | M | 3 | | | 20.0–24.0 | | India | 3837.0 |
| | F | 3 | 23.3 | | 21.0–27.0 | | 5 | |
| Dendrocopos kizuki | M | | 20.1 | | 18.5–21.7 | | | 3838.0 |
| | F | | 23.4 | | 22.0–25.9 | | 149 | |
| Dendrocopos minor | U | 50 | 19.8 | | 18.0–22.0 | Y | Britain 258 | 3839.0 |
| Dendrocopos auriceps | M | 3 | | | 38.0–44.0 | | India | 3840.0 |
| | F | 3 | | | 38.0–40.0 | | 5 | |

## Body Masses of World Birds (continued)

| Species | Sex | N | Mean | Std dev | Range | Sn | Location | Number |
|---|---|---|---|---|---|---|---|---|
| Dendrocopos macei | M | | | | 43.0–48.0 | | India | 3841.0 |
| | F | | | | 38.0–52.0 | | 5 | |
| Dendrocopos atratus | B | | | | 42.0–52.0 | | India | 3842.0 |
| | | | | | | | 5 | |
| Dendrocopos mahrattensis | B | 13 | 34.4 | | 32.0–38.0 | | India | 3843.0 |
| | | | | | | | 5 | |
| Dendrocopos hyperythrus | M | 10 | | | 42.0–53.0 | | India | 3845.0 |
| | F | 6 | | | 42.0–50.0 | | 5 | |
| Dendrocopos cathparius | M | 1 | 35.0 | | | | India | 3846.0 |
| | F | 2 | 27.0 | | 26.0–28.0 | | 5 | |
| Dendrocopos darjellensis | M | 5 | | | 70.0–80.0 | | India | 3847.0 |
| | F | 5 | | | 61.0–73.0 | | 5 | |
| Dendrocopos medius | B | 31 | 59.0 | | 50.0–80.0 | Y | central Europe | 3848.0 |
| | | | | | | | 118 | |
| Dendrocopos leucotos | B | 16 | 108.0 | 7.98 | 99.0–126.0 | | | 3849.0 |
| | | | | | | | 118 | |
| Dendrocopos major | U | 50 | 81.6 | | 71.0–83.0 | Y | Britain | 3850.0 |
| | | | | | | | 258 | |
| Dendrocopos major pinetorum | M | 9 | 76.0 | | 70.0–87.0 | B | Netherlands | 3850.0 |
| | F | 7 | 72.7 | | 68.0–79.0 | | 118 | |
| Dendrocopos syriacus | M | 9 | 79.5 | | 76.0–82.0 | | SE Europe | 3851.0 |
| | F | 11 | 74.0 | 3.20 | 70.0–81.0 | | 118 | |
| Dendrocopos leucopterus | U | | 67.0 | | | | | 3852.0 |
| | | | | | | | 560 | |
| Dendrocopos assimilis | U | | | | 42.0–64.0 | | | 3853.0 |
| | | | | | | | 560 | |
| Dendrocopos himalayensis | M | 3 | | | 72.0–74.0 | | India | 3854.0 |
| | F | 6 | | | 59.0–67.0 | | 5 | |
| Picoides lignarius | U | | 39.1 | | | | Chile | 3855.0 |
| | | | | | | | 292 | |
| Picoides mixtus | M | 2 | 33.5 | | 33.0–34.0 | | Brazil | 3856.0 |
| | F | 3 | | | 32.0–33.0 | | 37 | |
| Picoides nuttallii | U | 9 | 38.3 | | 32.8–43.1 | | California, USA | 3857.0 |
| | | | | | | | 591 | |
| Picoides scalaris | U | 21 | 30.3 | 3.55 | 25.0–41.4 | W | W. USA; Mexico | 3858.0 |
| | | | | | | | 594 | |
| Picoides pubescens | U | 383 | 27.0 | 0.19 | 20.7–32.2 | Y | Pennsylvania,USA | 3859.0 |
| | | | | | | | 101 | |
| Picoides borealis | U | 12 | 43.6 | 3.90 | | | Florida, USA | 3860.0 |
| | | | | | | | 244 | |
| Picoides stricklandi | M | 11 | 49.4 | 2.32 | | B | Arizona, USA | 3861.0 |
| | F | 5 | 44.1 | | | | 343 | |
| Picoides villosus | M | 27 | 70.0 | 3.20 | 60.8–79.6 | Y | Pennsylvania,USA | 3862.0 |
| | F | 11 | 62.5 | 1.98 | 59.3–65.9 | | 101 | |

## Body Masses of World Birds (continued)

| Species | Sex | N | Mean | Std dev | Range | Sn | Location | Number |
|---|---|---|---|---|---|---|---|---|
| Picoides albolarvatus | M | 18 | 63.0 | 3.40 | 55.6–68.0 | PB | Oregon, USA | 3863.0 |
| | F | 17 | 59.2 | 3.82 | 52.6–66.4 | | 177 | |
| Picoides tridactylus | M | 10 | 70.1 | 3.76 | 65.0–74.0 | | Norway | 3864.0 |
| | F | 5 | 61.2 | | 57.0–66.0 | | 265 | |
| Picoides arcticus | M | 6 | 72.0 | | 61.3–88.1 | | California, USA | 3865.0 |
| | F | 3 | 66.6 | | 63.4–71.3 | | 229 | |
| Veniliornis callonotus | M | 2 | 25.0 | | | | Peru | 3866.0 |
| | | | | | | | 672 | |
| Veniliornis dignus | B | 5 | 37.6 | | 35.7–40.0 | | Columbia | 3867.0 |
| | | | | | | | 387 | |
| Veniliornis nigriceps | M | 1 | 45.3 | | | | Peru; Ecuador | 3868.0 |
| | F | 1 | 39.0 | | | | 320, 668 | |
| Veniliornis fumigatus | B | 23 | 34.7 | | | | | 3869.0 |
| | | | | | | | 244, 387, 457, 510 | |
| Veniliornis passerinus | B | 5 | 31.4 | | | | | 3870.0 |
| | | | | | | | 111, 668 | |
| Veniliornis frontalis | U | | | | 30.0–36.0 | | | 3871.0 |
| | | | | | | | 560 | |
| Veniliornis spilogaster | M | 6 | | | 34.5–45.0 | | Brazil | 3872.0 |
| | | | | | | | 37, 510a | |
| Veniliornis sanguineus | B | 6 | 26.2 | | 23.0–30.0 | | Surinam | 3873.0 |
| | | | | | | | 245 | |
| Veniliornis kirkii | B | 7 | 36.7 | | 32.0–42.0 | | Trinidad | 3874.0 |
| | | | | | | | 188 | |
| Veniliornis cassini | M | 4 | 30.6 | | 27.0–35.0 | | Surinam; Brazil | 3876.0 |
| | F | 2 | 32.5 | | 32.0–33.0 | | 245, 246, 610a | |
| Veniliornis affinis | M | 7 | 37.4 | | | | Peru; Brazil | 3877.0 |
| | F | 2 | 34.0 | | 32.5–35.5 | | 193, 610a | |
| Piculus simplex | U | | 55.0 | | | | | 3879.0 |
| | | | | | | | 603a | |
| Piculus leucolaemus | U | | 69.0 | | | | Peru | 3882.0 |
| | | | | | | | 623 | |
| Piculus flavigula | B | 6 | 52.9 | | 44.0–59.5 | | Surinam; Brazil | 3883.0 |
| | | | | | | | 246, 610a | |
| Piculus chrysochloros | U | | 88.0 | | | | Peru | 3884.0 |
| | | | | | | | 623 | |
| Piculus aurulentus | B | 4 | 60.6 | | 52.0–64.1 | | Brazil | 3885.0 |
| | | | | | | | 37, 46, 610a | |
| Piculus auricularis | M | 3 | 65.4 | | 63.9–68.3 | | Oaxaca, Mexico | 3886.0 |
| | | | | | | | 584 | |
| Piculus rubiginosus | M | 14 | 56.2 | | 51.0–68.0 | | Trinidad | 3887.0 |
| | F | 9 | 55.4 | | 51.0–61.0 | | 188 | |
| Piculus rivolii | F | 1 | 86.0 | | | | Peru | 3888.0 |
| | | | | | | | 668 | |

## Body Masses of World Birds (continued)

| Species | Sex | N | Mean | Std dev | Range | Sn | Location | Number |
|---|---|---|---|---|---|---|---|---|
| Colaptes atricollis | U | | | | 73.0–90.0 | | 560 | 3889.0 |
| Colaptes punctigula | B | 7 | 65.7 | | 61.0–72.0 | | 245, 439, 625 | 3890.0 |
| Colaptes melanochloros | M | 7 | | | 120.0–137.0 | | Brazil | 3891.0 |
| | F | 11 | | | 114.0–140.0 | | 37 | |
| Colaptes melanolaimus | M | 1 | 129.0 | | | | Paraguay | 3892.0 |
| | F | 3 | 113.0 | | 104.0–120.0 | | 610 | |
| Colaptes auratus auratus | M | 94 | 135.0 | 6.37 | 114.0–160.0 | Y | Pennsylvania,USA | 3893.0 |
| | F | 65 | 129.0 | 7.67 | 106.0–164.0 | | 101 | |
| Colaptes auratus collaris | U | 82 | 142.0 | 2.20 | 121.0–167.0 | | Oregon, USA | 3893.1 |
| | | | | | | | 177 | |
| Colaptes auratus mearnsi | U | 100 | 111.0 | 7.36 | 92.2–129.0 | Y | Arizona, USA | 3893.2 |
| | | | | | | | 177 | |
| Colaptes pitius | U | | 100.0 | | | | Chile | 3896.0 |
| | | | | | | | 293 | |
| Colaptes rupicola | B | 5 | 180.0 | | 161.0–190.0 | | Peru | 3897.0 |
| | | | | | | | 451 | |
| Colaptes campestris | B | 5 | 165.0 | | 149.0–180.0 | | 37, 111 | 3898.0 |
| Celeus brachyurus phaioceps | B | 2 | 108.0 | | 102.0–113.0 | | India | 3899.0 |
| | | | | | | | 5 | |
| Celeus brachyurus | B | 3 | 68.4 | | 65.4–73.2 | | Borneo | 3899.0 |
| | | | | | | | 627 | |
| Celeus loricatus | B | 4 | 76.6 | | 74.0–79.8 | | Panama | 3900.0 |
| | | | | | | | 86, 497, 611 | |
| Celeus grammicus | M | 1 | 87.0 | | | | Peru | 3901.0 |
| | F | 4 | 76.6 | | | | 193 | |
| Celeus undatus | B | 3 | 63.2 | | 61.0–64.6 | | Surinam; Brazil | 3902.0 |
| | | | | | | | 246, 437 | |
| Celeus castaneus | B | 10 | 86.6 | 8.33 | 80.0–95.0 | | Mexico; Belize | 3903.0 |
| | | | | | | | 457, 510 | |
| Celeus elegans | B | 9 | 117.0 | | 93.5–133.0 | | Trinidad | 3904.0 |
| | | | | | | | 576 | |
| Celeus lugubris | M | 3 | 143.0 | | 134.0–157.0 | | Paraguay | 3905.0 |
| | | | | | | | 610 | |
| Celeus flavescens | B | 3 | 150.0 | | 141.0–165.0 | | Brazil | 3906.0 |
| | | | | | | | 37, 441 | |
| Celeus flavus | U | | 201.0 | | | | Peru | 3907.0 |
| | | | | | | | 623 | |
| Celeus spectabilis | U | 1 | 111.0 | | | | Peru | 3908.0 |
| | | | | | | | 193 | |
| Celeus torquatus | U | | 134.0 | | | | Peru | 3909.0 |
| | | | | | | | 623 | |

## Body Masses of World Birds (continued)

| Species | Sex | N | Mean | Std dev | Range | Sn | Location | Number |
|---|---|---|---|---|---|---|---|---|
| Dryocopus galeatus | F | 1 | 124.0 | | | | Paraguay 610 | 3910.0 |
| Dryocopus pileatus | M | 2 | 308.0 | | 308.0–309.0 | | Pennsylvania,USA | 3911.0 |
| | F | 4 | 266.0 | | 250.0–284.0 | | 177 | |
| Dryocopus lineatus | M | 12 | 194.0 | | | | Panama; Surinam | 3912.0 |
| | F | 7 | 173.0 | | | | 244, 245 | |
| Dryocopus javensis | M | 4 | 290.0 | | 240.0–347.0 | | Philippines | 3914.0 |
| | F | 6 | 254.0 | | 244.0–267.0 | | 475 | |
| Dryocopus martius | B | 18 | 321.0 | 30.30 | 255.0–361.0 | | 118 | 3916.0 |
| Campephilus haematogaster | B | 3 | 237.0 | | 225.0–250.0 | | Panama 50, 497 | 3918.0 |
| Campephilus rubricollis | B | 9 | 214.0 | | 193.0–239.0 | | 161, 245, 246, 610a, 668 | 3919.0 |
| Campephilus robustus | U | | 200.0 | | | | Brazil 564 | 3920.0 |
| Campephilus guatemalensis | B | 17 | 242.0 | | | | Panama 244 | 3921.0 |
| Campephilus melanoleucos | B | 14 | 243.0 | 25.80 | 181.0–279.0 | | 86,244,245,246, 311,384,497,611 | 3922.0 |
| Campephilus gayaquilensis | U | | | | 230.0–253.0 | | 560 | 3923.0 |
| Campephilus leucopogon | M | 1 | 205.0 | | | | Paraguay | 3924.0 |
| | F | 1 | 237.0 | | | | 610 | |
| Campephilus magellanicus | U | | | | 276.0–363.0 | | 560 | 3925.0 |
| Campephilus principalis | U | 2 | 511.0 | | 454.0–568.0 | | 616 | 3927.0 |
| Picus mineaceus | U | 2 | 90.8 | | 79.5–102.0 | | Malaysia; Borneo 627, 690 | 3928.0 |
| Picus chlorolophus chlorigaster | M | 4 | | | 57.0–74.0 | | India | 3929.0 |
| | F | 3 | | | 62.0–70.0 | | 5 | |
| Picus puniceus | U | 2 | 71.4 | | 65.8–77.0 | | Malaysia;Borneo 627, 690 | 3930.0 |
| Picus flavinucha | M | 5 | | | 172.0–198.0 | | India | 3931.0 |
| | F | 3 | | | 165.0–174.0 | | 5 | |
| Picus mentalis | U | 6 | 109.0 | | | | Malaysia 690 | 3932.0 |
| Picus vittatus | U | | | | 94.0–132.0 | | 560 | 3934.0 |
| Picus xanthopygaeus | B | 3 | 99.7 | | 90.0–111.0 | | India 5, 361 | 3935.0 |

## Body Masses of World Birds (continued)

| Species | Sex | N | Mean | Std dev | Range | Sn | Location | Number |
|---|---|---|---|---|---|---|---|---|
| Picus squamatus | B | 4 | 170.0 | | 159.0–185.0 | | 5, 149 | 3936.0 |
| Picus awokera | U | | | | 120.0–138.0 | | 560 | 3937.0 |
| Picus viridis | B | 69 | 176.0 | | 138.0–190.0 | | France 118 | 3938.0 |
| Picus erythropygius | U | | | | 100.0–135.0 | | 560 | 3940.0 |
| Picus canus | B | 28 | 137.0 | | 125.0–165.0 | | Switzerland 118 | 3941.0 |
| Dinopium rafflesii | U | 1 | 98.0 | | | | Malaysia 690 | 3942.0 |
| Dinopium shorii | F | 1 | 101.0 | | | | 5 | 3943.0 |
| Dinopium javanense | U | | | | 67.0–90.0 | | 560 | 3944.0 |
| Dinopium benghalense benghalense | U | 11 | 100.0 | | 89.0–121.0 | | India 5 | 3945.0 |
| Chrysocolaptes lucidus | B | 16 | 142.0 | | 125.0–164.0 | | Philippincs 475 | 3946.0 |
| Chrysocolaptes festivus | U | | 213.0 | | | | 560 | 3947.0 |
| Gecinulus grantia | U | | | | 68.0–85.0 | | 560 | 3948.0 |
| Blythipicus rubiginosus | U | 8 | 84.2 | | | | Malaysia 690 | 3951.0 |
| Blythipicus pyrrhotis | B | 6 | 132.0 | | 114.0–170.0 | | 5, 371 | 3952.0 |
| Reinwardtipicus validus | B | 2 | 172.0 | | 168.0–176.0 | | Borneo 627 | 3953.0 |
| Meiglyptes tristis | F | 1 | 31.7 | | | | Borneo 627 | 3954.0 |
| Meiglyptes jugularis | U | | | | 50.0–57.0 | | 560 | 3955.0 |
| Meiglyptes tukki | U | 24 | 53.1 | | | | Malaysia 690 | 3956.0 |
| Hemicircus concretus | U | | | | 27.0–32.0 | | 560 | 3957.0 |
| Hemicircus canente | U | | | | 37.0–50.0 | | 560 | 3958.0 |
| Mulleripicus funebris | B | 6 | 163.0 | | 143.0–180.0 | | Philippines 475 | 3960.0 |
| Mulleripicus pulverulentus | U | | | | 360.0–563.0 | | 560 | 3961.0 |

## Body Masses of World Birds (continued)

| Species | Sex | N | Mean | Std dev | Range | Sn | Location | Number |
|---|---|---|---|---|---|---|---|---|
| | | | **ORDER: PASSERIFORMES** | | | | **FAMILY: ACANTHISITTIDAE** | |
| Acanthisitta chloris | M | 21 | 6.3 | 0.60 | 5.0–7.0 | | New Zealand | 3962.0 |
| | F | 6 | 7.6 | | 7.0–8.5 | | 500 | |
| Xenicus longipes | M | 1 | 15.0 | | | | New Zealand | 3963.0 |
| | F | 1 | 17.0 | | | | 597 | |
| Xenicus gilviventris | B | 17 | 16.1 | | | | New Zealand | 3964.0 |
| | | | | | | | 249, 250 | |
| | | | **ORDER: PASSERIFORMES** | | | | **FAMILY: PITTIDAE** | |
| Pitta cyanea | U | | | | 99.0–120.0 | | India | 3972.0 |
| | | | | | | | 5 | |
| Pitta guajana | M | 1 | 81.5 | | | | Borneo | 3973.0 |
| | | | | | | | 627 | |
| Pitta sordida | U | 5 | 68.4 | | | | New Guinea | 3977.0 |
| | | | | | | | 35, 217, 218 | |
| Pitta steerii | B | 8 | 97.2 | | 83.8–113.0 | | Philippines | 3980.0 |
| | | | | | | | 475 | |
| Pitta erythrogaster | U | 16 | 87.0 | | | | New Guinea | 3982.0 |
| | | | | | | | 35 | |
| Pitta brachyura | B | 25 | 55.5 | | 47.0–66.0 | | India | 3989.0 |
| | | | | | | | 5 | |
| Pitta moluccensis | U | | 113.0 | | | | India | 3991.0 |
| | | | | | | | 5 | |
| | | | **ORDER: PASSERIFORMES** | | | | **FAMILY: EURYLAIMIDAE** | |
| Smithornis capensis | M | 7 | 23.7 | | 21.0–26.1 | | | 3997.0 |
| | F | 3 | 21.0 | | 17.4–23.9 | | 166, 280, 281 | |
| Eurylaimus javanicus | M | 1 | 81.8 | | | | Borneo | 4003.0 |
| | | | | | | | 627 | |
| Eurylaimus ochromalus | B | 5 | 33.3 | | 32.0–34.7 | | Borneo | 4004.0 |
| | | | | | | | 627 | |
| Eurylaimus steerii | B | 18 | 33.0 | | 33.4–41.5 | | Philippines | 4005.0 |
| | | | | | | | 475 | |
| Serilophus lunatus | F | 2 | 34.0 | | 33.0–35.0 | | India | 4006.0 |
| | | | | | | | 5 | |
| Psarisomus dalhousiae | F | 1 | 67.0 | | | | India | 4007.0 |
| | | | | | | | 5 | |
| Calyptomena viridis | F | 2 | 60.3 | | 53.2–67.4 | | Borneo | 4008.0 |
| | | | | | | | 627 | |
| | | | **ORDER: PASSERIFORMES** | | | | **FAMILY: PHILEPITTIDAE** | |
| Philepitta castanea | B | 6 | 34.4 | | 31.5–37.0 | | Madagascar | 4011.0 |
| | | | | | | | 39 | |

## Body Masses of World Birds (continued)

| Species | Sex | N | Mean | Std dev | Range | Sn | Location | Number |
|---|---|---|---|---|---|---|---|---|
| ORDER: PASSERIFORMES | | | | | FAMILY: DENDROCOLAPTIDAE | | | |
| Dendrocincla tyrannina | U | 3 | 60.2 | | | | Peru 668 | 4015.0 |
| Dendrocincla fuliginosa | B | 47 | 37.2 | | 31.0–43.5 | | Trinidad 576 | 4016.0 |
| Dendrocincla turdina | U | 22 | 39.0 | | 31.0–45.3 | | Brazil 441 | 4017.0 |
| Dendrocincla anabatina | M | 8 | 37.4 | | 35.2–42.4 | | Yucatan, Mexico 457 | 4018.0 |
| | F | 6 | 31.3 | | 28.8–34.1 | | | |
| Dendrocincla merula | U | 10 | 41.0 | | 38.5–44.0 | | 161, 437 | 4019.0 |
| Dendrocincla homochroa | M | 6 | 45.3 | | | | Panama 244 | 4020.0 |
| | F | 4 | 36.8 | | | | | |
| Deconychura longicauda | U | 11 | 23.8 | 2.00 | | | 315 | 4021.0 |
| Deconychura stictolaema | B | 6 | 18.1 | | 14.5–20.5 | | Brazil 136, 610a | 4022.0 |
| Sittasomus griseicapillus | M | 21 | 14.3 | 0.60 | | | Panama 244 | 4023.0 |
| | F | 11 | 12.9 | 0.73 | | | | |
| Glyphorhynchus spirurus | U | 113 | 14.6 | | 12.2–16.9 | | Brazil 437 | 4024.0 |
| Drymornis bridgesii | M | 1 | 80.0 | | | | 37, 411 | 4025.0 |
| | F | 1 | 110.0 | | | | | |
| Nasica longirostris | U | | 92.0 | | | | Peru 623 | 4026.0 |
| Dendrexetastes rufigula | U | | 70.0 | | | | Peru 623 | 4027.0 |
| Hylexetastes perrotii | U | 66 | 114.0 | 7.70 | | | Brazil 45 | 4029.0 |
| Hylexetastes promeropirhynchus | U | | 136.0 | | | | Peru 623 | 4030.0 |
| Xiphocolaptes albicollis | U | 2 | 116.0 | | 113.0–118.0 | | Brazil 37, 71 | 4031.0 |
| Xiphocolaptes major | B | 3 | 155.0 | | 147.0–162.0 | | Paraguay 610 | 4034.0 |
| Dendrocolaptes certhia | B | 20 | 64.2 | 6.48 | 52.7–73.9 | | 161, 244, 437, 457, 510, 611 | 4035.0 |
| Dendrocolaptes concolor | B | 3 | 62.7 | | 51.0–71.0 | | Brazil; Bolivia 22, 610a | 4036.0 |
| Dendrocolaptes hoffmannsi | F | 1 | 89.0 | | | | Brazil 610a | 4037.0 |
| Dendrocolaptes picumnus | U | 6 | 88.7 | | | | Surinam 678 | 4038.0 |

## Body Masses of World Birds (continued)

| Species | Sex | N | Mean | Std dev | Range | Sn | Location | Number |
|---|---|---|---|---|---|---|---|---|
| Dendrocolaptes picumnus | B | 4 | 73.8 | | | | north Andes 678 | 4038.0 |
| Dendrocolaptes picumnus | B | 6 | 61.0 | | | | Nicaragua 678 | 4038.0 |
| Dendrocolaptes platyrostris | B | 10 | 61.9 | | 50.5–69.0 | | Brazil 37, 440, 441 | 4039.0 |
| Xiphorhynchus picus | U | 14 | 41.6 | 3.16 | 35.3–46.8 | | Venezuela 625 | 4040.0 |
| Xiphorhynchus obsoletus | U | | 39.0 | | | | Peru 623 | 4042.0 |
| Xiphorhynchus ocellatus | U | 7 | 35.1 | 1.90 | | | Peru 668 | 4043.0 |
| Xiphorhynchus spixii | U | 17 | 31.2 | | 27.2–36.2 | | Brazil 437 | 4044.0 |
| Xiphorhynchus elegans | B | 13 | 34.1 | 2.58 | 30.5–38.5 | | Brazil 610a | 4045.0 |
| Xiphorhynchus pardalotus | U | 6 | 40.4 | | 32.0–44.0 | | Brazil 439, 610a | 4046.0 |
| Xiphorhynchus guttatus | B | 34 | 49.5 | | 36.5–59.0 | | Trinidad 188 | 4047.0 |
| Xiphorhynchus flavigaster | M | 9 | 47.2 | 0.88 | 43.6–52.6 | | Yucatan, Mexico 457 | 4047.5 |
| | F | 12 | 40.0 | 1.70 | 35.0–44.1 | | | |
| Xiphorhynchus eytoni | B | 11 | 58.8 | 7.19 | 45.0–69.0 | | Brazil 226, 610a | 4048.0 |
| Xiphorhynchus lachrymosus | B | 4 | 56.4 | | 50.7–66.0 | | Panama 436, 497, 611 | 4050.0 |
| Xiphorhynchus erythropygius | U | 13 | 46.8 | 2.36 | 42.6–50.0 | | Panama 50 | 4051.0 |
| Xiphorhynchus triangularis | B | 6 | 48.4 | | 46.3–51.8 | | Columbia 387 | 4052.0 |
| Lepidocolaptes leucogaster | B | 12 | 36.0 | 1.95 | 32.1–39.2 | | Mexico 584 | 4053.0 |
| Lepidocolaptes souleyetii | B | 10 | 25.7 | | | | Peru 672 | 4054.0 |
| Lepidocolaptes angustirostris | B | 11 | 31.3 | | 27.8–37.0 | | 37, 111, 440 | 4055.0 |
| Lepidocolaptes affinis | M | 16 | 35.4 | 2.32 | | | Panama 244 | 4056.0 |
| | F | 20 | 34.6 | 2.24 | | | | |
| Lepidocolaptes squamatus | U | 1 | 27.0 | | | | Brazil 440 | 4057.0 |
| Lepidocolaptes fuscus | U | 20 | 21.8 | 1.36 | | | Brazil 71 | 4058.0 |
| Lepidocolaptes albolineatus | U | | 33.0 | | | | Peru 623 | 4059.0 |

## Body Masses of World Birds (continued)

| Species | Sex | N | Mean | Std dev | Range | Sn | Location | Number |
|---|---|---|---|---|---|---|---|---|
| Campylorhamphus pucherani | M | 1 | 70.5 | | | | Peru 668 | 4060.0 |
| Campylorhamphus trochilirostris | B | 6 | 39.1 | | 35.0–46.0 | | 87, 385, 672 | 4061.0 |
| Campylorhamphus falcularius | U | 5 | 44.4 | | | | Brazil 37, 71, 441 | 4062.0 |
| Campylorhamphus pusillus | B | 9 | 39.6 | | 33.0–48.0 | | 451 | 4063.0 |
| Campylorhamphus procurvoides | M | 1 | 33.0 | | | | Fr. Guiana 161 | 4064.0 |

### ORDER: PASSERIFORMES     FAMILY: FURNARIIDAE

| Species | Sex | N | Mean | Std dev | Range | Sn | Location | Number |
|---|---|---|---|---|---|---|---|---|
| Geositta cunicularia | M | 1 | 30.0 | | | | Brazil 37 | 4066.0 |
| Geositta rufipennis harrisoni | B | 12 | 26.6 | | 24.0–29.0 | | Chile 363a | 4073.0 |
| Geositta rufipennis fasciata | B | 18 | 39.4 | | 36.0–44.0 | | Chile 363a | 4073.0 |
| Upucerthia ruficauda | U | 12 | 30.1 | | | | Peru 538 | 4078.0 |
| Upucerthia dumetaria | B | 7 | 49.3 | | 47.2–51.3 | | Argentina 411 | 4081.0 |
| Upucerthia albigula | U | 7 | 39.9 | | | | Peru 538 | 4082.0 |
| Upucerthia jelskii | U | 23 | 40.1 | | | | Peru 538 | 4083.0 |
| Cinclodes fuscus | B | 5 | 31.8 | | 29.0–33.0 | | 37, 644 | 4085.0 |
| Cinclodes pabsti | U | | 53.0 | | | | Brazil 564 | 4087.0 |
| Cinclodes nigrofumosus | F | 1 | 66.7 | | | | Peru 562 | 4093.0 |
| Cinclodes atacamensis | F | 1 | 54.7 | | | | Peru 562 | 4095.0 |
| Chilia melanura | U | | 40.0 | | | | Chile 292 | 4097.0 |
| Furnarius figulus | M | 1 | 28.0 | | | | Brazil 226 | 4099.0 |
| Furnarius leucopus | U | | 44.0 | | | | Peru 623 | 4100.0 |
| Furnarius rufus | B | 9 | 56.5 | | 48.5–65.0 | | 37, 111, 411 | 4102.0 |
| Sylviorthorhynchus desmursii | U | | 10.9 | | | | Chile 52 | 4104.0 |

## Body Masses of World Birds (continued)

| Species | Sex | N | Mean | Std dev | Range | Sn | Location | Number |
|---|---|---|---|---|---|---|---|---|
| Aphrastura spinicauda | U | | 12.2 | | | | Chile 52 | 4105.0 |
| Leptasthenura platensis | B | 3 | 10.2 | | 9.8–10.8 | | Argentina 411 | 4109.0 |
| Leptasthenura aegithaloides | B | 4 | 9.0 | | 8.0–10.0 | | Argentina 411 | 4110.0 |
| Leptasthenura striolata | U | | 10.5 | | | | Brazil 564 | 4111.0 |
| Leptasthenura setaria | B | 2 | 11.0 | | 11.0–11.0 | | Brazil 37 | 4116.0 |
| Schizoeaca coryi | B | 4 | 16.8 | | 15.5–17.5 | | Venezuela 644 | 4118.0 |
| Schizoeaca fuliginosa | B | 9 | 18.3 | | | | 644, 668 | 4119.0 |
| Schizoeaca griseomurina | B | 6 | 17.6 | | 15.5–19.0 | | 451 | 4120.0 |
| Schoeniophylax phryganophila | U | 9 | 18.4 | | 15.0–22.5 | | 37, 111 | 4126.0 |
| Synallaxis ruficapilla | B | 10 | 13.9 | | 11.5–15.0 | | Brazil 37, 441 | 4127.0 |
| Synallaxis frontalis | B | 8 | 15.0 | | 13.0–16.3 | | 37, 111 | 4128.0 |
| Synallaxis azarae | U | 31 | 16.9 | 0.87 | 15.8–19.3 | | Ecuador 320 | 4129.0 |
| Synallaxis albescens | B | 18 | 15.0 | 1.97 | 12.5–20.0 | | 244, 246, 311, 384, 611 | 4131.0 |
| Synallaxis spixi | B | 3 | 12.6 | | 12.0–13.0 | | Brazil 37 | 4132.0 |
| Synallaxis brachyura | B | 13 | 18.3 | | | | Panama; Peru 244, 497, 672 | 4133.0 |
| Synallaxis albigularis | U | 4 | 16.6 | | | | Peru 668 | 4134.0 |
| Synallaxis cabanisi | U | 22 | 22.7 | 1.49 | | | Peru 668 | 4139.0 |
| Synallaxis tithys | B | 15 | 15.8 | | | | Peru 672 | 4141.0 |
| Synallaxis cinerascens | B | 4 | 13.2 | | 12.5–14.5 | | Brazil 37, 441 | 4142.0 |
| Synallaxis gujanensis | U | 12 | 18.1 | 1.28 | | | Peru 668 | 4146.0 |
| Synallaxis rutilans | B | 6 | 17.6 | | 15.0–19.9 | | Surinam; Brazil 245, 437 | 4148.0 |
| Synallaxis cherriei | M | 1 | 16.0 | | | | Brazil 226 | 4149.0 |

## Body Masses of World Birds (continued)

| Species | Sex | N | Mean | Std dev | Range | Sn | Location | Number |
|---|---|---|---|---|---|---|---|---|
| Synallaxis unirufa | U | 7 | 17.6 | | | | Peru 668 | 4150.0 |
| Synallaxis erythrothorax | B | 13 | 17.1 | 1.18 | 15.0–18.8 | | Mexico; Belize 457, 510, 584 | 4154.0 |
| Synallaxis cinnamomea carri | B | 15 | 16.8 | | 15.5–18.0 | | Trinidad 311 | 4155.0 |
| Synallaxis candei | M | 2 | 15.3 | | 14.5–16.0 | | Venezuela 57 | 4157.0 |
| Synallaxis scutata | U | 2 | 12.3 | | 12.0–12.5 | | Brazil 440 | 4159.0 |
| Hellmayrea gularis | B | 29 | 12.4 | | 10.5–14.5 | | Peru 451 | 4160.0 |
| Cranioleuca erythrops | B | 18 | 16.9 | | | | Panama 244 | 4161.0 |
| Cranioleuca pallida | F | 1 | 11.5 | | | | Brazil 610a | 4163.0 |
| Cranioleuca curtata | M | 1 | 16.8 | | | | Peru 668 | 4164.0 |
| Cranioleuca demissa | U | 1 | 14.9 | | | | Venezuela 626 | 4165.0 |
| Cranioleuca pyrrhophia | B | 28 | 14.9 | | 13.0–17.0 | | Brazil 37 | 4168.0 |
| Cranioleuca obsoleta | B | 24 | 13.4 | | 12.0–16.0 | | Brazil 37 | 4169.0 |
| Cranioleuca marcapatae | F | 2 | 19.8 | | 19.1–20.5 | | Peru 193 | 4170.0 |
| Cranioleuca albiceps | U | 9 | 20.4 | | | | Peru 668 | 4171.0 |
| Cranioleuca semicinerea | B | 3 | 15.0 | | 14.5–16.0 | | Brazil 620 | 4172.0 |
| Cranioleuca vulpina | B | 4 | 16.0 | | 15.0–17.4 | | Brazil; Columbia 439, 589 | 4175.0 |
| Cranioleuca gutturata | F | 1 | 13.8 | | | | Peru 193 | 4177.0 |
| Cranioleuca sulphurifera | B | 3 | 13.5 | | 12.5–14.0 | | Brazil 37 | 4178.0 |
| Certhiaxis cinnamomea | B | 17 | 14.8 | 1.99 | 12.0–18.6 | | 37, 111, 245, 311, 385, 576 | 4179.0 |
| Asthenes baeri | B | 11 | 17.8 | | 15.0–21.0 | | 37, 411 | 4182.0 |
| Asthenes humicola humicola | B | 9 | 21.8 | | 20.0–24.0 | | Chile 363a | 4191.0 |
| Asthenes humicola goodalli | B | 4 | 19.3 | | 18.0–20.0 | | Chile 363a | 4191.0 |

## Body Masses of World Birds (continued)

| Species | Sex | N | Mean | Std dev | Range | Sn | Location | Number |
|---|---|---|---|---|---|---|---|---|
| Asthenes patagonica | B | 4 | 16.4 | | 15.5–17.1 | | Argentina 411 | 4192.0 |
| Asthenes flammulata | B | 19 | 22.5 | | 17.0–27.0 | | Peru 451 | 4199.0 |
| Phacellodomus rufifrons | B | 36 | 24.6 | | 21.5–27.8 | | Venezuela 625 | 4206.0 |
| Phacellodomus sibilatrix | M | 1 | 14.0 | | | | Paraguay 610 | 4207.0 |
| Phacellodomus striaticollis | B | 3 | 25.7 | | 24.5–26.5 | | Brazil 37 | 4209.0 |
| Phacellodomus ruber | B | 9 | 38.8 | | 35.8–45.0 | | 37, 111 | 4210.0 |
| Phacellodomus erythropthalmus | M | 2 | 24.5 | | 24.0–25.0 | | Brazil 37 | 4212.0 |
| Clibanornis dendrocolaptoides | F | 1 | 52.5 | | | | Brazil 37 | 4215.0 |
| Spartonoica maluroides | B | 2 | 11.0 | | 10.0–12.0 | | Brazil 37 | 4216.0 |
| Phleocryptes melanops | U | 4 | 14.2 | | 13.0–16.0 | | Brazil 37 | 4217.0 |
| Limnornis curvirostris | B | 3 | 28.6 | | 27.0–30.0 | | Brazil 37 | 4218.0 |
| Limnornis rectirostris | B | 3 | 18.5 | | 16.0–21.0 | | Brazil 37 | 4219.0 |
| Anumbius annumbi | U | 2 | 41.5 | | 38.0–45.0 | | Brazil 37 | 4220.0 |
| Coryphistera alaudina | U | 4 | 36.9 | | 31.0–42.0 | | Brazil 37 | 4221.0 |
| Siptornis striaticolis | F | 1 | 13.0 | | | | Ecuador 589 | 4223.0 |
| Metopothrix aurantiacus | F | 1 | 11.0 | | | | Peru 193 | 4224.0 |
| Xenerpestes minlosi | F | 1 | 11.0 | | | | Ecuador 589 | 4225.0 |
| Premnornis guttuligera | U | 15 | 15.9 | 1.12 | | | Peru 668 | 4228.0 |
| Premnoplex brunnescens | U | 18 | 16.3 | 0.98 | | | Peru 668 | 4229.0 |
| Margarornis rubiginosus | U | 2 | 17.8 | | 16.1–19.5 | | Panama 244 | 4231.0 |
| Margarornis bellulus | M | 1 | 18.5 | | | | Panama 497 | 4233.0 |
| Margarornis squamiger | U | 17 | 17.3 | 1.40 | | | Peru 668 | 4234.0 |

## Body Masses of World Birds (continued)

| Species | Sex | N | Mean | Std dev | Range | Sn | Location | Number |
|---|---|---|---|---|---|---|---|---|
| Lochmias nematura | B | 8 | 24.8 | | | | Peru; Brazil<br>37, 441, 668 | 4235.0 |
| Pseudoseisura lophotes | B | 6 | 86.0 | | 78.0–92.5 | | 37, 411 | 4237.0 |
| Pseudoseisura gutturalis | M<br>F | 1<br>1 | 78.2<br>66.6 | | | | Argentina<br>411 | 4238.0 |
| Pseudocolaptes lawrencii | B | 5 | 51.6 | | | | Panama<br>244 | 4239.0 |
| Pseudocolaptes boissonneautii | U | 6 | 47.7 | | | | Peru<br>668 | 4240.0 |
| Ancistrops strigilatus | B | 8 | 35.9 | | | | Peru<br>193 | 4242.0 |
| Hyloctistes subulatus | U | 7 | 33.4 | | | | Panama; Peru<br>37, 50, 315, 668 | 4244.0 |
| Syndactyla subalaris | U | 14 | 28.9 | 1.25 | | | Peru<br>668 | 4246.0 |
| Syndactyla rufosuperciliata | U | 9 | 27.7 | | | | Peru<br>668 | 4247.0 |
| Anabacerthia striaticollis | U | 18 | 24.8 | 1.26 | | | Peru<br>668 | 4249.0 |
| Philydor ruficaudatus | U | 7 | 24.9 | | | | Fr. Guiana<br>161, 624 | 4250.0 |
| Philydor pyrrhodes | U | | 29.0 | | | | Peru<br>623 | 4251.0 |
| Philydor erythrocercus | U | 12 | 26.4 | 4.60 | | | Fr. Guiana<br>624 | 4254.0 |
| Philydor ochrogaster | B | 41 | 24.2 | | | | Peru<br>193 | 4255.0 |
| Philydor erythropterus | U | | 30.0 | | | | Peru<br>623 | 4256.0 |
| Philydor amaurotis | U | 4 | 18.8 | | 16.8–20.0 | | Brazil<br>441 | 4257.0 |
| Philydor lichtensteini | U | 4 | 21.0 | | 20.0–22.5 | | Brazil<br>37, 441 | 4258.0 |
| Philydor rufus | B | 3 | 26.7 | | 25.0–28.1 | | Brazil<br>37, 135 | 4259.0 |
| Philydor atricapillus | U | 6 | 22.8 | | 20.5–24.0 | | Brazil<br>441 | 4260.0 |
| Simoxenops ucayalae | B | 11 | 49.8 | | | | Peru; Brazil<br>193, 226, 549 | 4262.0 |
| Thripadectes rufobrunneus | B | 4 | 56.2 | | 51.5–64.8 | | Panama<br>244 | 4266.0 |
| Thripadectes melanorhynchus | U | 5 | 48.8 | | | | Peru<br>668 | 4268.0 |

## Body Masses of World Birds (continued)

| Species | Sex | N | Mean | Std dev | Range | Sn | Location | Number |
|---|---|---|---|---|---|---|---|---|
| Thripadectes holostictus | U | 2 | 43.9 | | | | Peru 668 | 4269.0 |
| Thripadectes flammulatus | B | 5 | 55.0 | | 50.0–62.0 | | Peru 451 | 4270.0 |
| Thripadectes scrutator | U | 1 | 66.3 | | | | Peru 668 | 4271.0 |
| Automolus ruficollis | F | 5 | 31.0 | | 24.0–34.0 | | Peru 451 | 4272.0 |
| Automolus ochrolaemus | U | 25 | 40.2 | 2.50 | | | 315 | 4273.0 |
| Automolus dorsalis | U | 7 | 35.4 | | | | Peru 668 | 4274.0 |
| Automolus infuscatus | U | 11 | 32.9 | | 30.2–36.2 | | Brazil 437 | 4275.0 |
| Automolus leucophthalmus | U | 7 | 35.5 | | 32.0–37.5 | | Brazil 441 | 4276.0 |
| Automolus melanopezus | B | 3 | 31.2 | | 28.7–32.9 | | Peru 193 | 4278.0 |
| Automolus rubiginosus | B | 4 | 37.4 | | 32.0–43.4 | | 50, 161, 315 | 4279.0 |
| Automolus rufipileatus | U | 2 | 33.5 | | 32.0–35.0 | | 161, 437 | 4280.0 |
| Hylocryptus rectirostris | U | | 48.0 | | | | Brazil 564 | 4281.0 |
| Hylocryptus erythrocephalus | B | 10 | 46.5 | | | | Peru 672 | 4282.0 |
| Sclerurus mexicanus | B | 7 | 25.0 | | 21.0–27.6 | | Panama; Mexico 46, 50, 314, 497 | 4283.0 |
| Sclerurus rufigularis | U | 10 | 21.6 | | 20.2–23.1 | | 161, 246, 437 | 4284.0 |
| Sclerurus scansor | B | 5 | 37.4 | | 35.0–40.0 | | 37, 135, 441 | 4287.0 |
| Sclerurus guatemalensis | U | 25 | 34.7 | 3.30 | | | 315 | 4288.0 |
| Heliobletus contaminatus | B | 4 | 14.2 | | 13.0–15.0 | | Brazil 37, 441, 610a | 4289.0 |
| Xenops milleri | U | 3 | 12.1 | | | | Brazil 45, 610a | 4290.0 |
| Xenops tenuirostris | F | 4 | 10.3 | | 10.3–10.3 | | Peru 193 | 4291.0 |
| Xenops minutus | U | 16 | 10.6 | | 8.2–11.4 | | Brazil 437 | 4292.0 |
| Xenops rutilans | B | 12 | 12.2 | | | | Panama 244 | 4293.0 |

## Body Masses of World Birds (continued)

| Species | Sex | N | Mean | Std dev | Range | Sn | Location | Number |
|---|---|---|---|---|---|---|---|---|
| Megaxenops parnaguae | M | 1 | 25.0 | | | | Brazil 621 | 4294.0 |

### ORDER: PASSERIFORMES                    FAMILY: THAMNOPHILIDAE

| Species | Sex | N | Mean | Std dev | Range | Sn | Location | Number |
|---|---|---|---|---|---|---|---|---|
| Cymbilaimus lineatus | B | 13 | 34.7 | | | | 244, 624 | 4296.0 |
| Cymbilaimus sanctaemariae | B | 9 | 28.7 | | | | Peru 193 | 4297.0 |
| Hypoehaleus guttatus | F | 2 | 38.8 | | 36.0–41.5 | | Paraguay 610 | 4298.0 |
| Batara cinerea | U | 5 | 149.0 | | 134.0–156.0 | | Brazil 38, 441, 564 | 4299.0 |
| Mackenziaena severa | B | 3 | 51.3 | | 50.0–53.0 | | Brazil 38, 610a | 4300.0 |
| Mackenziaena leachii | B | 4 | 67.8 | | 59.0–76.1 | | 38, 135 | 4301.0 |
| Frederickena viridis | U | 1 | 71.0 | | | | Surinam 246 | 4302.0 |
| Frederickena unduligera | B | 6 | 83.0 | | | | Peru 193 | 4303.0 |
| Taraba major | B | 10 | 67.5 | | | | Panama 244 | 4304.0 |
| Sakesphorus canadensis | B | 14 | 24.5 | | 20.5–29.5 | | Trinidad 188 | 4305.0 |
| Sakesphorus bernardi | B | 2 | 30.4 | | 28.9–32.0 | | Peru 672 | 4307.0 |
| Sakesphorus luctuosus | B | 7 | 31.4 | | 28.0–34.0 | | Brazil 226 | 4310.0 |
| Thamnophilus doliatus | B | 29 | 27.9 | | 25.0–30.5 | | Trinidad 576 | 4312.0 |
| Thamnophilus multistriatus | B | 7 | 27.6 | | 25.6–30.4 | | Columbia 387 | 4313.0 |
| Thamnophilus palliatus | U | 1 | 23.3 | | | | Brazil 437 | 4314.0 |
| Thamnophilus bridgesi | U | | 27.0 | | | | 603a | 4315.0 |
| Thamnophilus nigriceps | U | 6 | 22.9 | | 21.8–26.2 | | Panama 87 | 4316.0 |
| Thamnophilus cryptoleucus | B | 2 | 27.5 | | 27.5–27.5 | | Columbia 512 | 4319.0 |
| Thamnophilus aethiops | U | 11 | 22.7 | | 20.0–25.5 | | Brazil 437, 610a | 4320.0 |
| Thamnophilus unicolor | B | 7 | 24.2 | | 22.2–27.2 | | Columbia 387 | 4321.0 |

## Body Masses of World Birds (continued)

| Species | Sex | N | Mean | Std dev | Range | Sn | Location | Number |
|---|---|---|---|---|---|---|---|---|
| Thamnophilus schistaceus | U | | 21.0 | | | | Peru 623 | 4322.0 |
| Thamnophilus murinus | U | 13 | 19.3 | 1.70 | | | Fr. Guiana 624 | 4323.0 |
| Thamnophilus punctatus | U | 11 | 23.6 | | 20.7–26.8 | | Panama 436 | 4325.0 |
| Thamnophilus amazonicus | U | 5 | 21.1 | | 19.3–24.4 19.3–24.4 | | Brazil 437 | 4327.0 |
| Thamnophilus caerulescens | B | 12 | 21.4 | | | | Brazil 441 | 4328.0 |
| Thamnophilus ruficapillus | B | 4 | 21.2 | | 18.0–23.0 | | Brazil 38 | 4330.0 |
| Pygiptila stellaris | U | 5 | 24.7 | | 23.8–25.8 | | Surinam; Brazil 246, 610a | 4331.0 |
| Neoctantes niger | U | | 32.0 | | | | Peru 623 | 4333.0 |
| Clytoctantes atrogularis | F | 1 | 31.0 | | | | Brazil 333 | 4334.1 |
| Xenornis setifrons | F | 1 | 24.6 | | | | Panama 50 | 4335.0 |
| Thamnistes anabatinus | B | 10 | 20.7 | 1.78 | 19.0–23.6 | | Panama; Belize 244, 314, 497, 510 | 4336.0 |
| Dysithamnus mentalis andrei | B | 14 | 12.8 | | 11.5–14.5 | | Trinidad 188 | 4338.0 |
| Dysithamnus striaticeps | U | 1 | 12.0 | | | | Costa Rica 315 | 4339.0 |
| Thamnomanes ardesiacus | U | 22 | 17.7 | 1.50 | | | Fr. Guiana 624 | 4345.0 |
| Thamnomanes saturninus | B | 13 | 20.4 | | 18.0–22.5 | | Bolivia 610a | 4345.1 |
| Thamnomanes caesius | B | 12 | 16.0 | | 13.0–18.0 | | Fr. Guiana 161 | 4346.0 |
| Thamnomanes schistogynus | U | | 17.0 | | | | Peru 623 | 4347.0 |
| Myrmotherula brachyura | U | 9 | 7.3 | | | | 461 | 4348.0 |
| Myrmotherula obscura | M | 5 | 7.1 | | | | 232 | 4349.0 |
| Myrmotherula sclateri | M | 7 | 8.4 | | | | 232, 610a | 4350.0 |
| Myrmotherula ambigua | M | 2 | 7.0 | | 6.4–7.6 | | Brazil 610a | 4352.0 |
| Myrmotherula surinamensis | B | 10 | 9.7 | | | | Panama 244 | 4353.0 |

## Body Masses of World Birds (continued)

| Species | Sex | N | Mean | Std dev | Range | Sn | Location | Number |
|---|---|---|---|---|---|---|---|---|
| Myrmotherula longicauda | M | 5 | 8.3 | | | | 232 | 4355.0 |
| Myrmotherula hauxwelli | U | 23 | 10.7 | | 8.0–16.0 | | Brazil 437 | 4356.0 |
| Myrmotherula guttata | U | 10 | 11.0 | 1.30 | | | Fr. Guiana 624 | 4357.0 |
| Myrmotherula gularis | U | 7 | 11.2 | | 10.6–11.7 | | Brazil 441 | 4358.0 |
| Myrmotherula gutturalis | U | 16 | 9.2 | 1.00 | | | Fr. Guiana 624 | 4359.0 |
| Myrmotherula fulviventris | U | 19 | 10.4 | 0.70 | | | 315 | 4360.0 |
| Myrmotherula leucophthalma | M | 12 | 9.4 | 0.90 | 8.0–10.7 | | Brazil 610a | 4361.0 |
| Myrmotherula haematonota | M | 20 | 8.7 | | | | 232 | 4362.0 |
| Myrmotherula ornata | U | 12 | 9.4 | | | | 226, 461 | 4364.0 |
| Myrmotherula erythrura | U | 8 | 11.4 | | | | 232, 461 | 4365.0 |
| Myrmotherula axillaris | B | 25 | 8.4 | | | | Peru 668 | 4366.0 |
| Myrmotherula schisticolor | B | 16 | 9.6 | | | | Panama 244 | 4368.0 |
| Myrmotherula longipennis | U | 33 | 9.4 | 1.10 | | | Fr. Guiana 624 | 4370.0 |
| Myrmotherula minor | F | 1 | 6.4 | | | | Brazil 610a | 4371.0 |
| Myrmotherula iheringi | U | | 8.0 | | | | Peru 623 | 4372.0 |
| Myrmotherula grisea | M | 5 | 8.4 | | | | 232 | 4373.0 |
| Myrmotherula menetriesii | U | 13 | 8.6 | 0.50 | | | Fr. Guiana 624 | 4377.0 |
| Myrmotherula assimilis | M | 5 | 9.3 | | | | 232 | 4378.0 |
| Dichrozona cincta | U | | 16.0 | | | | Peru 623 | 4379.0 |
| Myrmorchilus strigilatus | U | 1 | 20.0 | | | | Paraguay 610 | 4380.0 |
| Herpsilochmus sticturus | M | 1 | 11.0 | | | | Surinam 246 | 4385.0 |
| Herpsilochmus axillaris | F | 1 | 11.0 | | | | Peru 450 | 4391.0 |

## Body Masses of World Birds (continued)

| Species | Sex | N | Mean | Std dev | Range | Sn | Location | Number |
|---|---|---|---|---|---|---|---|---|
| Herpsilochmus rufimarginatus | M<br>F | 2<br>1 | 11.2<br>10.0 | | 11.0–11.5 | | Panama; Brazil 497, 610a | 4392.0 |
| Microrhopias quixensis | B | 11 | 7.9 | 0.65 | 7.5–9.7 | | Belize 510 | 4393.0 |
| Formicivora grisea intermedia | B | 12 | 9.3 | | 9.0–10.5 | | Trinidad 188 | 4395.0 |
| Formicivora grisea tobagensis | B | 7 | 10.9 | | 9.5–14.0 | | Tobago 188 | 4395.0 |
| Drymophila ferruginea | U | 4 | 10.2 | | 9.4–11.0 | | Brazil 441, 610a | 4399.0 |
| Drymophila devillei | M | 1 | 10.3 | | | | Peru 668 | 4403.0 |
| Drymophila caudata | B | 5 | 11.9 | | | | Peru 668 | 4404.0 |
| Drymophila malura | B | 4 | 12.2 | | 11.0–13.0 | | Brazil 38, 441 | 4405.0 |
| Drymophila squamata | M | 2 | 10.8 | | 10.5–11.0 | | Brazil 622 | 4406.0 |
| Terenura maculata | M | 1 | 6.5 | | | | Brazil 621 | 4407.0 |
| Terenura callinota | F | 1 | 7.0 | | | | Panama 497 | 4409.0 |
| Terenura humeralis | U | | 13.0 | | | | Peru 623 | 4410.0 |
| Terenura spodioptila | B | 3 | 6.9 | | 6.4–7.3 | | Brazil 610a | 4412.0 |
| Cercomacra cinerascens | U | | 20.0 | | | | Peru 623 | 4413.0 |
| Cercomacra tyrannina | B | 17 | 16.3 | | | | Panama 244 | 4415.0 |
| Cercomacra nigrescens | B | 12 | 16.5 | 1.93 | 14.0–21.0 | | Brazil 226 | 4416.0 |
| Cercomacra serva | U | | 17.0 | | | | Peru 623 | 4418.0 |
| Cercomacra nigricans | M<br>F | 7<br>4 | 17.6<br>15.6 | | | | Panama 244 | 4419.0 |
| Cercomacra carbonaria | F | 1 | 14.5 | | | | Brazil 610a | 4420.0 |
| Cercomacra melanaria | U | 2 | 19.0 | | 18.5–19.5 | | Paraguay 610 | 4421.0 |
| Cercomacra manu | M<br>F | 12<br>9 | 19.7<br>16.6 | | | | Peru 192 | 4422.0 |
| Pyriglena leuconota | U | 78 | 32.3 | | 26.2–37.5 | | Brazil 437 | 4423.0 |

## Body Masses of World Birds (continued)

| Species | Sex | N | Mean | Std dev | Range | Sn | Location | Number |
|---|---|---|---|---|---|---|---|---|
| Pyriglena leucoptera | U | 21 | 28.8 | | 25.5–35.6 | | Brazil 441 | 4424.0 |
| Pyriglena atra | M | 1 | 32.0 | | | | Brazil 621 | 4425.0 |
| Rhopornis ardesiaca | B | 6 | 26.3 | | 23.5–28.0 | | Brazil 619 | 4426.0 |
| Myrmoborus leucophrys | M | 11 | 22.0 | 1.30 | | | Peru 668 | 4427.0 |
| | F | 12 | 20.5 | 1.40 | | | | |
| Myrmoborus myotherinus | U | | 20.0 | | | | Peru 623 | 4429.0 |
| Hypocnemis cantator | U | 10 | 11.6 | 0.70 | | | Fr. Guiana 624 | 4431.0 |
| Hypocnemoides melanopogon | U | 3 | 14.7 | | 14.0–16.0 | | Surinam 246 | 4433.0 |
| Hypocnemoides maculicauda | U | | 13.0 | | | | Peru 623 | 4434.0 |
| Gymnocichla nudiceps | B | 13 | 30.6 | 2.70 | 27.0–35.0 | | 86, 497, 510, 611 | 4436.0 |
| Sclateria naevia | B | 5 | 21.6 | | 20.0–23.0 | | 245, 311, 437, 576 | 4437.0 |
| Percnostola rufifrons | U | 13 | 27.1 | | | | 161, 245, 246, 668 | 4438.0 |
| Percnostola lophotes | U | | 28.0 | | | | Peru 623 | 4443.0 |
| Myrmeciza nigricauda | M | 1 | 24.0 | | | | Columbia 260 | 4445.0 |
| Myrmeciza longipes | U | 10 | 27.9 | | | | Panama 314 | 4446.0 |
| Myrmeciza exsul | B | 12 | 26.5 | 2.11 | 22.5–30.0 | | Panama 59, 87, 436, 611 | 4447.0 |
| Myrmeciza ferruginea | U | 5 | 26.6 | | 25.0–29.0 | | 161, 245, 246 | 4448.0 |
| Myrmeciza squamosa | U | 4 | 18.5 | | 16.5–20.0 | | Brazil 441 | 4451.0 |
| Myrmeciza laemosticta | U | | 25.0 | | | | 603a | 4452.0 |
| Myrmeciza hemimelaena | B | 18 | 16.0 | | | | Peru 668 | 4455.0 |
| Myrmeciza hyperythra | U | | 41.0 | | | | Peru 623 | 4456.0 |
| Myrmeciza goeldii | U | | 42.0 | | | | Peru 623 | 4458.0 |
| Myrmeciza fortis | U | | 46.0 | | | | Peru 623 | 4459.0 |

## Body Masses of World Birds (continued)

| Species | Sex | N | Mean | Std dev | Range | Sn | Location | Number |
|---|---|---|---|---|---|---|---|---|
| Myrmeciza immaculata | U | 2 | 48.8 | | 42.0–55.5 | | Panama 315, 497 | 4460.0 |
| Myrmeciza atrothorax | U | 4 | 15.3 | | 12.0–18.0 | | Surinam 161, 245 | 4462.0 |
| Pithys albifrons | U | 19 | 20.8 | 1.17 | | | Peru 668 | 4464.0 |
| Gymnopithys rufigula | U | 834 | 29.1 | 2.08 | | | Brazil 45 | 4466.0 |
| Gymnopithys leucaspis | U | 32 | 31.1 | 1.90 | 28.0–37.0 | | Panama 50 | 4468.0 |
| Gymnopithys salvini | U | | 25.0 | | | | Peru 623 | 4470.0 |
| Myrmornis torquata | U | 6 | 48.0 | | | | Panama 314 | 4471.0 |
| Rhegmatorhina melanosticta | U | | 32.0 | | | | Peru 623 | 4472.0 |
| Rhegmatorhina hoffmannsi | B | 7 | 31.4 | | 27.0–34.0 | | Brazil 610a | 4474.0 |
| Hylophylax naevioides | U | 34 | 17.8 | | 15.8–20.0 | | Panama 436 | 4477.0 |
| Hylophylax naevia | B | 24 | 14.2 | | | | Peru 668 | 4478.0 |
| Hylophylax punctulata | B | 5 | 12.4 | | 12.0 13.0 | | Bolivia 22 | 4479.0 |
| Hylophylax poecilonota | U | 24 | 16.7 | 1.10 | | | Fr. Guiana 624 | 4480.0 |
| Phlegopsis nigromaculata | U | 45 | 44.5 | | 38.0–53.0 | | Brazil 437 | 4481.0 |
| Phaenostictus mcleannani | U | 12 | 51.1 | 2.70 | | | 315 | 4485.0 |

### ORDER: PASSERIFORMES  FAMILY: FORMICARIIDAE

| Species | Sex | N | Mean | Std dev | Range | Sn | Location | Number |
|---|---|---|---|---|---|---|---|---|
| Formicarius colma | U | 7 | 46.4 | | 42.0–49.6 | | 37, 161, 441 | 4486.0 |
| Formicarius analis | B | 16 | 62.2 | 3.54 | 56.0–67.8 | | Belize 510 | 4487.0 |
| Formicarius nigricapillus | B | 2 | 60.2 | | 59.0–61.5 | | 589 | 4489.0 |
| Formicarius rufipectus | F | 2 | 78.5 | | 75.0–82.0 | | Panama 497 | 4490.0 |
| Chamaeza campanisona | U | 4 | 97.2 | | 86.0–107.0 | | Brazil 441 | 4491.0 |
| Chamaeza nobilis | U | | 123.0 | | | | Peru 623 | 4492.0 |

## Body Masses of World Birds (continued)

| Species | Sex | N | Mean | Std dev | Range | Sn | Location | Number |
|---|---|---|---|---|---|---|---|---|
| Chamaeza ruficauda | M | 2 | 72.8 | | 70.0–75.5 | | Brazil 38 | 4493.0 |
| Pittasoma michleri | U | 1 | 109.0 | | | | Panama 314 | 4495.0 |
| Grallaria varia | U | | 125.0 | | | | Brazil 564 | 4500.0 |
| Grallaria guatimalensis | B | 6 | 94.1 | | 77.0–116.0 | | 315, 387, 584 | 4501.0 |
| Grallaria eludens | M | 1 | 115.0 | | | | Peru 352 | 4506.0 |
| Grallaria ruficapilla | U | 13 | 71.9 | | | | Columbia; Peru 387, 672 | 4508.0 |
| Grallaria nuchalis | B | 4 | 117.0 | | 110.0–122.0 | | Peru 451 | 4512.0 |
| Grallaria hypoleuca | F | 1 | 82.0 | | | | Peru 451 | 4516.0 |
| Grallaria rufula | M | 1 | 38.0 | | | | Peru 668 | 4521.0 |
| Grallaria quitensis | B | 25 | 65.7 | | 57.0–83.0 | | Peru 451 | 4524.0 |
| Hylopezus perspicillatus | U | 11 | 43.0 | | | | 87, 314, 315 | 4526.0 |
| Hylopezus malcularius | M | 2 | 40.5 | | 38.0–43.0 | | Fr. Guiana 161 | 4527.0 |
| Hylopezus fulviventris | M | 2 | 41.0 | | 38.0–44.0 | | Panama 497 | 4528.0 |
| Hylopezus berlepschi | U | | 48.0 | | | | Peru 623 | 4529.0 |
| Hylopezus ochroleucus | U | | 28.0 | | | | Brazil 564 | 4530.0 |
| Myrmothera campanisona | U | | 47.0 | | | | Peru 623 | 4532.0 |
| Grallaricula flavirostris | U | 9 | 17.1 | | | | Peru 668 | 4534.0 |
| Grallaricula ferrugineipectus | B | 43 | 17.2 | | 13.0–21.0 | | Peru 451 | 4535.0 |
| Grallaricula nana | B | 8 | 19.6 | | 17.5–21.5 | | Peru 451 | 4536.0 |
| Grallaricula peruviana | M | 2 | 17.3 | | 17.0–17.5 | | Peru 451 | 4538.0 |
| | F | 3 | 20.0 | | 19.5–21.0 | | 451 | |

### ORDER: PASSERIFORMES          FAMILY: CONOPOPHAGIDAE

| Species | Sex | N | Mean | Std dev | Range | Sn | Location | Number |
|---|---|---|---|---|---|---|---|---|
| Conopophaga lineata | B | 11 | 22.1 | | 20.0–25.0 | | Brazil 38, 135, 441 | 4542.0 |

## Body Masses of World Birds (continued)

| Species | Sex | N | Mean | Std dev | Range | Sn | Location | Number |
|---|---|---|---|---|---|---|---|---|
| Conopophaga aurita | B | 2 | 12.6 | | 12.0–13.2 | B | Fr. Guiana 161 | 4543.0 |
| Conopophaga roberti | U | 1 | 20.8 | | | | Brazil 437 | 4544.0 |
| Conopophaga peruviana | U | | 23.0 | | | | Peru 623 | 4545.0 |
| Conopophaga ardesiaca | B | 57 | 26.3 | | | | Peru 193 | 4546.0 |
| Conopophaga castaneiceps | B | 7 | 27.6 | | | | Peru 668 | 4547.0 |
| Conopophaga melanops | U | 3 | 20.1 | | 19.2–21.5 | | Brazil 441 | 4548.0 |
| Conopophaga melanogaster | M | 2 | 42.5 | | 42.0–43.0 | | Brazil 226, 610a | 4549.0 |

### ORDER: PASSERIFORMES    FAMILY: RHINOCRYPTIDAE

| Species | Sex | N | Mean | Std dev | Range | Sn | Location | Number |
|---|---|---|---|---|---|---|---|---|
| Pteroptochos tarnii | U | | 163.0 | | | | Chile 52 | 4550.0 |
| Pteroptochos megapodius | U | | 119.0 | | | | Chile 292 | 4551.0 |
| Scelorchilus albicollis | M | 6 | 54.8 | | 52.0–60.0 | | Chile 363a | 4552.0 |
| | F | 2 | 46.5 | | 45.0–48.0 | | | |
| Scelorchilus albicollis atacamae | B | 2 | 41.5 | | 39.0–44.0 | | Chile 363a | 4552.0 |
| Scelorchilus rubecula | U | | 76.0 | | | | Chile 52 | 4553.0 |
| Rhinocrypta lanceolata | B | 4 | 63.6 | | 56.4–72.3 | | Argentina 411 | 4554.0 |
| Liosceles thoracicus | U | | 81.0 | | | | Peru 623 | 4556.0 |
| Melanopareia elegans | F | 1 | 16.0 | | | | Peru 672 | 4559.0 |
| Eugralla paradoxa | U | | 44.0 | | | | Chile 52 | 4563.0 |
| Myornis senilis | M | 1 | 23.6 | | | | Peru 451 | 4564.0 |
| | F | 6 | 20.7 | | 18.1–24.5 | | | |
| Scytalopus unicolor | U | 4 | 17.6 | | | | Peru 668 | 4565.0 |
| Scytalopus femoralis | U | 8 | 22.6 | | | | Peru 668 | 4567.0 |
| Scytalopus argentifrons | U | | 17.0 | | | | 603a | 4570.0 |
| Scytalopus latebricola | B | 3 | 22.7 | | 21.0–24.0 | | Peru 451 | 4571.0 |

## Body Masses of World Birds (continued)

| Species | Sex | N | Mean | Std dev | Range | Sn | Location | Number |
|---|---|---|---|---|---|---|---|---|
| Scytalopus magellanicus | U | 5 | | | 10.4–13.9 | | Argentina 559 | 4572.0 |
| Scytalopus speluncae | M | 1 | 13.0 | | | | Brazil 38 | 4574.0 |
| Scytalopus novacapitalis | U | | 19.2 | | | | Brazil 564 | 4575.0 |
| Acropternis orthonyx | B | 5 | 91.0 | | 90.0–100.0 | | Peru 451 | 4577.0 |

### ORDER: PASSERIFORMES
### FAMILY: COTINGIDAE

| Species | Sex | N | Mean | Std dev | Range | Sn | Location | Number |
|---|---|---|---|---|---|---|---|---|
| Phoenicircus carnifex | M | 1 | 70.0 | | | | 575 | 4579.0 |
| Laniisoma elegans | B | 5 | 47.1 | | 45.8–51.0 | | 574, 575, 668 | 4580.0 |
| Phibalura flavirostris | B | 3 | 46.5 | | 43.5–52.0 | | 38, 575 | 4581.0 |
| Tijuca condita | F | 1 | 80.0 | | | | 575 | 4583.0 |
| Carpornis cucullatus | B | 9 | 75.7 | | 67.0–84.4 | | 38, 575 | 4584.0 |
| Carpornis melanocephalus | B | 3 | 64.2 | | 62.7–66.0 | | 575, 622 | 4585.0 |
| Ampelion rubrocristata | B | 21 | 66.3 | | 45.0–80.0 | | 575 | 4586.0 |
| Ampelion rufaxilla | B | 7 | 73.9 | | 69.0–77.0 | | 575 | 4587.0 |
| Ampelion sclateri | B | 8 | 60.5 | | 54.0–69.0 | | 575 | 4588.0 |
| Ampelion stresemanni | B | 16 | 52.2 | | 46.0–57.0 | | 448, 575 | 4589.0 |
| Pipreola riefferii | B | 8 | 50.4 | | 49.0–51.6 | | 575 | 4590.0 |
| Pipreola intermedia intermedia | B | 7 | 47.9 | | 44.0–58.0 | | 575 | 4591.0 |
| Pipreola intermedia signata | M | 3 | 54.7 | | 50.8–58.9 | | 575 | 4591.0 |
| Pipreola arcuata | B | 6 | 120.0 | | 112.0–128.0 | | 575 | 4592.0 |
| Pipreola aureopectus | B | 2 | 46.0 | | 46.0–46.0 | | 575 | 4593.0 |
| Pipreola pulchra | B | 6 | 58.2 | | 51.2–63.0 | | 575, 668 | 4596.0 |
| Pipreola chlorolepidota | B | 2 | 29.5 | | 28.0–31.0 | | 575 | 4597.0 |

## Body Masses of World Birds (continued)

| Species | Sex | N | Mean | Std dev | Range | Sn | Location | Number |
|---|---|---|---|---|---|---|---|---|
| Pipreola frontalis | B | 7 | 42.4 | | 39.5–45.3 | | 575 | 4598.0 |
| Pipreola formosa | B | 5 | 45.0 | | 41.0–49.0 | | 575 | 4599.0 |
| Ampelioides tschudii | B | 7 | 80.8 | | 71.5–95.0 | | 575 | 4601.0 |
| Iodopleura isabellae | B | 3 | 20.0 | | 19.8–20.2 | | Peru 193 | 4603.0 |
| Iodopleura fusca | M | 1 | 15.3 | | | | 575 | 4604.0 |
| Lipaugus subalaris | B | 3 | 83.4 | | 81.8–86.3 | | 575 | 4606.0 |
| Lipaugus fuscocinereus | F | 1 | 138.0 | | | | Peru 451 | 4608.0 |
| Lipaugus uropygialis | M | 1 | 116.0 | | | | 575 | 4609.0 |
| Lipaugus vociferans | B | 22 | 82.2 | | 77.0–87.0 | | Surinam 575 | 4610.0 |
| Lipaugus vociferans | B | 10 | 72.2 | | 68.0–74.0 | | Peru 575 | 4610.0 |
| Lipaugus unirufus | B | 9 | 82.1 | | 69.0–87.2 | | 575 | 4611.0 |
| Lipaugus lanioides | B | 12 | 94.8 | | 85.0–110.0 | | 575 | 4612.0 |
| Porphyrolaema porphyrolaema | U | | 60.0 | | | | Peru 623 | 4614.0 |
| Cotinga amabilis | B | 5 | 71.5 | | 66.0–75.0 | | 575 | 4615.0 |
| Cotinga ridgwayi | M | 16 | 53.6 | 3.60 | | | Panama 244 | 4616.0 |
| | F | 6 | 61.0 | | | | | |
| Cotinga maynana | U | | 69.0 | | | | Peru 623 | 4618.0 |
| Cotinga cotinga | B | 2 | 54.0 | | 53.0–55.0 | | 575 | 4619.0 |
| Cotinga maculata | M | 1 | 65.0 | | | | 575 | 4620.0 |
| Cotinga cayana | B | 13 | 64.3 | | 56.0–72.5 | | 575 | 4621.0 |
| Xipholena punicea | B | 15 | 68.1 | | 58.0–76.0 | | 575 | 4622.0 |
| Carpodectes nitidus | M | 2 | 116.0 | | | | | 4625.0 |
| | F | 1 | 89.0 | | | | 269a | |
| Carpodectes antoniae | U | | 98.0 | | | | 603a | 4626.0 |

## Body Masses of World Birds (continued)

| Species | Sex | N | Mean | Std dev | Range | Sn | Location | Number |
|---|---|---|---|---|---|---|---|---|
| Conioptilon mcilhennyi | U | | 90.0 | | | | Peru 623 | 4628.0 |
| Gymnoderus foetidus | U | | 275.0 | | | | Peru 623 | 4629.0 |
| Querula purpurata | B | 34 | 106.0 | | 91.0–133.0 | | Peru, Guianas 575 | 4631.0 |
| Pyroderus scutatus scutatus | B | 8 | 357.0 | | 293.0–419.0 | | 575 | 4632.0 |
| Cephalopterus glabricollis | M | | 450.0 | | | | | 4633.0 |
| | F | | 320.0 | | | | 603a | |
| Cephalopterus penduliger | M | 1 | 338.0 | | | | Ecuador 589 | 4634.0 |
| Cephalopterus ornatus | F | 1 | 380.0 | | | | 575 | 4635.0 |
| Perissocephalus tricolor | M | 5 | 360.0 | | 320.0–395.0 | | | 4636.0 |
| | F | 6 | 319.0 | | 267.0–367.0 | | 575 | |
| Procnias tricarunculata | M | 6 | 210.0 | | 194.0–233.0 | | 575 | 4637.0 |
| Procnias alba | M | 1 | 210.0 | | | | 575 | 4638.0 |
| Procnias averano | M | 1 | 178.0 | | | | 575 | 4639.0 |
| | F | 3 | 131.0 | | 127.0–135.0 | | 575 | |
| Procnias nudicollis | M | 4 | 200.0 | | 177.0–225.0 | | 575 | 4640.0 |
| | F | 5 | 148.0 | | 140.0–150.0 | | 575 | |
| Rupicola rupicola | F | 1 | 140.0 | | | | 575 | 4641.0 |
| Rupicola peruviana sanguinolenta | M | 1 | 266.0 | | | | 575 | 4642.0 |
| | F | 3 | 221.0 | | 213.0–226.0 | | 575 | |

ORDER: PASSERIFORMES     FAMILY: OXYRUNCIDAE

| Species | Sex | N | Mean | Std dev | Range | Sn | Location | Number |
|---|---|---|---|---|---|---|---|---|
| Oxyruncus cristatus | B | 6 | 42.0 | | 37.0–46.0 | | 138, 497, 564 | 4643.0 |

ORDER: PASSERIFORMES     FAMILY: PHYTOTOMIDAE

| Species | Sex | N | Mean | Std dev | Range | Sn | Location | Number |
|---|---|---|---|---|---|---|---|---|
| Phytotoma raimondii | B | 4 | 39.8 | | 36.0–44.0 | | Peru 584 | 4644.0 |
| Phytotoma rutila | U | | 32.0 | | | | Brazil 564 | 4645.0 |
| Phytotoma rara | U | | 40.0 | | | | Chile 52 | 4646.0 |

ORDER: PASSERIFORMES     FAMILY: PIPRIDAE

| Species | Sex | N | Mean | Std dev | Range | Sn | Location | Number |
|---|---|---|---|---|---|---|---|---|
| Pipra aureola | U | 5 | 15.7 | | 14.0–18.0 | | 245, 246, 455 | 4647.0 |

## Body Masses of World Birds (continued)

| Species | Sex | N | Mean | Std dev | Range | Sn | Location | Number |
|---|---|---|---|---|---|---|---|---|
| Pipra fasciicauda | B | 130 | 15.9 | | | | Peru 668 | 4648.0 |
| Pipra filicauda | B | 6 | 13.9 | | | | 455 | 4649.0 |
| Pipra mentalis | U | 24 | 15.0 | | 12.5–17.2 | | Panama 436 | 4650.0 |
| Pipra erythrocephala | M | 291 | 12.8 | 0.97 | 10.5–17.0 | | Trinidad 576 | 4651.0 |
| | F | 214 | 14.1 | 0.97 | 12.0–16.5 | | | |
| Pipra rubrocapilla | U | 17 | 12.0 | | 10.2–13.0 | | Brazil 437 | 4652.0 |
| Pipra chloromeros | B | 16 | 16.5 | | | | 455 | 4653.0 |
| Pipra pipra | M | 11 | 10.3 | | | | 455 | 4655.0 |
| | F | 10 | 12.0 | | | | | |
| Pipra coronata | B | 133 | 8.5 | | | | 455 | 4656.0 |
| Pipra serena | B | 10 | 11.4 | | 10.8–12.0 | | Fr. Guiana 161 | 4657.0 |
| Pipra nattereri | M | 2 | 8.0 | | 7.8–8.1 | | Brazil 610a | 4660.0 |
| | F | 8 | 8.8 | | 8.1–9.4 | | | |
| Pipra coeruleocapilla | M | 25 | 9.1 | | | | Peru 668 | 4662.0 |
| Chiroxiphia linearis | M | 41 | 16.8 | | | | 455 | 4664.0 |
| | F | 48 | 19.1 | | | | | |
| Chiroxiphia lanceolata | U | 27 | 18.5 | 1.40 | | | Panama 315 | 4665.0 |
| Chiroxiphia pareola | B | 16 | 16.5 | | | | 455 | 4666.0 |
| Chiroxiphia boliviana | M | 9 | 16.4 | | | | Peru 193 | 4667.0 |
| | F | 20 | 17.9 | 1.10 | | | | |
| Chiroxiphia caudata | U | 43 | 25.6 | | 20.5–30.0 | | Brazil 441 | 4668.0 |
| Masius chrysopterus | M | 8 | 10.2 | | 9.0–12.5 | | Ecuador 470 | 4669.0 |
| | F | 1 | 13.5 | | | | | |
| Ilicura militaris | U | 8 | 14.6 | | 12.5–19.5 | | Brazil 440, 441 | 4670.0 |
| Corapipo gutturalis | U | 65 | 8.2 | 1.02 | | | Brazil 45 | 4671.0 |
| Corapipo leucorrhoa | M | 10 | 10.9 | 0.41 | 10.5–11.7 | | Panama 50 | 4673.0 |
| | F | 10 | 13.9 | 1.20 | 11.8–15.7 | | | |
| Manacus candei | M | 21 | 18.8 | 3.10 | | | Panama 315 | 4674.0 |
| | F | 8 | 20.9 | 1.00 | | | | |
| Manacus aurantiacus | U | | 15.5 | | | | 603a | 4675.0 |

## Body Masses of World Birds (continued)

| Species | Sex | N | Mean | Std dev | Range | Sn | Location | Number |
|---------|-----|---|------|---------|-------|----|----------|--------|
| Manacus vitellinus | M | 17 | 19.3 | 0.21 | | | Panama | 4676.0 |
| | F | 16 | 17.1 | 0.20 | | | 244 | |
| Manacus manacus trinitatis | M | 191 | 18.5 | 1.30 | 16.0–23.0 | | Trinidad | 4677.0 |
| | F | 344 | 16.8 | 1.30 | 14.0–21.5 | | 576 | |
| Manacus manacus | U | 44 | 15.0 | | 12.2–18.1 | | Brazil | 4677.0 |
| | | | | | | | 437 | |
| Machaeropterus pyrocephalus | B | 14 | 9.8 | | | | 455, 668 | 4678.0 |
| Machaeropterus regulus | B | 28 | 9.4 | | | | 455 | 4679.0 |
| Machaeropterus deliciosus | F | 1 | 12.0 | | | | 455 | 4680.0 |
| Xenopipo atronitens | M | 2 | 15.5 | | | | 455 | 4681.0 |
| Chloropipo unicolor | U | 27 | 15.5 | 0.91 | | | Peru | 4682.0 |
| | | | | | | | 668 | |
| Chloropipo holochlora | B | 10 | 15.4 | | 11.0–18.7 | | Ecuador; Panama | 4684.0 |
| | | | | | | | 50, 589 | |
| Chloropipo flavicapilla | F | 5 | 17.8 | | 16.8–19.5 | | Columbia | 4685.0 |
| | | | | | | | 387 | |
| Neopipo cinnamomea | U | 1 | 7.0 | | | | Brazil | 4686.0 |
| | | | | | | | 45 | |
| Heterocercus flavivertex | U | 3 | 21.0 | | | | Brazil | 4687.0 |
| | | | | | | | 439 | |
| Heterocercus linteatus | U | | 24.0 | | | | Brazil | 4689.0 |
| | | | | | | | 564 | |
| Neopelma chrysocephalum | U | 3 | 15.9 | | 15.7–16.0 | | Surinam | 4690.0 |
| | | | | | | | 246, 610a | |
| Neopelma sulphureiventer | M | 2 | 18.0 | | | | Bolivia | 4691.0 |
| | F | 2 | 13.7 | | | | 22 | |
| Neopelma aurifrons | U | 5 | 14.0 | | 13.5–14.9 | | Brazil | 4693.0 |
| | | | | | | | 441, 610a | |
| Tyranneutes stolzmanni | U | | 9.0 | | | | Peru | 4694.0 |
| | | | | | | | 623 | |
| Tyranneutes virescens | U | | 7.0 | | | | Brazil | 4695.0 |
| | | | | | | | 564 | |
| Piprites pileatus | M | 1 | 15.0 | | | | Brazil | 4696.0 |
| | | | | | | | 38 | |
| Piprites griseiceps | U | 4 | 16.0 | | | | Costa Rica | 4697.0 |
| | | | | | | | 50a | |
| Piprites chloris | U | | 20.0 | | | | Peru | 4698.0 |
| | | | | | | | 623 | |
| Sapayoa aenigma | U | 9 | 20.8 | | | | Panama | 4699.0 |
| | | | | | | | 50, 314, 497 | |

## Body Masses of World Birds (continued)

| Species | Sex | N | Mean | Std dev | Range | Sn | Location | Number |
|---|---|---|---|---|---|---|---|---|
| ORDER: PASSERIFORMES | | | | | FAMILY: TYRANNIDAE | | | |
| Mionectes striaticollis | U | 89 | 15.0 | 1.69 | | | Peru 668 | 4700.0 |
| Mionectes olivaceus | B | 20 | 16.4 | 1.22 | 14.4–18.5 | | Panama 50 | 4701.0 |
| Mionectes oleagineus | B | 261 | 12.1 | | 9.0–14.5 | | Trinidad 576 | 4702.0 |
| Mionectes oleagineus | U | 47 | 9.6 | 0.67 | 7.5–10.8 | | Peru 93 | 4702.0 |
| Mionectes macconnelli | U | 39 | 13.5 | 1.32 | 10.0–16.0 | | Bolivia 93 | 4703.0 |
| Mionectes rufiventris | U | 14 | 13.3 | | 7.5–16.8 | | Brazil 38, 441 | 4704.0 |
| Leptopogon rufipectus | B | 6 | 13.3 | | 10.0–18.0 | | Peru 451 | 4705.0 |
| Leptopogon amaurocephalus | U | 27 | 11.7 | 0.87 | | | Peru 668 | 4707.0 |
| Leptopogon superciliaris | B | 30 | 11.7 | | 9.5–13.0 | | Trinidad 188 | 4708.0 |
| Pseudotriccus pelzelni | U | 22 | 10.9 | 0.82 | | | Peru 668 | 4709.0 |
| Pseudotriccus simplex | B | 8 | 9.6 | | | | Peru 193 | 4710.0 |
| Pseudotriccus ruficeps | U | 3 | 12.0 | | | | Peru 668 | 4711.0 |
| Poecilotriccus capitale | U | 2 | 8.0 | | 8.0–8.0 | | Peru 450 | 4713.0 |
| Taeniotriccus andrei | U | 1 | 9.4 | | | | Brazil 437 | 4715.0 |
| Hemitriccus minor | U | 2 | 6.0 | | | | Brazil 439 | 4716.0 |
| Hemitriccus flammulatus | B | 7 | 10.0 | | 8.8–11.7 | | Peru; Bolivia 22, 193 | 4718.0 |
| Hemitriccus diops | U | 9 | 9.9 | | 8.0–11.5 | | Brazil 38, 441 | 4719.0 |
| Hemitriccus obsoletus | B | 2 | 10.0 | | 9.0–11.0 | | Brazil 38 | 4720.0 |
| Hemitriccus zosterops | U | | 9.0 | | | | Peru 623 | 4721.0 |
| Hemitriccus orbitatus | B | 3 | 9.7 | | 9.0–10.5 | | Brazil 38 | 4723.0 |
| Hemitriccus iohannis | U | | 10.0 | | | | Peru 623 | 4724.0 |

## Body Masses of World Birds (continued)

| Species | Sex | N | Mean | Std dev | Range | Sn | Location | Number |
|---|---|---|---|---|---|---|---|---|
| Hemitriccus nidipendulus | B | 2 | 7.5 | | 7.0–8.0 | | Brazil 603a | 4726.0 |
| Hemitriccus margaritaceiventer | B | 24 | 8.4 | 0.97 | 6.6–11.5 | | Paraguay 610 | 4728.0 |
| Hemitriccus granadensis | U | 9 | 7.9 | | | | Peru 668 | 4730.0 |
| Hemitriccus rufigularis | U | 2 | 9.2 | | 9.0–9.5 | | Peru 138 | 4731.0 |
| Hemitriccus cinnamomeipectus | U | 7 | 7.5 | | 6.5–8.5 | | Peru 191 | 4732.0 |
| Todirostrum plumbeiceps | B | 7 | 5.7 | | 5.4–6.0 | | Brazil 38, 136, 441 | 4738.0 |
| Todirostrum latirostre | U | 22 | 8.1 | 0.71 | | | Peru 668 | 4739.0 |
| Todirostrum sylvia | U | 17 | 7.1 | | | | 244, 314, 315, 384, 510, 625 | 4741.0 |
| Todirostrum maculatum | U | 4 | 7.0 | | 6.4–7.6 | | 246, 483 | 4742.0 |
| Todirostrum cinereum | B | 23 | 6.4 | | | | Panama 244 | 4744.0 |
| Todirostrum nigriceps | M | 1 | 6.5 | | | | Panama 497 | 4746.0 |
| Todirostrum pictum | M | 1 | 6.9 | | | | Brazil 610a | 4747.0 |
| Todirostrum chrysocrotaphum | U | | 7.0 | | | | Peru 623 | 4748.0 |
| Todirostrum calopterum | B | 2 | 8.0 | | 7.2–8.7 | | Peru 193 | 4749.0 |
| Corythopis torquata | U | 5 | 14.7 | | 14.0–15.0 | | 161, 437, 610a | 4751.0 |
| Corythopis delalandi | B | 3 | 14.9 | | 13.0–16.0 | | 38, 136 | 4752.0 |
| Phyllomyias fasciatus | B | 3 | 10.5 | | 10.0–11.0 | | Brazil 38 | 4753.0 |
| Phyllomyias zeledoni | U | 1 | 10.3 | | | | Panama 244 | 4754.0 |
| Phyllomyias burmeisteri | B | 3 | 10.8 | | 9.8–11.5 | | Brazil; Peru 38, 138 | 4755.0 |
| Phyllomyias virescens | M | 3 | | | 10.0–12.0 | | Brazil 38 | 4756.0 |
| Phyllomyias griseiceps | B | 2 | 7.2 | | 7.0–7.4 | | Ecuador 589, 668 | 4760.0 |
| Phyllomyias uropygialis | F | 1 | 8.7 | | | | Peru 668 | 4764.0 |

## Body Masses of World Birds (continued)

| Species | Sex | N | Mean | Std dev | Range | Sn | Location | Number |
|---------|-----|---|------|---------|-------|-----|----------|--------|
| Zimmerius vilissimus | U | 15 | 9.7 | | | | Panama 337 | 4765.0 |
| Zimmerius bolivianus | B | 2 | 12.0 | | 10.0–11.9 | | Peru 668 | 4767.0 |
| Zimmerius cinereicapillus | B | 6 | 11.8 | | | | Peru 193 | 4768.0 |
| Zimmerius gracilipes | M | 1 | 8.1 | | | | Brazil 610a | 4769.0 |
| Zimmerius viridiflavus | M | 1 | 10.0 | | | | Peru 672 | 4770.0 |
| Zimmerius chrysops | B | 4 | 10.5 | | 9.5–12.0 | | Columbia 387 | 4771.0 |
| Ornithion inerme | U | | 7.0 | | | | Peru 623 | 4772.0 |
| Ornithion semiflavum | B | 5 | 7.0 | | 6.5–7.4 | | Mexico 584 | 4773.0 |
| Ornithion brunneicapillum | U | 3 | 7.1 | | 6.9–7.2 | | Panama 611 | 4774.0 |
| Camptostoma imberbe | B | 13 | 7.4 | 0.80 | 6.3–9.5 | B | Mexico 584 | 4775.0 |
| Camptostoma obsoletum | U | 18 | 8.0 | 1.18 | 6.0–10.0 | | 246, 311, 387, 611 | 4776.0 |
| Phaeomyias murina | U | 37 | 10.0 | | 8.0–12.0 | | Trinidad 188 | 4777.0 |
| Nesotriccus ridgwayi | U | | 11.0 | | | | 603a | 4778.0 |
| Capsiempis flaveola | B | 12 | 7.7 | | | | 244, 314, 611, 625 | 4779.0 |
| Sublegatus modestus | U | 23 | 12.3 | | 10.5–15.0 | | Trinidad 188 | 4781.0 |
| Suiriri suiriri | B | 6 | | | 14.0–16.5 | | Brazil 38 | 4783.0 |
| Tyrannulus elatus | U | 7 | 7.8 | | 7.2–8.6 | | 244, 246, 611, 668 | 4784.0 |
| Myiopagis gaimardii | B | 13 | 12.6 | | 9.5–14.0 | | Trinidad 188 | 4785.0 |
| Myiopagis caniceps | B | 6 | 10.3 | | 9.5–11.0 | | Peru; Brazil 38, 136, 589 | 4786.0 |
| Myiopagis subplacens | M | 1 | 14.9 | | | | Peru 672 | 4787.0 |
| | F | 4 | 16.9 | | | | | |
| Myiopagis flavivertex | F | 1 | 11.0 | | | | Surinam 246 | 4788.0 |
| Myiopagis cotta | U | | 13.0 | | | | Jamaica 185 | 4789.0 |

## Body Masses of World Birds (continued)

| Species | Sex | N | Mean | Std dev | Range | Sn | Location | Number |
|---|---|---|---|---|---|---|---|---|
| Myiopagis viridicata viridicata | B | 17 | 12.3 | 0.78 | 11.4–13.6 | | Argentina 136 | 4790.0 |
| Myiopagis viridicata minima | B | 11 | 10.1 | 0.39 | 9.6–10.8 | | W. Mexico 593 | 4790.0 |
| Elaenia martinica | B | 23 | 22.6 | | 18.5–28.5 | | Cayman Is. 432 | 4792.0 |
| Elaenia flavogaster | B | 29 | 24.7 | | 21.0–27.5 | | Trinidad 576 | 4793.0 |
| Elaenia spectabilis | B | 6 | 28.2 | | | | 38, 200 | 4794.0 |
| Elaenia albiceps | U | 9 | 15.9 | | | | Peru 668 | 4796.0 |
| Elaenia parvirostris | U | 10 | 17.0 | | | | Paraguay 200 | 4797.0 |
| Elaenia strepera | M | 2 | 16.7 | | 13.3–20.1 | | Peru 193 | 4798.0 |
| Elaenia mesoleuca | B | 17 | | | 15.5–20.0 | | Brazil 38 | 4799.0 |
| Elaenia gigas | U | 1 | 31.0 | | | | Peru 668 | 4800.0 |
| Elaenia cristata | U | 8 | 18.0 | | 14.0–23.0 | | Surinam 245, 246 | 4802.0 |
| Elaenia chiriquensis | U | 13 | 16.1 | | | | 188, 244, 314, 611 | 4804.0 |
| Elaenia frantzii | U | 10 | 19.4 | 1.40 | 18.1–20.8 | | Panama 337 | 4805.0 |
| Elaenia obscura | B | 9 | 22.4 | | 14.7–34.5 | | Columbia; Brazil 38, 387, 441 | 4806.0 |
| Elaenia pallatangae | B | 8 | 14.9 | | | | Peru 193 | 4808.0 |
| Elaenia fallax | U | | 14.0 | | | | 185 | 4809.0 |
| Mecocerculus leucophrys | U | 6 | 13.9 | | | | Peru 668 | 4810.0 |
| Mecocerculus calopterus | F | 3 | 7.0 | | | | Peru 672 | 4813.0 |
| Mecocerculus stictopterus | U | 1 | 10.7 | | | | Ecuador 320 | 4815.0 |
| Serpophaga cinerea | U | 4 | 8.1 | | | | Panama 244 | 4816.0 |
| Serpophaga nigricans | B | 2 | 9.0 | | 9.0–9.0 | | Brazil 38 | 4817.0 |
| Serpophaga subcristata | B | 5 | 7.1 | | 6.5–8.6 | | 38, 111 | 4819.0 |

## Body Masses of World Birds (continued)

| Species | Sex | N | Mean | Std dev | Range | Sn | Location | Number |
|---------|-----|---|------|---------|-------|----|----------|--------|
| Serpophaga munda | U | 2 | 4.6 | | 4.5–4.8 | | Argentina 111 | 4820.0 |
| Inezia subflava | U | 14 | 8.4 | 0.63 | 7.5–10.2 | | Venezuela 625 | 4823.0 |
| Stigmatura budytoides | B | 4 | 11.4 | | 10.9–12.2 | | Argentina 411 | 4825.0 |
| Anairetes agraphia | B | 2 | 9.9 | | 9.9–9.9 | | Peru 668 | 4828.0 |
| Anairetes reguloides | B | 11 | 5.9 | | 5.0–8.0 | | Chile 363a | 4830.0 |
| Anairetes flavirostris | B | 5 | 5.6 | | 5.5–5.8 | | Argentina 411 | 4831.0 |
| Anairetes parulus | B | 11 | 5.9 | | 5.4–6.1 | | Argentina 411 | 4833.0 |
| Tachuris rubrigastra | U | 2 | 7.2 | | 6.5–8.0 | | Brazil 38 | 4834.0 |
| Pseudocolopteryx sclateri | B | 3 | 7.3 | | 6.8–8.0 | | 38, 111 | 4838.0 |
| Pseudocolopteryx flaviventris | U | 2 | 7.5 | | 7.0–8.0 | | Brazil 38 | 4841.0 |
| Euscarthmus meloryphus | U | 10 | 6.8 | | 5.3–8.0 | | 22, 38, 384, 625, 672 | 4842.0 |
| Phylloscartes ophthalmicus | B | 9 | 9.2 | | | | Peru 193 | 4844.0 |
| Phylloscartes orbitalis | M | 6 | 7.6 | | | | Peru 193 | 4847.0 |
| | F | 8 | 6.6 | | | | | |
| Phylloscartes poecilotis | U | 3 | 8.0 | | 7.0–9.5 | | Columbia; Peru 387, 668 | 4848.0 |
| Phylloscartes eximius | U | | 11.4 | | | | Brazil 564 | 4849.0 |
| Phylloscartes oustaleti | B | 2 | 10.0 | | 9.4–10.5 | | Brazil 610a | 4856.0 |
| Phylloscartes difficilis | U | 2 | 7.0 | | 6.0–8.0 | | Brazil 38, 441 | 4857.0 |
| Phylloscartes ventralis | B | 6 | 8.1 | | | | Brazil; Peru 38, 668 | 4859.0 |
| Phylloscartes flavovirens | U | 1 | 8.8 | | | | Panama 611 | 4860.0 |
| Phylloscartes virescens | U | 2 | 9.0 | | 6.9–11.0 | | 101a, 161 | 4861.0 |
| Phylloscartes superciliaris | M | 2 | 8.5 | | 8.0–9.0 | | Ecuador 589 | 4862.0 |
| | F | 1 | 7.0 | | | | | |
| Myiornis auricularis | B | 17 | 5.3 | | | | 136, 668 | 4865.0 |

## Body Masses of World Birds (continued)

| Species | Sex | N | Mean | Std dev | Range | Sn | Location | Number |
|---|---|---|---|---|---|---|---|---|
| Myiornis ecaudatus | U | | 6.0 | | | | Peru 623 | 4867.0 |
| Lophotriccus pileatus | M | 9 | 8.1 | | | | Panama 244 | 4868.0 |
| | F | 7 | 6.9 | | | | | |
| Lophotriccus eulophotes | F | 1 | 6.0 | | | | Peru 549 | 4870.0 |
| Lophotriccus galeatus | U | 5 | 6.9 | | 6.2–7.4 | | Brazil; Surinam 246, 625 | 4871.0 |
| Atalotriccus pilaris | U | 6 | 6.8 | | 5.3–9.1 | | 314, 384, 625 | 4872.0 |
| Oncostoma cinereigulare | B | 15 | 6.1 | | | | Mexico; Belize 457, 510 | 4873.0 |
| Oncostoma olivaceum | U | 12 | 6.6 | | | | Panama 87, 314, 315, 436 | 4874.0 |
| Cnipodectes subbrunneus | U | 9 | 23.2 | | | | Panama 315 | 4875.0 |
| Ramphotrigon megacephala | U | 7 | 13.8 | | 7.4–17.0 | | 441, 449 | 4876.0 |
| Ramphotrigon fuscicauda | U | 12 | 18.6 | | 16.5–20.5 | | Brazil; Peru 449 | 4877.0 |
| Ramphotrigon ruficauda | U | 10 | 17.9 | | 15.0–20.0 | | 161, 246, 449 | 4878.0 |
| Rhynchocyclus brevirostris | B | 10 | 22.8 | | | | Panama 244 | 4879.0 |
| Rhynchocyclus olivaceus | B | 26 | 21.4 | | | | 50,87,246,314, 315,437,497,611 | 4880.0 |
| Rhynchocyclus fulvipectus | U | 3 | 26.2 | | | | Peru 450, 668 | 4881.0 |
| Tolmomyias sulphurescens | B | 28 | 14.9 | 1.50 | 11.6–18.5 | | 244,311,384,457, 510,611,672 | 4882.0 |
| Tolmomyias assimilis | U | | 17.0 | | | | Peru 623 | 4883.0 |
| Tolmomyias poliocephalus | U | 16 | 11.5 | | | | Surinam; Brazil 246, 668 | 4884.0 |
| Tolmomyias flaviventris | B | 24 | 12.2 | | 9.5–14.5 | | Trinidad 188 | 4885.0 |
| Platyrinchus saturatus | U | 5 | 10.6 | | 8.8–15.0 | | Brazil 437, 610a | 4886.0 |
| Platyrinchus mystaceus | B | 23 | 9.7 | | 7.5–11.0 | | Trinidad 188 | 4888.0 |
| Platyrinchus coronatus | U | 13 | 9.4 | 0.60 | | | Costa Rica 315 | 4889.0 |

## Body Masses of World Birds (continued)

| Species | Sex | N | Mean | Std dev | Range | Sn | Location | Number |
|---|---|---|---|---|---|---|---|---|
| Platyrinchus coronatus | U | 14 | 8.9 | 0.40 | | | Panama 315 | 4889.0 |
| Platyrinchus platyrhynchos | U | | 12.0 | | | | Peru 623 | 4891.0 |
| Platyrinchus leucoryphus | F | 1 | 17.0 | | | | Brazil 38 | 4892.0 |
| Onchorhynchus coronatus | U | 13 | 14.0 | | 9.7–16.2 | | Brazil 437 | 4893.0 |
| Myiotriccus ornatus | U | 1 | 13.5 | | | | Peru 668 | 4894.0 |
| Myiophobus flavicans | U | 15 | 12.7 | 0.94 | | | Peru 668 | 4895.0 |
| Myiophobus phoenicomitra | U | 8 | 10.6 | | 9.0–11.5 | | Peru 138, 450 | 4896.0 |
| Myiophobus roraimae | U | 6 | 13.2 | | | | Peru 450, 668 | 4898.0 |
| Myiophobus pulcher | U | 1 | 9.1 | | | | Peru 193 | 4899.0 |
| Myiophobus lintoni | F | 2 | 9.8 | | 9.5–10.0 | | Peru 451 | 4900.0 |
| Myiophobus fasciatus | B | 20 | 9.9 | | 8.5–12.0 | | Trinidad 188 | 4902.0 |
| Myiophobus cryptoxanthus | U | 1 | 11.0 | | | | Peru 450 | 4903.0 |
| Myiobius erythrurus | U | 14 | 7.4 | 0.26 | 6.9–8.0 | | Panama 50 | 4904.0 |
| Myiobius villosus | B | 6 | 13.7 | | 12.0–15.0 | | 589, 668 | 4905.0 |
| Myiobius barbatus | U | 29 | 11.9 | 1.10 | | | 315 | 4906.0 |
| Myiobius atricaudus | U | 13 | 10.0 | 1.10 | | | Panama 315 | 4907.0 |
| Pyrrhomyias cinnamomea | U | 4 | 10.2 | | | | 105, 386, 668 | 4908.0 |
| Hirundinea ferruginea | U | 2 | 24.0 | | 23.0–25.0 | | Brazil 38 | 4909.0 |
| Hirundinea bellicosa | M | 1 | 42.0 | | | | Argentina 111 | 4910.0 |
| Cnemotriccus fuscatus | M | 5 | | | 12.0–14.0 | | Brazil 38 | 4911.0 |
| | F | 4 | | | 12.0–16.0 | | | |
| Lathrotriccus euleri | B | 28 | 12.0 | | 9.5–14.0 | | Trinidad 188 | 4912.0 |
| Aphanotriccus capitalis | U | 6 | 11.8 | | | | Costa Rica 50a | 4913.0 |

## Body Masses of World Birds (continued)

| Species | Sex | N | Mean | Std dev | Range | Sn | Location | Number |
|---------|-----|---|------|---------|-------|----|----------|--------|
| Aphanotriccus audax | B | 4 | | | 10.5–12.0 | | Panama 497 | 4914.0 |
| Xenotriccus callizonus | B | 3 | 11.2 | | 10.5–11.5 | | Chiapas, Mexico 584 | 4915.0 |
| Xenotriccus mexicanus | M | 1 | 13.8 | | | | Oaxaca, Mexico 584 | 4916.0 |
| Mitrephanes phaeocercus | M | 8 | 8.9 | | | | Panama 244 | 4917.0 |
| | F | 8 | 8.3 | | | | | |
| Contopus borealis | B | 26 | 32.1 | 2.27 | 26.7–42.2 | M | Pennsylvania,USA 101 | 4919.0 |
| Contopus pertinax | B | 27 | 27.2 | 3.30 | 21.6–35.8 | | Mexico 585 | 4920.0 |
| Contopus lugubris | U | 23 | 21.5 | | | | Panama 244 | 4921.0 |
| Contopus fumigatus | U | 4 | 20.6 | | 17.8–23.2 | | 105, 387, 672 | 4922.0 |
| Contopus ochraceus | U | | 23.0 | | | | 603a | 4923.0 |
| Contopus sordidulus | B | 15 | 12.8 | 1.45 | 12.4–15.2 | S | California, USA 107 | 4924.0 |
| Contopus virens | B | 135 | 14.1 | 0.91 | 10.4–18.2 | | Pennsylvania,USA 101 | 4925.0 |
| Contopus cinereus | B | 16 | 11.7 | | 8.5–13.5 | | Trinidad 311 | 4926.0 |
| Contopus nigrescens | B | 2 | 9.5 | | 9.0–10.0 | | Peru 138 | 4927.0 |
| Contopus caribaeus | U | 8 | 11.2 | | 9.3–13.0 | | 429, 600 | 4929.0 |
| Contopus latirostris | M | 4 | 11.4 | | 11.0–11.7 | | Puerto Rico 431 | 4930.0 |
| | F | 5 | 9.7 | | 9.2–10.0 | | | |
| Empidonax flaviventris | B | 169 | 11.6 | 0.63 | 9.2–15.5 | M | Pennsylvania,USA 101 | 4932.0 |
| Empidonax virescens | B | 62 | 12.9 | 0.77 | 9.9–16.1 | S | Pennsylvania,USA 101 | 4933.0 |
| Empidonax alnorum | M | 42 | 12.7 | 1.03 | 10.0–14.4 | B | Ontario, Canada 590 | 4934.0 |
| | F | 12 | 13.1 | 1.56 | 10.1–15.6 | | | |
| Empidonax traillii | M | 13 | 13.1 | 1.37 | 12.0–15.7 | B | Connecticut,USA 222a | 4935.0 |
| | F | 11 | 13.7 | 1.46 | 11.3–16.4 | | | |
| Empidonax albigularis | B | 3 | 10.8 | | 10.3–11.0 | | 584 | 4936.0 |
| Empidonax minimus | B | 413 | 10.3 | 0.47 | 8.2–14.9 | | Pennsylvania,USA 101 | 4937.0 |
| Empidonax hammondii | B | 10 | 10.1 | 1.19 | 7.7–12.1 | S | Arizona, USA 594 | 4938.0 |

## Body Masses of World Birds (continued)

| Species | Sex | N | Mean | Std dev | Range | Sn | Location | Number |
|---|---|---|---|---|---|---|---|---|
| Empidonax wrightii | B | 16 | 12.5 | 0.82 | 11.3–14.5 | B | Nevada, USA 300 | 4939.0 |
| Empidonax oberholseri | B | 16 | 10.4 | 0.64 | 9.3–11.4 | M | Arizona, USA 177 | 4940.0 |
| Empidonax affinis | M | 1 | 11.3 | | | | Mexico 139 | 4941.0 |
| Empidonax difficilis | B | 32 | 10.0 | 0.80 | 8.4–11.8 | S | California, USA 103 | 4942.0 |
| Empidonax occidentalis | B | 17 | 11.4 | 0.52 | 10.8–12.6 | B | Arizona, USA 509 | 4943.0 |
| Empidonax flavescens | B | 11 | 12.2 | | | | Panama 244 | 4944.0 |
| Empidonax fulvifrons | B | 23 | 7.9 | 0.70 | 6.9–9.3 | Y | Mexico 586 | 4945.0 |
| Empidonax atriceps | U | 3 | 9.0 | | 8.9–9.3 | | Costa Rica 584 | 4946.0 |
| Sayornis phoebe | B | 334 | 19.8 | 7.47 | 11.4–24.4 | | Pennsylvania,USA 101 | 4947.0 |
| Sayornis saya | B | 8 | 21.2 | | 17.5–24.1 | B | New Mexico, USA 177 | 4948.0 |
| Sayornis nigricans | M F | 78 154 | 19.5 17.8 | 0.90 1.07 | 16.9–22.0 | B | California, USA 177 | 4949.0 |
| Pyrocephalus rubinus | B | 11 | 14.4 | | 12.0–16.1 | | Nicaragua 272 | 4950.0 |
| Silvicultrix frontalis | U | 5 | 10.7 | | | | Peru 668 | 4951.0 |
| Silvicultrix pulchella | U | 6 | 12.3 | | | | Peru 668 | 4954.0 |
| Ochthoeca cinnamomeiventris | U | 1 | 11.9 | | | | Peru 668 | 4955.0 |
| Ochthoeca rufipectoralis | U | 6 | 10.8 | | | | Peru 668 | 4956.0 |
| Ochthoeca fumicolor superciliosa | B | 3 | 14.7 | | | | Venezuela 644 | 4957.0 |
| Colorhamphus parvirostris | U | | 8.5 | | | | Chile 292 | 4962.0 |
| Ochthornis littoralis | U | 3 | 13.3 | | 12.3–14.9 | | Peru 193 | 4963.0 |
| Myiotheretes striaticollis | M | 1 | 64.0 | | | | Peru 668 | 4965.0 |
| Myiotheretes fuscorufus | M | 1 | 32.2 | | | | Peru 668 | 4968.0 |
| Xolmis pyrope | U | | 38.3 | | | | Chile 292 | 4969.0 |

## Body Masses of World Birds (continued)

| Species | Sex | N | Mean | Std dev | Range | Sn | Location | Number |
|---|---|---|---|---|---|---|---|---|
| Xolmis cinerea | U | 4 | 59.6 | | 54.8–61.8 | | 38, 111 | 4970.0 |
| Xolmis coronata | U | 2 | 47.0 | | 43.0–51.0 | | Brazil 38 | 4971.0 |
| Xolmis irupero | U | 3 | 29.8 | | 29.0–31.0 | | 38, 111 | 4973.0 |
| Heteroxolmis dominicana | U | 2 | 42.8 | | 42.0–43.5 | | Brazil 38 | 4976.0 |
| Neoxolmis rufiventris | U | | 77.0 | | | | Brazil 564 | 4977.0 |
| Agriornis livida | U | | 99.2 | | | | Chile 52 | 4980.0 |
| Agriornis microptera | M | 1 | 72.9 | | | | Argentina 411 | 4981.0 |
| Polioxolmis rufipennis | M | 1 | 34.5 | | | | Bolivia 643 | 4983.0 |
| Muscigralla brevicauda | U | 2 | 12.6 | | 11.8–13.5 | | Peru 562 | 4996.0 |
| Lessonia rufa | M | 2 | 15.0 | | 15.0–15.0 | | Brazil 38 | 4998.0 |
| | F | 2 | 12.0 | | 10.0–14.0 | | | |
| Knipolegus poecilocercus | U | 3 | 14.3 | | | | Brazil 439 | 5001.0 |
| Knipolegus cyanirostris | B | 4 | 15.4 | | 14.0–17.0 | | Brazil 38, 440 | 5003.0 |
| Knipolegus poecilurus | B | 7 | 14.6 | | 13.5–15.3 | | Columbia 387 | 5004.0 |
| Knipolegus aterrimus | M | 1 | 23.5 | | | | Peru 193 | 5006.0 |
| Knipolegus nigerrimus | M | 1 | 20.0 | | | | Brazil 38 | 5007.0 |
| Knipolegus lophotes | B | 3 | 31.8 | | 30.0–34.0 | | Brazil 38 | 5008.0 |
| Hymenops perspicillatus | B | 2 | 24.0 | | 24.0–24.0 | | Brazil 38 | 5009.0 |
| Fluvicola pica | B | 20 | 12.2 | | | | 87,244,311,384,576 | 5010.0 |
| Arundinicola leucocephala | U | 8 | 14.1 | | 12.0–16.5 | | 38, 111, 188, 245 | 5013.0 |
| Gubernetes yetapa | B | 2 | 67.0 | | 64.0–70.0 | | Brazil 38 | 5016.0 |
| Satrapa icterophrys | B | 3 | 20.3 | | 19.5–21.5 | | Brazil 38 | 5017.0 |
| Colonia colonus | U | 4 | 16.1 | | 14.7–17.0 | | Panama; Brazil 38, 87, 611 | 5018.0 |

## Body Masses of World Birds (continued)

| Species | Sex | N | Mean | Std dev | Range | Sn | Location | Number |
|---|---|---|---|---|---|---|---|---|
| Machetornis rixosus | B | 6 | 33.4 | | 28.3–37.0 | | 38, 111, 625 | 5019.0 |
| Muscipipra vetula | B | 2 | 27.0 | | 26.0–28.0 | | Brazil 38 | 5020.0 |
| Attila phoenicurus | U | 2 | 34.5 | | 34.0–35.0 | | Brazil 38, 441 | 5021.0 |
| Attila cinnamomeus | M F | 4 1 | 39.5 28.0 | | 37.0–41.0 | | Surinam 245 | 5022.0 |
| Attila bolivianus | U | | 45.0 | | | | Peru 623 | 5025.0 |
| Attila rufus | U | 6 | 43.4 | | 36.5–51.5 | | Brazil 38, 441 | 5026.0 |
| Attila spadiceus | B | 16 | 39.1 | 3.26 | 35.0–44.1 | | 457, 497, 510 | 5027.0 |
| Casiornis rufa | U | 15 | 24.7 | 2.70 | | | Peru 668 | 5028.0 |
| Casiornis fusca | U | 1 | 19.5 | | | | Brazil 440 | 5029.0 |
| Rhytipterna holerythra | M | 18 | 36.8 | | | | Panama 244 | 5030.0 |
| Rhytipterna simplex | B | 5 | 35.8 | | 33.0–38.0 | | Fr. Guiana 161 | 5031.0 |
| Laniocera rufescens | B | 8 | 48.1 | | 38.6–56.0 | | Panama; Belize 243, 314, 497, 510, 611 | 5033.0 |
| Laniocera hypopyrra | U | | 51.0 | | | | Peru 623 | 5034.0 |
| Sirystes sibilator | U | 5 | 33.3 | | 27.5–36.0 | | 38, 87, 497, 611 | 5035.0 |
| Myiarchus yucatanensis | B | 4 | 21.4 | | 19.4–23.0 | | Yucatan, Mexico 457 | 5037.0 |
| Myiarchus tuberculifer | B | 25 | 19.9 | 1.06 | | | Panama 244 | 5038.0 |
| Myiarchus barbirostris | B | 66 | 13.4 | 0.87 | 11.5–16.2 | | Jamaica 613a | 5039.0 |
| Myiarchus swainsoni | B | 9 | 23.5 | | 18.0–27.5 | | 188, 441 | 5040.0 |
| Myiarchus panamensis | M | 1 | 28.1 | | | | Panama 611 | 5042.0 |
| Myiarchus ferox | U | 18 | 28.3 | 1.80 | 23.0–30.2 | | Surinam 246 | 5043.0 |
| Myiarchus cephalotes | U | 2 | 27.0 | | | | Peru 668 | 5044.0 |
| Myiarchus phaeocephalus | M | 1 | 27.2 | | | | Peru 672 | 5045.0 |

**Body Masses of World Birds (continued)**

| Species | Sex | N | Mean | Std dev | Range | Sn | Location | Number |
|---|---|---|---|---|---|---|---|---|
| Myiarchus apicalis | M | 3 | 29.3 | | 28.0–30.0 384 | | Columbia | 5046.0 |
| Myiarchus cinerascens | B | 32 | 27.2 | | 24.0–31.0 334 | | | 5047.0 |
| Myiarchus nuttingi | U | 9 | 23.0 | | 21.0–23.9 | | Mexico 593, 594 | 5048.0 |
| Myiarchus crinitus | B | 70 | 33.5 | 3.01 | 27.2–39.6 | | Pennsylvania,USA 177 | 5049.0 |
| Myiarchus tyrannulus | B | 59 | 43.8 | 3.37 | 37.9–53.7 | B | Arizona, USA 177 | 5050.0 |
| Myiarchus nugator | U | | 37.0 | | | | 185 | 5051.0 |
| Myiarchus validus | U | 4 | 40.6 | | 38.6–43.2 | | Jamaica 600 | 5053.0 |
| Myiarchus sagrae | B | 7 | 18.6 | | 17.0–20.0 | | Cayman Is. 432 | 5054.0 |
| Myiarchus stolidus | U | 46 | 22.9 | 1.30 | 20.5–26.3 | W | Puerto Rico 186 | 5055.0 |
| Myiarchus antillarum | B | 10 | 23.3 | 0.91 | 22.0–25.3 | | Puerto Rico 431 | 5056.0 |
| Myiarchus oberi | U | | 37.0 | | | | 185 | 5057.0 |
| Deltarhynchus flammulatus | M | 1 | 17.2 | | | | Oaxaca, Mexico 584 | 5058.0 |
| Tyrannus melancholicus | M | 20 | 36.5 | 2.47 | 32.0–40.6 | | | 5061.0 |
| | F | 15 | 38.6 | 2.86 | 32.7–42.5 | | 593 | |
| Tyrannus couchii | M | 2 | 45.0 | | | | Mexico 139 | 5062.0 |
| | F | | | | | | | |
| Tyrannus vociferans | B | 14 | 45.6 | 3.44 | 37.4–50.0 | | 585 | 5063.0 |
| Tyrannus crassirostris | B | 7 | 56.0 | | 53.0–59.0 | B | Jalisco, Mexico 548 | 5064.0 |
| Tyrannus verticalis | B | 17 | 39.6 | 2.75 | 34.9–44.1 | | 177 | 5065.0 |
| Tyrannus forficatus | B | 33 | 43.2 | 4.50 | 36.3–56.3 | | Oklahoma, USA 590a | 5066.0 |
| Tyrannus savana | B | 13 | 28.6 | | | | 38, 244, 311 | 5067.0 |
| Tyrannus tyrannus | B | 22 | 43.6 | 3.40 | | B | Manitoba, Canada 44 | 5068.0 |
| Tyrannus dominicensis | B | 25 | 43.8 | 3.50 | 37.0–51.2 | W | Puerto Rico 186 | 5069.0 |
| Tyrannus caudifasciatus | B | 13 | 44.1 | | 34.5–52.0 | | Cayman Is. 432 | 5070.0 |

## Body Masses of World Birds (continued)

| Species | Sex | N | Mean | Std dev | Range | Sn | Location | Number |
|---|---|---|---|---|---|---|---|---|
| Empidonomus varius | B | 12 | 27.1 | | | | Paraguay 200 | 5072.0 |
| Griseotyrannus aurantioatrocristatus | B | 3 | 27.2 | | 22.5–32.0 | | Brazil 38, 610a | 5073.0 |
| Tyrannopsis sulphurea | B | | | | 52.0–61.0 | | Trinidad 188 | 5074.0 |
| Megarhynchus pitangua | M F | 6 4 | 77.0 70.0 | | | | Panama 269a | 5075.0 |
| Conopias albovittata | M | 4 | 24.4 | | 23.5–26.0 | | Panama 497 | 5076.0 |
| Conopias parva | M | 1 | 24.0 | | | | Fr. Guiana 161 | 5077.0 |
| Myiodynastes hemichrysus | U | 1 | 43.4 | | | | Panama 244 | 5080.0 |
| Myiodynastes chrysocephalus | B | 3 | 40.5 | | 39.0–43.0 | | 105, 497 | 5081.0 |
| Myiodynastes bairdii | F | 1 | 45.0 | | | | Peru 672 | 5082.0 |
| Myiodynastes maculatus | U | 28 | 45.9 | 4.32 | 39.0–52.5 | | 87,105,245,246, 311,436,510,576,611 | 5083.0 |
| Myiodynastes luteiventris | B | 8 | 45.9 | | 44.2–49.2 | | Guatemala 571, 593 | 5084.0 |
| Myiozetetes cayanensis | U | 14 | 25.9 | 2.10 | 22.0–29.5 | | Venezuela 625 | 5085.0 |
| Myiozetetes similis | B | 24 | 28.0 | | | | Panama 269a | 5086.0 |
| Myiozetetes granadensis | B | 10 | 29.3 | | | | Panama 269a | 5087.0 |
| Legatus leucophaius | B | 15 | 24.4 | 3.49 | 19.0–31.5 | | 38, 244, 245, 510, 611 | 5089.0 |
| Philohydor lictor | U | 7 | 24.2 | | 23.0–27.0 | | 245, 611, 625 | 5090.0 |
| Pitangus sulphuratus | U | 15 | 61.0 | 4.07 | 53.5–67.5 | | Venezuela 625 | 5091.0 |
| Phelpsia inornata | U | 35 | 29.4 | 1.71 | 26.8–33.5 | | Venezuela 625 | 5092.0 |
| Schiffornis major | U | | 31.0 | | | | Peru 623 | 5093.0 |
| Schiffornis turdinus | B | 13 | 31.7 | | | | 455 | 5094.0 |
| Schiffornis virescens | U | 12 | 24.8 | | 23.0–26.5 | | Brazil 441 | 5095.0 |

## Body Masses of World Birds (continued)

| Species | Sex | N | Mean | Std dev | Range | Sn | Location | Number |
|---|---|---|---|---|---|---|---|---|
| Xenopsaris albinucha | M | 1 | 10.2 | | | | Venezuela 626 | 5096.0 |
| Pachyramphus viridis | M | 1 | 21.0 | | | | Brazil 38 | 5097.0 |
| Pachyramphus versicolor | B | 3 | 17.0 | | 15.8–18.4 | | Columbia 387 | 5098.0 |
| Pachyramphus cinnamomeus | B | 14 | 20.3 | 0.83 | 18.7–22.0 | | Panama; Belize 87, 244, 510, 611 | 5099.0 |
| Pachyramphus castaneus | U | 8 | 19.5 | | 18.0–21.0 | | Brazil 38, 441, 610a | 5100.0 |
| Pachyramphus polychopterus | B | 18 | 21.6 | | | | 246, 668 | 5101.0 |
| Pachyramphus major | U | 2 | 25.2 | | 22.1–28.4 | | Belize; Mexico 323, 510 | 5102.0 |
| Pachyramphus albogriseus | B | 9 | 17.4 | | 12.5–20.0 | | Peru 584 | 5103.0 |
| Pachyramphus marginatus | U | | 18.0 | | | | Peru 623 | 5104.0 |
| Pachyramphus surinamus | M | 4 | 18.6 | | 17.5–21.0 | | Surinam | 5105.0 |
| | F | 3 | 20.9 | | 18.9–23.0 | | 248 | |
| Pachyramphus rufus | U | 8 | 17.3 | | 14.0–19.9 | | 245, 246, 384 | 5106.0 |
| Pachyramphus spodiurus | B | 4 | 18.2 | | | | Peru 672 | 5107.0 |
| Pachyramphus aglaiae yucatensis | B | 9 | 30.0 | | 25.8–33.3 | | Yucatan, Mexico 457 | 5108.0 |
| Pachyramphus homochrous | M | 1 | 35.0 | | | | Panama 497 | 5109.0 |
| Pachyramphus minor | U | 8 | 39.9 | | 36.0–44.0 | | 245, 246, 610a, 668 | 5110.0 |
| Pachyramphus validus | F | 2 | 45.0 | | 41.0–49.0 | | Brazil 38, 610a | 5112.0 |
| Tityra cayana | B | 13 | 73.9 | 7.91 | 60.0–87.0 | | 38, 161, 188, 245, 311 | 5113.0 |
| Tityra semifasciata | B | 25 | 79.3 | | | | Panama 244 | 5114.0 |
| Tityra inquisitor | B | 21 | 43.3 | 3.06 | 36.0–47.0 | | 38, 244, 245, 457, 510, 611 | 5115.0 |

### ORDER: PASSERIFORMES    FAMILY: CLIMACTERIDAE

| Species | Sex | N | Mean | Std dev | Range | Sn | Location | Number |
|---|---|---|---|---|---|---|---|---|
| Cormobates leucophaea | U | 31 | 22.6 | 1.89 | | | | 5117.0 |

## Body Masses of World Birds (continued)

| Species | Sex | N | Mean | Std dev | Range | Sn | Location | Number |
|---|---|---|---|---|---|---|---|---|
| Climacteris erythrops | U | 18 | 22.8 | 1.40 | | | 423 | 5119.0 |
| Climacteris picumnus | U | 32 | 37.0 | 2.55 | | | 423 | 5120.0 |

**ORDER: PASSERIFORMES**   **FAMILY: MENURIDAE**

| | | | | | | | | |
|---|---|---|---|---|---|---|---|---|
| Menura novaehollandiae | U | | 746.0 | | | | 480a | 5124.0 |

**ORDER: PASSERIFORMES**   **FAMILY: ATRICHORNITHIDAE**

| | | | | | | | | |
|---|---|---|---|---|---|---|---|---|
| Atrichornis clamosus | U | | 57.0 | | | | W. Australia 550 | 5126.0 |

**ORDER: PASSERIFORMES**   **FAMILY: PTILONORHYNCHIDAE**

| | | | | | | | | |
|---|---|---|---|---|---|---|---|---|
| Ailuroedus buccoides | U | 23 | 138.0 | | | | New Guinea 35 | 5127.0 |
| Ailuroedus melanotis | F | 1 | 228.0 | | | | New Guinea 218 | 5128.0 |
| Ailuroedus crassirostris | M | 1 | 205.0 | | | | New Guinea 154 | 5129.0 |
| Ailuroedus dentirostris | U | 11 | 157.0 | | | | Australia 205 | 5130.0 |
| Amblyornis inornatus | U | 40 | 125.0 | | 105.0–155.0 | | New Guinea 157 | 5133.0 |
| Amblyornis macgregoriae | B | 30 | 126.0 | | 110.0–140.0 | | New Guinea 531 | 5134.0 |
| Amblyornis subalaris | M | 7 | 103.0 | | 96.0–107.0 | | New Guinea 531 | 5135.0 |
| | F | 3 | 114.0 | | 109.0–122.0 | | | |
| Sericulus bakeri | B | 8 | 176.0 | | 164.0–184.0 | | New Guinea 218 | 5139.0 |
| Chlamydera maculata | U | | 128.0 | | | | W. Australia 550 | 5143.0 |
| Chlamydera lauterbachi | M | 1 | 133.0 | | | | New Guinea 217 | 5145.0 |
| | F | 1 | 112.0 | | | | | |

**ORDER: PASSERIFORMES**   **FAMILY: MALURIDAE**

| | | | | | | | | |
|---|---|---|---|---|---|---|---|---|
| Clytomyias insignis | B | 4 | 13.2 | | 12.0–15.2 | | New Guinea 154 | 5147.0 |
| Sipodotus wallacii | B | | | | 6.5–8.0 | | 525 | 5148.0 |
| Malurus grayi | M | 5 | 16.0 | | 14.0–17.0 | | New Guinea 155 | 5149.0 |
| | F | 7 | 14.2 | | 12.0–16.0 | | | |
| Malurus alboscapulatus | B | 14 | 10.9 | | 9.3–12.3 | | New Guinea 154 | 5150.0 |

## Body Masses of World Birds (continued)

| Species | Sex | N | Mean | Std dev | Range | Sn | Location | Number |
|---|---|---|---|---|---|---|---|---|
| Malurus campbelli | B | 3 | 9.0 | | 8.0–10.0 | | 526 | 5150.1 |
| Malurus melanocephalus | U | | | | 6.0–6.5 | | W. Australia 550 | 5151.0 |
| Malurus leucopterus | U | 8 | 6.1 | | | | W. Australia 695 | 5152.0 |
| Malurus cyaneus | B | 100 | 9.0 | 0.60 | 8.0–10.5 | | Australia 235 | 5153.0 |
| Malurus splendens | U | | 10.0 | | | | 77 | 5154.0 |
| Malurus lamberti | B | 29 | 8.0 | 0.60 | 7.0–9.5 | | Australia 235 | 5155.0 |
| Malurus amabilis | B | | | | 8.0–11.0 | | 525 | 5156.0 |
| Malurus elegans | B | | | | 9.0–11.0 | | 525 | 5157.0 |
| Malurus pulcherrimus | B | | | | 9.0–10.0 | | 525 | 5158.0 |
| Malurus coronatus | B | | | | 9.0–12.0 | | 525 | 5159.0 |
| Malurus cyanocephalus | U | 8 | 13.0 | | | | New Guinea 35 | 5160.0 |
| Stipiturus ruficeps | B | 7 | 5.1 | | 4.5–6.3 | | Australia 199 | 5161.0 |
| Amytornis textilis | M | 17 | 23.8 | 1.30 | | | west Australia 76 | 5164.0 |
| | F | 18 | 21.6 | 1.20 | | | | |
| Amytornis purnelli | B | | | | 18.0–25.0 | | 525 | 5165.0 |
| Amytornis housei | B | | | | 25.0–35.0 | | 525 | 5166.0 |
| Amytornis striatus | M | 2 | 16.3 | | 16.0–16.6 | | Australia 199 | 5167.0 |
| Amytornis woodwardi | M | | | | 33.0–40.0 | | | 5168.0 |
| | F | | | | 30.0–37.0 | | 525 | |
| Amytornis dorotheae | M | | | | 24.0–25.0 | | | 5169.0 |
| | F | | | | 21.0–23.0 | | 525 | |
| Amytornis barbatus | F | 1 | 21.0 | | | | Australia 527 | 5170.0 |
| Amytornis goyderi | B | | | | 16.0–19.0 | | 525 | 5171.0 |

### ORDER: PASSERIFORMES                    FAMILY: PARDILOTIDAE

| Species | Sex | N | Mean | Std dev | Range | Sn | Location | Number |
|---|---|---|---|---|---|---|---|---|
| Pardalotus punctatus | U | 15 | 9.2 | 0.46 | | | Tasmania 687 | 5172.0 |

## Body Masses of World Birds (continued)

| Species | Sex | N | Mean | Std dev | Range | Sn | Location | Number |
|---|---|---|---|---|---|---|---|---|
| Pardalotus quadragintus | U | 52 | 10.7 | 0.94 | | | Tasmania 687 | 5174.0 |
| Pardalotus striatus | B | 290 | 12.2 | | | | Australia 688 | 5176.0 |
| Pardalotus substriatus | U | | 11.0 | | | | Australia 454 | 5176.1 |
| Dasyornis longirostris | U | 1 | 37.0 | | | | Australia 568 | 5177.0 |
| Crateroscelis murina | B | 17 | 15.0 | | 13.0–17.0 | | New Guinea 154 | 5183.0 |
| Crateroscelis robusta | B | 10 | 16.5 | 1.04 | 14.3–18.0 | | New Guinea 154 | 5185.0 |
| Amalocichla incerta | M | 1 | 30.7 | | | | New Guinea 154 | 5187.0 |
| Sericornis frontalis | U | | 12.8 | | | | 266 | 5190.0 |
| Sericornis keri | U | 3 | 12.3 | | 12.1–12.4 | | Australia 209, 551 | 5192.0 |
| Sericornis beccarii | U | | 12.7 | | | | New Guinea 123 | 5193.0 |
| Sericornis virgatus | U | | 12.7 | | | | New Guinea 123 | 5194.0 |
| Sericornis nouhuysi | M | 9 | 16.2 | | 13.0–17.7 | | New Guinea 154 | 5195.0 |
| | F | 7 | 14.6 | | 14.0–15.8 | | | |
| Sericornis magnirostris | B | 26 | 10.3 | | | | Australia 209 | 5196.0 |
| Sericornis perspicillatus | B | 6 | 9.0 | | 8.0–10.2 | | New Guinea 154 | 5198.0 |
| Sericornis arfakianus | B | 20 | 8.7 | 0.84 | 6.5–10.3 | | New Guinea 154, 216 | 5199.0 |
| Sericornis papuensis | B | 13 | 10.6 | | 9.8–11.7 | | New Guinea 154 | 5200.0 |
| Sericornis spilodera | B | 13 | 11.7 | | 10.3–13.2 | | New Guinea 154 | 5201.0 |
| Chthonicola sagittatus | U | | 13.0 | | | | Australia 194a | 5204.0 |
| Acanthiza murina | U | 3 | 8.4 | | 7.8–9.0 | | New Guinea 216 | 5209.0 |
| Acanthiza reguloides | U | 10 | 7.5 | | | | Australia 480 | 5210.0 |
| Acanthiza inornata | U | | 7.0 | | | | 77 | 5211.0 |
| Acanthiza pusilla | U | 17 | 6.2 | 0.82 | | | Australia 481 | 5215.0 |

## Body Masses of World Birds (continued)

| Species | Sex | N | Mean | Std dev | Range | Sn | Location | Number |
|---|---|---|---|---|---|---|---|---|
| Acanthiza ewingii | U | | 10.0 | | | | Tasmania 134 | 5216.0 |
| Acanthiza chrysorrhoa | U | | 8.8 | | | | 266 | 5217.0 |
| Acanthiza robustirostris | U | 1 | 6.5 | | | | Australia 199 | 5219.0 |
| Acanthiza nana | U | 10 | 5.9 | | | | Australia 480 | 5220.0 |
| Acanthiza lineata | U | 29 | 6.7 | 0.54 | | | Australia 481 | 5221.0 |
| Smicrornis brevirostris | U | 10 | 5.1 | | | | Australia 480 | 5222.0 |
| Gerygone cinerea | B | 5 | 7.5 | | 7.0–8.0 | | New Guinea 154 | 5223.0 |
| Gerygone chloronota | U | | 6.3 | | | | New Guinea 123 | 5224.0 |
| Gerygone palpebrosa | U | 13 | 8.0 | | | | New Guinea 35 | 5225.0 |
| Gerygone olivacea | U | 10 | 7.7 | | | | Australia 480 | 5226.0 |
| Gerygone chrysogaster | U | 6 | 8.0 | | | | New Guinea 35 | 5227.0 |
| Gerygone magnirostris | U | 15 | 8.0 | | | | New Guinea 35 | 5228.0 |
| Gerygone sulphurea | U | 2 | 5.6 | | 5.5–5.8 | | Thailand 378 | 5231.0 |
| Gerygone igata | U | 112 | 6.4 | 0.72 | 5.0–8.0 | | New Zealand 500 | 5239.0 |
| Gerygone albofrontata | M | 39 | 11.1 | 0.87 | 9.8–13.8 | | Chatham Is. 499 | 5240.0 |
| | F | 11 | 9.5 | 1.02 | 8.3–11.5 | | | |
| Gerygone flavolateralis | U | 12 | 6.3 | 0.30 | 5.5–6.5 | | New Caledonia 506 | 5241.0 |

## ORDER: PASSERIFORMES          FAMILY: MELIPHAGIDAE

| Species | Sex | N | Mean | Std dev | Range | Sn | Location | Number |
|---|---|---|---|---|---|---|---|---|
| Timeliopsis fulvigula | B | 5 | 18.2 | | 16.0–20.0 | | New Guinea 154, 216 | 5245.0 |
| Timeliopsis griseigula | U | 6 | 32.0 | | | | New Guinea 35 | 5246.0 |
| Melilestes megarhynchus | M | 10 | 46.6 | 1.70 | 43.0–49.5 | | New Guinea 154 | 5247.0 |
| | F | 10 | 41.1 | 3.30 | 36.3–47.5 | | | |
| Glycichaera fallax | U | 12 | 10.0 | | | | New Guinea 35 | 5249.0 |
| Lichmera indistincta | U | | 13.0 | | | | Australia 501 | 5252.0 |

## Body Masses of World Birds (continued)

| Species | Sex | N | Mean | Std dev | Range | Sn | Location | Number |
|---|---|---|---|---|---|---|---|---|
| Lichmera incana | U | 8 | 13.1 | | 9.5–14.5 | | New Caledonia 506 | 5253.0 |
| Lichmera alboauricularis | M | 12 | | | 13.5–19.0 | | New Guinea | 5255.0 |
| | F | 3 | 13.3 | | 12.5–14.5 | | 217 | |
| Myzomela blasii | M | 8 | 14.9 | 1.80 | | | Ceram 160 | 5261.0 |
| Myzomela albigula | M | 3 | 16.3 | | | | | 5262.0 |
| | F | 1 | 13.0 | | | | 160 | |
| Myzomela eques eques | M | 22 | 13.8 | 2.40 | | | | 5263.0 |
| | | | | | | | 160 | |
| Myzomela eques nymanni | M | 21 | 15.8 | 2.00 | | | New Guinea | 5263.0 |
| | F | 8 | 12.4 | 1.20 | | | 160 | |
| Myzomela obscura | M | 7 | 13.5 | | | | Australia | 5264.0 |
| | F | 2 | 11.3 | | | | 160 | |
| Myzomela cruentata | M | 14 | 8.6 | 0.60 | | | New Britain | 5265.0 |
| | F | 9 | 7.9 | | | | 160 | |
| Myzomela nigrita | M | 26 | 10.1 | 0.50 | | | New Guinea | 5266.0 |
| | F | 9 | 8.5 | 0.60 | | | 160 | |
| Myzomela pulchella | B | 5 | 10.0 | | | | New Ireland 30 | 5267.0 |
| Myzomela erythrocephala | B | 16 | 8.2 | | | | Australia 160 | 5269.0 |
| Myzomela adolphinae | B | 10 | 7.4 | | | | New Guinea 160 | 5270.0 |
| Myzomela sanguinolenta | M | 2 | 7.8 | | 7.5–8.2 | | New Caledonia | 5274.0 |
| | F | 2 | 6.8 | | 6.6–6.9 | | 506 | |
| Myzomela cardinalis | M | 6 | 15.4 | | 14.5–16.0 | | San Cristobal | 5277.0 |
| | F | 6 | 12.3 | | 11.5–13.0 | | 160 | |
| Myzomela sclateri | B | 26 | 10.5 | | | | 160 | 5279.0 |
| Myzomela pammelaena | M | 28 | 17.0 | 1.20 | | | | 5280.0 |
| | F | 28 | 14.6 | 1.00 | | | 160 | |
| Myzomela tristrami | M | 21 | 14.4 | | 12.5–16.5 | | San Cristobal | 5285.0 |
| | F | 10 | 12.3 | | 11.5–13.0 | | 160 | |
| Myzomela jugularis | M | 10 | 9.0 | 0.60 | 8.0–10.0 | | Fiji | 5286.0 |
| | F | 9 | 8.4 | | 8.0–9.0 | | 332 | |
| Myzomela erythromelas | B | 46 | 8.0 | | | | New Britain 160 | 5287.0 |
| Myzomela rosenbergii | M | 15 | 11.0 | 2.50 | 9.3–12.7 | | New Guinea | 5289.0 |
| | F | 4 | 9.7 | | 9.0–10.7 | | 154 | |
| Myzomela cineracea | M | 39 | 15.6 | 1.60 | | | New Britain | 5290.0 |
| | F | 19 | 12.2 | 1.20 | | | 160 | |
| Certhionyx pectoralis | U | | 9.2 | | | | 453 | 5291.0 |

## Body Masses of World Birds (continued)

| Species | Sex | N | Mean | Std dev | Range | Sn | Location | Number |
|---|---|---|---|---|---|---|---|---|
| Certhionyx niger | U | | 7.8 | | | | | 5292.0 |
| | | | | | | | 453 | |
| Meliphaga montana auga | M | 14 | 30.7 | 2.30 | 27.0–34.0 | | New Guinea | 5294.0 |
| | F | 13 | 25.6 | 1.80 | 22.0–29.0 | | 154 | |
| Meliphaga mimikae | M | 25 | 28.5 | 2.00 | 24.0–33.0 | | New Guinea | 5295.0 |
| | F | 41 | 25.4 | 1.70 | 22.0–29.0 | | 154 | |
| Meliphaga orientalis | B | 20 | 18.0 | | 16.0–20.0 | | New Guinea | 5296.0 |
| | | | | | | | 154 | |
| Meliphaga albonotata | U | 5 | 22.4 | | | | New Guinea | 5297.0 |
| | | | | | | | 35, 267 | |
| Meliphaga aruensis | M | 3 | 27.8 | | 27.0–28.7 | | New Guinea | 5298.0 |
| | F | 3 | 24.7 | | 23.0–26.5 | | 154 | |
| Meliphaga analoga | U | 9 | 21.0 | | | | New Guinea | 5299.0 |
| | | | | | | | 35 | |
| Meliphaga gracilis | U | 3 | 22.0 | | | | New Guinea | 5301.0 |
| | | | | | | | 35 | |
| Meliphaga flavirictus | U | 2 | 18.5 | | 17.0–20.0 | | New Guinea | 5303.1 |
| Meliphaga lewinii | U | | 33.3 | | | | | 5304.0 |
| | | | | | | | 453 | |
| Meliphaga albilineata | U | | 25.8 | | | | | 5305.0 |
| | | | | | | | 453 | |
| Foulehaio carunculata | U | 10 | 34.6 | 4.40 | 29.0–43.0 | | Fiji | 5308.0 |
| | | | | | | | 332 | |
| Lichenostomus subfrenatus | M | 9 | 32.6 | | 29.0–37.0 | | New Guinea | 5309.0 |
| | F | 9 | 27.0 | | 25.0–29.0 | | 154, 216 | |
| Lichenostomus obscurus | B | 11 | 26.4 | 2.29 | 24.0–31.0 | | New Guinea | 5310.0 |
| | | | | | | | 154 | |
| Lichenostomus hindwoodi | M | 1 | 19.7 | | | | Australia | 5312.0 |
| | | | | | | | 348 | |
| Lichenostomus chrysops | U | | 17.1 | | | | | 5313.0 |
| | | | | | | | 266 | |
| Lichenostomus fasciogularis | U | | 27.5 | | | | | 5315.0 |
| | | | | | | | 453 | |
| Lichenostomus virescens | U | 22 | 24.6 | | | | W. Australia | 5316.0 |
| | | | | | | | 695 | |
| Lichenostomus flavus | U | | 23.5 | | | | | 5317.0 |
| | | | | | | | 453 | |
| Lichenostomus unicolor | U | | 32.4 | | | | | 5318.0 |
| | | | | | | | 453 | |
| Lichenostomus leucotis | U | 8 | 23.1 | | | | | 5319.0 |
| | | | | | | | 95 | |
| Lichenostomus flavicollis | U | | 29.0 | | | | | 5320.0 |
| | | | | | | | 134 | |

## Body Masses of World Birds (continued)

| Species | Sex | N | Mean | Std dev | Range | Sn | Location | Number |
|---|---|---|---|---|---|---|---|---|
| Lichenostomus melanops | U | | 20.7 | | | | 453 | 5321.0 |
| Lichenostomus cratitius | U | | 19.6 | | | | 454 | 5323.0 |
| Lichenostomus flavescens | U | | 13.3 | | | | 453 | 5325.0 |
| Lichenostomus fuscus | U | | 17.9 | | | | 194 | 5326.0 |
| Lichenostomus plumulus | U | | 18.2 | | | | 453 | 5327.0 |
| Lichenostomus ornatus | U | | 17.8 | | | | 454 | 5328.0 |
| Lichenostomus penicillatus | U | | 19.8 | | | | 454 | 5329.0 |
| Xanthotis flaviventer | M | 12 | 47.3 | 4.20 | 40.0–55.0 | | New Guinea | 5330.0 |
| | F | 10 | 42.7 | 3.10 | 38.0–49.0 | | 154 | |
| Xanthotis polygramma | M | 3 | 21.1 | | 19.7–23.5 | | New Guinea | 5331.0 |
| | F | 3 | 19.0 | | 18.0–19.5 | | 154 | |
| Xanthotis macleayana | U | | 33.1 | | | | 453 | 5332.0 |
| Melithreptus lunatus | U | | 14.7 | | | | 266 | 5336.0 |
| Melithreptus affinis | U | 8 | 16.4 | | | | Tasmania 687 | 5337.0 |
| Melithreptus albogularis | U | | 11.0 | | | | 453 | 5338.0 |
| Melithreptus gularis | U | | 21.8 | | | | 453 | 5340.0 |
| Melithreptus validirostris | U | | 23.5 | | | | 453 | 5341.0 |
| Melithreptus brevirostris | U | | 14.6 | | | | 266 | 5342.0 |
| Notiomystis cincta | M | 59 | 37.3 | 2.70 | 29.0–42.0 | | New Zealand | 5343.0 |
| | F | 50 | 30.1 | 1.72 | 26.0–32.0 | | 215 | |
| Pycnopygius ixoides | M | 4 | 29.5 | | 28.0–33.0 | | New Guinea | 5344.0 |
| | F | 8 | 25.8 | | 22.0–31.0 | | 154 | |
| Pycnopygius cinereus | M | 7 | 48.8 | | 40.0–58.0 | | New Guinea | 5345.0 |
| | F | 4 | 41.2 | | 36.0–46.0 | | 154 | |
| Pycnopygius stictocephalus | M | 1 | 38.0 | | | | New Guinea 267 | 5346.0 |
| Philemon meyeri | B | 11 | 52.9 | | | | New Guinea 35, 154, 218 | 5348.0 |
| Philemon citreogularis | U | | 62.8 | | | | 453 | 5352.0 |

## Body Masses of World Birds (continued)

| Species | Sex | N | Mean | Std dev | Range | Sn | Location | Number |
|---|---|---|---|---|---|---|---|---|
| Philemon buceroides | M | 23 | 119.0 | | | | Australia | 5356.0 |
| | F | 13 | 106.0 | | | | 530 | |
| Philemon novaeguineae | B | 9 | 147.0 | | 133.0–177.0 | | New Guinea | 5357.0 |
| | | | | | | | 154, 218, 219, 267 | |
| Philemon eichhorni | M | 1 | 116.0 | | | | New Ireland | 5360.0 |
| | F | 2 | 88.0 | | 82.0–94.0 | | 30 | |
| Philemon argenticeps | U | | 86.5 | | | | | 5361.0 |
| | | | | | | | 453 | |
| Philemon gordoni | U | | 91.6 | | | | | 5361.1 |
| | | | | | | | 453 | |
| Philemon corniculatus | U | | 107.0 | | | | Australia | 5362.0 |
| | | | | | | | 194a | |
| Philemon diemenensis | M | 9 | 74.5 | | 69.5–84.0 | | New Caledonia | 5363.0 |
| | F | 6 | 58.3 | | 53.0–62.0 | | 506 | |
| Ptiloprora plumbea | M | 1 | 16.5 | | | | New Guinea | 5364.0 |
| | | | | | | | 216 | |
| Ptiloprora erythropleura | F | 1 | 21.0 | | | | New Guinea | 5366.0 |
| | | | | | | | 157 | |
| Ptiloprora guisei | M | 10 | 24.2 | 1.70 | 21.3–27.7 | | New Guinea | 5368.0 |
| | F | 10 | 20.4 | 1.40 | 17.6–24.0 | | 154 | |
| Ptiloprora perstriata | M | 13 | 27.1 | 1.82 | 23.5–30.0 | | New Guinea | 5369.0 |
| | F | 7 | 23.1 | | 21.0–25.5 | | 216 | |
| Melidectes whitemanensis | F | 2 | 48.0 | | 47.0–49.0 | | New Britain | 5371.0 |
| | | | | | | | 219 | |
| Melidectes leucostephes | M | 8 | 80.0 | | 63.0–97.0 | | New Guinea | 5375.0 |
| | F | 5 | 70.8 | | 61.0–78.0 | | 157 | |
| Melidectes belfordi | M | 3 | 74.7 | | 70.0–80.0 | | New Guinea | 5376.0 |
| | F | 4 | 60.5 | | 55.0–67.0 | | 216 | |
| Melidectes rufocrissalis | M | 2 | 79.0 | | 74.0–84.0 | | New Guinea | 5377.0 |
| | | | | | | | 216 | |
| Melidectes torquatus | M | 7 | 52.4 | | 47.0–58.0 | | New Guinea | 5379.0 |
| | F | 3 | 38.3 | | 34.0–44.0 | | 154 | |
| Melipotes fumigatus | B | 20 | 53.0 | | 44.0–68.0 | | New Guinea | 5382.0 |
| | | | | | | | 154 | |
| Gymnomyza aubryana | M | 4 | 244.0 | | 211.0–284.0 | | New Caledonia | 5388.0 |
| | F | 2 | 156.0 | | 152.0–159.0 | | 506, 604 | |
| Phylidonyris pyrrhoptera | U | | 16.4 | | | | | 5394.0 |
| | | | | | | | 266 | |
| Phylidonyris novaehollandiae | U | | 20.0 | | | | | 5395.0 |
| | | | | | | | 454 | |
| Phylidonyris nigra | U | | 18.3 | | | | | 5396.0 |
| | | | | | | | 453 | |
| Phylidonyris albifrons | U | | 18.0 | | | | | 5397.0 |
| | | | | | | | 454 | |

## Body Masses of World Birds (continued)

| Species | Sex | N | Mean | Std dev | Range | Sn | Location | Number |
|---|---|---|---|---|---|---|---|---|
| Phylidonyris undulata | F | 1 | 16.0 | | | | New Caledonia 506 | 5398.0 |
| Gliciphila melanops | U | | 18.5 | | | | 454 | 5400.0 |
| Ramsayornis modestus | U | 17 | 12.1 | 1.37 | 8.5–14.0 | | Australia 360 | 5401.0 |
| Ramsayornis fasciatus | U | | 13.3 | | | | 453 | 5402.0 |
| Conopophila albogularis | F | 2 | 13.2 | | 12.0–14.5 | | New Guinea 217 | 5404.0 |
| Conopophila rufogularis | U | | 11.5 | | | | 453 | 5405.0 |
| Grantiella picta | U | | 24.0 | | | | 453 | 5407.0 |
| Acanthorhynchus tenuirostris | M F | 33 20 | 14.1 11.5 | 1.30 1.00 | | | Australia 95 | 5409.0 |
| Acanthorhynchus superciliosus | U | | 10.8 | | | | 453 | 5410.0 |
| Entomyzon cyanotis | U | | 107.0 | | | | 453 | 5411.0 |
| Manorina melanocephala | U | | 68.0 | | | | 194 | 5413.0 |
| Manorina flavigula | U | | | | 64.0–71.0 | | W. Australia 550 | 5414.0 |
| Anthornis melanura | M F | 202 94 | 30.7 23.6 | 2.42 2.16 | 21.0–38.0 20.0–32.0 | | New Zealand 500 | 5415.0 |
| Acanthagenys rufogularis | M F | | 47.0 41.0 | | | | Australia 482 | 5416.0 |
| Anthochaera chrysoptera | U | 6 | 74.3 | | | | Australia 95 | 5417.0 |
| Anthochaera carunculata | M | | 125.0 | | | | 194a | 5418.0 |
| Anthochaera paradoxa | U | | 152.0 | | | | 453 | 5419.0 |
| Prosthemadera novaeseelandiae | M F | 75 46 | 125.0 89.6 | 10.00 6.54 | 97.0–150.0 70.0–105.0 | | New Zealand 500 | 5420.0 |
| Epthianura crocea | B | 3 | 10.1 | | 9.5–11.2 | | Australia 199 | 5423.0 |
| Epthianura albifrons | U | | 12.0 | | | | 134 | 5424.0 |

### ORDER: PASSERIFORMES    FAMILY: CALLAEIDAE

| Species | Sex | N | Mean | Std dev | Range | Sn | Location | Number |
|---|---|---|---|---|---|---|---|---|
| Callaeas cinerea | U | | | | 200.0–250.0 | | 43 | 5426.0 |

## Body Masses of World Birds (continued)

| Species | Sex | N | Mean | Std dev | Range | Sn | Location | Number |
|---------|-----|---|------|---------|-------|----|----------|--------|
| ORDER: PASSERIFORMES | | | | | FAMILY: EOPSALTRIIDAE | | | |
| Culicicapa ceylonensis | U | 20 | | | 7.0–9.0 | | India 5 | 5429.0 |
| Culicicapa helianthea | M | 2 | 8.6 | | 8.2–9.0 | | Philippines 475 | 5430.0 |
| Monachella muelleriana | B | 4 | 24.5 | | 23.0–27.5 | | New Guinea 154 | 5431.0 |
| Microeca flavigaster | U | 3 | 12.7 | | 12.5–13.0 | | New Guinea 216 | 5434.0 |
| Microeca griseoceps | U | | 12.8 | | | | 123 | 5435.0 |
| Microeca flavovirescens | B | 14 | 15.0 | | 13.0–17.0 | | New Guinea 154 | 5436.0 |
| Microeca papuana | B | 14 | 14.1 | | 13.0–16.0 | | New Guinea 154 | 5437.0 |
| Eugerygone rubra | U | | 8.9 | | | | 123 | 5438.0 |
| Petroica bivittata | U | | 9.0 | | | | 123 | 5439.0 |
| Petroica multicolor | U | 10 | 9.6 | 1.00 | 8.0–11.0 | | Fiji 332 | 5441.0 |
| Petroica macrocephala | B | 84 | 10.7 | | 9.0–14.0 | | New Zealand 500 | 5442.0 |
| Petroica goodenovii | U | | | | 7.0–9.0 | | 55 | 5443.0 |
| Petroica phoenicea | U | | 13.3 | | | | 266 | 5444.0 |
| Petroica rosea | U | | 8.7 | | | | 266 | 5445.0 |
| Petroica rodinogaster | U | | | | 8.0–10.0 | | 55 | 5446.0 |
| Petroica australis | U | 2 | 29.5 | | 29.0–30.0 | | New Zealand 215 | 5447.0 |
| Melanodryas cucullata | U | | 21.2 | | | | 194 | 5449.0 |
| Melanodryas vittata | U | | 20.0 | | | | 134 | 5450.0 |
| Tregellasia capito | U | | | | 15.0–18.0 | | 55 | 5451.0 |
| Tregellasia leucops | M | 10 | 17.4 | 1.00 | 15.3–19.4 | | New Guinea 154 | 5452.0 |
| | F | 7 | 15.4 | | 14.0–18.0 | | | |
| Eopsaltria australis | U | | 19.6 | | | | 194 | 5453.0 |

## Body Masses of World Birds (continued)

| Species | Sex | N | Mean | Std dev | Range | Sn | Location | Number |
|---------|-----|---|------|---------|-------|----|----------|--------|
| Eopsaltria flaviventris | B | 3 | 13.0 | | 10.5–14.5 | | New Caledonia 506 | 5455.0 |
| Eopsaltria pulverulenta | B | 2 | 23.2 | | 20.0–26.5 | | New Guinea 217 | 5457.0 |
| Poecilodryas brachyura | U | | 25.1 | | | | 123 | 5458.0 |
| Poecilodryas hypoleuca | U | 6 | 18.0 | | | | New Guinea 35 | 5459.0 |
| Poecilodryas superciliosa | U | | | | 16.0–22.0 | | 55 | 5460.0 |
| Poecilodryas placens | M | 2 | 27.0 | | 26.0–28.0 | | New Guinea 154 | 5461.0 |
| | F | 3 | 24.0 | | 23.0–25.0 | | | |
| Poecilodryas albonotata | B | 8 | 38.2 | | 34.5–43.0 | | New Guinea 154, 216 | 5462.0 |
| Peneothello sigillatus | U | | 22.2 | | | | 123 | 5463.0 |
| Peneothello cryptoleucus | U | | 18.9 | | | | 123 | 5464.0 |
| Peneothello cyanus | M | 10 | 26.9 | 1.50 | 24.0–30.0 | | New Guinea 154 | 5465.0 |
| | F | 10 | 23.2 | 1.70 | 20.7–26.0 | | | |
| Peneothello bimaculatus | M | 7 | 26.3 | | 25.0–28.0 | | New Guinea 154 | 5466.0 |
| | F | 2 | 21.5 | | 21.0–22.0 | | | |
| Heteromyias albispecularis | B | 11 | 32.0 | | 30.0–36.0 | | New Guinea 154, 157, 216 | 5467.0 |
| Pachycephalopsis hattamensis | U | | 34.6 | | | | 123 | 5469.0 |
| Pachycephalopsis poliosoma | M | 10 | 40.5 | 2.60 | 35.0–42.0 | | New Guinea 154 | 5470.0 |
| | F | 10 | 35.6 | 2.70 | 32.0–40.0 | | | |
| Drymodes superciliaris | B | 4 | 46.8 | | 36.0–58.0 | | New Guinea 154 | 5471.0 |
| Drymodes brunneopygia | U | | | | 36.0–38.0 | | Australia 55 | 5472.0 |

### ORDER: PASSERIFORMES      FAMILY: PACHYCEPHALIDAE

| Species | Sex | N | Mean | Std dev | Range | Sn | Location | Number |
|---------|-----|---|------|---------|-------|----|----------|--------|
| Daphoenosita chrystoptera | U | | 11.9 | | | | 266 | 5473.0 |
| Mohoua albicilla | M | 20 | 18.3 | 1.27 | 16.0–21.0 | | New Zealand 500 | 5475.0 |
| | F | 27 | 17.0 | 2.42 | 12.0–20.0 | | | |
| Mohoua novaeseelandiae | M | 51 | 13.4 | 0.69 | 12.0–15.0 | | New Zealand 129 | 5477.0 |
| | F | 24 | 11.0 | 0.43 | 10.5–12.0 | | | |
| Falcunculus frontatus | U | | 28.6 | | | | 266 | 5478.0 |
| Oreoica gutturalis | U | | | | 57.0–67.0 | | 55 | 5479.0 |

## Body Masses of World Birds (continued)

| Species | Sex | N | Mean | Std dev | Range | Sn | Location | Number |
|---------|-----|---|------|---------|-------|----|----------|--------|
| Rhagologus leucostigma | B | 7 | 26.8 | | 23.0–29.5 | | New Guinea 154 | 5480.0 |
| Pachycare flavogrisea | B | 17 | 16.2 | | 12.5–19.0 | | New Guinea 154 | 5481.0 |
| Aleadryas rufinucha | B | 7 | 39.9 | | 37.5–42.5 | | New Guinea 154, 216 | 5484.0 |
| Pachycephala tenebrosa | U | | 46.0 | | | | 123 | 5485.0 |
| Pachycephala olivacea | U | | | | 31.0–45.0 | | 55 | 5486.0 |
| Pachycephala rufogularis | U | 8 | 36.1 | | | | Australia 686 | 5487.0 |
| Pachycephala inornata | U | 7 | 32.7 | | | | Australia 686 | 5488.0 |
| Pachycephala grisola | B | 4 | 21.0 | | | | India 5 | 5489.0 |
| Pachycephala hyperythra | B | 8 | 27.7 | | 25.5–29.0 | | New Guinea 154 | 5492.0 |
| Pachycephala modesta | B | 8 | 18.6 | | 16.5–20.1 | | New Guinea 154, 216 | 5493.0 |
| Pachycephala philippinensis | B | 67 | 22.3 | | 18.0–28.8 | | Philippines 475 | 5496.0 |
| Pachycephala meyeri | U | 5 | 19.0 | | | | New Guinea 157 | 5497.0 |
| Pachycephala griseiceps | B | 20 | 21.8 | | 20.0–24.3 | | New Guinea 154 | 5498.0 |
| Pachycephala simplex | U | | | | 16.0–25.0 | | 55 | 5499.0 |
| Pachycephala pectoralis | U | 26 | 32.7 | 2.40 | 28.5–38.0 | | Fiji 332 | 5501.0 |
| Pachycephala pectoralis | M | 4 | 24.4 | | 22.0–26.5 | | New Ireland 30 | 5501.0 |
|  | F | 4 | 22.2 | | 20.4–24.0 | | | |
| Pachycephala soror | B | 18 | 24.8 | | 22.5–27.2 | | New Guinea 154 | 5502.0 |
| Pachycephala lorentzi | U | 5 | 19.4 | | 18.0–20.3 | | New Guinea 216 | 5503.0 |
| Pachycephala caledonica | U | 17 | 20.9 | 1.30 | 18.0–23.0 | | New Caledonia 506 | 5505.0 |
| Pachycephala schlegelii | B | 20 | 22.4 | | 19.4–24.8 | | New Guinea 154 | 5508.0 |
| Pachycephala rufiventris | B | 38 | 18.1 | | 15.5–23.0 | | New Caledonia 506 | 5516.0 |
| Pachycephala lanioides | U | | 43.0 | | | | W. Australia 550 | 5517.0 |

## Body Masses of World Birds (continued)

| Species | Sex | N | Mean | Std dev | Range | Sn | Location | Number |
|---|---|---|---|---|---|---|---|---|
| Colluricincla megarhyncha | U | 81 | 32.0 | | | | New Guinea 35 | 5519.0 |
| Colluricincla megarhyncha | M | 10 | 43.4 | 2.34 | 40.0–45.0 | | Australia 196 | 5519.0 |
| | F | 11 | 40.9 | 2.60 | 38.0–45.0 | | | |
| Colluricincla boweri | U | | | | 39.0–48.0 | | 55 | 5520.0 |
| Colluricincla woodwardi | U | | | | 47.0–60.0 | | 55 | 5521.0 |
| Colluricincla harmonica | U | | 75.6 | | | | 266 | 5522.0 |
| Pitohui kirhocephalus | B | 17 | 92.9 | 4.00 | 85.0–100.0 | | New Guinea 154 | 5524.0 |
| Pitohui dichrous | B | 20 | 72.0 | | 67.0–79.0 | | New Guinea 154 | 5525.0 |
| Pitohui ferrugineus | M | 10 | 98.0 | 8.00 | 86.0–110.0 | | New Guinea 154 | 5527.0 |
| | F | 7 | 90.0 | | 77.0–99.0 | | | |
| Pitohui cristatus | U | 3 | 79.0 | | | | New Guinea 35 | 5528.0 |
| Pitohui nigrescens | B | 9 | 78.4 | | 73.0–86.0 | | New Guinea 35, 216 | 5529.0 |
| Eulacestoma nigropectus | B | 6 | 21.0 | | 19.7–22.0 | | New Guinea 154 | 5530.0 |

**ORDER: PASSERIFORMES**          **FAMILY: ORTHONYCHIDAE**

| Species | Sex | N | Mean | Std dev | Range | Sn | Location | Number |
|---|---|---|---|---|---|---|---|---|
| Orthonyx temminckii | U | | | | 46.0–69.0 | | Australia 55 | 5532.0 |
| Orthonyx spaldingii | U | | | | 126.0–213.0 | | Australia 55 | 5533.0 |

**ORDER: PASSERIFORMES**          **FAMILY: POMATOSTOMIDAE**

| Species | Sex | N | Mean | Std dev | Range | Sn | Location | Number |
|---|---|---|---|---|---|---|---|---|
| Pomatostomus isidorei | U | 11 | 70.0 | | | | New Guinea 35 | 5534.0 |
| Pomatostomus temporalis | B | 321 | 75.0 | 4.93 | | | Australia 214 | 5535.0 |
| Pomatostomus superciliosus | B | 2 | 35.0 | | 35.0–35.0 | | Australia 199 | 5536.0 |
| Pomatostomus halli | U | 15 | 41.5 | 1.78 | | | Australia 214 | 5537.0 |
| Pomatostomus ruficeps | U | | | | 50.0–62.0 | | Australia 55 | 5538.0 |

**ORDER: PASSERIFORMES**          **FAMILY: CINCLOSOMATIDAE**

| Species | Sex | N | Mean | Std dev | Range | Sn | Location | Number |
|---|---|---|---|---|---|---|---|---|
| Psophodes olivaceus | U | | 62.2 | | | | Australia 266 | 5540.0 |

## Body Masses of World Birds (continued)

| Species | Sex | N | Mean | Std dev | Range | Sn | Location | Number |
|---|---|---|---|---|---|---|---|---|
| Psophodes nigrogularis | M | 11 | 46.0 | 2.20 | | | W. Australia | 5541.0 |
| | F | 12 | 43.1 | 2.69 | | | 569 | |
| Psophodes occidentalis | U | | | | 35.0−49.0 | | Australia | 5542.0 |
| | | | | | | | 55 | |
| Psophodes cristatus | U | | | | 31.0−46.0 | | Australia | 5543.0 |
| | | | | | | | 55 | |
| Cinclosoma punctatum | U | | | | 67.0−87.0 | | Australia | 5544.0 |
| | | | | | | | 55 | |
| Cinclosoma castanotus | M | 72 | 74.6 | 4.14 | 67.0−87.0 | | Australia | 5545.0 |
| | F | 44 | 72.1 | 6.35 | 58.0−85.0 | | 197 | |
| Cinclosoma castaneothorax | M | 13 | 69.2 | 3.20 | 63.0−75.0 | | Australia | 5546.0 |
| | F | 14 | 61.5 | 3.94 | 56.0−67.0 | | 197 | |
| Cinclosoma castaneothorax marginatus | M | 62 | 60.8 | 4.45 | 54.0−74.0 | | Australia | 5546. |
| | F | 41 | 58.7 | 4.13 | 51.0−70.0 | | 197 | |
| Cinclosoma cinnamomeum | M | 19 | 58.8 | 3.14 | 53.0−67.0 | | Australia | 5547.0 |
| | F | 6 | 58.4 | | 56.0−62.0 | | 197 | |
| Ptilorrhoa leucosticta | B | 2 | 50.0 | | 49.0−51.0 | | New Guinea | 5549.0 |
| | | | | | | | 154 | |
| Ptilorrhoa caerulescens | U | 21 | 49.0 | | | | New Guinea | 5550.0 |
| | | | | | | | 35 | |
| Ptilorrhoa castanonota | B | 5 | 72.0 | | 70.0−74.0 | | New Guinea | 5551.0 |
| | | | | | | | 154 | |
| Ifrita kowaldi | B | 15 | 32.1 | 2.14 | 28.8−36.0 | | New Guinea | 5553.0 |
| | | | | | | | 154, 216 | |

### ORDER: PASSERIFORMES                    FAMILY: CORCORACIDAE

| Species | Sex | N | Mean | Std dev | Range | Sn | Location | Number |
|---|---|---|---|---|---|---|---|---|
| Corcorax melanorhamphos | U | 191 | 364.0 | 22.10 | | W | Australia | 5555.0 |
| | | | | | | | 508 | |

### ORDER: PASSERIFORMES                    FAMILY: MONARCHIDAE

| Species | Sex | N | Mean | Std dev | Range | Sn | Location | Number |
|---|---|---|---|---|---|---|---|---|
| Rhipidura hypoxantha | B | 10 | | | 5.0−6.0 | | India | 5557.0 |
| | | | | | | | 5 | |
| Rhipidura superciliaris | B | 47 | 12.5 | | 9.5−16.8 | | Philippines | 5558.0 |
| | | | | | | | 475 | |
| Rhipidura nigrocinnamomea | B | 22 | 13.0 | | 10.5−14.5 | | Philippines | 5561.0 |
| | | | | | | | 475 | |
| Rhipidura albicollis | U | 24 | 12.9 | | 11.0−15.0 | | Malaysia | 5562.0 |
| | | | | | | | 371 | |
| Rhipidura aureola | B | 5 | 11.2 | | 10.0−13.0 | | India | 5564.0 |
| | | | | | | | 5 | |
| Rhipidura javanica | U | 13 | 12.5 | 0.90 | 10.7−13.9 | | Thailand | 5565.0 |
| | | | | | | | 378 | |
| Rhipidura perlata | M | 1 | 15.1 | | | | Borneo | 5566.0 |
| | F | 2 | 12.8 | | 12.4−13.2 | | 627 | |

## Body Masses of World Birds (continued)

| Species | Sex | N | Mean | Std dev | Range | Sn | Location | Number |
|---|---|---|---|---|---|---|---|---|
| Rhipidura leucophrys | B | 17 | 27.7 | 3.53 | 23.0–34.0 | | New Guinea 154, 216, 217, 219, 267 | 5567.0 |
| Rhipidura rufiventris | U | 9 | 15.0 | | | | New Guinea 35 | 5569.0 |
| Rhipidura albolimbata | B | 14 | 10.2 | 0.83 | 9.0–11.5 | | New Guinea 154, 216 | 5572.0 |
| Rhipidura hyperythra | B | 13 | 11.2 | | | | New Guinea 35, 154, 216 | 5573.0 |
| Rhipidura threnothorax | B | 10 | 17.6 | 1.81 | 14.7–20.0 | | New Guinea 154 | 5574.0 |
| Rhipidura maculipectus | U | 7 | 18.0 | | | | New Guinea 35 | 5575.0 |
| Rhipidura leucothorax | B | 6 | 16.9 | | 11.0–20.0 | | New Guinea 154, 217 | 5576.0 |
| Rhipidura atra | B | 14 | 12.2 | | 10.5–14.0 | | New Guinea 154 | 5577.0 |
| Rhipidura fuliginosa | U | 92 | 8.0 | 1.00 | 5.5–10.0 | | New Zealand 500 | 5579.0 |
| Rhipidura spilodera | U | 21 | 12.9 | 1.10 | 11.5–15.0 | | Fiji 332 | 5583.0 |
| Rhipidura brachyrhyncha | M | 2 | 10.5 | | 10.3–10.7 | | New Guinea 154 | 5586.0 |
| | F | 2 | 9.0 | | 8.3–9.7 | | | |
| Rhipidura rufidorsa | U | 5 | 10.0 | | | | New Guinea 35 | 5592.0 |
| Rhipidura dahli | B | | | | 9.0–11.0 | | New Britain 219 | 5593.0 |
| Rhipidura rufifrons | U | | 10.4 | | | | Australia 266 | 5596.0 |
| Erythrocercus holochlorus | U | 1 | 5.0 | | | | Somalia 691 | 5599.0 |
| Elminia longicauda | U | 11 | 9.8 | 2.79 | 6.3–17.0 | | 228, 349 | 5601.0 |
| Trochocercus nigromitratus | U | 5 | 9.8 | | 8.9–10.8 | | Liberia 314a | 5603.0 |
| Trochocercus albiventris | B | 5 | 9.6 | | 9.0–10.0 | | Cameroon 349 | 5604.0 |
| Trochocercus albonotatus | B | 44 | 9.2 | | 7.9–11.6 | | Malawi 168 | 5605.0 |
| Trochocercus cyanomelas | B | 24 | 10.2 | | 9.3–12.1 | | Somalia 691 | 5607.0 |
| Hypothymis helenae | B | 4 | 9.9 | | 9.2–10.3 | | Philippines 475 | 5608.0 |
| Hypothymis coelestis | M | 1 | 13.6 | | | | Philippines 475 | 5609.0 |

**Body Masses of World Birds (continued)**

| Species | Sex | N | Mean | Std dev | Range | Sn | Location | Number |
|---|---|---|---|---|---|---|---|---|
| Hypothymis azurea | B | 21 | 11.1 | | 8.9–13.4 | | Philippines 475 | 5610.0 |
| Terpsiphone rufiventer | U | 10 | 15.1 | 1.60 | 12.5–17.3 | | Liberia 314a | 5612.0 |
| Terpsiphone rufocinerea | U | 1 | 16.5 | | | | Gabon Karr 1976 | 5614.0 |
| Terpsiphone viridis | B | 16 | 14.4 | | 11.3–17.0 | | Ghana 228 | 5615.0 |
| Terpsiphone mutata | B | 17 | 12.8 | | 11.0–15.5 | | Madagascar 152 | 5617.0 |
| Terpsiphone bourbonnensis | M | 12 | 11.7 | 0.38 | 11.0–12.5 | | | 5619.0 |
| | F | 10 | 11.3 | 0.51 | 10.4–12.1 | | 152 | |
| Terpsiphone paradisi | U | 50 | 18.5 | | 16.0–21.0 | F | India 5 | 5620.0 |
| Terpsiphone atrocaudata | M | 2 | 18.7 | | | | Thailand 378 | 5621.0 |
| Terpsiphone cinnamomea | M | 3 | 22.4 | | 21.2–23.1 | | Philippines 475 | 5622.0 |
| | F | 1 | 19.4 | | | | | |
| Chapsiempis sandwichensis | U | | 14.0 | | | | Hawaiian Is. 400 | 5624.0 |
| Mayrornis lessoni | U | 13 | 11.2 | 1.50 | 9.5–14.5 | | Fiji 332 | 5631.0 |
| Clytorhynchus pachycephaloides | M | 2 | 24.8 | | 24.5–25.0 | | New Caledonia 506 | 5631.0 |
| Clytorhynchus vitiensis | U | 74 | 29.2 | 3.30 | | | Fiji 332 | 5635.0 |
| Monarcha axillaris | B | 14 | 15.6 | | 14.3–17.0 | | New Guinea 154 | 5639.0 |
| Monarcha frater | B | 11 | 22.0 | 1.06 | 20.8–24.0 | | New Guinea 154 | 5642.0 |
| Monarcha melanopsis | U | | 23.6 | | | | Australia 266 | 5643.0 |
| Monarcha leucotis | U | | | | 10.0–13.0 | | 55 | 5649.0 |
| Monarcha guttulus | B | 14 | 16.4 | | 14.0–18.3 | | New Guinea 154 | 5650.0 |
| Monarcha trivirgatus | U | | | | 11.0–16.0 | | 55 | 5652.0 |
| Monarcha manadensis | U | | 22.5 | | | | New Guinea 123 | 5659.0 |
| Monarcha verticalis | B | 11 | 21.7 | 2.61 | 19.2–29.0 | | 30, 219 | 5663.0 |
| Monarcha chrysomela | B | 9 | 14.6 | | | | 30, 35, 154 | 5669.0 |

## Body Masses of World Birds (continued)

| Species | Sex | N | Mean | Std dev | Range | Sn | Location | Number |
|---------|-----|---|------|---------|-------|----|----------|--------|
| Arses telescopthalmus | B | 11 | 17.0 | | 15.7–19.0 | | New Guinea 154 | 5670.0 |
| Arses kaupi | U | | | | 12.0–15.0 | | 55 | 5672.0 |
| Myiagra rubecula | U | | 14.0 | | | | Australia 194 | 5679.0 |
| Myiagra caledonica | U | 11 | 10.8 | 0.70 | 10.0–12.0 | | New Caledonia 506 | 5682.0 |
| Myiagra vanikorensis | U | 3 | 13.2 | | 13.0–13.5 | | Fiji 332 | 5683.0 |
| Myiagra azureocapilla | B | 17 | 12.9 | | 9.5–16.0 | | Fiji 332 | 5685.0 |
| Myiagra ruficollis | U | | | | 10.0–13.0 | | 55 | 5686.0 |
| Myiagra cyanoleuca | U | | 17.5 | | | | Australia 266 | 5687.0 |
| Myiagra inquieta | U | | 24.0 | | | | Australia 194a | 5688.0 |
| Piezorhynchus alecto | U | 8 | 24.0 | | | | New Guinea 35 | 5689.0 |
| Piezorhynchus hebetior | B | 4 | 20.2 | | 18.5–21.5 | | New Britain 219 | 5690.0 |
| Lamprolia victoriae victoriae | U | 8 | 19.0 | | 16.0–21.0 | | Fiji 332 | 5691.0 |
| Lamprolia victoriae kleinschmidti | U | 5 | 11.3 | | 10.0–12.5 | | Fiji 332 | 5691.0 |
| Machaerirhynchus flaviventer | U | 9 | 11.2 | | | | New Guinea 35, 154 | 5692.0 |
| Machaerirhynchus nigripectus | B | 7 | 11.6 | | 11.0–12.5 | | New Guinea 154, 216 | 5693.0 |

### ORDER: PASSERIFORMES      FAMILY: DICRURIDAE

| Species | Sex | N | Mean | Std dev | Range | Sn | Location | Number |
|---------|-----|---|------|---------|-------|----|----------|--------|
| Chaetorhynchus papuensis | M | 10 | 42.0 | 2.00 | 36.0–45.0 | | New Guinea 154 | 5694.0 |
| | F | 10 | 33.0 | 4.00 | 27.0–39.0 | | | |
| Dicrurus ludwigii | U | 3 | 26.7 | | 24.5–28.5 | | Ghana 228 | 5695.0 |
| Dicrurus adsimilis | B | 15 | 45.7 | | 40.0–51.0 | | India 5 | 5697.0 |
| Dicrurus aldabranus | M | 1 | 51.0 | | | | Aldabra Is. 40 | 5699.0 |
| | F | 1 | 46.0 | | | | | |
| Dicrurus forficatus | B | 5 | 43.8 | | 39.0–48.0 | | Madagascar 39 | 5701.0 |
| Dicrurus macrocercus | U | 3 | 49.8 | | 38.0–59.0 | | 78 | 5703.0 |

## Body Masses of World Birds (continued)

| Species | Sex | N | Mean | Std dev | Range | Sn | Location | Number |
|---|---|---|---|---|---|---|---|---|
| Dicrurus leucophaeus | B | 40 | 37.6 | | 32.0–45.0 | | India 5 | 5704.0 |
| Dicrurus caerulescens | U | 2 | 40.0 | | 39.0–41.0 | | India 5 | 5705.0 |
| Dicrurus annectans | F | 1 | 44.0 | | | | India 5 | 5706.0 |
| Dicrurus aeneus | M | 5 | | | 26.0–28.0 | | India 5 | 5707.0 |
| | F | 2 | 26.0 | | 22.0–30.0 | | | |
| Dicrurus remifer | U | 19 | 43.1 | | 40.0–50.0 | | Malaysia 371 | 5708.0 |
| Dicrurus megarhynchus | F | 2 | 130.0 | | 129.0–130.0 | | New Ireland 30 | 5713.0 |
| Dicrurus paradiseus | B | 8 | 83.3 | | 70.0–124.0 | | India 5 | 5717.0 |

### ORDER: PASSERIFORMES          FAMILY: CORVIDAE

| Species | Sex | N | Mean | Std dev | Range | Sn | Location | Number |
|---|---|---|---|---|---|---|---|---|
| Platysmurus leucopterus | B | 2 | 180.0 | | 178.0–182.0 | | Borneo 627 | 5719.0 |
| Gymnorhinus cyanocephalus | B | 14 | 103.0 | | | | Arizona, USA 21 | 5720.0 |
| Cyanocitta cristata | B | 462 | 86.8 | 8.08 | 64.1–109.0 | Y | Pennsylvania,USA 101 | 5721.0 |
| Cyanocitta stelleri | B | 67 | 128.0 | | 111.0–142.0 | | western Canada 694 | 5722.0 |
| Aphelocoma californica nevadae | M | 56 | 80.3 | | 69.0–98.0 | | western USA 13 | 5724.0 |
| | F | 37 | 73.8 | | 65.0–83.0 | | | |
| Aphelocoma californica insularis | M | 42 | 123.0 | | 100.0–147.0 | | California, USA 13 | 5724.0 |
| | F | 44 | 109.0 | | 99.0–117.0 | | | |
| Aphelocoma coerulescens | B | 100 | 80.2 | | | | Florida, USA 698 | 5725.0 |
| Aphelocoma ultramarina | B | 20 | 124.0 | 10.60 | 105.0–144.0 | | Arizona, USA 594 | 5726.0 |
| Aphelocoma unicolor | M | 3 | 124.0 | | 121.0–130.0 | | Mexico 584 | 5727.0 |
| Cyanolyca viridicyana | M | 8 | 106.0 | | 87.0–127.0 | | | 5730.0 |
| | F | 4 | 93.2 | | 82.0–113.0 | | 584, 668 | |
| Cyanolyca cucullata | M | 1 | 109.0 | | | | | 5731.0 |
| | F | 4 | 94.2 | | | | 584 | |
| Cyanolyca pumilo | M | 1 | 47.0 | | | | Guatemala 584 | 5733.0 |
| Cyanolyca nana | B | 2 | 41.0 | | 41.0–41.0 584 | | Mexico | 5734.0 |
| Cyanolyca mirabilis | M | 3 | 52.4 | | 49.9–54.0 | | Mexico 584 | 5735.0 |

## Body Masses of World Birds (continued)

| Species | Sex | N | Mean | Std dev | Range | Sn | Location | Number |
|---|---|---|---|---|---|---|---|---|
| Cyanolyca argentigula | B | 2 | 200.0 | | 190.0–210.0 | | Panama 244 | 5736.0 |
| Cyanocorax melanocyaneus | B | 9 | 108.0 | | 103.0–115.0 | | 584 | 5737.0 |
| Cyanocorax sanblasianus | U | 3 | 109.0 | | 101.0–114.0 | | Mexico 593 | 5738.0 |
| Cyanocorax yucatanicus | B | 13 | 118.0 | 7.00 | 105.0–128.0 | | Mexico; Belize 323, 457, 510 | 5739.0 |
| Cyanocorax beecheii | U | 18 | 193.0 | | | | W. Mexico 683 | 5740.0 |
| Cyanocorax cyanomelas | B | 7 | 222.0 | | | | Peru 193 | 5741.0 |
| Cyanocorax caeruleus | M | 1 | 272.0 | | | | Brazil 38 | 5742.0 |
| Cyanocorax violaceus | U | | 262.0 | | | | Peru 623 | 5743.0 |
| Cyanocorax cayanus | F | 1 | 175.0 | | | | Surinam 245 | 5746.0 |
| Cyanocorax affinis | U | 11 | 212.0 | 10.77 | 194.0–232.0 | | 87, 244, 384, 497 | 5747.0 |
| Cyanocorax dickeyi | U | 3 | 176.0 | | | | Mexico 121, 594 | 5748.0 |
| Cyanocorax chrysops | B | 13 | 157.0 | | | | 38, 111, 200 | 5749.0 |
| Cyanocorax yncas | B | 122 | 78.5 | 5.60 | 66.0–92.0 | | Texas, USA 212 | 5752.0 |
| Psilorhinus morio | B | 13 | 204.0 | | 173.0–224.0 | | Mexico 457 | 5753.0 |
| Calocitta colliei | B | 4 | 234.0 | | 225.0–251.0 | PB | Mexico 548 | 5754.0 |
| Calocitta formosa | B | 2 | 210.0 | | 206.0–213.0 | | Mexico 458 | 5755.0 |
| Garrulus glandarius | U | 50 | 161.0 | | 140.0–187.0 | S | Britain 258 | 5756.0 |
| Garrulus lanceolatus | B | 25 | 97.7 | | 84.0–104.0 | | India 5 | 5757.0 |
| Perisoreus infaustus | U | 56 | 84.4 | | 72.0–97.0 | | 78 | 5759.0 |
| Perisoreus canadensis | B | 20 | 73.0 | | 62.0–73.0 | | western Canada 694 | 5761.0 |
| Urocissa ornata | F | 1 | 196.0 | | | | Sri Lanka 5 | 5762.0 |
| Urocissa flavirostris flavirostris | M F | 5 3 | | | 168.0–180.0 132.0–143.0 | | India 5 | 5764.0 |

## Body Masses of World Birds (continued)

| Species | Sex | N | Mean | Std dev | Range | Sn | Location | Number |
|---|---|---|---|---|---|---|---|---|
| Urocissa flavirostris cucullata | M | 6 | | | 130.0–165.0 | | India | 5764.0 |
| | F | 5 | | | 123.0–163.0 | | 5 | |
| Urocissa erythrorhyncha | B | 8 | | | 196.0–232.0 | | India 5 | 5765.0 |
| Cissa chinensis | M | 4 | | | 130.0–133.0 | | India | 5767.0 |
| | F | 3 | | | 120.0–124.0 | | 5 | |
| Cyanopica cyana | F | 2 | 72.0 | | 62.0–82.0 | | 149 | 5769.0 |
| Dendrocitta vagabunda pallida | U | 13 | 100.0 | | 90.0–118.0 | | India 5 | 5770.0 |
| Dendrocitta formosae | M | 38 | | | 90.0–121.0 | | India | 5771.0 |
| | F | 21 | | | 89.0–115.0 | | 5 | |
| Dendrocitta leucogastra | F | 1 | 99.2 | | | | India 5 | 5773.0 |
| Dendrocitta bayleyi | U | | | | 92.0–113.0 | | Andaman Is. 5 | 5775.0 |
| Pica pica | M | 81 | 189.0 | 10.30 | 159.0–209.0 | Y | Colorado, USA | 5779.0 |
| | F | 39 | 166.0 | 14.30 | 135.0–197.0 | | 177 | |
| Pica nuttalli | M | 32 | 174.0 | 8.75 | 158.0–189.0 | Y | California, USA | 5780.0 |
| | F | 31 | 144.0 | 8.69 | 126.0–158.0 | | 638 | |
| Podeces panderi | U | 4 | 91.1 | | 86.0–96.0 | | 149 | 5784.0 |
| Pseudopodoces humilis | M | 4 | | | 42.5–48.5 | | Tibet 5 | 5786.0 |
| Nucifraga columbiana | M | 180 | 141.0 | 8.10 | 123.0–161.0 | | Montana, USA | 5787.0 |
| | F | 92 | 129.0 | 8.90 | 106.0–155.0 | | 382 | |
| Nucifraga caryocatactes | M | 40 | 176.0 | | 153.0–190.0 | | | 5788.0 |
| | F | 28 | 169.0 | | 124.0–184.0 | | 149 | |
| Pyrrhocorax pyrrhocorax | U | 20 | 324.0 | | 260.0–375.0 | F | Britain 258 | 5789.0 |
| Pyrrhocorax graculus | B | 8 | | | 203.0–244.0 | | India 5 | 5790.0 |
| Ptilostomus afer | U | 1 | 128.0 | | | | Sudan 680 | 5791.0 |
| Corvus monedula | U | 196 | 246.0 | | | S | Britain 258 | 5792.0 |
| Corvus dauuricus | M | 1 | 123.0 | | | | 149 | 5793.0 |
| Corvus splendens | B | 6 | 296.0 | | 252.0–362.0 | | India 5 | 5794.0 |
| Corvus moneduloides | B | 22 | 275.0 | 31.50 | 230.0–330.0 | | New Caledonia 506 | 5795.0 |
| Corvus enca | B | 5 | 238.0 | | 222.0–255.0 | | Philippines 475 | 5796.0 |

## Body Masses of World Birds (continued)

| Species | Sex | N | Mean | Std dev | Range | Sn | Location | Number |
|---|---|---|---|---|---|---|---|---|
| Corvus tristis | M | 1 | 635.0 | | | | New Guinea 35 | 5805.0 |
| Corvus capensis | U | | 697.0 | | | | 78 | 5806.0 |
| Corvus frugilegus | U | 75 | 488.0 | | | S | Britain 258 | 5807.0 |
| Corvus caurinus | M | 19 | 415.0 | | 389.0–486.0 | | Washington, USA 306 | 5808.0 |
| | F | 8 | 368.0 | | 315.0–421.0 | | | |
| Corvus brachyrhynchos | M | 6 | 458.0 | | | | | 5809.0 |
| | F | 6 | 438.0 | | | | 244 | |
| Corvus sinaloae | U | 2 | 244.0 | | 229.0–258.0 | | W. Mexico 594 | 5811.0 |
| Corvus ossifragus | M | 20 | 300.0 | 20.30 | 260.0–332.0 | | Florida, USA 25 | 5812.0 |
| | F | 19 | 270.0 | 23.00 | 195.0–304.0 | | | |
| Corvus palmarum | B | 2 | 289.0 | | 263.0–315.0 | | Cuba 496 | 5813.0 |
| Corvus nasicus | B | 6 | 360.0 | | 330.0–385.0 | | Cuba 429 | 5815.0 |
| Corvus corone | U | 126 | 570.0 | | | S | Britain 258 | 5817.0 |
| Corvus macrorhynchos intermedius | M | 7 | | | 460.0–582.0 | | India 5 | 5818.0 |
| | F | 10 | | | 395.0–495.0 | | | |
| Corvus macrorhynchos tibetosinensis | M | 4 | | | 560.0–650.0 | | India 5 | 5818.0 |
| | F | 7 | | | 450.0–565.0 | | | |
| Corvus orru | B | 3 | 433.0 | | 428.0–444.0 | | New Britain 219 | 5820.0 |
| Corvus bennetti | U | | | | 312.0–450.0 | | W. Australia 550 | 5821.0 |
| Corvus coronoides | M | 216 | 675.0 | | 540.0–820.0 | | Australia 146 | 5822.0 |
| Corvus mellori | M | 138 | 567.0 | | 407.0–660.0 | | Australia 146 | 5823.0 |
| Corvus tasmanicus | M | 5 | 712.0 | | 500.0–800.0 | | 146 | 5824.0 |
| Corvus cryptoleucus | M | 76 | 556.0 | 43.60 | 442.0–667.0 | | | 5827.0 |
| | F | 68 | 512.0 | 42.70 | 378.0–607.0 | | 24 | |
| Corvus albus | B | 3 | 529.0 | | 491.0–594.0 | | Sudan 680 | 5828.0 |
| Corvus corax | M | 5 | 1240.0 | | 1100.0–1400.0 | B | Alaska, USA 278 | 5830.0 |
| | F | 3 | 1158.0 | | 1050.0–1300.0 | | | |
| Corvus rhipidurus | M | 1 | 745.0 | | | | Sudan 680 | 5831.0 |
| Corvus albicollis | U | | 900.0 | | | | 78 | 5832.0 |

## Body Masses of World Birds (continued)

| Species | Sex | N | Mean | Std dev | Range | Sn | Location | Number |
|---------|-----|---|------|---------|-------|-----|----------|--------|
| ORDER: PASSERIFORMES | | | | | FAMILY: PARADISAEIDAE | | | |
| Melampitta lugubris | M | 8 | 44.2 | | 41.0–47.0 | | | 5834.0 |
| | F | 6 | 41.2 | | 38.0–45.0 | | 206 | |
| Melampitta gigantea | M | 1 | 205.0 | | | | New Guinea 156 | 5835.0 |
| Loboparadisaea sericea | B | 6 | 73.2 | | 70.0–77.0 | | New Guinea 154 | 5836.0 |
| Cnemophilus macgregorii | M | 1 | 94.0 | | | | New Guinea 154 | 5837.0 |
| Cnemophilus loria | B | 10 | 90.1 | | 77.0–101.0 | | New Guinea 154, 216 | 5838.0 |
| Manucodia atra | U | 11 | 208.0 | | | | New Guinea 35 | 5841.0 |
| Manucodia chalybata | B | 10 | 192.0 | | | | New Guinea 31, 154, 218 | 5842.0 |
| Manucodia keraudrenii | M | 9 | 175.0 | | 166.0–190.0 | | New Guinea | 5845.0 |
| | F | 6 | 152.0 | | 131.0–172.0 | | 154, 218 | |
| Paradigalla brevicauda | M | 1 | 175.0 | | | | New Guinea 216 | 5848.0 |
| Epimachus fastuosus | M | 1 | 313.0 | | | | New Guinea | 5849.0 |
| | F | 1 | 218.0 | | | | 216 | |
| Epimachus meyeri | M | 1 | 189.0 | | | | New Guinea | 5850.0 |
| | F | 1 | 145.0 | | | | 216 | |
| Epimachus albertisi | U | | 110.0 | | | | New Guinea 158 | 5851.0 |
| Epimachus bruijnii | U | | 159.0 | | | | New Guinea 158 | 5852.0 |
| Lophorina superba | M | 4 | 89.0 | | 80.0–93.0 | | New Guinea | 5853.0 |
| | F | 4 | 67.0 | | 60.0–75.0 | | 154 | |
| Parotia sefilata | U | | 182.0 | | | | New Guinea 158 | 5854.0 |
| Parotia carolae | M | 2 | 208.0 | | 205.0–211.0 | | New Guinea | 5855.0 |
| | F | 1 | 163.0 | | | | 216 | |
| Parotia lawesii | M | 1 | 175.0 | | | | New Guinea | 5856.0 |
| | F | 5 | 151.0 | | 139.0–166.0 | | 154 | |
| Ptiloris magnificus | U | 11 | 175.0 | | | | New Guinea 35 | 5859.0 |
| Cicinnurus magnificus | M | 12 | 90.0 | 8.00 | | | New Guinea 31 | 5862.0 |
| Cicinnurus regius | U | 11 | 52.0 | | | | New Guinea 35 | 5864.0 |
| Astrapia splendidissima | M | 6 | 136.0 | | 124.0–151.0 | | New Guinea | 5866.0 |
| | F | 8 | 123.0 | | 108.0–133.0 | | 216 | |

## Body Masses of World Birds (continued)

| Species | Sex | N | Mean | Std dev | Range | Sn | Location | Number |
|---|---|---|---|---|---|---|---|---|
| Astrapia stephaniae | B | 2 | 149.0 | | 145.0–153.0 | | New Guinea 154 | 5868.0 |
| Pteridophora alberti | M | 8 | 75.2 | | 68.0–83.0 | | New Guinea 216 | 5870.0 |
| Seleucidis melanoleuca | U | 9 | 178.0 | | | | New Guinea 35 | 5871.0 |
| Paradisaea rubra | M | 8 | 204.0 | | 158.0–224.0 | | | 5872.0 |
| | F | 5 | 157.0 | | 137.0–182.0 | | 338 | |
| Paradisaea minor finschi | M | 5 | 279.0 | | 264.0–291.0 | | | 5873.0 |
| | F | 1 | 185.0 | | | | 338 | |
| Paradisaea minor jobiensis | M | 2 | 296.0 | | 293.0–300.0 | | | 5873.0 |
| | F | 2 | 168.0 | | 152.0–182.0 | | 338 | |
| Paradisaea minor minor | M | 5 | 250.0 | | 225.0–285.0 | | | 5873.0 |
| | F | 2 | 168.0 | | 165.0–170.0 | | 338 | |
| Paradisaea apoda | F | 2 | 172.0 | | 170.0–173.0 | | 338 | 5874.0 |
| Paradisaea raggiana | F | 4 | 184.0 | | 164–203 | | 154, 338 | 5875.0 |
| Paradisaea decora | M | 1 | 237.0 | | | | 338 | 5876.0 |
| Paradisaea rudolphi | M | 2 | 172.0 | | 165.0–178.0 | | 338 | 5878.0 |

### ORDER: PASSERIFORMES          FAMILY: ARTAMIDAE

| Species | Sex | N | Mean | Std dev | Range | Sn | Location | Number |
|---|---|---|---|---|---|---|---|---|
| Artamus fuscus | B | 4 | 39.8 | | 37.0–42.0 | | India 5 | 5879.0 |
| Artamus leucorhynchus | B | 11 | 47.4 | | 36.4–49.6 | | Philippines 475 | 5880.0 |
| Artamus maximus | B | 12 | 58.2 | | 52.0–64.0 | | New Guinea 154 | 5883.0 |
| Artamus insignus | B | 3 | 52.0 | | 49.0–54.0 | | New Britain 219 | 5884.0 |
| Artamus superciliosus | U | | 40.0 | | | | 194a | 5886.0 |
| Artamus cinereus | U | | 35.0 | | | | 133 | 5887.0 |
| Artamus cyanopterus | U | | 40.0 | | | | 194a | 5888.0 |

### ORDER: PASSERIFORMES          FAMILY: CRACTICIDAE

| Species | Sex | N | Mean | Std dev | Range | Sn | Location | Number |
|---|---|---|---|---|---|---|---|---|
| Cracticus torquatus latens | B | 22 | 92.4 | | 79.0–110.0 | | Australia 195 | 5891.0 |
| Cracticus torquatus argenteus | B | 8 | 79.0 | | 74.0–84.0 | | Australia 195 | 5891.0 |

## Body Masses of World Birds (continued)

| Species | Sex | N | Mean | Std dev | Range | Sn | Location | Number |
|---|---|---|---|---|---|---|---|---|
| Cracticus cassicus | B | 7 | 142.0 | | 130.0–155.0 | | New Guinea 154 | 5892.0 |
| Cracticus nigrogularis | U | | 156.0 | | | | W. Australia 550 | 5894.0 |
| Cracticus quoyi | U | 15 | 159.0 | | | | New Guinea 35 | 5895.0 |
| Gymnorhina tibicen | U | | 314.0 | | | | Australia 266 | 5896.0 |
| Gymnorhina dorsalis | U | | | | 284.0–340.0 | | W. Australia 550 | 5897.0 |
| Strepera graculina | U | | 300.0 | | | | Australia 194 | 5899.0 |
| Strepera fuliginosa | M | | 300.0 | | | | estimated 424 | 5900.0 |

ORDER: PASSERIFORMES                    FAMILY: GRALLINIDAE

| Species | Sex | N | Mean | Std dev | Range | Sn | Location | Number |
|---|---|---|---|---|---|---|---|---|
| Grallina cyanoleuca | U | | 89.0 | | | | Australia 194 | 5902.0 |
| Grallina bruijni | M | 1 | 38.4 | | | | New Guinea 154 | 5903.0 |

ORDER: PASSERIFORMES                    FAMILY: AEGITHINIDAE

| Species | Sex | N | Mean | Std dev | Range | Sn | Location | Number |
|---|---|---|---|---|---|---|---|---|
| Aegithina tiphia | B | 7 | 10.9 | | 10.0–12.0 | | India 361 | 5904.0 |
| Aegithina nigrolutea | B | 17 | | | 10.0–14.0 | | India 5 | 5905.0 |
| Aegithina viridissima | B | 4 | 13.0 | | 12.5–13.5 | | Borneo 627 | 5906.0 |

ORDER: PASSERIFORMES                    FAMILY: ORIOLIDAE

| Species | Sex | N | Mean | Std dev | Range | Sn | Location | Number |
|---|---|---|---|---|---|---|---|---|
| Oriolus szalayi | B | 18 | 95.2 | 8.23 | 91.0–115.0 | | New Guinea 154 | 5912.0 |
| Oriolus sagittatus | U | | 96.0 | | | | Australia 194 | 5913.0 |
| Oriolus oriolus | U | 51 | 79.0 | 7.43 | 59.7–96.0 | F | Egypt 392 | 5919.0 |
| Oriolus auratus | U | 1 | 79.4 | | | | Zimbabwe 285 | 5920.0 |
| Oriolus chinensis | B | 5 | | | 72.0–92.0 | | India 5 | 5921.0 |
| Oriolus brachyrhynchus | U | 1 | 45.5 | | | | Gabon 314a | 5925.0 |
| Oriolus larvatus | U | 3 | 61.3 | | 59.6–63.3 | | 51, 284 | 5927.0 |

## Body Masses of World Birds (continued)

| Species | Sex | N | Mean | Std dev | Range | Sn | Location | Number |
|---|---|---|---|---|---|---|---|---|
| Oriolus nigripennis | B | 4 | 58.0 | | 56.0–60.0 | | Cameroon 349 | 5929.0 |
| Oriolus xanthornus | B | 7 | 56.3 | | 48.3–79.0 | | 5, 361, 475 | 5930.0 |
| Oriolus traillii | B | 10 | | | 67.0–81.0 | | India 5 | 5933.0 |
| Specothercs viridis | U | | 135.0 | | | | 134 | 5936.0 |

### ORDER: PASSERIFORMES         FAMILY: CAMPEPHAGIDAE

| Species | Sex | N | Mean | Std dev | Range | Sn | Location | Number |
|---|---|---|---|---|---|---|---|---|
| Coracina caledonica | B | 38 | 140.0 | | 117.0–180.0 | | New Caledonia 506 | 5945.0 |
| Coracina novaehollandiae | U | 8 | 93.3 | | 87.0–102.0 | | Malaysia 371 | 5946.0 |
| Coracina caeruleogrisea | M | 10 | 148.0 | 11.40 | 127.0–170.0 | | New Guinea 154, 216, 218 | 5947.0 |
| | F | 12 | 130.0 | 8.75 | 116.0–142.0 | | | |
| Coracina striata | B | 18 | 111.0 | | 98.2–128.0 | | Philippines 475 | 5949.0 |
| Coracina lineata | B | 5 | 69.8 | | 66.0–75.0 | | 218, 219 | 5951.0 |
| Coracina boyeri | B | 15 | 66.4 | | 61.0–74.0 | | New Guinea 154 | 5952.0 |
| Coracina papuensis | U | 6 | 80.0 | | | | New Guinea 35 | 5954.0 |
| Coracina longicauda | M | 2 | 96.5 | | 92.0–101.0 | | New Guinea 216 | 5956.0 |
| Coracina analis | M | 1 | 97.0 | | | | New Caledonia 506 | 5959.0 |
| | F | 1 | 70.0 | | | | | |
| Coracina pectoralis | B | 2 | 58.0 | | 58.0–58.0 | | Cameroon 349 | 5960.0 |
| Coracina caesia | U | 2 | 42.8 | | | | 51 | 5961.0 |
| Coracina cinerea | B | 2 | 43.2 | | 41.5–45.0 | | Madagascar 39 | 5964.0 |
| Coracina typica | U | 1 | 42.9 | | | | Mauritius 152 | 5965.0 |
| Coracina tenuirostris | B | 4 | 61.0 | | 54.0–67.0 | | New Britain 219 | 5968.0 |
| Coracina morio | B | 10 | 58.1 | 4.75 | 50.0–67.0 | | 154, 216, 218, 475 | 5973.0 |
| Coracina schisticeps | B | 13 | 52.8 | | 48.0–57.0 | | New Guinea 154 | 5976.0 |
| Coracina montana | B | 19 | 64.6 | | 57.0–70.0 | | New Guinea 154 | 5978.0 |

## Body Masses of World Birds (continued)

| Species | Sex | N | Mean | Std dev | Range | Sn | Location | Number |
|---------|-----|---|------|---------|-------|----|----------|--------|
| Coracina mcgregori | B | 21 | 42.6 | | 39.0–47.5 | | Philippines 475 | 5980.0 |
| Coracina polioptera | U | 1 | 35.8 | | | | Thailand 378 | 5982.0 |
| Coracina melaschistos | U | 4 | | | 35.0–42.0 | | India 5 | 5983.0 |
| Coracina fimbriata | U | 3 | 29.3 | | 28.4–30.0 | | Borneo 627 | 5984.0 |
| Coracina melanoptera | U | 19 | | | 24.0–36.0 | | India 5 | 5985.0 |
| Coracina melaena | U | 15 | 55.0 | | | | New Guinea 35 | 5985.1 |
| Campochaera sloetii | B | 4 | 41.0 | | 36.0–46.0 | | New Guinea 154 | 5986.0 |
| Lalage melanoleuca | B | 4 | 46.1 | | 41.9–48.6 | | Philippines 475 | 5987.0 |
| Lalage nigra | B | 12 | 28.4 | | 24.0–31.3 | | Philippines 475 | 5988.0 |
| Lalage sueurii | U | | 26.0 | | | | Australia 194 | 5989 0 |
| Lalage atrovirens | M | 1 | 31.5 | | | | New Guinea 218 | 5993.0 |
| Lalage leucomela | B | 6 | 26.5 | | 24.0–29.0 | | New Guinea 154 | 5994.0 |
| Lalage maculosa | U | 19 | 30.3 | | | | Tonga 491 | 5995.0 |
| Lalage leucopyga | M | 2 | 18.8 | | 16.5–21.0 | | New Caldonia 506 | 5997.0 |
| Campephaga petiti | U | 4 | 28.0 | | 26.0–30.0 | | Cameroon 349 | 5998.0 |
| Campephaga flava | M | 2 | 32.2 | | 31.6–32.9 | | Zimbabwe 285 | 5999.0 |
| Campephaga phoenicea | B | 7 | 28.5 | | 25.5–35.5 | | Ghana 228 | 6000.0 |
| Pericrocotus roseus | B | 4 | 17.2 | | 14.0–19.0 | | India 5 | 6004.0 |
| Pericrocotus cinnamomeus | U | 5 | 8.2 | | 6.0–12.0 | | India 5, 361 | 6008.0 |
| Pericrocotus igneus | B | 3 | 15.0 | | 14.0–15.8 | | Borneo 627 | 6009.0 |
| Pericrocotus solaris | B | 4 | 14.0 | | 11.0–17.2 | | India 5 | 6012.0 |
| Pericrocotus ethologus | B | 2 | 18.0 | | 18.0–18.0 | | India 5 | 6013.0 |

## Body Masses of World Birds (continued)

| Species | Sex | N | Mean | Std dev | Range | Sn | Location | Number |
|---|---|---|---|---|---|---|---|---|
| Pericrocotus brevirostris | B | | | | 16.0–17.0 | | India 5 | 6014.0 |
| Pericrocotus flammeus | M | 3 | 19.6 | | 19.0–20.0 | | Borneo 627 | 6016.0 |
| Hemipus picatus | U | | | | 8.5–9.5 | | India 5 | 6017.0 |
| Hemipus hirundinaccus | F | 1 | 10.2 | | | | Borneo 627 | 6018.0 |
| Pityriasis gymnocephala | M | 2 | 118.0 | | 115.0–121.0 | | Borneo 627 | 6019.0 |
| | F | 1 | 140.0 | | | | | |
| Peltops montanus | B | | 31.6 | 2.59 | 27.0–36.0 | | New Guinea 154 | 6020.0 |
| Peltops blainvillii | U | 11 | 30.0 | | | | New Guinea 35 | 6021.0 |

### ORDER: PASSERIFORMES                 FAMILY: IRENIDAE

| Species | Sex | N | Mean | Std dev | Range | Sn | Location | Number |
|---|---|---|---|---|---|---|---|---|
| Irena puella | U | 7 | 62.5 | | 51.0–70.3 | | Malaysia;Borneo 371, 627 | 6022.0 |
| Irena cyanogaster | B | 29 | 81.4 | | 76.0–96.1 | | Philippines 475 | 6023.0 |
| Chloropsis sonnerati | M | 1 | 45.8 | | | | Borneo 627 | 6026.0 |
| Chloropsis cyanopogon | B | 3 | 22.4 | | 19.8–25.3 | | Borneo 627 | 6027.0 |
| Chloropsis cochinchinensis | M | 1 | 24.0 | | | | India 361 | 6028.0 |
| Chloropsis aurifrons | M | 2 | 35.5 | | 35.0–36.0 | | | 6029.0 |
| | F | 1 | 29.0 | | | | 5 | |
| Chloropsis hardwickii | M | | | | 32.0–40.0 | | India 5 | 6030.0 |
| | F | | | | 25.0–34.0 | | | |

### ORDER: PASSERIFORMES                 FAMILY: VIREONIDAE

| Species | Sex | N | Mean | Std dev | Range | Sn | Location | Number |
|---|---|---|---|---|---|---|---|---|
| Cyclarhis gujanensis | U | 26 | 28.8 | | 22.5–35.0 | | Trinidad 188 | 6032.0 |
| Cyclarhis nigrirostris | U | 3 | 31.7 | | 29.6–33.1 | | Columbia 387 | 6033.0 |
| Vireolanius melitophrys | M | 1 | 34.7 | | | | Oaxaca, Mexico 584 | 6034.0 |
| Vireolanius pulchellus | B | 4 | 24.0 | | 22.0–26.1 | | Belize; Mexico 510, 584 | 6035.0 |
| Vireolanius leucotis | U | | 26.0 | | | | Peru 623 | 6037.0 |
| Vireo brevipennis | M | 1 | 11.8 | | | | Oaxaca, Mexico 584 | 6038.0 |

## Body Masses of World Birds (continued)

| Species | Sex | N | Mean | Std dev | Range | Sn | Location | Number |
|---------|-----|---|------|---------|-------|----|----------|--------|
| Vireo bellii | B | 33 | 8.5 | 0.55 | 7.4–9.8 | | Arizona, USA 177 | 6039.0 |
| Vireo atricapillus | B | 5 | 8.5 | | 8.0–9.2 | B | Coahuila,Mexico 386 | 6040.0 |
| Vireo nelsoni | M | 1 | 9.4 | | | | Oaxaca, Mexico 584 | 6041.0 |
| Vireo huttoni | B | 15 | 11.6 | 0.38 | 11.0–12.6 | B | Coahuila,Mexico 386 | 6042.0 |
| Vireo carmioli | B | 5 | 12.9 | | | | Panama 244, 315 | 6043.0 |
| Vireo griseus | B | 153 | 11.4 | | 10.0–14.3 | F | Florida, USA 190 | 6044.0 |
| Vireo pallens | U | 8 | 11.6 | | 8.9–13.0 | | Yucatan, Mexico 323, 510 | 6045.0 |
| Vireo bairdi | B | 9 | 12.0 | | 11.2–12.7 | | Mexico 323, 457, 584 | 6046.0 |
| Vireo gundlachii | B | 4 | 13.0 | | 12.5–13.3 | | Cuba 429 | 6047.0 |
| Vireo crassirostris | B | 19 | 13.1 | 1.19 | 11.0–15.0 | | 432, 600 | 6048.0 |
| Vireo caribaeus | M | 4 | 9.3 | | 8.6–10.0 | | San Andres Is. 19, 511 | 6049.0 |
| Vireo vicinior | B | 19 | 12.8 | 1.06 | 11.0–14.9 | | Sonora, Mexico 177 | 6050.0 |
| Vireo hypochryseus | B | 22 | 12.0 | | 8.5–14.1 | | W. Mexico 593 | 6051.0 |
| Vireo modestus | B | 8 | 9.6 | | 9.2–10.4 | | Jamaica 600 | 6052.0 |
| Vireo nanus | U | | 11.0 | | | | 185 | 6053.0 |
| Vireo latimeri | U | 10 | 11.2 | 0.60 | 10.2–12.2 | W | Puerto Rico 186 | 6054.0 |
| Vireo osburni | U | 4 | 19.9 | | | | Jamaica 127, 600 | 6055.0 |
| Vireo solitarius | B | 32 | 16.6 | 1.17 | 14.1–19.3 | M | Pennsylvania,USA 177 | 6057.0 |
| Vireo flavifrons | B | 16 | 18.0 | 1.63 | 15.6–21.4 | B | Pennsylvania,USA 177 | 6058.0 |
| Vireo philadelphicus | B | 116 | 12.2 | 0.53 | 10.3–16.1 | M | Pennsylvania,USA 101 | 6059.0 |
| Vireo olivaceus | B | 1405 | 16.7 | 1.57 | 12.0–25.1 | | Pennsylvania,USA 101 | 6060.0 |
| Vireo olivaceus chivi | U | 7 | 13.8 | | | | Peru 193 | 6060.0 |

## Body Masses of World Birds (continued)

| Species | Sex | N | Mean | Std dev | Range | Sn | Location | Number |
|---|---|---|---|---|---|---|---|---|
| Vireo flavoviridis | U | 33 | 17.9 | 0.66 | 16.0–21.1 | B | Panama 397 | 6061.0 |
| Vireo altiloquus | B | 27 | 17.9 | | | B | Jamaica 127 | 6063.0 |
| Vireo magister | B | 12 | 15.4 | | 13.5–17.0 | | Cayman Is. 432 | 6064.0 |
| Vireo gilvus | B | 80 | 14.8 | 1.20 | | B | Manitoba,Canada 44 | 6066.0 |
| Vireo leucophrys | B | 9 | 12.5 | | 11.4–13.4 | | 46, 387 | 6067.0 |
| Hylophilus poicilotis | B | 4 | 10.0 | | 9.0–10.8 | | Brazil 38, 441 | 6068.0 |
| Hylophilus thoracicus | M | 1 | 13.8 | | | | Peru 193 | 6069.0 |
| Hylophilus semicinereus | M | 1 | 13.0 | | | | Bolivia 22 | 6070.0 |
| Hylophilus pectoralis | U | 3 | 11.7 | | 11.7–11.8 | | Surinam 246 | 6071.0 |
| Hylophilus muscicapinus | U | 2 | 12.0 | | 10.5–13.5 | | Brazil 45, 610a | 6073.0 |
| Hylophilus brunneiceps | B | 4 | 9.5 | | 8.0–11.0 | | Brazil 226, 610a | 6074.0 |
| Hylophilus hypoxanthus | U | | 17.0 | | | | Peru 623 | 6075.0 |
| Hylophilus semibrunneus | B | 2 | 11.4 | | 10.2–12.6 | | Columbia 387 | 6076.0 |
| Hylophilus aurantiifrons | B | 26 | 9.5 | | 7.5–12.0 | | Trinidad 188 | 6077.0 |
| Hylophilus flavipes | B | 14 | 12.6 | 0.79 | 11.3–14.0 | | 384, 611 | 6078.0 |
| Hylophilus flavipes insularis | B | 5 | 10.8 | | 9.5–12.0 | | Trinidad 311 | 6078.0 |
| Hylophilus olivaceus | U | 2 | 11.5 | | 11.0–12.0 | | Peru 584 | 6079.0 |
| Hylophilus ochraceiceps | B | 16 | 10.4 | 0.97 | 8.5–12.5 | | 161, 457, 497, 510 | 6080.0 |
| Hylophilus decurtatus | U | 8 | 8.8 | | 7.8–10.3 | | Belize 510 | 6081.0 |

### ORDER: PASSERIFORMES          FAMILY: LANIIDAE

| Species | Sex | N | Mean | Std dev | Range | Sn | Location | Number |
|---|---|---|---|---|---|---|---|---|
| Lanius tigrinus | U | 6 | 25.0 | | | W | Malaysia 420 | 6082.0 |
| Lanius collurio | U | 27 | 29.9 | | 22.5–47.3 | M | Britain 258 | 6084.0 |

## Body Masses of World Birds (continued)

| Species | Sex | N | Mean | Std dev | Range | Sn | Location | Number |
|---|---|---|---|---|---|---|---|---|
| Lanius cristatus | B | 14 | 28.0 | | 21.0–35.0 | W | India 5 | 6086.0 |
| Lanius vittatus | U | 17 | 20.8 | | 18.0–24.0 | | India 5 | 6090.0 |
| Lanius schach | B | 26 | 38.4 | | 30.0–45.0 | | Philippines 475 | 6091.0 |
| Lanius tephronotus | M | 8 | | | 39.0–51.0 | | India | 6092.0 |
| | F | 4 | | | 43.0–54.0 | | 5 | |
| Lanius minor | B | 11 | 48.6 | 3.34 | 42.0–53.0 | | 5, 149 | 6094.0 |
| Lanius ludovicianus | U | 38 | 47.4 | 3.26 | 40.5–54.1 | | Florida, USA 177 | 6095.0 |
| Lanius excubitor | B | 6 | 65.6 | | 61.8–69.0 | W | western USA 594 | 6096.0 |
| Lanius sphenocercus | M | 2 | 93.6 | | 87.2–100.0 | | | 6097.0 |
| | F | 1 | 87.2 | | | | 149 | |
| Lanius mackinnoni | B | 2 | 36.0 | | 35.6–36.4 | | 51 | 6102.0 |
| Lanius collaris | B | 33 | 41 5 | | 34.7–50.0 | | South Africa 256 | 6103.0 |
| Lanius collaris | U | 60 | 30.2 | | 21.7–37.5 | F | Uganda | 6103.0 |
| Lanius senator | B | 65 | 29.1 | | 21.0–41.0 | S | Morocco 8 | 6106.0 |
| Lanius nubicus | U | 32 | 20.4 | | 16.5–28.5 | | 78 | 6107.0 |
| Corvinella melanoleuca | F | 3 | 98.3 | | 90.0–110.0 | | South Africa 256 | 6109.0 |

### ORDER: PASSERIFORMES       FAMILY: MALACONOTIDAE

| Species | Sex | N | Mean | Std dev | Range | Sn | Location | Number |
|---|---|---|---|---|---|---|---|---|
| Nilaus afer | U | 3 | 18.7 | | 17.3–19.9 | | Ghana 228 | 6113.0 |
| Dryoscopus gambensis | B | 7 | 33.5 | | 27.2–39.0 | | 51, 228 | 6114.0 |
| Dryoscopus cubla | M | 12 | 27.4 | 1.83 | 23.6–29.8 | | Zimbabwe | 6116.0 |
| | F | 11 | 25.5 | 2.30 | 23.2–30.8 | | 280, 281, 282, 283, 284 | |
| Dryoscopus senegalensis | M | 1 | 28.0 | | | | Cameroon 349 | 6117.0 |
| Tchagra minuta | B | 3 | 31.4 | | 30.0–32.2 | | 228, 349 | 6120.0 |
| Tchagra senegala | B | 15 | 49.2 | | 43.3–53.0 | | Kenya 67 | 6122.0 |
| Tchagra australis | U | 22 | 36.3 | | 31.2–42.0 | | Kenya 67 | 6123.0 |

## Body Masses of World Birds (continued)

| Species | Sex | N | Mean | Std dev | Range | Sn | Location | Number |
|---|---|---|---|---|---|---|---|---|
| Tchagra tchagra | U | 2 | 52.4 | | 50.6–54.3 | | 78 | 6125.0 |
| Laniarius luehderi | M | 3 | 54.0 | | 52.0–55.0 | | Cameroon 349 | 6127.0 |
| Laniarius aethiopicus | B | 11 | 50.5 | 6.08 | 40.1–61.5 | | Zimbabwe 314a, 282, 283, 284 | 6131.0 |
| Laniarius ferrugineus | U | 32 | 48.7 | | 38.8–66.0 | | Kenya 67 | 6133.0 |
| Laniarius barbarus | U | 12 | 47.0 | | 43.0–52.0 | | Ghana 228 | 6134.0 |
| Laniarius atrococcineus | U | 3 | 49.2 | | 46.7–52.0 | | South Africa 256 | 6136.0 |
| Laniarius mufumbiri | U | 12 | 41.5 | 3.62 | 34.4–46.0 | | Kenya 68 | 6137.0 |
| Laniarius atroflavus | B | 2 | 40.0 | | 40.0–40.0 | | Cameroon 349 | 6138.0 |
| Laniarius fuelleborni | B | 10 | 47.4 | | 43.0–53.0 | | Malawi 168 | 6142.0 |
| Telophorus zeylonus | B | 12 | 62.7 | | 43.2–73.1 | | South Africa 256 | 6144.0 |
| Telophorus bocagei | M | 1 | 29.0 | | | | Cameroon 349 | 6145.0 |
| Telophorus sulfureopectus | B | 12 | 29.8 | 1.79 | 27.8–34.0 | | 228, 282, 283, 284, 314a | 6146.0 |
| Telophorus olivaceus | B | 9 | 33.1 | | 26.2–37.9 | | Zimbabwe 281 | 6147.0 |
| Telophorus multicolor | F | 1 | 37.0 | | | | Cameroon 349 | 6148.0 |
| Telophorus quadricolor | B | 8 | 36.9 | | 29.6–40.5 | | Zimbabwe 280, 281, 283 | 6152.0 |
| Malaconotus blanchoti | B | 2 | 74.1 | | 72.3–75.9 | | Zimbabwe 280, 284 | 6157.0 |
| Prionops plumatus | B | 23 | 33.0 | | 27.3–39.8 | | Zimbabwe 280 | 6160.0 |
| Prionops retzii | M | 2 | 40.2 | | 37.9–42.5 | | Zimbabwe 280 | 6165.0 |
| | F | 1 | 46.7 | | | | | |
| Pseudobias wardi | M | 1 | 12.5 | | | | Madagascar 39 | 6170.0 |
| Batis diops | U | 4 | 12.7 | | | | 51 | 6171.0 |
| Batis mixta | U | | 11.8 | | | | Tanzania 417a | 6173.0 |

## Body Masses of World Birds (continued)

| Species | Sex | N | Mean | Std dev | Range | Sn | Location | Number |
|---------|-----|---|------|---------|-------|----|----------|--------|
| Batis capensis | B | 70 | 12.8 | | 11.3–15.6 | | Malawi 168 | 6176.0 |
| Batis molitor | B | 15 | 10.2 | 0.94 | 9.3–12.0 | | Zimbabwe 280, 281, 282, 283, 284 | 6178.0 |
| Batis pririt | B | 10 | 8.7 | | 7.5–10.1 | | South Africa 256 | 6180.0 |
| Batis senegalensis | B | 20 | 9.7 | | 8.0–11.4 | | Ghana 228 | 6181.0 |
| Batis minor | B | 4 | 12.5 | | 11.0–14.0 | | Cameroon 349 | 6183.0 |
| Platysteira cyanea | B | 12 | 13.4 | | 11.6–15.0 | | Ghana 228 | 6190.0 |
| Platysteira peltata | B | 2 | 16.0 | | 15.8–16.3 | | 51 | 6193.0 |
| Platysteira castanea | U | 4 | 12.7 | | 12.0–13.5 | | 314a, 349 | 6194.0 |
| Platysteira blissetti | U | 4 | 11.6 | | 11.0–12.0 | | Ghana; Kenya 314a | 6196.0 |
| Platysteira concreta | U | 5 | 12.8 | | 12.0–13.5 | | Liberia 314a | 6199.0 |
| Philentoma velatum | B | 2 | 26.1 | | 25.0–27.2 | | Borneo 627 | 6201.0 |
| Tephrodornis gularis | U | 4 | | | 38.0–46.0 | | India 5 | 6202.0 |
| Tephrodornis pondicerianus | B | 17 | 20.2 | | 18.0–27.0 | | India 5, 361 | 6203.0 |

### ORDER: PASSERIFORMES          FAMILY: VANGIDAE

| Species | Sex | N | Mean | Std dev | Range | Sn | Location | Number |
|---------|-----|---|------|---------|-------|----|----------|--------|
| Calicalicus madagascariensis | U | | 16.0 | | | | Madagascar 39 | 6204.0 |
| Vanga curvirostris | M | 1 | 72.0 | | | | Madagascar 39 | 6206.0 |
| | F | 1 | 65.0 | | | | | |
| Falculea palliata | U | 3 | 119.0 | | 115.0–124.0 | | Madagascar 39 | 6210.0 |
| Artamella viridis | M | 2 | 47.2 | | 46.5–48.0 | | Madagascar 39 | 6211.0 |
| | F | 1 | 57.0 | | | | | |
| Leptopterus chabert | M | 2 | 23.2 | | 20.5–26.0 | | Madagascar 39 | 6212.0 |
| Cyanolanius madagascarinus | U | 4 | 21.8 | | 19.5–23.5 | | Madagascar 39 | 6213.0 |
| Hypositta corallirostris | M | 1 | 15.2 | | | | Madagascar 39 | 6217.0 |
| | F | 1 | 13.0 | | | | | |

## Body Masses of World Birds (continued)

| Species | Sex | N | Mean | Std dev | Range | Sn | Location | Number |
|---|---|---|---|---|---|---|---|---|
| **ORDER: PASSERIFORMES** | | | | | | **FAMILY: BOMBYCILLIDAE** | | |
| Ptilogonys cinereus | B | 6 | 33.6 | | 32.3–34.9 | | Mexico 456, 593 | 6218.0 |
| Ptilogonys caudatus | M | 4 | 39.1 | | | | Panama | 6219.0 |
| | F | 1 | 33.5 | | | | 244 | |
| Phainopepla nitens | B | 33 | 24.0 | | 22.0–28.0 | | 651 | 6220.0 |
| Phainoptila melanoxantha | U | | 56.0 | | | | 603a | 6221.0 |
| Bombycilla garrulus | B | 45 | 56.4 | 5.22 | 46.5–69.0 | | Colorado, USA 177 | 6222.0 |
| Bombycilla cedorum | M | 58 | 30.6 | 1.72 | 25.5–39.6 | | Pennsylvania,USA | 6224.0 |
| | F | 190 | 33.1 | 1.07 | 28.0–40.2 | | 101 | |
| **ORDER: PASSERIFORMES** | | | | | | **FAMILY: DULIDAE** | | |
| Dulus dominicus | U | | 42.0 | | | | 185 | 6225.0 |
| **ORDER: PASSERIFORMES** | | | | | | **FAMILY: CINCLIDAE** | | |
| Cinclus cinclus | M | 178 | 64.2 | 3.80 | 53.0–76.0 | | Germany | 6226.0 |
| | F | 180 | 55.4 | 5.50 | 46.0–72.0 | | 119 | |
| Cinclus cinclus gularis | M | 124 | 69.1 | 3.00 | 60.0–76.0 | | Scotland | 6226.0 |
| | F | 101 | 58.1 | 3.30 | 50.0–67.0 | | 119 | |
| Cinclus pallasii | B | 11 | 76.0 | 8.59 | 66.0–88.0 | | 5, 149 | 6227.0 |
| Cinclus mexicanus | M | 21 | 61.0 | 2.19 | 57.0–66.0 | B | Utah, USA | 6228.0 |
| | F | 18 | 54.6 | 4.76 | 43.0–65.0 | | 177 | |
| Cinclus leucocephalus | U | 7 | 43.9 | | 38.0–59.0 | | Ecuador 320 | 6229.0 |
| **ORDER: PASSERIFORMES** | | | | | | **FAMILY: TURDIDAE** | | |
| Neocossyphus fraseri | U | 4 | 37.1 | | 32.5–41.5 | | 314a | 6232.0 |
| Neocossyphus poensis | U | 4 | 54.5 | | 47.5–58.0 | | Liberia 314a | 6234.0 |
| Pseudocossyphus sharpei | F | 2 | 26.5 | | 25.5–27.5 | | Madagascar 39 | 6235.0 |
| Monticola rupestris | F | 1 | 60.0 | | | | South Africa 256 | 6238.0 |
| Monticola saxatilis | B | 22 | 48.5 | | 40.0–65.0 | S | 119 | 6243.0 |
| Monticola cinclorhynchus | B | 20 | 36.0 | | 30.0–41.0 | | India 5 | 6245.0 |

## Body Masses of World Birds (continued)

| Species | Sex | N | Mean | Std dev | Range | Sn | Location | Number |
|---|---|---|---|---|---|---|---|---|
| Monticola gularis | M | 1 | 34.0 | | | | USSR 149 | 6246.0 |
| Monticola rufiventris | M | 7 | | | 50.0–61.0 | Y | India 5 | 6247.0 |
| | F | 5 | | | 48.0–56.0 | | | |
| Monticola solitarius solitarius | U | 26 | 57.0 | 3.50 | 50.5–63.5 | | Malta 119 | 6248.0 |
| Monticola solitarius philippensis | M | | 50.5 | | 39.0–60.0 | | China 119 | 6248.0 |
| Myiophonus robinsoni | U | 5 | 99.2 | | 87.0–105.0 | | Malaysia 371 | 6252.0 |
| Myiophonus horsfieldii | U | 10 | 117.0 | | 101.0–130.0 | | India 5 | 6253.0 |
| Myiophonus caeruleus | M | 11 | | | 136.0–199.0 | | India 5 | 6255.0 |
| | F | 6 | | | 154.0–181.0 | | | |
| Zoothera wardii | U | 6 | 58.6 | | 52.0–72.0 | | India 5 | 6262.0 |
| Zoothera citrina | U | 20 | 53.3 | | 47.0–60.0 | | India 5 | 6265.0 |
| Zoothera sibirica | U | 76 | 75.5 | | 62.0–90.0 | | Malaysia 371 | 6267.0 |
| Zoothera naevia | B | 50 | 78.4 | | 69.6–90.0 | F | California, USA 581 | 6268.0 |
| Zoothera pinicola | M | 1 | 88.0 | | | | Mexico 634 | 6269.0 |
| Zoothera piaggiae | U | 2 | 57.0 | | | | 51 | 6270.0 |
| Zoothera gurneyi | M | 2 | 54.5 | | 53.0–56.0 | | Zambia 166 | 6273.0 |
| Zoothera spiloptera | M | 1 | 70.0 | | | | Sri Lanka 5 | 6279.0 |
| Zoothera andromedae | M | 1 | 108.0 | | | | Philippines 475 | 6280.0 |
| Zoothera mollissima | U | 4 | 98.2 | | 89.6–112.0 | | India 5 | 6281.0 |
| Zoothera dixoni | B | 6 | 90.0 | | 71.5–103.0 | | India 5 | 6282.0 |
| Zoothera dauma | B | 9 | 104.0 | | 92.0–130.0 | | India 5 | 6283.0 |
| Zoothera monticola | B | 2 | 126.0 | | 122.0–131.0 | | India 5 | 6290.0 |
| Zoothera marginata | F | 1 | 81.0 | | | | Thailand 378 | 6291.0 |
| Nesocichla eremita | U | | 91.0 | | | | Tristan de Cunha 513 | 6294.0 |

## Body Masses of World Birds (continued)

| Species | Sex | N | Mean | Std dev | Range | Sn | Location | Number |
|---|---|---|---|---|---|---|---|---|
| Cichlherminia lherminieri | U | | 100.0 | | | | 185 | 6295.0 |
| Myadestes obscurus | U | | 50.0 | | | | Hawaiian Is. 400 | 6299.0 |
| Myadestes townsendi | B | 11 | 34.0 | 3.73 | 28.5–43.3 | | Arizona, USA 594 | 6301.0 |
| Myadestes occidentalis | B | 4 | 41.6 | | 38.2–44.1 | | Mexico 456, 593 | 6302.0 |
| Myadestes elisabeth | U | | 27.0 | | | | Cuba 185 | 6303.0 |
| Myadestes genibarbis | U | 10 | 27.1 | | 24.4–30.0 | | Jamaica 126 | 6304.0 |
| Myadestes melanops | B | 9 | 32.1 | | | | Panama 244 | 6305.0 |
| Myadestes coloratus | M F | 5 9 | | | 26.0–34.0 24.5–32.0 | | Panama 497 | 6306.0 |
| Myadestes ralloides | U | 17 | 27.7 | 1.60 | | | Peru 668 | 6307.0 |
| Myadestes unicolor | B | 12 | 36.7 | 4.27 | 30.0–44.5 | | Mexico 584 | 6308.0 |
| Cichlopsis leucogenys | U | | 61.0 | | | | Brazil 564 | 6309.0 |
| Entomodestes leucotis | U | 6 | 57.9 | | | | Peru 668 | 6310.0 |
| Catharus gracilirostris | U | | 21.0 | | | | 603a | 6312.0 |
| Catharus aurantiirostris | B | 15 | 27.0 | | 21.5–32.4 | | 311, 387, 548, 593, 649 | 6313.0 |
| Catharus fuscater | U | 4 | 40.3 | | 36.9–44.8 | | Panama 50 | 6314.0 |
| Catharus occidentalis | M | 1 | 26.2 | | | | Mexico 139 | 6315.0 |
| Catharus frantzii | U | 1 | 31.4 | | | | Panama 244 | 6316.0 |
| Catharus mexicanus | M | 9 | 33.0 | | 30.4–34.1 | | Mexico 584 | 6317.0 |
| Catharus dryas | B | 12 | 35.1 | | | | 48, 668, 672 | 6318.0 |
| Catharus fuscescens | B | 72 | 31.2 | 2.90 | 26.2–41.7 | M | Pennsylvania,USA 101 | 6319.0 |
| Catharus minimus | B | 206 | 32.8 | 2.68 | 26.4–50.5 | M | Pennsylvania,USA 101 | 6320.0 |
| Catharus ustulatus | B | 1022 | 30.8 | 1.83 | 21.9–50.7 | M | Pennsylvania,USA 101 | 6321.0 |

## Body Masses of World Birds (continued)

| Species | Sex | N | Mean | Std dev | Range | Sn | Location | Number |
|---|---|---|---|---|---|---|---|---|
| Catharus guttatus | B | 70 | 31.0 | 1.50 | 26.6–37.4 | M | Pennsylvania,USA 101 | 6322.0 |
| Catharus mustelinus | B | 308 | 47.4 | 4.17 | 39.2–57.7 | | Pennsylvania,USA 101 | 6323.0 |
| Platycichla flavipes | B | 18 | 60.0 | | 52.0–67.0 | | Trinidad 188 | 6324.0 |
| Platycichla leucops | U | 5 | 61.5 | | | | Peru 668 | 6325.0 |
| Psophocichla litsipsirupa | U | 5 | 74.8 | | 68.0–78.5 | | Ethiopia 633 | 6326.0 |
| Turdus abyssinicus | M | 1 | 65.0 | | | | Zambia 166 | 6329.0 |
| Turdus olivaceus | B | 42 | 73.8 | | 60.8–88.0 | | South Africa 256 | 6330.0 |
| Turdus bewsheri | U | 3 | 63.3 | | 57.0–68.0 | | Comoro Is. 97 | 6333.0 |
| Turdus libonyanus | B | 6 | 55.4 | | 46.2–64.5 | | Zimbabwe 280, 282, 284 | 6335.0 |
| Turdus hortulorum | M | 1 | 68.0 | | | | 149 | 6337.0 |
| Turdus unicolor | U | 15 | 64.2 | | 57.0–75.1 | | India 5 | 6338.0 |
| Turdus albocinctus | B | 5 | 99.4 | | 90.0–107.0 | | India 5 | 6341.0 |
| Turdus torquatus | U | 50 | 109.0 | | 88.4–132.0 | S | Britain 258 | 6342.0 |
| Turdus boulboul | B | 18 | 97.1 | | 85.0–111.0 | | India 5 | 6343.0 |
| Turdus merula merula | B | 522 | 113.0 | | 82.0–149.0 | F | Britain 119 | 6344.0 |
| Turdus merula nigropileus | U | 19 | 74.9 | | 60.0–94.0 | | India 5 | 6344.0 |
| Turdus merula intermedius | B | 13 | 93.4 | | 84.0–102.0 | | 119 | 6344.0 |
| Turdus poliocephalus | U | 38 | 61.7 | 3.90 | 53.5–69.3 | | Fiji 332 | 6345.0 |
| Turdus poliocephalus rennellianus | B | 20 | 58.1 | | 52.0–67.0 | | Solomon Is. 159 | 6345.0 |
| Turdus poliocephalus keysseri | B | 6 | 72.0 | | 69.0–78.0 | | New Guinea 159 | 6345.0 |
| Turdus rubrocanus | M | 2 | 92.2 | | 84.6–99.7 | | India 5 | 6346.0 |
| Turdus obscurus | U | 136 | 62.6 | | 45.0–80.0 | | Malaysia 371 | 6349.0 |

## Body Masses of World Birds (continued)

| Species | Sex | N | Mean | Std dev | Range | Sn | Location | Number |
|---------|-----|---|------|---------|-------|-----|----------|--------|
| Turdus pallidus | B | 7 | 72.7 | | 64.0–90.0 | | 149 | 6350.0 |
| Turdus ruficollis | U | 25 | 78.1 | | 57.0–96.0 | | India 5 | 6353.0 |
| Turdus naumanni | B | 26 | 71.5 | | 50.0–88.0 | M | NE China 119 | 6354.0 |
| Turdus pilaris | M | 31 | 108.0 | 9.30 | 81.0–120.0 | F | Netherlands 119 | 6355.0 |
| | F | 38 | 104.0 | 10.20 | 83.0–128.0 | | | |
| Turdus iliacus | B | 282 | 61.2 | | 46.0–80.0 | F | Netherlands 119 | 6356.0 |
| Turdus philomelos | M | 129 | 68.9 | 6.00 | 56.0–89.0 | F | Netherlands 119 | 6357.0 |
| | F | 140 | 66.6 | 6.60 | 52.0–89.0 | | | |
| Turdus viscivorus | B | 19 | 115.0 | | 96.0–140.0 | | 119 | 6359.0 |
| Turdus aurantius | U | 8 | 81.9 | | | | Jamaica 125a, 600 | 6360.0 |
| Turdus plumbeus | U | 14 | 74.5 | 4.30 | 67.0–82.4 | W | Puerto Rico 186 | 6362.0 |
| Turdus plumbeus rubripes | U | 12 | 79.3 | 7.84 | 69.0–96.0 | | Cuba 374 | 6362.0 |
| Turdus chiguanco | M | 1 | 107.0 | | | | Peru 193 | 6363.0 |
| Turdus nigrescens | U | | 96.0 | | | | 603a | 6364.0 |
| Turdus fuscater | B | 5 | 140.0 | | 128.0–154.0 | | 644, 649 | 6365.0 |
| Turdus infuscatus | M | 3 | 73.8 | | 71.5–81.0 | | 584 | 6366.0 |
| Turdus serranus | U | 5 | 84.5 | | 79.0–89.8 | | Columbia; Peru 387, 668 | 6367.0 |
| Turdus nigriceps | B | 9 | 51.2 | | | | Peru 668 | 6368.0 |
| Turdus reevei | M | 1 | 61.0 | | | | Peru 672 | 6370.0 |
| Turdus olivater | U | 2 | 71.9 | | 69.8–74.0 | | Venezuela 105 | 6371.0 |
| Turdus rufiventris | M | 13 | 66.8 | 5.68 | | | Paraguay 200 | 6374.0 |
| | F | 10 | 72.2 | 5.01 | | | | |
| Turdus falcklandii | U | | 94.3 | | | | Chile 292 | 6375.0 |
| Turdus leucomelas | B | 17 | 68.6 | | | | Paraguay 200 | 6376.0 |
| Turdus amaurochalinus | B | 8 | 63.1 | | | | Paraguay 200 | 6377.0 |

## Body Masses of World Birds (continued)

| Species | Sex | N | Mean | Std dev | Range | Sn | Location | Number |
|---|---|---|---|---|---|---|---|---|
| Turdus plebejus | M | 16 | 81.1 | 9.56 | | | Panama | 6378.0 |
| | F | 5 | 91.9 | | | | 244 | |
| Turdus ignobilis | U | 12 | 59.6 | | | | Peru | 6379.0 |
| | | | | | | | 668 | |
| Turdus lawrencii | M | 1 | 72.6 | | | | Peru | 6380.0 |
| | | | | | | | 193 | |
| Turdus obsoletus | U | 9 | 76.7 | | 66.0–98.0 | | Panama | 6381.0 |
| | | | | | | | 50 | |
| Turdus fumigatus | B | 124 | 71.4 | | 56.5–83.0 | | Trinidad | 6382.0 |
| | | | | | | | 576 | |
| Turdus hauxwelli | U | 44 | 69.0 | 4.70 | | | Peru | 6383.0 |
| | | | | | | | 193 | |
| Turdus grayi | U | 25 | 73.8 | 5.80 | | | Panama | 6384.0 |
| | | | | | | | 315 | |
| Turdus nudigenis | B | 88 | 63.9 | | 55.0–74.5 | | Trinidad | 6385.0 |
| | | | | | | | 576 | |
| Turdus jamaicensis | U | 7 | 59.1 | | | | Jamaica | 6388.0 |
| | | | | | | | 125a, 600 | |
| Turdus assimilis | U | 10 | 67.5 | 6.01 | | | Panama | 6389.0 |
| | | | | | | | 244 | |
| Turdus albicollis | U | 45 | 54.1 | | 45.0–62.5 | | Trinidad | 6390.0 |
| | | | | | | | 576 | |
| Turdus rufopalliatus | B | 8 | 76.8 | | 72.0–85.0 | | Jalisco, Mexico | 6391.0 |
| | | | | | | | 548 | |
| Turdus graysoni | B | 8 | 76.7 | | 63.5–101.0 | | W. Mexico | 6392.0 |
| | | | | | | | 593 | |
| Turdus swalesi | U | | 75.0 | | | | | 6393.0 |
| | | | | | | | 185 | |
| Turdus migratorius | B | 401 | 77.3 | 0.36 | 63.5–103.0 | Y | Pennsylvania,USA | 6394.0 |
| | | | | | | | 101 | |
| Turdus rufitorques | B | 3 | 72.0 | | 70.0–74.0 | | | 6395.0 |
| | | | | | | | 584 | |
| Brachypteryx leucophrys | U | 15 | 15.8 | | 13.0–18.0 | | Malaysia | 6400.0 |
| | | | | | | | 371 | |
| Brachypteryx montana | B | 4 | 21.6 | | 20.0–23.8 | | | 6401.0 |
| | | | | | | | 5, 475 | |
| Alethe poliocephala | U | 14 | 36.2 | | 27.5–42.0 | | | 6405.0 |
| | | | | | | | 314a, 349 | |
| Alethe poliophrys | U | 29 | 35.6 | 2.33 | | | | 6406.0 |
| | | | | | | | 51 | |
| Alethe fuelleborni | B | 27 | 49.3 | | 41.6–58.0 | | Malawi | 6407.0 |
| | | | | | | | 168 | |
| Alethe diademata | U | 9 | 33.6 | | 31.5–35.5 | | | 6409.0 |
| | | | | | | | 314a | |

## Body Masses of World Birds (continued)

| Species | Sex | N | Mean | Std dev | Range | Sn | Location | Number |
|---|---|---|---|---|---|---|---|---|
| ORDER: PASSERIFORMES | | | | | FAMILY: MIMIDAE | | | |
| Dumetella carolinensis | B | 1736 | 36.9 | 3.12 | 26.6−56.5 | | Pennsylvania,US 101 | 6411.0 |
| Melanoptila glabrirostris | B | 8 | 36.3 | | 31.6−42.0 | | Yucatan, Mexico 323, 457 | 6412.0 |
| Melanotis caerulescens caerulescens | B | 8 | 63.1 | | 58.0−69.0 | | Jalisco, Mexico 548 | 6413. |
| Melanotis caerulescens longirostris | B | 18 | 59.7 | 4.53 | 50.2−68.0 50.2−68.0 | | W. Mexico 593 | |
| Melanotis hypoleucus | B | 2 | 62.0 | | 60.0−64.0 | | Guatemala 584 | 6414.0 |
| Mimus polyglottos | B | 221 | 48.5 | | 36.2−55.7 | | Florida, USA 190 | 6415.0 |
| Mimus gilvus | U | 16 | 58.4 | 4.55 | 52.5−66.0 | | Venezuela 625 | 6416.0 |
| Mimus gundlachii | B | 10 | 66.8 | 8.81 | 57.0−85.0 | | Bahamas 600 | 6417.0 |
| Mimus saturninus | U | | 73.0 | | | | Brazil 564 | 6418.0 |
| Mimus patagonicus | M | 1 | 57.0 | | | | Argentina 411 | 6419.0 |
| Mimus dorsalis | B | 7 | 58.7 | | 52.0−65.0 | | Bolivia 584 | 6420.0 |
| Mimus triurus | U | 2 | 51.5 | | 49.0−54.0 | | Brazil 38 | 6421.0 |
| Mimus longicaudatus | B | 19 | 66.6 | 5.72 | 54.0−79.0 | | Peru 584 | 6422.0 |
| Mimus thenca | U | | 66.0 | | | | Chile 292 | 6423.0 |
| Nesomimus parvulus | M F | 150 178 | 56.2 51.2 | 4.90 4.00 | | | Galapagos Is. 130 | 6424.0 |
| Nesomimus trifasciatus | M F | 34 61 | 65.7 59.8 | 4.08 3.91 | | | Galapagos Is. 130 | 6425.0 |
| Nesomimus macdonaldi | M F | 140 77 | 76.1 64.8 | 4.73 5.26 | | | Galapagos Is. 130 | 6426.0 |
| Nesomimus melanotis | M F | 27 18 | 53.2 48.0 | 3.12 2.97 | | | Galapagos Is. 130 | 6427.0 |
| Mimodes graysoni | B | 3 | 75.5 | | 74.8−76.2 | | W. Mexico 593 | 6428.0 |
| Oreoscoptes montanus | M F | 14 8 | 41.1 45.5 | 3.10 | 36.8−46.3 41.6−49.6 | B | Oregon, USA 177 | 6429.0 |
| Toxostoma rufum | B | 177 | 68.8 | 2.96 | 57.6−89.0 | | Pennsylvania,USA 101 | 6430.0 |

## Body Masses of World Birds (continued)

| Species | Sex | N | Mean | Std dev | Range | Sn | Location | Number |
|---|---|---|---|---|---|---|---|---|
| Toxostoma longirostre | B | 19 | 69.9 | 3.86 | 63.0–79.9 | | Texas, USA 585 | 6431.0 |
| Toxostoma guttatum | B | 3 | 52.8 | | 49.0–59.8 | | Yucatan, Mexico 323, 457 | 6432.0 |
| Toxostoma bendirei | B | 23 | 62.2 | 5.06 | 53.5–74.6 | Y | Arizona, USA 177 | 6433.0 |
| Toxostoma cinereum | M | 4 | 63.9 | | 58.6–69.8 | | W. Mexico | 6434.0 |
| | F | 1 | 54.4 | | | | 561, 593 | |
| Toxostoma curvirostre | B | 70 | 79.4 | 5.16 | 67.2–90.5 | Y | Arizona, USA 177 | 6435.0 |
| Toxostoma ocellatum | U | 3 | 84.3 | | 77.7–88.9 | | Mexico 584 | 6436.0 |
| Toxostoma lecontei | B | 174 | 61.9 | 3.81 | 54.5–75.5 | Y | SW USA, Mexico 556 | 6437.0 |
| Toxostoma redivivum | B | 21 | 84.4 | 3.86 | 78.0–93.0 | | California, USA 177 | 6438.0 |
| Toxostoma crissale | B | 34 | 62.7 | 4.27 | 53.2–70.0 | Y | Arizona, USA 177 | 6439.0 |
| Cinclocerthia ruficauda | U | | 50.0 | | | | 185 | 6440.0 |
| Ramphocinclus brachyurus | U | | 50.0 | | | | estimated 185 | 6442.0 |
| Margarops fuscus | U | 311 | 100.0 | | 81.0–128.0 | W | Puerto Rico 186 | 6443.0 |
| Margarops fuscatus | U | | 75.0 | | | | 185 | 6444.0 |

### ORDER: PASSERIFORMES          FAMILY: STURNIDAE

| Species | Sex | N | Mean | Std dev | Range | Sn | Location | Number |
|---|---|---|---|---|---|---|---|---|
| Aplonis tabuensis | B | 2 | 67.8 | | 67.0–68.6 | | Tonga 490 | 6452.0 |
| Aplonis striata | U | 3 | 54.2 | | 49.5–60.0 | | New Caledonia 506 | 6453.0 |
| Aplonis opaca | B | 11 | 83.4 | | 71.5–93.0 | | Guam 290 | 6455.0 |
| Aplonis cantoroides | U | 7 | 55.0 | | | | New Guinea 35 | 6457.0 |
| Aplonis panayensis | B | 10 | 56.2 | | 52.0–60.5 | | Philippines 475 | 6464.0 |
| Aplonis metallica | U | 10 | 61.0 | | | | New Guinea 35 | 6465.0 |
| Grafisia torquata | B | 2 | 64.0 | | 61.0–67.0 | | Cameroon 349 | 6472.0 |
| Onychognathus morio | M | 33 | 138.0 | 9.10 | 115.0–155.0 | | South Africa | 6476.0 |
| | F | 22 | 128.0 | 10.70 | 110.0–151.0 | | 256 | |

## Body Masses of World Birds (continued)

| Species | Sex | N | Mean | Std dev | Range | Sn | Location | Number |
|---|---|---|---|---|---|---|---|---|
| Onychognathus fulgidus | M | 1 | 54.0 | | | | Cameroon 349 | 6479.0 |
| Lamprotornis nitens | U | 11 | 77.6 | | 65.3–91.0 | | 78, 256 | 6488.0 |
| Lamprotornis chalybaeus | U | 37 | 100.0 | 11.10 | 84.0–127.0 | | Ethiopia 633 | 6490.0 |
| Lamprotornis chloropterus | B | 2 | 67.5 | | 65.0–70.0 | | Cameroon 349 | 6491.0 |
| Lamprotornis splendidus | M | 2 | 135.0 | | 120.0–150.0 | | Cameroon 349 | 6494.0 |
| Cinnyricinclus leucogaster | M | 1 | 39.5 | | | | Zimbabwe 282 | 6506.0 |
| | F | 1 | 45.5 | | | | | |
| Spreo bicolor | U | 38 | 105.0 | | 88.0–146.0 | | 78 | 6510.0 |
| Saroglossa aurata | M | 1 | 40.5 | | | | Madagascar 39 | 6514.0 |
| Saroglossa spiloptera | M | 1 | 47.5 | | | | India 5 | 6515.0 |
| Creatophora cinerea | F | 2 | 67.0 | | 65.0–69.1 | | South Africa 256 | 6516.0 |
| Sturnus malabaricus | U | 20 | 39.6 | | 32.0–44.0 | | India 5 | 6520.0 |
| Sturnus pagodarum | U | 9 | 49.0 | | 40.0–54.0 | | India 5 | 6522.0 |
| Sturnus sinensis | U | | 61.0 | | | | 78 | 6526.0 |
| Sturnus roseus | B | 15 | | | 53.0–80.0 | | India 5 | 6527.0 |
| Sturnus vulgaris | M | 1942 | 84.7 | | | W | Ohio, USA 259 | 6528.0 |
| | F | 915 | 79.9 | | | | | |
| Sturnus contra | M | 4 | 84.0 | | 76.0–90.0 | | India 361 | 6531.0 |
| Sturnus nigricollis | U | | 166.0 | | 122.0–210.0 | | 78 | 6532.0 |
| Acridotheres tristis | M | 17 | 110.0 | | 82.0–130.0 | | India 5 | 6536.0 |
| | F | 3 | | | 120.0–138.0 | | | |
| Acridotheres ginginianus | U | 6 | 72.0 | | 64.0–76.0 | | India 5 | 6537.0 |
| Acridotheres fuscus | U | 15 | 82.8 | | 72.0–98.0 | | India 5 | 6538.0 |
| Acridotheres grandis | U | | 90.0 | | | | 78 | 6539.0 |
| Acridotheres cristatellus | U | 3 | 113.0 | | 92.0–140.0 | | 78 | 6542.0 |

## Body Masses of World Birds (continued)

| Species | Sex | N | Mean | Std dev | Range | Sn | Location | Number |
|---|---|---|---|---|---|---|---|---|
| Ampeliceps coronatus | U | | | | 78.0–99.0 | | India 5 | 6543.0 |
| Mino anais | U | 6 | 152.0 | | | | New Guinea 35, 218 | 6544.0 |
| Mino dumontii | U | 20 | 217.0 | | | | New Guinea 35 | 6545.0 |
| Sarcops calvus | B | 31 | 142.0 | | –170.0 | | Philippines 475 | 6552.0 |
| Gracula religiosa | B | 9 | 192.0 | | 161.0–229.0 | | India 5, 227, 361 | 6554.0 |
| **ORDER: PASSERIFORMES** | | | | | **FAMILY: MUSCICAPIDAE** | | | |
| Bradornis pallidus | B | 11 | 22.4 | 4.34 | 18.0–34.3 | | 228, 280, 282, 349, 680 | 6560.0 |
| Bradornis mariquensis | U | 4 | 24.7 | | 20.8–28.5 | | 78, 256 | 6562.0 |
| Dioptrornis chocolatinus | U | 8 | 23.4 | | 22.0–27.0 | | Ethiopia 633 | 6565.0 |
| Melaenornis edolioides | U | 9 | 30.8 | | 28.1  33.9 | | Ghana 228 | 6568.0 |
| Sigelus silens | B | 40 | 25.6 | 1.95 | 22.5–37.0 | | South Africa 256 | 6574.0 |
| Rhinomyias umbratilis | M | 3 | 18.4 | | 16.7–20.7 | | Borneo 627 | 6579.0 |
| Rhinomyias ruficauda | B | 19 | 17.3 | | 15.1–20.8 | | Philippines 475 | 6580.0 |
| Muscicapa striata | U | 76 | 14.6 | | 12.2–17.0 | | Wales 84 | 6586.0 |
| Muscicapa griseisticta | B | 6 | 16.2 | | 15.1–17.4 | | Philippines 475 | 6588.0 |
| Muscicapa sibirica | B | 18 | 9.7 | | 8.5–11.5 | | India 5 | 6589.0 |
| Muscicapa dauurica | U | 13 | 9.8 | 0.87 | | | Malaysia 420 | 6590.0 |
| Muscicapa ruficauda | U | 13 | 13.3 | | 12.0–16.0 | | India 5 | 6591.0 |
| Muscicapa muttui | U | 13 | 12.0 | | 10.0–14.0 | | India 5 | 6592.0 |
| Muscicapa ferruginea | F | 1 | 12.0 | | | | India 5 | 6593.0 |
| Muscicapa aquatica | U | 2 | 10.6 | | 10.2–11.0 | | Ghana 228 | 6597.0 |
| Muscicapa olivascens | U | 2 | 15.8 | | 15.7–16.0 | | Liberia 314a | 6598.0 |

## Body Masses of World Birds (continued)

| Species | Sex | N | Mean | Std dev | Range | Sn | Location | Number |
|---|---|---|---|---|---|---|---|---|
| Muscicapa adusta | B | 6 | 11.5 | | 9.6–16.0 | | 166, 281, 349 | 6601.0 |
| Muscicapa sethsmithi | U | 2 | 8.4 | | 7.9–8.8 | | Gabon 314a | 6603.0 |
| Muscicapa caerulescens | B | 10 | 17.1 | 1.11 | 15.2–18.4 | | Zimbabwe 280, 282, 284 | 6607.0 |
| Myioparus griseigularis | U | 2 | 11.4 | | 11.2–11.7 | | Liberia 314a | 6608.0 |
| Myioparus plumbeus | U | 4 | 12.2 | | 11.5–13.0 | | Ghana 228 | 6609.0 |
| Ficedula hypoleuca | U | 60 | 11.6 | | 9.7–14.3 | S | Morocco 8 | 6611.0 |
| Ficedula albicollis | U | 1 | 10.3 | | | | 78 | 6612.0 |
| Ficedula zanthopygia | U | 9 | 11.2 | | | F | Malaysia 420 | 6614.0 |
| Ficedula narcissina | U | | 14.0 | | | | 149 | 6615.0 |
| Ficedula mugimaki | U | 9 | 11.7 | | 10.0–14.0 | | Malaysia 371 | 6616.0 |
| Ficedula hodgsonii | B | 2 | 10.0 | | 9.0–11.0 | | India 5 | 6617.0 |
| Ficedula strophiata | B | 11 | | | 11.0–15.0 | | India 5 | 6618.0 |
| Ficedula parva | U | 7 | 10.7 | | 8.0–13.5 | | 149 | 6619.0 |
| Ficedula subrubra | U | 15 | 10.4 | | 9.0–12.0 | | India 5 | 6620.0 |
| Ficedula monileger | F | 1 | 11.0 | | | | India 5 | 6621.0 |
| Ficedula hyperythra | U | 19 | 8.2 | | 6.0–12.0 | | Malaysia 371 | 6623.0 |
| Ficedula dumetoria | M | 1 | 11.8 | | | | Borneo 627 | 6624.0 |
| Ficedula basilanica | M | 3 | 14.8 | | 14.0–16.1 | | Philippines 475 | 6627.0 |
| | F | 1 | 12.4 | | | | | |
| Ficedula westermanni | B | 13 | 7.8 | | 6.8–10.0 | | Philippines 475 | 6633.0 |
| Ficedula superciliaris | B | 2 | 8.0 | | 8.0–8.0 | | 5 | 6634.0 |
| Ficedula tricolor | M | 13 | 9.1 | | 7.5–10.0 | | India 5 | 6635.0 |
| | F | 11 | 7.5 | | 7.0–8.0 | | | |
| Ficedula sapphira | M | 2 | 7.8 | | 7.5–8.0 | | India 5 | 6636.0 |

## Body Masses of World Birds (continued)

| Species | Sex | N | Mean | Std dev | Range | Sn | Location | Number |
|---------|-----|---|------|---------|-------|----|----------|--------|
| Ficedula nigrorufa | M | 5 | | | 10.0–11.0 | | India | 6637.0 |
| | F | 5 | | | 7.0–10.0 | | 5 | |
| Cyanoptila cyanomelana | M | 1 | 25.0 | | | | Thailand | 6639.0 |
| | | | | | | | 378 | |
| Eumyias thalassina | B | 27 | 18.1 | | 15.0–20.0 | | India | 6640.0 |
| | | | | | | | 5 | |
| Eumyias panayensis | B | 23 | 20.0 | | 17.7–21.7 | | Philippines | 6642.0 |
| | | | | | | | 475 | |
| Eumyias albicaudata | B | 14 | 16.4 | | 12.0–19.0 | | India | 6643.0 |
| | | | | | | | 5 | |
| Niltava grandis | U | 8 | 34.5 | | 24.0–40.0 | | | 6645.0 |
| | | | | | | | 5, 371 | |
| Niltava macgrigoriae | B | | | | 11.0–13.0 | | India | 6646.0 |
| | | | | | | | 5 | |
| Niltava sundara | U | 28 | 21.1 | | 19.0–24.0 | | India | 6648.0 |
| | | | | | | | 5 | |
| Niltava vivida | B | 3 | 33.0 | | | | India | 6650.0 |
| | | | | | | | 5 | |
| Cyornis herioti | F | 1 | 18.0 | | | | Philippines | 6656.0 |
| | | | | | | | 221 | |
| Cyornis pallipes | B | 21 | 19.0 | | 14.0–23.0 | | India | 6658.0 |
| | | | | | | | 5 | |
| Cyornis unicolor | B | 2 | 21.0 | | 21.0–21.1 | | India | 6660.0 |
| | | | | | | | 5 | |
| Cyornis rubeculoides | B | 28 | 14.2 | | 10.0–19.0 | | India | 6661.0 |
| | | | | | | | 5 | |
| Cyornis banyumas | U | 5 | 14.3 | | 12.8–15.3 | | Thailand | 6662.0 |
| | | | | | | | 378 | |
| Cyornis tickelliae | U | 19 | 14.6 | | 12.0–17.0 | | India | 6667.0 |
| | | | | | | | 5 | |
| Cyornis rufigastra | M | 2 | 17.0 | | 16.8–17.3 | | Philippines | 6668.0 |
| | | | | | | | 475 | |
| Pogonocichla stellata | B | 207 | 18.6 | | 15.5–26.3 | | Malawi | 6671.0 |
| | | | | | | | 168 | |
| Sheppardia cyornithopis | U | 4 | 16.8 | | 13.8–21.5 | | Liberia | 6676.0 |
| | | | | | | | 314a | |
| Sheppardia aequatorialis | U | 21 | 16.8 | 1.16 | | | | 6677.0 |
| | | | | | | | 51 | |
| Sheppardia sharpei | U | | 14.0 | | | | Tanzania | 6678.0 |
| | | | | | | | 417a | |
| Sheppardia gunningi | U | 1 | 15.9 | | | | Kenya | 6679.0 |
| | | | | | | | 314a | |
| Erithacus rubecula | B | 50 | 18.2 | 0.50 | 14.2–22.5 | Y | Britain | 6683.0 |
| | | | | | | | 119 | |

## Body Masses of World Birds (continued)

| Species | Sex | N | Mean | Std dev | Range | Sn | Location | Number |
|---|---|---|---|---|---|---|---|---|
| Luscinia luscinia | U | 28 | 23.8 | 2.90 | 18.2–34.1 | W | Ethiopia 9 | 6687.0 |
| Luscinia megarhynchos megarhynchos | U | 276 | 18.3 | | 12.0–25.0 | S | Morocco 391 | 6688.0 |
| Luscinia megarhynchos africana | U | 207 | 19.8 | 1.65 | 16.0–25.0 | | Kenya 119 | 6688.0 |
| Luscinia megarhynchos hafizi | U | 116 | 20.8 | 1.80 | 16.0–26.0 | | Kenya 119 | 6688.0 |
| Luscinia calliope | B | 20 | 18.5 | | 16.0–27.0 | M | China 119 | 6689.0 |
| Luscinia pectoralis | B | 13 | | | 21.5–25.2 | | India 5 | 6690.0 |
| Luscinia svecica svecica | B | 142 | 18.2 | | 14.7–25.5 | B | Russia 119 | 6691.0 |
| Luscinia svecica pallidogularis | M | 132 | 16.1 | 1.10 | 14.0–19.0 | F | Siberia 119 | 6691.0 |
| | F | 64 | 15.6 | 1.20 | 13.0–20.0 | | | |
| Luscinia ruficeps | U | 1 | 18.0 | | | | Malaysia 371 | 6692.0 |
| Luscinia brunnea | B | 6 | 17.5 | | 14.0–20.0 | | India 5 | 6695.0 |
| Luscinia cyane | U | 52 | 14.8 | | | W | Malaysia 420 | 6696.0 |
| Tarsiger cyanurus cyanurus | B | 25 | 12.1 | | 10.0–16.0 | | China 119 | 6697.0 |
| Tarsiger cyanurus pallidior | U | 12 | 12.9 | | 11.0–14.0 | | India 5 | 6697.0 |
| Tarsiger chrysaeus | B | 12 | 13.8 | 1.39 | 11.9–16.4 | | India 5 | 6698.0 |
| Tarsiger indicus | B | 6 | 14.6 | | 13.2–15.5 | | India 5 | 6699.0 |
| Tarsiger hyperythrus | M | 1 | 12.0 | | | | India 5 | 6700.0 |
| Irania gutturalis | B | 16 | 21.4 | | 16.0–24.0 | S | Kuwait 119 | 6702.0 |
| Cossypha isabellae | B | 15 | 24.0 | | 21.0–28.0 | | Cameroon 349 | 6703.0 |
| Cossypha anomala | B | 67 | 24.5 | | 20.6–27.4 | | Malawi 168 | 6706.0 |
| Cossypha caffra | B | 58 | 28.5 | | 25.4–34.0 | | Malawi 168 | 6707.0 |
| Cossypha humeralis | B | 15 | 21.5 | 2.31 | 19.0–25.8 | | Zimbabwe 281, 282, 283, 284 | 6708.0 |
| Cossypha cyanocampter | U | 1 | 26.5 | | | | Kenya 314a | 6709.0 |

## Body Masses of World Birds (continued)

| Species | Sex | N | Mean | Std dev | Range | Sn | Location | Number |
|---|---|---|---|---|---|---|---|---|
| Cossypha polioptera | F | 1 | 24.0 | | | | Cameroon 349 | 6710.0 |
| Cossypha semirufa | U | 11 | 27.3 | | 22.5–31.0 | | Ethiopia 633, 635 | 6711.0 |
| Cossypha heuglini | U | 45 | 41.7 | 4.00 | 34.5–51.0 | | Kenya 67 | 6712.0 |
| Cossypha natalensis | U | 69 | 29.1 | 2.30 | 23.7–33.4 | | Somalia 691 | 6713.0 |
| Cossypha niveicapilla | U | 14 | 33.8 | 3.10 | 27.3–38.5 | | Ghana 228 | 6716.0 |
| Cossypha albicapilla | U | 11 | 59.2 | | 53.0–65.0 | | Ghana 228 | 6717.0 |
| Cercotrichas leucosticta | U | 1 | 24.0 | | | | Liberia 314a | 6722.0 |
| Cercotrichas quadrivirgata | U | 20 | 24.1 | 1.70 | 20.3–26.3 | | Somalia 691 | 6723.0 |
| Cercotrichas hartlaubi | F | 2 | 19.0 | | 18.0–20.0 | | Cameroon 349 | 6726.0 |
| Cercotrichas leucophrys | B | 19 | 16.7 | | 12.9–20.3 | | Zimbabwe 280, 282, 283, 284 | 6727.0 |
| Cercotrichas galactotes | U | 26 | 20.3 | | 17.5–24.0 | W | Kenya 119 | 6728.0 |
| Cercotrichas paena | B | 10 | 19.4 | | 17.7–21.0 | | South Africa 256 | 6729.0 |
| Cercotrichas coryphaeus | U | 5 | 21.4 | | 19.7–23.5 | | South Africa 256 | 6730.0 |
| Cercotrichas podobe | B | 7 | 25.3 | | 24.0–27.0 | | Niger; Chad 119 | 6731.0 |
| Copsychus albospecularis | B | 14 | 22.3 | | 18.5–27.0 | | Madagascar 39 | 6734.0 |
| Copsychus saularis | B | 24 | 36.0 | | 31.0–42.5 | | India 5 | 6735.0 |
| Copsychus malabaricus | U | 4 | | | 28.0–32.0 | | India 5 | 6736.0 |
| Saxicoloides fulicata | B | 10 | 15.9 | | 17.0–21.0 | | India 5 | 6742.0 |
| Saxicoloides fulicata cambaiensis | B | 14 | 16.6 | | 17.0–21.0 | | India 5 | 6742.0 |
| Phoenicurus erythronota | B | 23 | 18.5 | | 15.0–22.0 | W | USSR 119 | 6744.0 |
| Phoenicurus caeruleocephalus | U | 30 | 15.5 | 0.15 | | M | Gibraltar 189 | 6745.0 |
| Phoenicurus ochruros | U | 50 | 16.5 | 1.50 | 13.0–20.0 | W | Malta 119 | 6746.0 |

## Body Masses of World Birds (continued)

| Species | Sex | N | Mean | Std dev | Range | Sn | Location | Number |
|---|---|---|---|---|---|---|---|---|
| Phoenicurus phoenicurus | M | 1016 | 14.7 | 1.40 | 10.0–19.0 | F | Poland | 6747.0 |
| | F | 1282 | 14.5 | 1.40 | 9.0–21.0 | | 119 | |
| Phoenicurus hodgsoni | M | 7 | | | 16.0–19.5 | W | India | 6748.0 |
| | F | 6 | | | 14.5–18.5 | | 5 | |
| Phoenicurus schisticeps | B | 7 | 17.7 | | 16.0–20.6 | | India | 6749.0 |
| | | | | | | | 5 | |
| Phoenicurus auroreus | B | 6 | 15.2 | | 13.9–17.0 | | | 6750.0 |
| | | | | | | | 149 | |
| Phoenicurus moussieri | B | 4 | 14.9 | | 14.5–15.0 | F | Algeria | 6751.0 |
| | | | | | | | 119 | |
| Phoenicurus erythrogaster | M | 367 | 25.2 | 2.30 | 22.0–29.0 | F | India | 6752.0 |
| | F | 346 | 23.9 | 1.30 | 21.0–28.0 | | 119 | |
| Phoenicurus frontalis | M | 25 | | | 14.5–19.0 | | India | 6753.0 |
| | F | 18 | | | 12.5–18.0 | | 5 | |
| Chimarraiornis leucocephalus | M | 16 | | | 30.0–37.0 | | | 6754.0 |
| | F | 12 | | | 24.0–30.6 | | 5 | |
| Rhyacornis fuliginosus | M | 13 | | | 14.0–21.0 | | | 6755.0 |
| | F | 8 | | | 14.8–18.0 | | 5 | |
| Hodgsonius phaenicuroides | B | 5 | 23.2 | | 22.0–24.8 | | India | 6757.0 |
| | | | | | | | 5 | |
| Cinclidium leucurum | B | 6 | | | 24.0–30.0 | | India | 6758.0 |
| | | | | | | | 5 | |
| Cinclidium frontale | B | 4 | 25.4 | | 25.0–26.2 | | India | 6760.0 |
| | | | | | | | 5 | |
| Grandala coelicolor | B | 5 | 45.6 | | 38.0–52.0 | | India | 6761.0 |
| | | | | | | | 5 | |
| Sialia sialis | B | 33 | 31.6 | 0.92 | | | | 6762.0 |
| | | | | | | | 244 | |
| Sialia mexicana | M | 22 | 29.0 | 1.79 | 26.0–32.5 | PB | California, USA | 6763.0 |
| | F | 19 | 27.1 | 2.01 | 22.5–320.5 | | 177 | |
| Sialia currucoides | B | 42 | 29.6 | 3.00 | | | western USA | 6764.0 |
| | | | | | | | 257 | |
| Enicurus scouleri | M | 2 | 17.8 | | 17.0–18.6 | | India | 6765.0 |
| | F | 3 | | | 12.0–16.0 | | 5 | |
| Enicurus immaculatus | B | 2 | 25.5 | | 25.0–26.0 | | India | 6768.0 |
| | | | | | | | 5 | |
| Enicurus schistaceus | U | 13 | 31.0 | | 25.0–38.0 | | Malaysia | 6769.0 |
| | | | | | | | 371 | |
| Enicurus leschenaulti | M | 1 | 53.5 | | | | India | 6770.0 |
| | | | | | | | 5 | |
| Enicurus maculatus | U | 14 | | | 34.0–48.0 | | India | 6771.0 |
| | | | | | | | 5 | |
| Cochoa purpurea | B | 2 | 103.0 | | 100.0–106.0 | | India | 6772.0 |
| | | | | | | | 5 | |

## Body Masses of World Birds (continued)

| Species | Sex | N | Mean | Std dev | Range | Sn | Location | Number |
|---|---|---|---|---|---|---|---|---|
| Cochoa viridis | M | 5 | | | 88.0–99.0 | | India | 6773.0 |
| | F | 2 | 120.0 | | 117.0–122.0 | | 5 | |
| Saxicola rubetra | U | 72 | 16.6 | 1.50 | 13.0–22.0 | S | Britain | 6776.0 |
| | | | | | | | 119 | |
| Saxicola torquata rubicola | U | 50 | 15.3 | 1.40 | 13.0–19.0 | W | Malta | 6780.0 |
| | | | | | | | 119 | |
| Saxicola torquata indica | U | 17 | 11.2 | | 8.0–16.0 | | India | 6780.0 |
| | | | | | | | 5 | |
| Saxicola maura | B | 29 | 12.6 | | 11.0–15.0 | | India | 6781.0 |
| | | | | | | | 5 | |
| Saxicola tectes | M | 13 | 12.3 | 0.72 | 11.2–13.8 | | Reunion | 6782.0 |
| | F | 8 | 13.0 | | 11.9–13.6 | | 152 | |
| Saxicola caprata | B | 19 | 15.2 | | 13.0–17.0 | | India | 6784.0 |
| | | | | | | | 5 | |
| Saxicola ferrea | M | 17 | | | 13.8–16.2 | | India | 6786.0 |
| | F | 3 | 14.7 | | 13.1–16.0 | | 5 | |
| Oenanthe leucopyga | B | 8 | 27.9 | | 25.0–39.0 | W | Algeria | 6789.0 |
| | | | | | | | 119 | |
| Oenanthe monacha | M | 1 | 23.0 | | | | Iran | 6790.0 |
| | F | 2 | 20.2 | | 18.0–22.5 | | 119 | |
| Oenanthe alboniger | B | 27 | 25.2 | | 22.0–28.5 | | Iran | 6791.0 |
| | | | | | | | 119 | |
| Oenanthe leucura | M | 2 | 41.0 | | 38.0–44.0 | | | 6792.0 |
| | | | | | | | 119 | |
| Oenanthe monticola | F | 2 | 37.0 | | 35.5–38.5 | | South Africa | 6793.0 |
| | | | | | | | 256 | |
| Oenanthe oenanthe oenanthe | M | 22 | 24.0 | 1.50 | 19.0–27.0 | S | Netherlands | 6795.0 |
| | F | 45 | 22.3 | 2.50 | 18.0–29.0 | | 119 | |
| Oenanthe oenanthe leucorhoa | B | 54 | 30.6 | | | S | Scotland | 6795.0 |
| | | | | | | | 119 | |
| Oenanthe lugens | M | 3 | 24.0 | | 22.0–25.0 | F | Algeria | 6796.0 |
| | F | 3 | 21.0 | | 19.0–22.0 | | 119 | |
| Oenanthe finschii | M | 29 | 25.6 | | 21.0–28.5 | S | USSR | 6798.0 |
| | F | 10 | 29.3 | | 23.5–32.5 | | 119 | |
| Oenanthe picata | M | 41 | | | 21.0–25.0 | | India | 6799.0 |
| | F | 12 | | | 18.0–23.0 | | 5 | |
| Oenanthe pleschanka | M | 21 | 17.4 | 1.10 | 15.0–20.0 | S | | 6801.0 |
| | F | 9 | 19.9 | | 18.0–25.0 | | 119 | |
| Oenanthe hispanica | M | 15 | 16.4 | 1.70 | 13.0–19.0 | S | | 6803.0 |
| | F | 3 | 13.8 | | 12.0–15.0 | | 119 | |
| Oenanthe xanthoprymna | B | 9 | 22.7 | | 20.0–27.0 | F | | 6804.0 |
| | | | | | | | 119 | |
| Oenanthe deserti | M | 19 | 19.5 | 1.70 | 17.0–22.0 | S | | 6805.0 |
| | F | 7 | 17.1 | | 15.0–19.0 | | 119 | |

## Body Masses of World Birds (continued)

| Species | Sex | N | Mean | Std dev | Range | Sn | Location | Number |
|---|---|---|---|---|---|---|---|---|
| Oenanthe pileata | U | 3 | 26.8 | | 23.9–32.5 | | 78, 256 | 6806.0 |
| Oenanthe isabellina | B | 18 | 28.7 | | 25.0–34.0 | B | 119 | 6807.0 |
| Cercomela sinuata | F | 2 | 18.6 | | 17.4–19.7 | | South Africa 256 | 6809.0 |
| Cercomela familiaris | B | 19 | 22.0 | | 17.3–28.0 | | South Africa 256 | 6812.0 |
| Cercomela fusca | F | 1 | 12.7 | | | | India 5 | 6814.0 |
| Cercomela melanura | B | 4 | 14.8 | | 14.0–15.0 | | Chad; Niger 119 | 6816.0 |
| Cercomela sordida | U | 1 | 20.0 | | | | Ethiopia 635 | 6817.0 |
| Myrmococichla aethiops | U | 5 | 58.0 | | | | Kenya 119 | 6819.0 |
| Myrmococichla formicivora | U | 65 | 41.6 | | 22.0–49.0 | | 78 | 6820.0 |
| Myrmococichla nigra | B | 9 | 40.1 | | 37.0–46.0 | | Cameroon 349 | 6821.0 |
| Myrmococichla albifrons | B | 2 | 19.5 | | 19.0–20.0 | | Cameroon 349 | 6823.0 |
| Thamnolaea cinnamomeiventris | M | 1 | 46.7 | | | | Zimbabwe 285 | 6825.0 |
| Pinarornis plumosus | M | 1 | 65.8 | | | | Zimbabwe 285 | 6828.0 |

### ORDER: PASSERIFORMES    FAMILY: SITTIDAE

| Species | Sex | N | Mean | Std dev | Range | Sn | Location | Number |
|---|---|---|---|---|---|---|---|---|
| Sitta europaea | U | 30 | 22.0 | | 19.5–24.0 | S | Britain 258 | 6829.0 |
| Sitta nagaensis | U | 2 | 14.7 | | 13.9–15.5 | | Thailand 378 | 6830.0 |
| Sitta castanea | M | 8 | | | 17.4–20.9 | | India | 6832.0 |
| | F | 5 | | | 19.9–20.5 | | 5 | |
| Sitta himalayensis | B | 31 | | | 12.0–17.0 | | India 5 | 6833.0 |
| Sitta pygmaea | B | 267 | 10.6 | | | | 422 | 6835.0 |
| Sitta pusilla | B | 123 | 10.2 | | | | 422 | 6836.0 |
| Sitta whiteheadi | U | | 12.5 | | | | 78 | 6837.0 |
| Sitta canadensis | B | 310 | 9.8 | 0.70 | 8.0–12.7 | F | New Jersey, USA 407 | 6842.0 |

## Body Masses of World Birds (continued)

| Species | Sex | N | Mean | Std dev | Range | Sn | Location | Number |
|---|---|---|---|---|---|---|---|---|
| Sitta leucopsis | M | 10 | | | 13.5–15.9 | | India | 6843.0 |
| | F | 3 | | | 14.7–15.1 | | 5 | |
| Sitta carolinensis | B | 266 | 21.1 | 2.39 | 18.5–26.7 | Y | Pennsylvania,USA | 6844.0 |
| | | | | | | | 101 | |
| Sitta neumayer | U | 5 | 31.0 | | 24.0–37.6 | | | 6845.0 |
| | | | | | | | 78 | |
| Sitta tephronota | M | 3 | | | 33.0–35.0 | | India | 6846.0 |
| | F | 3 | | | 31.0–32.0 | | 5 | |
| Sitta frontalis | M | 13 | | | 11.0–14.5 | | India | 6847.0 |
| | F | 4 | | | 10.0–13.4 | | 5 | |
| Tichodroma muraria | M | 6 | | | 13.0–19.0 | | India | 6853.0 |
| | F | 2 | 17.2 | | 16.5–18.0 | | 5 | |

### ORDER: PASSERIFORMES — FAMILY: CERTHIIDAE

| Species | Sex | N | Mean | Std dev | Range | Sn | Location | Number |
|---|---|---|---|---|---|---|---|---|
| Certhia familiaris | U | 50 | 9.0 | | 7.8–11.5 | S | Britain | 6854.0 |
| | | | | | | | 258 | |
| Certhia americana | B | 112 | 8.4 | 0.21 | 7.2–9.9 | M | Pennsylvania,USA | 6855.0 |
| | | | | | | | 101 | |
| Certhia brachydactyla | U | 1 | 8.5 | | | | | 6856.0 |
| | | | | | | | 227 | |
| Certhia himalayana | B | 10 | 8.8 | 0.84 | 7.8–10.3 | | India | 6857.0 |
| | | | | | | | 5 | |
| Certhia nipalensis | B | 6 | 11.1 | | 10.0–12.0 | | India | 6858.0 |
| | | | | | | | 5 | |
| Certhia discolor | B | 8 | 10.4 | | 9.6–11.0 | | India | 6859.0 |
| | | | | | | | 5 | |
| Salpornis spilonotus | U | 1 | 14.0 | | | | India | 6860.0 |
| | | | | | | | 5 | |

### ORDER: PASSERIFORMES — FAMILY: TROGLODYTIDAE

| Species | Sex | N | Mean | Std dev | Range | Sn | Location | Number |
|---|---|---|---|---|---|---|---|---|
| Donacobius atricapillus | B | 6 | 34.8 | | 31.0–42.0 | | Surinam; Panama | 6861.0 |
| | | | | | | | 87, 245 | |
| Campylorhynchus gularis | B | 3 | 30.1 | | 28.3–31.0 | | Jalisco, Mexico | 6862.0 |
| | | | | | | | 548 | |
| Campylorhynchus brunneicapillus | B | 42 | 38.9 | | 33.4–46.9 | | Arizona, USA | 6863.0 |
| | | | | | | | 6 | |
| Campylorhynchus jocosus | B | 5 | 27.6 | | 23.8–29.8 | | Oaxaca, Mexico | 6864.0 |
| | | | | | | | 584 | |
| Campylorhynchus yucatanicus | U | 2 | 35.5 | | 31.0–40.0 | | Yucatan, Mexico | 6865.0 |
| | | | | | | | 323 | |
| Campylorhynchus chiapensis | B | 8 | 50.9 | | 43.4–57.0 | | Chiapas, Mexico | 6866.0 |
| | | | | | | | 584 | |
| Campylorhynchus griseus | U | 6 | 42.4 | | 37.0–46.5 | | Venezuela | 6867.0 |
| | | | | | | | 625 | |

## Body Masses of World Birds (continued)

| Species | Sex | N | Mean | Std dev | Range | Sn | Location | Number |
|---|---|---|---|---|---|---|---|---|
| Campylorhynchus rufinucha | B | 2 | 30.4 | | 28.9–31.8 | | Mexico 139 | 6868.0 |
| Campylorhynchus turdinus | U | | 39.0 | | | | Brazil 564 | 6869.0 |
| Campylorhynchus megalopterus | U | 3 | 33.1 | | 32.8–33.5 | | Mexico 456 | 6870.0 |
| Campylorhynchus zonatus | B | 7 | 34.6 | | 28.3–39.5 | | 510, 584 | 6871.0 |
| Campylorhynchus albobrunneus | U | 6 | 32.9 | | 27.5–39.0 | | Panama 497, 611 | 6872.0 |
| Campylorhynchus nuchalis | U | 10 | 23.2 | 1.60 | 21.1–25.5 | | Venezuela 625 | 6873.0 |
| Campylorhynchus fasciatus | F | 1 | 24.9 | | | | Peru 672 | 6874.0 |
| Odontorchilus branickii | B | 2 | 9.6 | | 9.0–10.2 | | 260, 668 | 6875.0 |
| Odontorchilus cinereus | M | 1 | 11.0 | | | | Brazil 610a | 6876.0 |
| Salpinctes obsoletus | B | 31 | 16.5 | 0.63 | | | California, USA 465 | 6877.0 |
| Catherpes mexicanus | B | 7 | 12.6 | | 9.0–18.0 | | 368, 594 | 6878.0 |
| Catherpes sumichrasti | F | 1 | 28.4 | | | | 590 | 6879.0 |
| Cinnycerthia unirufa | B | 17 | 24.9 | | 21.0–29.0 | | Peru 451 | 6880.0 |
| Cinnycerthia peruana | U | 8 | 19.6 | | | | Peru 668 | 6881.0 |
| Cistothorus platensis | B | 28 | 9.0 | 0.69 | 7.2  10.3 | Y | 584 | 6882.0 |
| Cistothorus palustris | M | 38 | 11.9 | 0.72 | 10.5–13.5 | B | New York, USA 177 | 6885.0 |
| | F | 38 | 10.6 | 0.99 | 9.0–13.5 | | | |
| Thryomanes bewickii | B | 56 | 9.9 | 0.77 | 7.8–11.8 | Y | Arizona, USA 177 | 6886.0 |
| Thryomanes sissonii | B | 7 | 7.8 | | 5.6–9.8 | | Mexico 593 | 6887.0 |
| Thryothorus atrogularis | M | 6 | 26.3 | | 24.3–27.3 | | Costa Rica 584 | 6889.0 |
| | F | 1 | 22.5 | | | | | |
| Thryothorus fasciatoventris | U | 12 | 24.0 | | | | 87, 244, 315, 497, 584, 611 | 6891.0 |
| Thryothorus euophrys | B | 8 | 29.2 | | 26.0–34.0 | | Peru 584 | 6892.0 |
| Thryothorus genibarbis | U | 8 | 19.3 | | 16.2–22.8 | | Brazil 437 | 6895.0 |

## Body Masses of World Birds (continued)

| Species | Sex | N | Mean | Std dev | Range | Sn | Location | Number |
|---|---|---|---|---|---|---|---|---|
| Thryothorus coraya | U | 10 | 23.8 | 2.52 | | | Peru 668 | 6896.0 |
| Thryothorus felix | B | 24 | 13.2 | 1.91 | 9.0–16.4 | | Mexico 593 | 6897.0 |
| Thryothorus maculipectus | M | 9 | 15.8 | | 14.3–16.8 | | Belize 510 | 6898.0 |
| | F | 11 | 13.9 | 1.15 | 12.4–16.2 | | | |
| Thryothorus rutilus | B | 26 | 16.4 | | 13.5–18.5 | | Trinidad 576 | 6899.0 |
| Thryothorus semibadius | U | 1 | 17.0 | | | | Costa Rica 315 | 6901.0 |
| Thryothorus nigricapillus | U | 8 | 21.9 | | 17.7–26.3 | | Panama 87, 611 | 6902.0 |
| Thryothorus thoracicus | U | 3 | 17.6 | | | | Costa Rica 315 | 6903.0 |
| Thryothorus pleurostictus | B | 13 | 18.1 | 2.73 | 14.0–23.8 | | 315, 458, 584 | 6905.0 |
| Thryothorus ludovicianus | U | 20 | 18.7 | 2.26 | 15.0–22.0 | W | Georgia, USA 176 | 6906.0 |
| Thryothorus rufalbus | M | 25 | 28.0 | | | | | 6907.0 |
| | F | 25 | 25.0 | | | | 186a | |
| Thryothorus sinaloa | M | 17 | 15.4 | 1.02 | 13.1–16.6 | | Mexico 593 | 6909.0 |
| | F | 7 | 14.3 | | 13.5–16.4 | | | |
| Thryothorus modestus | M | 6 | 19.7 | | | | Panama 244 | 6910.0 |
| | F | 7 | 17.9 | | | | | |
| Thryothorus zeledoni | U | | 23.0 | | | | 603a | 6910.1 |
| Thryothorus leucotis | M | 35 | 21.0 | | | | | 6911.0 |
| | F | 35 | 18.0 | | | | 186a | |
| Thryothorus guarayanus | B | 2 | 13.5 | | 13.0–14.0 | | Bolivia 22 | 6913.0 |
| Troglodytes troglodytes | B | 54 | 8.9 | 0.20 | 7.5–10.5 | F | Pennsylvania,USA 101 | 6916.0 |
| Troglodytes troglodytes fridariensi | U | 50 | 12.2 | | | | Fair Is. 119 | 6916.0 |
| Troglodytes troglodytes troglodytes | U | 272 | 9.3 | 1.10 | 6.0–12.0 | | Germany 119 | 6916.0 |
| Troglodytes troglodytes | U | 50 | 9.9 | | 8.0–12.7 | Y | Britain 258 | 6916.0 |
| Troglodytes aedon | B | 346 | 10.9 | 0.80 | 8.9–14.2 | | Pennsylvania,USA 101 | 6917.0 |
| Troglodytes aedon brunneicollis | B | 15 | 10.4 | 0.56 | 9.7–11.8 | | Coahuila, Mexico 386 | 691 |
| Troglodytes rufociliatus | F | 1 | 11.0 | | | | 584 | 6919.0 |

## Body Masses of World Birds (continued)

| Species | Sex | N | Mean | Std dev | Range | Sn | Location | Number |
|---|---|---|---|---|---|---|---|---|
| Troglodytes ochraceus | U | 4 | 9.4 | | 8.0–10.0 | | Panama 244, 497 | 6920.0 |
| Troglodytes solstitialis | U | 9 | 11.8 | | | | Peru 668 | 6922.0 |
| Thryorchilus browni | U | | 14.0 | | | | 603a | 6924.0 |
| Uropsila leucogastra | M | 10 | 9.6 | 0.30 | 9.8–10.5 | | Yucatan, Mexico | 6925.0 |
| | F | 7 | 8.5 | | 8.0–9.1 | | 457 | |
| Henicorhina leucosticta | U | 26 | 15.7 | 1.20 | | | Peru 668 | 6926.0 |
| Henicorhina leucophrys | M | 17 | 18.1 | 1.03 | | | Panama | 6927.0 |
| | F | 6 | 16.4 | | | | 244 | |
| Henicorhina leucoptera | M | 10 | | | 12.0–16.5 | | Peru | 6928.0 |
| | F | 7 | | | 12.0–14.5 | | 138 | |
| Microcerculus philomela | M | 7 | 18.6 | | 17.4–21.5 | | | 6929.0 |
| | F | 4 | 17.0 | | 16.4–17.4 | | 603 | |
| Microcerculus luscinia | M | 5 | 20.1 | | 18.2–22.0 | | | 6929.1 |
| | F | 4 | 17.4 | | 17.0–18.0 | | 603 | |
| Microcerculus marginatus | U | 17 | 17.3 | 0.81 | | | Peru 668 | 6930.0 |
| Microcerculus bambla | U | 2 | 17.2 | | 17.0–17.5 | | 549, 161 | 6932.0 |
| Cyphorhinus phaeocephalus | U | 21 | 24.6 | 2.20 | | | Panama 315 | 6933.0 |
| Cyphorhinus thoracicus | M | 5 | 33.4 | | 30.5–35.0 | | Columbia | 6934.0 |
| | F | 4 | 29.3 | | 29.0–29.6 | | 387 | |
| Cyphorhinus aradus | U | 5 | 21.8 | | 18.0–24.0 | | 161, 610a | 6935.0 |

**ORDER: PASSERIFORMES**  **FAMILY: POLIOPTILIDAE**

| Species | Sex | N | Mean | Std dev | Range | Sn | Location | Number |
|---|---|---|---|---|---|---|---|---|
| Auriparus flaviceps | B | 49 | 6.8 | 0.69 | 5.5–8.5 | Y | SW USA; Mexico 594 | 6936.0 |
| Microbates collaris | U | 3 | 10.8 | | 10.0–11.5 | | Fr. Guiana 161 | 6937.0 |
| Microbates cinereiventris | U | 21 | 11.9 | 0.80 | | | Panama 315 | 6938.0 |
| Ramphocaenus melanurus | U | 17 | 9.7 | 0.90 | | | Panama 315 | 6939.0 |
| Polioptila caerulea | B | 184 | 6.0 | 0.13 | 4.8–8.9 | | Pennsylvania,USA 101 | 6940.0 |
| Polioptila californica | M | 5 | 6.0 | | 5.9–6.3 | | California, USA 12 | 6941.0 |
| Polioptila melanura | B | 8 | 5.1 | | 4.5–5.6 | | Mexico 584, 634 | 6942.0 |

## Body Masses of World Birds (continued)

| Species | Sex | N | Mean | Std dev | Range | Sn | Location | Number |
|---|---|---|---|---|---|---|---|---|
| Polioptila lembeyei | B | 5 | 4.5 | | 4.5–4.6 | | Cuba 429 | 6943.0 |
| Polioptila nigriceps | U | 9 | 5.6 | | 4.1–6.6 | | W. Mexico 593, 594 | 6944.0 |
| Polioptila albiloris albiventris | B | 9 | 5.6 | | 5.0–6.2 | | Yucatan, Mexico 457 | 6945.0 |
| Polioptila albiloris vanrossemi | B | 6 | 7.0 | | 6.5–8.1 | | Yucatan, Mexico 457 | 6945.0 |
| Polioptila plumbea | B | 25 | 6.0 | 0.67 | 4.8–7.0 | | 46,87,244,246,384 439,457,510,611,668 | 6946.0 |
| Polioptila lactea | U | | 7.0 | | | | Brazil 564 | 6947.0 |
| Polioptila guianensis | B | 2 | 5.8 | | 5.8–5.8 | | Brazil 610a | 6948.0 |
| Polioptila dumicola | M | 2 | 7.2 | | 7.0–7.5 | | Brazil 38 | 6950.0 |
| | F | 1 | 9.5 | | | | | |

### ORDER: PASSERIFORMES                                    FAMILY: AEGITHALIDAE

| Species | Sex | N | Mean | Std dev | Range | Sn | Location | Number |
|---|---|---|---|---|---|---|---|---|
| Aegithalos caudatus | U | 50 | 8.2 | | 7.8–9.5 | S | Britain 258 | 6951.0 |
| Aegithalos leucogenys | M | 9 | | | 6.0–8.0 | | India 5 | 6952.0 |
| | F | 7 | | | 6.0–7.0 | | | |
| Aegithalos concinnus | B | 29 | 6.1 | | 5.0–7.5 | | India 5 | 6953.0 |
| Aegithalos niveogularis | U | 3 | 7.8 | | | | India 467 | 6954.0 |
| Aegithalos iouschistos | M | 5 | 7.0 | | 6.5–7.5 | | India 5 | 6955.0 |
| Psaltriparus minimus | B | 53 | 5.3 | 0.45 | 4.5–6.0 | | California, USA 177 | 6957.0 |

### ORDER: PASSERIFORMES                                    FAMILY: HIRUNDINIDAE

| Species | Sex | N | Mean | Std dev | Range | Sn | Location | Number |
|---|---|---|---|---|---|---|---|---|
| Tachycineta bicolor | B | 82 | 20.1 | 1.58 | 15.6–25.4 | S | Pennsylvania,USA 177 | 6961.0 |
| Tachycineta albilinea | B | 9 | 13.9 | | | | 244, 457, 510, 611 | 6962.0 |
| Tachycineta albiventer | U | 6 | 17.7 | | 14.0–21.0 | | 38, 245, 311, 632 | 6963.0 |
| Tachycineta leucorrhoa | U | | 19.0 | | 17.0–21.0 | | 632 | 6964.0 |
| Tachycineta meyeni | M | | 17.0 | | 15.0–20.0 | | 632 | 6965.0 |
| Tachycineta thalassina | M | 16 | 14.4 | | 13.0–16.3 | | California, USA 106 | 6966.0 |
| | F | 15 | 13.9 | | 12.5–15.2 | | | |

## Body Masses of World Birds (continued)

| Species | Sex | N | Mean | Std dev | Range | Sn | Location | Number |
|---------|-----|---|------|---------|-------|-----|----------|--------|
| Tachycineta cyaneoviridis | B | 8 | 17.5 | | 16.3–19.5 | | Bahamas 600 | 6967.0 |
| Phaeoprogne tapera | U | | 36.1 | | 29.9–40.0 | | 632 | 6969.0 |
| Progne subis | B | 22 | 49.4 | 1.49 | | S | Maine, USA 243 | 6970.0 |
| Progne dominicensis | U | 5 | 39.6 | | 38.0–42.0 | | Trinidad 188 | 6972.0 |
| Progne chalybea | U | 115 | 42.9 | | 36.0–48.0 | | Trinidad 188 | 6974.0 |
| Notiochelidon cyanoleuca | U | 21 | 9.7 | 0.59 | 8.5–10.8 | | Venezuela 625 | 6977.0 |
| Notiochelidon flavipes | B | 8 | 9.4 | | 8.0–10.0 | | Peru 451 | 6978.0 |
| Notiochelidon pileata | B | 4 | 12.2 | | 12.0–13.0 | | Guatemala 584 | 6979.0 |
| Atticora fasciata | U | | | | 12.0–16.0 | | 632 | 6980.0 |
| Atticora melanoleuca | U | | | | 10.0–12.0 | | 632 | 6981.0 |
| Neochelidon tibialis | U | 3 | 9.1 | | 9.0–9.2 | | Panama 314, 611 | 6982.0 |
| Stelgidopteryx fucata | U | | | | 13.0–15.0 | | 632 | 6983.0 |
| Stelgidopteryx serripennis | B | 47 | 15.9 | 0.58 | 10.3–18.3 | S | Pennsylvania,USA 101 | 6984.0 |
| Stelgidopteryx ruficollis | U | | 15.2 | | 14.0–18.0 | | 632 | 6986.0 |
| Cheramoeca leucosternus | U | | 14.8 | | 12.0–15.8 | | 632 | 6987.0 |
| Riparia riparia | B | 249 | 14.6 | | 12.0–18.6 | B | New York, USA 606 | 6988.0 |
| Riparia paludicola | U | 61 | 13.4 | 1.00 | 11.0–15.0 | | South Africa 256 | 6989.0 |
| Riparia cincta | U | | | | 20.0–23.0 | | 632 | 6991.0 |
| Phedina borbonica | U | 19 | 23.9 | 1.78 | | | Mauritius 152 | 6992.0 |
| Phedina brazzae | U | | 13.0 | | | | 632 | 6993.0 |
| Hirundo griseopyga | U | | 9.5 | | 8.3–10.3 | | 632 | 6994.0 |
| Hirundo rupestris | U | 7 | 19.0 | | 17.0–21.0 | | India 5 | 6995.0 |

**Body Masses of World Birds (continued)**

| Species | Sex | N | Mean | Std dev | Range | Sn | Location | Number |
|---|---|---|---|---|---|---|---|---|
| Hirundo fuligula | U | 4 | 22.4 | | 16.0–30.0 | | 632 | 6997.0 |
| Hirundo concolor | U | 4 | | | 12.0–14.0 | | India 5 | 6998.0 |
| Hirundo rustica | M | 1337 | 16.2 | | 12.1–28.2 | S | Morocco | 6999.0 |
| | F | 994 | 15.8 | | 11.0–24.8 | | 8 | |
| Hirundo rustica rustica | B | 16 | 18.2 | | 16.0–22.0 | | Afghanistan 5 | 6999.0 |
| Hirundo lucida | U | | | | 12.0–14.0 | | 632 | 7000.0 |
| Hirundo aethiopica | U | | 13.0 | | 10.5–15.0 | | 632 | 7001.0 |
| Hirundo angolensis | U | 10 | 17.8 | | 15.7–19.0 | | Zambia 166 | 7002.0 |
| Hirundo albigularis | U | | 21.3 | | 16.0–28.0 | | 632 | 7003.0 |
| Hirundo tahitica | U | | 13.1 | | 11.1–15.6 | | 632 | 7005.0 |
| Hirundo neoxena | U | | 14.7 | | 12.5–17.3 | | 632 | 7006.0 |
| Hirundo smithii | U | | 13.9 | | 11.0–17.0 | | 632 | 7007.0 |
| Hirundo nigrita | U | | | | 16.0–19.0 | | 632 | 7008.0 |
| Hirundo nigrorufa | U | | 13.0 | | | | 632 | 7009.0 |
| Hirundo atrocaerulea | M | 1 | 13.1 | | | | Zambia 166 | 7010.0 |
| Hirundo dimidiata | U | | | | 10.0–12.0 | | 632 | 7013.0 |
| Hirundo cucullata | U | 3 | 27.0 | | 19.0–35.0 | | 632 | 7014.0 |
| Hirundo abyssinica | U | | 17.0 | | 15.0–21.0 | | 632 | 7015.0 |
| Hirundo semirufa | U | | 30.0 | | 28.0–34.0 | | 632 | 7016.0 |
| Hirundo senegalensis | U | | 42.5 | | 29.0–50.0 | | 632 | 7017.0 |
| Hirundo daurica | U | 7 | 16.7 | | 15.0–19.0 | | India 5 | 7018.0 |
| Hirundo striolata | F | 1 | 22.0 | | | | India 5 | 7020.0 |
| Hirundo spilodera | U | | 20.6 | | 16.0–26.0 | | 632 | 7024.0 |

## Body Masses of World Birds (continued)

| Species | Sex | N | Mean | Std dev | Range | Sn | Location | Number |
|---|---|---|---|---|---|---|---|---|
| Hirundo andecola | U | | 17.0 | | 14.0–19.0 | | 632 | 7025.0 |
| Hirundo pyrrhonota | B | 88 | 21.6 | 2.04 | 17.5–26.7 | B | California, USA 177 | 7026.0 |
| Hirundo fulva | B | 25 | 20.4 | | 18.4–22.3 | F | Texas, USA 547 | 7027.0 |
| Hirundo nigricans | U | | 15.9 | | 13.8–19.3 | | 632 | 7029.0 |
| Hirundo fluvicola | U | 14 | 9.7 | | 8.0–12.0 | | India 5 | 7030.0 |
| Hirundo ariel | U | | 11.2 | | 9.1–14.0 | | 632 | 7031.0 |
| Delichon urbica | U | 252 | 14.5 | | 10.3–19.8 | S | Morocco 8 | 7033.0 |
| Delichon dasypus | U | | 18.0 | | | | 632 | 7034.0 |
| Delichon nipalensis | U | | | | 14.0–16.0 | | 632 | 7035.0 |
| Psalidoprocne nitens | U | | 9.8 | | 8.3–11.0 | | 632 | 7036.0 |
| Psalidoprocne fuliginosa | B | 16 | 12.4 | | 11.0–14.0 | | Cameroon 349 | 7037.0 |
| Psalidoprocne albiceps | U | | | | 11.0–12.0 | | 632 | 7038.0 |
| Psalidoprocne pristoptera | B | 7 | 11.1 | | 10.0–13.0 | | Cameroon 349 | 7043.0 |
| Psalidoprocne orientalis | M | 1 | 11.9 | | | | Zimbabwe 285 | 7045.0 |
| Psalidoprocne obscura | U | | | | 8.8–10.0 | | 632 | 7047.0 |

### ORDER: PASSERIFORMES  FAMILY: REGULIDAE

| Species | Sex | N | Mean | Std dev | Range | Sn | Location | Number |
|---|---|---|---|---|---|---|---|---|
| Regulus calendula | M | 1424 | 6.9 | 0.36 | 5.0–9.7 | M | Pennsylvania,USA 101 | 7048.0 |
| | F | 1094 | 6.4 | 0.16 | 5.1–8.9 | | | |
| Regulus regulus | U | 50 | 5.7 | | 4.9–7.4 | S | Britain 258 | 7049.0 |
| Regulus ignicapillus | U | 44 | 5.6 | | 4.5–8.2 | Y | Britain 258 | 7052.0 |
| Regulus satrapa | M | 261 | 6.3 | 0.10 | 4.9–7.7 | M | Pennsylvania,USA 101 | 7053.0 |
| | F | 147 | 6.1 | 0.03 | 4.5–7.8 | | | |

### ORDER: PASSERIFORMES  FAMILY: PYCNONOTIDAE

| Species | Sex | N | Mean | Std dev | Range | Sn | Location | Number |
|---|---|---|---|---|---|---|---|---|
| Spizixos canifrons | M | 1 | 44.0 | | | | India 5 | 7054.0 |

## Body Masses of World Birds (continued)

| Species | Sex | N | Mean | Std dev | Range | Sn | Location | Number |
|---|---|---|---|---|---|---|---|---|
| Pycnonotus striatus | B | 10 | | | 45.0–60.0 | | India 5 | 7057.0 |
| Pycnonotus melanoleucus | U | 1 | 31.0 | | | | Malaysia 371 | 7060.0 |
| Pycnonotus melanicterus | M | 4 | | | 30.0–34.0 | | India | 7063.0 |
| | F | 2 | 29.5 | | 28.0–31.0 | | 5 | |
| Pycnonotus squamatus | B | 2 | 23.1 | | 22.2–24.0 | | Borneo 627 | 7064.0 |
| Pycnonotus cyaniventris | B | 2 | 21.0 | | 20.5–21.6 | | Borneo 627 | 7065.0 |
| Pycnonotus jocosus | U | 20 | 27.4 | | 25.0–31.0 | | India 5 | 7066.0 |
| Pycnonotus xanthorrhous | B | 12 | 26.9 | 1.93 | 24.4–31.1 | | Thailand 378 | 7067.0 |
| Pycnonotus barbatus | U | 44 | 35.9 | 1.90 | 30.8–42.4 | | Ghana 228 | 7070.0 |
| Pycnonotus nigricans | U | 359 | 30.8 | | 21.6–39.1 | | 78 | 7074.0 |
| Pycnonotus capensis | U | 49 | 39.5 | | 33.2  47.1 | | 78 | 7075.0 |
| Pycnonotus xanthopygos | U | 56 | 44.0 | | 35.0–46.0 | | Israel 119 | 7076.0 |
| Pycnonotus leucogenys leucotis | U | 10 | 23.0 | | 18.0–28.0 | | India 5 | 7078.0 |
| Pycnonotus leucogenys mesopotamiae | U | 5 | 32.0 | | 29.0–35.4 | | Kuwait; Iraq 119 | 7078.0 |
| Pycnonotus leucogenys leucogenys | U | 5 | | | 34.0–38.0 | | India 5 | 7078.0 |
| Pycnonotus cafer | U | 10 | 31.1 | | 28.0–40.0 | | India 5 | 7079.0 |
| Pycnonotus aurigaster | U | | 45.0 | | 40.0–50.0 | | 78 | 7080.0 |
| Pycnonotus urostictus | B | 35 | 24.8 | | 21.8–27.3 | | Philippines 475 | 7083.0 |
| Pycnonotus penicillatus | F | 2 | 36.5 | | 36.0–37.0 | | Sri Lanka 5 | 7087.0 |
| Pycnonotus flavescens | B | 13 | 29.8 | 2.14 | 27.1–35.0 | | Thailand 378 | 7088.0 |
| Pycnonotus luteolus | U | 10 | 34.7 | | 28.0–43.0 | | India 5 | 7089.0 |
| Pycnonotus goiavier | B | 24 | 27.8 | | 24.4–34.0 | | Philippines 475 | 7090.0 |
| Pycnonotus blanfordi | U | 8 | 33.4 | | 31.3–35.7 | | Thailand 378 | 7092.0 |

## Body Masses of World Birds (continued)

| Species | Sex | N | Mean | Std dev | Range | Sn | Location | Number |
|---|---|---|---|---|---|---|---|---|
| Pycnonotus brunneus | M | 2 | 36.6 | | 36.0–37.3 | | Borneo | 7094.0 |
| | F | 1 | 30.0 | | | | 627 | |
| Pycnonotus erythropthalmos | B | 3 | 20.2 | | 19.8–20.5 | | Borneo 627 | 7095.0 |
| Andropadus montanus | B | 6 | 31.8 | | 30.0–34.0 | | Cameroon 349 | 7096.0 |
| Andropadus masukucnsis | U | 18 | 24.9 | 1.32 | | | Kenya 362 | 7098.0 |
| Andropadus virens | U | 18 | 24.2 | 2.77 | 19.5–29.5 | | 314a | 7099.0 |
| Andropadus gracilis | U | 6 | 19.0 | | | | Kenya 362 | 7101.0 |
| Andropadus curvirostris | U | 5 | 23.0 | | 20.5–26.0 | | Ghana; Kenya 314a | 7103.0 |
| Andropadus gracilirostris | U | 1 | 27.0 | | | | 51 | 7104.0 |
| Andropadus importunus | U | 10 | 27.0 | 2.60 | 22.4–31.6 | | Somalia 691 | 7105.0 |
| Andropadus latirostris | U | 60 | 27.3 | 2.18 | 23.5–32.0 | | 314a | 7106.0 |
| Andropadus tephrolaemus | B | 130 | 37.3 | | 30.0–42.0 | | Malawi 168 | 7107.0 |
| Andropadus milanjensis | M | 39 | 38.5 | 2.50 | 34.6–45.9 | | Zimbabwe | 7110.0 |
| | F | 17 | 36.3 | 2.40 | 32.8–40.8 | | 281 | |
| Chlorocichla simplex | B | 4 | 46.6 | | 41.0–52.0 | | 314a, 349 | 7115.0 |
| Chlorocichla flavicollis | U | 5 | 45.3 | | 39.0–55.0 | | 228, 349 | 7116.0 |
| Chlorocichla flaviventris | M | 15 | 40.5 | 2.90 | 34.5–44.1 | | Zimbabwe | 7118.0 |
| | F | 7 | 36.8 | 2.10 | 33.6–39.4 | | 280 | |
| Chlorocichla laetissima | U | 4 | 48.5 | | | | Kenya 362 | 7119.0 |
| Phyllastrephus fischeri | U | 10 | 28.9 | 2.26 | 26.0–32.6 | | Kenya 314a | 7124.0 |
| Phyllastrephus placidus | U | | 28.2 | | | | Tanzania 417a | 7125.0 |
| Phyllastrephus terrestris | M | 22 | 34.1 | 2.60 | 30.1–40.5 | | Zimbabwe | 7126.0 |
| | F | 14 | 29.3 | 2.10 | 25.5–32.7 | | 280 | |
| Phyllastrephus strepitans | U | 45 | 27.1 | 3.50 | 21.4–32.9 | | Somalia 691 | 7127.0 |
| Phyllastrephus baumanni | U | 2 | 29.0 | | 27.5–30.5 | | Ghana 314a | 7130.0 |
| Phyllastrephus poensis | B | 3 | 31.0 | | 30.0–32.0 | | Cameroon 349 | 7132.0 |

## Body Masses of World Birds (continued)

| Species | Sex | N | Mean | Std dev | Range | Sn | Location | Number |
|---|---|---|---|---|---|---|---|---|
| Phyllastrephus flavostriatus | M | 11 | 31.8 | | 28.1–35.8 | | Malawi | 7135.0 |
| | F | 14 | 25.2 | | 22.0–32.4 | | 168 | |
| Phyllastrephus debilis | U | 5 | 13.6 | | 11.9–15.1 | | Kenya | 7137.0 |
| | | | | | | | 314a | |
| Phyllastrephus icterinus | U | 11 | 20.8 | 1.99 | 17.5–23.0 | | Liberia | 7139.0 |
| | | | | | | | 314a | |
| Phyllastrephus madagascariensis | M | 4 | 28.7 | | 25.5–31.2 | | Madagascar | 7142.0 |
| | F | 4 | 23.5 | | 20.7–26.5 | | 39 | |
| Phyllastrephus zosterops | M | 4 | 17.4 | | 16.0–18.2 | | Madagascar | 7143.0 |
| | F | 4 | 15.1 | | 13.5–16.7 | | 39 | |
| Bleda syndactyla | U | 13 | 47.9 | 4.36 | 42.5–58.5 | | | 7149.0 |
| | | | | | | | 314a | |
| Bleda eximia | U | 4 | 50.3 | | 49.5–51.5 | | Liberia | 7150.0 |
| | | | | | | | 314a | |
| Bleda eximia | U | 4 | 35.7 | | 34.5–37.5 | | Gabon | 7150.0 |
| | | | | | | | 314a | |
| Bleda canicapilla | B | 6 | 45.4 | | 36.5–52.0 | | Liberia | 7151.0 |
| | | | | | | | 314a | |
| Nicator chloris | U | 1 | 38.2 | | | | Liberia | 7152.0 |
| | | | | | | | 314a | |
| Criniger barbatus | U | 3 | 45.2 | | 44.5–45.7 | | Liberia | 7155.0 |
| | | | | | | | 314a | |
| Criniger chloronotus | U | 2 | 48.0 | | 46.5–49.5 | | Gabon | 7156.0 |
| | | | | | | | 314a | |
| Criniger calurus | U | 4 | 37.8 | | 33.2–43.0 | | Liberia | 7157.0 |
| | | | | | | | 314a | |
| Criniger olivaceus | U | 1 | 43.0 | | | | Malaysia | 7158.0 |
| | | | | | | | 371 | |
| Alophoixus finschii | B | 4 | 24.0 | | 23.1–24.5 | | Borneo | 7160.0 |
| | | | | | | | 627 | |
| Alophoixus flaveolus | M | 10 | 48.3 | | 38.0–54.0 | | India | 7161.0 |
| | F | 4 | | | 38.0–48.0 | | 5 | |
| Alophoixus pallidus | U | 10 | 46.0 | 3.00 | 41.0–52.0 | | Thailand | 7162.0 |
| | | | | | | | 378 | |
| Alophoixus bres | M | 2 | 43.7 | | 41.4–46.0 | | Borneo | 7164.0 |
| | F | 2 | 51.9 | | 51.2–52.6 | | 627 | |
| Alophoixus phaeocephalus | M | 1 | 35.0 | | | | Borneo | 7165.0 |
| | | | | | | | 627 | |
| Iole propinqua | F | 1 | 25.9 | | | | Thailand | 7170.0 |
| | | | | | | | 378 | |
| Iole olivacea | U | 3 | 75.0 | | 70.0–79.0 | | Mauritius | 7171.0 |
| | | | | | | | 152 | |
| Iole indica | U | 33 | 30.6 | | 27.0–34.0 | | India | 7172.0 |
| | | | | | | | 5 | |

## Body Masses of World Birds (continued)

| Species | Sex | N | Mean | Std dev | Range | Sn | Location | Number |
|---|---|---|---|---|---|---|---|---|
| Ixos philippinus | B | 44 | 38.6 | | 32.0–43.6 | | Philippines 475 | 7174.0 |
| Ixos everetti | B | 33 | 58.3 | | 51.9–66.9 | | Philippines 475 | 7178.0 |
| Ixos malaccensis | M | 1 | 41.0 | | | | Borneo 627 | 7179.0 |
| Hemixos flavala | M | 5 | 32.6 | | 31.0–34.0 | | Malaysia 371 | 7180.0 |
| Hypsipetes mcclellandii | U | 126 | 32.5 | | 25.0–40.0 | | Malaysia 371 | 7182.0 |
| Hypsipetes madagascariensis | U | 25 | 42.9 | | 35.0–49.0 | | India 5 | 7184.0 |
| Hypsipetes crassirostris | U | 32 | 79.5 | 1.10 | | | Seychelles 152 | 7185.0 |
| Hypsipetes borbonicus | B | 3 | 54.7 | | 51.0–57.2 | | Reunion 152 | 7187.0 |
| Hypsipetes thompsoni | M | 1 | 41.8 | | | | Thailand 378 | 7190.0 |

### ORDER: PASSERIFORMES      FAMILY: ZOSTEROPIDAE

| Species | Sex | N | Mean | Std dev | Range | Sn | Location | Number |
|---|---|---|---|---|---|---|---|---|
| Zosterops senegalensis | B | 193 | 10.9 | | 8.9–14.1 | | Malawi 168 | 7197.0 |
| Zosterops pallidus | U | 18 | 9.3 | | 8.4–10.5 | | 78 | 7203.0 |
| Zosterops maderaspatanus | B | 15 | 10.0 | | 8.5–10.7 | | Madagascar 39 | 7204.0 |
| Zosterops borbonicus | B | 13 | 8.2 | | 7.6–9.2 | | Mauritius 152 | 7208.0 |
| Zosterops olivaceus olivaceus | M<br>F | 9<br>11 | 8.6<br>9.7 | 0.86 | 7.7–9.8<br>8.6–11.4 | | Reunion 152 | 7209.0 |
| Zosterops olivaceus chloronotos | B | 5 | 8.1 | | 7.5–9.0 | | Mauritius 152 | 7209.0 |
| Zosterops ceylonensis | F | 1 | 12.6 | | | | Sri Lanka 5 | 7212.0 |
| Zosterops erythropleurus | U | 3 | 10.9 | | 9.9–11.5 | | Thailand 378 | 7213.0 |
| Zosterops palpebrosus | U | 20 | 8.6 | | 6.0–11.0 | | India 5 | 7214.0 |
| Zosterops japonicus | U | | 10.0 | | | | Hawaiian Is. 400 | 7215.0 |
| Zosterops conspicillatus | B | 9 | 7.9 | | | | Saipan 114 | 7218.0 |
| Zosterops everetti | B | 42 | 10.0 | | 8.1–12.0 | | Philippines 475 | 7222.0 |

## Body Masses of World Birds (continued)

| Species | Sex | N | Mean | Std dev | Range | Sn | Location | Number |
|---|---|---|---|---|---|---|---|---|
| Zosterops montanus | B | 20 | 10.5 | | 9.4–12.6 | | Philippines 475 | 7224.0 |
| Zosterops atrifrons | B | 8 | 11.1 | | 10.0–12.0 | | New Guinea 154, 216, 218 | 7234.0 |
| Zosterops fuscicapilla | M | 1 | 11.2 | | | | New Guinea 216 | 7240.0 |
| Zosterops novaeguineae | B | 3 | 12.5 | | 12.0–13.5 | | New Guinea 154 | 7243.0 |
| Zosterops lateralis | U | 1083 | 12.9 | 1.14 | 9.5–16.9 | | New Zealand 500 | 7256.0 |
| Zosterops explorator | U | 4 | 10.5 | | 8.5–11.5 | | Fiji 332 | 7262.0 |
| Zosterops xanthochrous | U | 46 | 10.3 | 0.90 | 8.5–12.0 | | New Caledonia 506 | 7264.0 |
| Zosterops virens | U | 14 | 13.4 | 1.09 | 12.0–16.0 | | Ethiopia | 7268.1 |
| Cleptornis marchei | B | 4 | 20.2 | | | | Saipan 114 | 7272.0 |
| Lophozosterops goodfellowi | B | 18 | 20.0 | | 18.2–22.3 | | Philippines 475 | 7277.0 |
| Hypocryptadius cinnamomeus | B | 22 | 30.0 | | 24.7–31.5 | | Philippines 475 | 7288.0 |

### ORDER: PASSERIFORMES          FAMILY: HYPOCOLIIDAE

| Species | Sex | N | Mean | Std dev | Range | Sn | Location | Number |
|---|---|---|---|---|---|---|---|---|
| Hypocolius ampelinus | B | 8 | 51.3 | | 48.0–57.0 | | 5, 119 | 7289.0 |

### ORDER: PASSERIFORMES          FAMILY: CISTICOLIDAE

| Species | Sex | N | Mean | Std dev | Range | Sn | Location | Number |
|---|---|---|---|---|---|---|---|---|
| Cisticola erythrops | U | 15 | 13.6 | 1.26 | 12.1–16.4 | | 228, 280, 284 | 7290.0 |
| Cisticola cantans | B | 23 | 11.9 | | 10.0–13.5 | | Zimbabwe 283 | 7292.0 |
| Cisticola lateralis | M | 12 | 18.9 | | 14.0–21.0 | | Cameroon 349 | 7293.0 |
| | F | 10 | 14.9 | | 12.0–19.0 | | | |
| Cisticola chubbi | U | 6 | 17.5 | | | | 51 | 7298.0 |
| Cisticola hunteri | M | 5 | 18.4 | | 17.0–20.0 | | Cameroon 349 | 7299.0 |
| | F | 7 | 15.1 | | 12.0–19.0 | | | |
| Cisticola aberrans | M | 7 | 15.4 | | 14.1–16.9 | | Zimbabwe 280, 284 | 7302.0 |
| | F | 3 | 13.0 | | 12.4–13.9 | | | |
| Cisticola chiniana | F | 1 | 12.8 | | | | Zimbabwe 284 | 7304.0 |
| Cisticola ruficeps | U | 1 | 9.8 | | | | Ghana 228 | 7306.0 |

## Body Masses of World Birds (continued)

| Species | Sex | N | Mean | Std dev | Range | Sn | Location | Number |
|---|---|---|---|---|---|---|---|---|
| Cisticola subruficapilla | F | 3 | 10.5 | | 9.6–12.0 | | South Africa 256 | 7309.0 |
| Cisticola lais | M | 4 | 14.4 | | 13.5–15.6 | | | 7310.0 |
| | F | 1 | 10.8 | | | | 166, 281 | |
| Cisticola njombe | M | 3 | 11.8 | | | | Zambia | 7313.0 |
| | F | 2 | 10.2 | | | | 166 | |
| Cisticola galactotes | M | 50 | 16.2 | 1.85 | 13.0–19.1 | | Kenya | 7314.0 |
| | F | 41 | 12.9 | 1.56 | 11.0–18.0 | | 68 | |
| Cisticola carruthersi | M | 37 | 11.7 | 0.97 | 10.1–14.4 | | Kenya | 7316.0 |
| | F | 34 | 10.8 | 0.82 | 9.5–12.4 | | 78 | |
| Cisticola tinniens | U | 12 | 12.9 | | 9.5–18.5 | | | 7317.0 |
| | | | | | | | 78, 256 | |
| Cisticola natalensis | B | 4 | 15.4 | | 14.5–16.1 | | Zimbabwe 283, 284 | 7321.0 |
| Cisticola fulvicapilla | U | 11 | 9.0 | | 8.0–10.3 | | | 7322.0 |
| | | | | | | | 256, 280 | |
| Cisticola brachyptera | M | 3 | 9.7 | | 9.0–10.0 | | Cameroon | 7325.0 |
| | F | 4 | 8.3 | | 7.0–9.0 | | 349 | |
| Cisticola rufa | M | 4 | 8.0 | | 7.4–9.1 | | Ghana | 7326.0 |
| | F | 3 | 6.5 | | 6.0–6.9 | | 228 | |
| Cisticola juncidis | M | 9 | | | 7.1–8.2 | | Australia | 7330.0 |
| | F | 8 | | | 5.0–7.9 | | 528 | |
| Cisticola cherina | B | 9 | 10.0 | | 8.2–11.2 | | Madagascar 39 | 7332.0 |
| Cisticola aridula | U | 1 | 8.1 | | | | 78 | 7333.0 |
| Cisticola exilis | U | 12 | 7.1 | | 6.5–7.5 | | Philippines 475 | 7339.0 |
| Scotocerca inquieta | B | | | | 8.1–9.4 | | 149 | 7340.0 |
| Prinia criniger criniger | U | 3 | 15.1 | | 13.2–17.0 | | India 5 | 7343.0 |
| Prinia criniger striatula | U | 3 | 12.5 | | 12.0–13.0 | | India 5 | 7343.0 |
| Prinia atrogularis | U | 11 | 11.8 | | 8.0–16.0 | | India 5 | 7345.0 |
| Prinia buchanani | U | 3 | | | 5.0–9.0 | | India 5 | 7347.0 |
| Prinia rufescens | U | 8 | | | 6.0–7.0 | | 5 | 7348.0 |
| Prinia hodgsonii | U | 8 | 6.4 | | 5.0–9.0 | | India 5 | 7349.0 |
| Prinia gracilis | U | 4 | 6.9 | | 6.5–7.1 | | 78 | 7350.0 |

## Body Masses of World Birds (continued)

| Species | Sex | N | Mean | Std dev | Range | Sn | Location | Number |
|---|---|---|---|---|---|---|---|---|
| Prinia sylvatica | U | 17 | 16.1 | | 10.0–21.0 | | India 5 | 7351.0 |
| Prinia flaviventris | M | 1 | 7.0 | | | | Pakistan 5 | 7353.0 |
| Prinia socialis | B | 12 | 8.0 | | 6.7–11.0 | | India 5 | 7354.0 |
| Prinia subflava | U | 11 | 9.2 | | | | Ghana 228 | 7355.0 |
| Prinia flavicans | U | 149 | 9.0 | | 6.8–12.0 | | 78 | 7359.0 |
| Prinia maculosa | B | 8 | 10.0 | | 8.2–11.2 | | 78, 256 | 7360.0 |
| Prinia substriata | M | 3 | 12.4 | | 11.5–13.2 | | South Africa 256 | 7361.0 |
| Prinia robertsi | M | 1 | 9.5 | | | | Zimbabwe 281 | 7363.0 |
| | F | 1 | 8.3 | | | | | |
| Prinia leucopogon | B | 8 | 13.8 | | 10.7–16.0 | | 314a, 349 | 7365.0 |
| Prinia bairdii | U | 4 | 13.9 | | 11.3–17.0 | | 314a, 349 | 7366.0 |
| Prinia erythroptera | B | 6 | 12.7 | | 10.3–16.0 | | 283, 349 | 7368.0 |
| Malcorus pectoralis | U | 10 | 10.0 | 1.00 | 8.0–11.5 | | South Africa 256 | 7369.0 |
| Urolais epichlora | B | 7 | 11.4 | | 10.0–12.0 | | Cameroon 349 | 7371.0 |
| Apalis pulchra | U | 17 | 10.3 | 3.22 | | | Kenya 362 | 7373.0 |
| Apalis ruwenzorii | U | 2 | 9.8 | | | | 51 | 7374.0 |
| Apalis thoracica | B | 74 | 12.1 | | 10.2–16.0 | | Malawi 168 | 7375.0 |
| Apalis flavida | U | 5 | 8.3 | | | | 280, 282, 314a | 7381.0 |
| Apalis melanocephala | U | 1 | 8.4 | | | | Somalia 691 | 7392.0 |
| Apalis chirindensis | M | 3 | 8.2 | | 7.5–8.9 | | Zimbabwe 281 | 7393.0 |
| Apalis cinerea | F | 1 | 10.0 | | | | Cameroon 349 | 7394.0 |
| Hypergerus atriceps | B | 2 | 28.0 | | 28.0–28.0 | | Cameroon 349 | 7398.0 |
| Camaroptera brevicaudata | B | 11 | 12.0 | | 11.0–13.0 | | Cameroon 349 | 7400.0 |

## Body Masses of World Birds (continued)

| Species | Sex | N | Mean | Std dev | Range | Sn | Location | Number |
|---|---|---|---|---|---|---|---|---|
| Camaroptera brachyura | U | 65 | 9.3 | 0.75 | 7.8–10.5 | | Kenya 67 | 7402.0 |
| Camaroptera chloronota | U | 15 | 11.4 | 0.95 | 8.7–12.3 | | 314a | 7404.0 |
| Calamonastes stierlingi | B | 4 | 13.2 | | 12.3–14.0 | | Zimbabwe 280 | 7406.0 |
| Calamonastes fasciolatus | F | 1 | 13.8 | | | | South Africa 256 | 7407.0 |
| Tesia castaneocoronata | U | | | | 8.0–10.0 | | India 5 | 7409.0 |
| Tesia olivea | B | 3 | 7.0 | | 6.0–9.0 | | India 5 | 7411.0 |
| Tesia cyaniventer | B | 5 | 9.7 | | 8.4–12.0 | | India 5 | 7412.0 |
| Urosphena squameiceps | U | | | | 9.0–10.0 | | 149 | 7416.0 |
| Cettia diphone | U | | | | 13.0–15.0 | | 149 | 7418.0 |
| Cettia ruficapilla | U | 18 | 12.7 | 1.60 | 9.0–15.5 | | Fiji 332 | 7421.0 |
| Cettia fortipes | U | 13 | | | 8.0–11.5 | | India 5 | 7423.0 |
| Cettia flavolivacea | B | 3 | 7.3 | | 6.0–9.0 | | India 5 | 7426.0 |
| Cettia brunnifrons | B | 16 | | | 6.0–9.0 | | India 5 | 7428.0 |
| Cettia cetti | M | 13 | 15.9 | | 14.5–17.5 | Y | Britain 258 | 7429.0 |
| | F | 18 | 12.6 | | 11.3–16.6 | | | |
| Bradypterus baboecala | U | 7 | 12.9 | | 11 7–15.0 | | Kenya 68 | 7430.0 |
| Bradypterus carpalis | M | 23 | 23.5 | 1.62 | 18.9–26.2 | | Kenya 68 | 7432.0 |
| | F | 18 | 21.3 | 1.02 | 19.0–23.0 | | | |
| Bradypterus barratti | M | 2 | 17.5 | | 16.0–19.0 | | Cameroon 349 | 7438.0 |
| Bradypterus cinnamomeus | U | 20 | 18.5 | | 16.3–22.0 | | 170 | 7439.0 |
| Bradypterus thoracicus | M | 1 | 10.0 | | | | India 5 | 7441.0 |
| Bradypterus palliseri | M | 1 | 9.0 | | | | Sri Lanka 5 | 7446.0 |
| Dromaeocercus seebohmi | M | 3 | 12.3 | | 11.2–13.2 | | Madagascar 39 | 7450.1 |
| Bathmocercus cerviniventris | U | 2 | 19.2 | | 18.7–19.7 | | Kenya 314a | 7451.0 |

## Body Masses of World Birds (continued)

| Species | Sex | N | Mean | Std dev | Range | Sn | Location | Number |
|---|---|---|---|---|---|---|---|---|
| Nesillas typica | U | 1 | 24.2 | | | | Comoro Is. 97 | 7455.0 |
| Melocichla mentalis | U | 9 | 33.5 | 1.80 | 30.6–37.0 | | Ghana 228 | 7458.0 |
| Sphenoeacus afer | B | 3 | 31.4 | | 28.3–33.7 | | Zimbabwe 281 | 7460.0 |
| Locustella lanceolata | M | 11 | 10.6 | | | | Malaysia 420 | 7461.0 |
| Locustella naevia | U | 50 | 13.3 | | 11.4–15.2 | B | Britain 258 | 7462.0 |
| Locustella certhiola | U | 136 | 14.4 | 1.52 | | W | Malaysia 420 | 7463.0 |
| Locustella ochotensis | U | 19 | 18.5 | | 16.0–24.0 | | 149 | 7464.0 |
| Locustella fluviatilis | M | 2 | 18.1 | | 16.7–19.5 | | 149 | 7466.0 |
| Locustella luscinioides | U | 1 | 15.0 | | | | Ghana 652 | 7467.0 |
| Locustella fasciolata | M | | 28.0 | | 25.0–32.0 | | | 7468.0 |
| | F | | 25.5 | | 21.0–28.0 | | 149 | |
| Acrocephalus melanopogon | U | 12 | 11.0 | | 8.5–13.0 | W | India 5 | 7470.0 |
| Acrocephalus paludicola | U | 47 | 11.6 | | 9.8–15.5 | F | Britain 258 | 7471.0 |
| Acrocephalus schoenobaenus | U | 232 | 11.2 | | 8.1–17.9 | | Wales 84 | 7472.0 |
| Acrocephalus bistrigiceps | M | 16 | 9.4 | | 8.0–11.0 | | India 5 | 7474.0 |
| | F | 8 | 8.0 | | 7.0–10.0 | | | |
| Acrocephalus agricola | U | 11 | 9.6 | | 8.0–11.0 | | India 5 | 7476.0 |
| Acrocephalus concinens | U | 4 | 8.3 | | 7.7–8.9 | | Thailand 378 | 7477.0 |
| Acrocephalus scirpaceus | U | 331 | 12.3 | 2.70 | 8.0–19.7 | S | Nigeria 169 | 7478.0 |
| Acrocephalus baeticatus | B | 23 | 10.3 | | 8.8–12.0 | | South Africa 256 | 7479.0 |
| Acrocephalus dumetorum | U | 25 | 11.2 | | 8.0–16.0 | | India 5 | 7480.0 |
| Acrocephalus palustris | U | 8 | 11.9 | | 10.4–12.9 | | Zimbabwe 282, 283, 284 | 7481.0 |
| Acrocephalus arundinaceus | U | 12 | 27.2 | | 22.0–31.0 | S | Morocco 8 | 7482.0 |
| Acrocephalus orientalis | U | 13 | 25.5 | | 22.0–29.0 | | India 5 | 7483.0 |

## Body Masses of World Birds (continued)

| Species | Sex | N | Mean | Std dev | Range | Sn | Location | Number |
|---|---|---|---|---|---|---|---|---|
| Acrocephalus stentoreus | U | 10 | 28.8 | | 23.0–24.0 | | India 5 | 7484.0 |
| Acrocephalus vaughani | M | 1 | 27.0 | | | | Pitcairn Is. 677 | 7496.0 |
| | F | 1 | 22.0 | | | | | |
| Acrocephalus rufescens | M | 30 | 23.9 | 1.95 | 21.7–27.2 | | Kenya 68 | 7497.0 |
| | F | 36 | 21.6 | 1.90 | 18.0–25.0 | | | |
| Acrocephalus gracilirostris | U | 9 | 14.4 | | 12.5–16.2 | | Kenya 68 | 7499.0 |
| Acrocephalus newtoni | B | 2 | 16.0 | | 15.5–16.5 | | Madagascar 39 | 7500.0 |
| Acrocephalus aedon | U | 10 | 26.1 | | 22.0–28.0 | | 5, 149 | 7501.0 |
| Bebrornis rodericanus | U | 1 | 11.4 | | | | Rodrigues Is. 152 | 7502.0 |
| Bebrornis sechellensis | M | 43 | 16.8 | 0.12 | | | Seychelles 152 | 7503.0 |
| | F | 25 | 15.0 | 0.20 | | | | |
| Hippolais caligata | U | 14 | 9.3 | | 8.0–11.0 | | India 5 | 7504.0 |
| Hippolais pallida | U | 26 | 10.6 | 0.80 | 9.1–12.3 | F | Nigeria 169 | 7505.0 |
| Hippolais languida | B | 7 | 10.0 | | 9.6–10.5 | | 149 | 7506.0 |
| Hippolais olivetorum | U | 17 | 18.1 | | 14.3–22.0 | | Kenya 460 | 7507.0 |
| Hippolais polyglotta | U | 54 | 11.0 | 0.95 | | S | Gibraltar 189 | 7508.0 |
| Hippolais icterina | U | 46 | 14.6 | | 9.5–22.8 | M | Britain 258 | 7509.0 |
| Chloropeta natalensis | B | 6 | 11.7 | | 9.6–13.0 | | 166, 283, 349 | 7510.0 |
| Chloropeta similis | B | 13 | 12.1 | | 10.9–14.0 | | Malawi 168 | 7511.0 |
| Chloropeta gracilirostris | U | 13 | 10.8 | 0.68 | 10.0–12.5 | | Kenya 68 | 7512.0 |
| Stenostira scita | M | 4 | 6.4 | | 5.6–6.7 | | South Africa 256 | 7513.0 |
| Orthotomus metopias | U | | 7.6 | | | | Tanzania 417a | 7515.0 |
| Orthotomus cuculatus | B | 4 | 6.0 | | 5.5–7.1 | | India 5 | 7517.0 |
| Orthotomus sutorius | U | 10 | 7.5 | | 6.0–10.0 | | India 5 | 7519.0 |
| Orthotomus atrogularis | M | 15 | 8.1 | | 6.8–8.7 | | Philippines 475 | 7520.0 |
| | F | 13 | 7.3 | | 6.2–8.2 | | | |

## Body Masses of World Birds (continued)

| Species | Sex | N | Mean | Std dev | Range | Sn | Location | Number |
|---|---|---|---|---|---|---|---|---|
| Orthotomus sericeus | F | 1 | 10.8 | | | | Borneo 627 | 7524.0 |
| Orthotomus sepium | B | 5 | 8.4 | | 7.8–8.8 | | Borneo 627 | 7526.0 |
| Orthotomus samarensis | M | 5 | 10.1 | | 10.0–10.2 | | Philippines | 7527.0 |
| | F | 2 | 7.9 | | 7.4–8.4 | | 475 | |
| Orthotomus cinereiceps | B | 2 | 9.0 | | 8.8–9.3 | | Philippines 475 | 7529.0 |
| Poliolais lopezi | B | 5 | 12.0 | | 11.0–13.0 | | Cameroon 349 | 7530.0 |
| Eremomela icteropygialis | B | 13 | 8.1 | 0.90 | 6.7–9.4 | | 256, 282, 284, 285 | 7532.0 |
| Eremomela pusilla | U | 26 | 6.1 | 0.20 | 5.0–8.4 | | Ghana 228 | 7535.0 |
| Eremomela scotops | M | 1 | 9.2 | | | | Zimbabwe 285 | 7537.0 |
| Newtonia amphichroa | B | 3 | 12.4 | | 11.7–13.0 | | Madagascar 39 | 7544.0 |
| Newtonia brunneicauda | B | 10 | 9 7 | 1.54 | 7.5–11.7 | | Madagascar 39 | 7545.0 |
| Newtonia fanovanae | M | 1 | 12.9 | | | | Madagascar 222 | 7547.0 |
| Sylvietta virens | B | 6 | 8.7 | | 7.6–10.0 | | 314a, 349 | 7548.0 |
| Sylvietta leucophrys | U | 7 | 10.4 | | | | 51 | 7551.0 |
| Sylvietta brachyura | U | 6 | 8.0 | | 7.4–8.4 | | Ghana 228 | 7552.0 |
| Sylvietta whytii | B | 11 | 9.9 | 0.49 | 9.1–10.5 | | Zimbabwe 280, 283, 284 | 7555.0 |
| Sylvietta rufescens | B | 12 | 11.3 | | 9.2–12.7 | | 256, 282, 284 | 7557.0 |
| Hemitesia neumanni | U | 1 | 11.3 | | | | 51 | 7558.0 |
| Macrosphenus kempi | U | 1 | 14.5 | | | | Ghana 314a | 7559.0 |
| Macrosphenus concolor | U | 1 | 13.7 | | | | Liberia 314a | 7561.0 |
| Hylia prasina | U | 11 | 12.5 | | 10.5–16.0 | | 314a, 349 | 7565.0 |
| Phylloscopus laetus | U | 3 | 8.9 | | | | 51 | 7568.0 |
| Phylloscopus ruficapilla | U | | 7.8 | | | | Tanzania 417a | 7570.0 |

## Body Masses of World Birds (continued)

| Species | Sex | N | Mean | Std dev | Range | Sn | Location | Number |
|---|---|---|---|---|---|---|---|---|
| Phylloscopus budongoensis | U | | | | 7.0–10.0 | | Kenya 495 | 7572.0 |
| Phylloscopus trochilus | U | 723 | 8.7 | | 6.5–11.8 | M | Wales 84 | 7574.0 |
| Phylloscopus collybita | U | 101 | 7.5 | | 5.7–9.2 | | Wales 84 | 7575.0 |
| Phylloscopus bonelli | U | 31 | 7.2 | | 5.5–8.4 | S | Morocco 8 | 7578.0 |
| Phylloscopus sibilatrix | U | 48 | 8.2 | 0.80 | 6.7–10.1 | | Nigeria 169 | 7579.0 |
| Phylloscopus fuscatus | U | 8 | 8.8 | | 7.0–10.2 | | India 5 | 7580.0 |
| Phylloscopus affinis | B | 17 | | | 6.3–7.7 | | India 5 | 7582.0 |
| Phylloscopus subaffinis | M | 1 | 6.2 | | | | Thailand 378 | 7583.0 |
| Phylloscopus griseolus | U | 6 | 7.6 | | 7.0–9.0 | | India 5 | 7584.0 |
| Phylloscopus armandii | B | 3 | 9.3 | | 9.0–9.6 | | Thailand 378 | 7585.0 |
| Phylloscopus schwarzi | M F | | | | 11.3–12.7 8.5–11.3 | | 149 | 7586.0 |
| Phylloscopus pulcher | U | 10 | 6.8 | | 5.4–8.1 | | India 5 | 7587.0 |
| Phylloscopus maculipennis | B | 18 | 5.1 | | 4.5–6.0 | | India 5 | 7588.0 |
| Phylloscopus proregulus | U | 8 | 5.1 | | 4.5–6.2 | | India 5 | 7589.0 |
| Phylloscopus subviridis | B | 8 | | | 5.0–6.0 | | India 5 | 7590.0 |
| Phylloscopus inornatus | M F | 18 3 | 6.0 | | 5.3–7.2 5.8–7.6 | | India 5 | 7591.0 |
| Phylloscopus borealis | B | 70 | 10.6 | | 8.0–15.0 | | India 5 | 7592.0 |
| Phylloscopus trochiloides | U | 25 | 9.5 | | 8.0–14.0 | | India 5 | 7593.0 |
| Phylloscopus nitidus | M F | 3 1 | 6.5 | | 7.2–7.5 | | India 5 | 7595.0 |
| Phylloscopus tenellipes | M | 2 | 12.1 | | 10.7–13.5 | | 149 | 7596.0 |
| Phylloscopus magnirostris | M | 2 | 11.6 | | 11.3–12.0 | | India 5 | 7597.0 |
| Phylloscopus tytleri | U | 30 | 7.2 | 0.50 | | | India 467 | 7598.0 |

### Body Masses of World Birds (continued)

| Species | Sex | N | Mean | Std dev | Range | Sn | Location | Number |
|---|---|---|---|---|---|---|---|---|
| Phylloscopus occipitalis | B | 13 | 8.5 | | 8.0–9.0 | B | India 5 | 7599.0 |
| Phylloscopus coronatus | M | 1 | 7.4 | | | | 149 | 7600.0 |
| Phylloscopus reguloides | B | 45 | | | 6.2–9.4 | | India 5 | 7602.0 |
| Phylloscopus davisoni | U | 9 | 6.4 | | 5.7–8.1 | | Thailand 378 | 7603.0 |
| Phylloscopus cantator | U | 3 | 6.0 | | 5.0–7.0 | | India 5 | 7604.0 |
| Phylloscopus olivaceus | B | 19 | 10.1 | | 8.6–11.6 | | Philippines 475 | 7606.0 |
| Phylloscopus trivirgatus | B | 20 | 9.0 | | 8.1–10.0 | | Philippines 475 | 7608.0 |
| Seicercus burkii | B | 13 | 7.3 | | 4.0–9.3 | | India | 7614.0 |
| Seicercus xanthoschistus | M | 27 | | | 6.0–8.5 | | India 5 | 7615.0 |
| | F | 10 | | | 6.0–7.6 | | | |
| Seicercus affinis | U | 6 | | | 6.0–8.0 | | India 5 | 7616.0 |
| Seicercus poliogenys | B | 3 | 6.3 | | 6.0–7.0 | | India 5 | 7617.0 |
| Seicercus castaniceps | U | 6 | 5.3 | | 4.0–6.0 | | India 5 | 7618.0 |
| Seicercus montis | U | 9 | 6.4 | | 5.0–7.0 | | Malaysia 371 | 7619.0 |
| Tickellia hodgsoni | U | 2 | 4.5 | | 4.0–5.0 | | India 5 | 7621.0 |
| Abroscopus schisticeps | U | 3 | 4.7 | | 4.0–6.0 | | India 5 | 7623.0 |
| Abroscopus superciliaris | M | 3 | | | 6.0–7.0 | | India 5 | 7624.0 |
| Hyliota flavigaster | F | 1 | 12.2 | | | | Ghana 228 | 7625.0 |
| Hyliota australis | B | 2 | 12.4 | | 12.3–12.4 | | Zimbabwe 285 | 7626.0 |
| Megalurus timoriensis | M | 10 | 32.9 | 6.06 | 21.0–40.0 | | | 7630.0 |
| | F | 8 | 28.1 | | 23.0–33.5 | | 154, 216, 217, 475 | |
| Megalurus palustris | M | 13 | 53.0 | | 48.0–56.8 | | Philippines | 7631.0 |
| | F | 12 | 36.7 | | 32.4–41.8 | | 475 | |
| Cincloramphus cruralis | U | | 28.0 | | | | W. Australia 550 | 7635.0 |
| Cincloramphus mathewsi | U | | 25.0 | | | | W. Australia 550 | 7636.0 |

## Body Masses of World Birds (continued)

| Species | Sex | N | Mean | Std dev | Range | Sn | Location | Number |
|---------|-----|---|------|---------|-------|----|----------|--------|
| Eremiornis carteri | U | 14 | 12.2 | | | | Australia 695 | 7637.0 |
| Megalurulus mariei | B | 28 | 25.0 | | 20.0–29.2 | | New Caledonia 506 | 7639.0 |
| Megalurulus rubiginosa | B | 2 | 41.2 | | 39.5–43.0 | | New Guinea 219 | 7643.0 |
| Graminicola bengalensis | B | 6 | | | 13.0–16.5 | | India 5 | 7646.0 |
| Schoenicola platyura | M | 1 | 16.0 | | | | Cameroon 349 | 7647.0 |
| Chamaea fasciata | M | 139 | 15.2 | | 13.0–17.8 | PB | California, USA 58 | 7649.0 |
|  | F | 113 | 14.1 | | 11.6–17.9 | | | |
| Sylvia lugens | B | 11 | 15.4 | | 14.0–18.0 | | Ethiopia 183 | 7651.0 |
| Sylvia layardi | U | 3 | 14.8 | | 14.0–15.5 | | South Africa 256 | 7653.0 |
| Sylvia subcaeruleum | U | 70 | 14.3 | | 11.6–17.0 | | 78 | 7654.0 |
| Sylvia atricapilla | U | 375 | 15.5 | | 11.8–20.9 | S | Jordan 391 | 7655.0 |
| Sylvia borin | U | 272 | 13.9 | | 8.5–20.7 | S | Cyprus 391 | 7656.0 |
| Sylvia communis | U | 1010 | 14.5 | 1.80 | 10.9–24.6 | S | Nigeria 169 | 7657.0 |
| Sylvia curruca | U | 227 | 10.1 | | 6.8–13.0 | S | Cyprus 391 | 7658.0 |
| Sylvia nana | U | 14 | 8.7 | | 7.0–10.5 | | India 5 | 7659.0 |
| Sylvia nisoria | U | 50 | 22.8 | | 18.8–31.0 | | India 5 | 7660.0 |
| Sylvia hortensis | U | 9 | 22.2 | | 22.0–24.0 | | India 5 | 7661.0 |
| Sylvia melanocephala | B | 12 | 11.3 | | 9.4–14.0 | S | Morocco 8 | 7664.0 |
| Sylvia cantillans | U | 36 | 10.8 | 1.40 | 8.8–14.8 | S | Nigeria 169 | 7666.0 |
| Sylvia mystacea | U | 30 | 9.9 | | 8.0–11.5 | | 78 | 7667.0 |
| Sylvia conspicillata | U | | 10.0 | | | | 78 | 7668.0 |
| Sylvia undata | U | 52 | 10.8 | | 9.7–11.8 | Y | Britain 258 | 7670.0 |
| Sylvia sarda | U | | 10.5 | | | | 78 | 7671.0 |

## Body Masses of World Birds (continued)

| Species | Sex | N | Mean | Std dev | Range | Sn | Location | Number |
|---------|-----|---|------|---------|-------|----|----------|--------|
| | | | | **ORDER: PASSERIFORMES** | | **FAMILY: TIMALIIDAE** | | |
| Garrulax cinereifrons | M | 1 | 70.0 | | | | Sri Lanka 5 | 7673.0 |
| Garrulax albogularis | M | 17 | | | 97.0–114.0 | | India 5 | 7677.0 |
| | F | 7 | | | 78.0–105.0 | | | |
| Garrulax leucolophus | M | 3 | | | 123.0–129.0 | | India 5 | 7678.0 |
| | F | 5 | | | 119.0–123.0 | | | |
| Garrulax monileger | U | 5 | | | 77.0–91.0 | | India 5 | 7679.0 |
| Garrulax pectoralis | M | 1 | 156.0 | | | | India 5 | 7680.0 |
| | F | 2 | 135.0 | | 135.0–135.0 | | | |
| Garrulax striatus | B | 6 | | | 126.0–148.0 | | India 5 | 7683.0 |
| Garrulax striatus sikkimensis | M | 1 | 92.0 | | | | India 5 | 7683.0 |
| | F | 3 | | | 99.0–106.0 | | | |
| Garrulax ruficollis | M | 6 | | | 60.0–73.0 | | India 5 | 7687.0 |
| Garrulax galbanus | B | 3 | 56.0 | | 55.0–57.0 | | India 5 | 7691.0 |
| Garrulax delesserti | M | 1 | 92.0 | | | | India 5 | 7692.0 |
| Garrulax cineraceus | M | 4 | | | 47.0–51.0 | | India 5 | 7696.0 |
| Garrulax rufogularis | B | 7 | | | 58.0–68.0 | | India 5 | 7697.0 |
| Garrulax ocellatus | B | 5 | 114.0 | | 110.0–121.0 | | India 5 | 7701.0 |
| Garrulax caerulatus | M | 4 | | | 82.0–99.0 | | | 7702.0 |
| | F | 6 | | | 79.0–84.0 | | 5 | |
| Garrulax mitratus | U | 17 | 62.0 | | 51.0–70.0 | | Malaysia 371 | 7704.0 |
| Garrulax canorus | U | | 55.0 | | | | Hawaiian Is. 400 | 7706.0 |
| Garrulax sannio | M | 2 | 68.0 | | 68.0–68.0 | | India 5 | 7707.0 |
| | F | 1 | 56.0 | | | | | |
| Garrulax lineatus | U | 25 | 40.7 | | 36.0–46.0 | | India | 7710.0 |
| Garrulax austeni | M | 4 | | | 63.0–74.0 | | India 5 | 7712.0 |
| | F | 1 | 59.0 | | | | | |
| Garrulax squamatus | F | 1 | 84.0 | | | | India 5 | 7713.0 |
| Garrulax subunicolor | M | 3 | | | 63.0–69.0 | | India 5 | 7714.0 |

## Body Masses of World Birds (continued)

| Species | Sex | N | Mean | Std dev | Range | Sn | Location | Number |
|---|---|---|---|---|---|---|---|---|
| Garrulax variegatus | U | 10 | 64.5 | | 59.0–72.0 | | India 5 | 7716.0 |
| Garrulax affinis | B | 14 | | | 66.0–80.0 | | India 5 | 7718.0 |
| Garrulax erythrocephalus | U | 68 | 71.7 | | 62.0–95.0 | | Malaysia 371 | 7720.0 |
| Liocichla phoenicea | U | 4 | 49.0 | | 45.0–52.0 | | India 5 | 7724.0 |
| Modulatrix stictigula | U | | 29.5 | | | | Tanzania 417a | 7727.0 |
| Malacocincla abbotti | M | 5 | | | 26.0–32.5 | | India 5 | 7733.0 |
| | F | 2 | 26.8 | | 26.5–27.0 | | | |
| Malacocincla sepiarium | B | 3 | 27.4 | | 25.7–28.2 | | Borneo 627 | 7734.0 |
| Malacocincla malaccensis | M | 1 | 24.5 | | | | Borneo 627 | 7737.0 |
| Pellorneum tickelli | B | 3 | 17.1 | | 16.2–17.7 | | Thailand 378 | 7739.0 |
| Pellorneum albiventre | M | 2 | 21.5 | | 21.0–22.0 | | India 5 | 7740.0 |
| Pellorneum ruficeps | U | 15 | 26.0 | | 21.0–30.0 21.0–30.0 | | India 5 | 7742.0 |
| Pellorneum fuscocapillum | M | 1 | 30.0 | | | | Sri Lanka 5 | 7743.0 |
| Pellorneum capistratum | B | 2 | 21.2 | | 21.1–21.4 | | Borneo 627 | 7745.0 |
| Malacopteron magnirostre | M | 4 | 20.3 | | 20.0–20.5 | | Borneo 627 | 7746.0 |
| Illadopsis cleaveri | U | 6 | 30.5 | | 26.0–36.0 | | 314a | 7752.0 |
| Illadopsis albipectus | U | 2 | 30.8 | | 29.0–32.5 | | Kenya 314a | 7753.0 |
| Illadopsis rufescens | U | 2 | 36.2 | | 34.5–38.0 | | Liberia 314a | 7754.0 |
| Illadopsis rufipennis | U | 7 | 22.2 | | 20.3–26.0 | | 314a, 349 | 7756.0 |
| Illadopsis fulvescens | U | 11 | 24.4 | 4.12 | 21.0–30.5 | | 314a | 7757.0 |
| Illadopsis pyrrhoptera | U | 8 | 26.9 | | | | 51 | 7758.0 |
| Illadopsis abyssinica | B | 13 | 18.2 | | 17.0–20.0 | | Cameroon 349 | 7760.0 |
| Kakamega poliothorax | U | 6 | 35.3 | | | | 51, 349 | 7761.0 |

## Body Masses of World Birds (continued)

| Species | Sex | N | Mean | Std dev | Range | Sn | Location | Number |
|---|---|---|---|---|---|---|---|---|
| Ptyrticus turdinus | M | 1 | 66.0 | | | | Cameroon 349 | 7762.0 |
| Pomatorhinus hypoleucos | U | 3 | 80.0 | | 79.0–82.0 | | Malaysia 371 | 7763.0 |
| Pomatorhinus erythrogenys | M | 8 | | | 62.0–70.0 | | India | 7765.0 |
| | F | 3 | | | 59.0–61.0 | | 5 | |
| Pomatorhinus horsfieldii | U | 12 | 43.0 | | 33.0–53.0 | | India 5 | 7766.0 |
| Pomatorhinus schisticeps | U | 12 | 43.0 | | 33.0–53.0 | | India 5 | 7767.0 |
| Pomatorhinus montanus | M | 2 | 30.6 | | 28.2–33.1 | | Borneo 627 | 7768.0 |
| Pomatorhinus ruficollis | M | 8 | 33.1 | | 30.0–39.0 | | India | 7769.0 |
| | F | 5 | 30.2 | | 26.0–35.0 | | 5 | |
| Pomatorhinus ochraceiceps | F | 1 | 34.0 | | | | India 5 | 7770.0 |
| Pomatorhinus ferruginosus | F | 1 | 40.0 | | | | India 5 | 7771.0 |
| Xiphirhynchus superciliaris | U | 3 | 28.0 | | 27.0–30.0 | | India 5 | 7772.0 |
| Rimator malacoptilus | M | 4 | | | 18.0–21.0 | | India 5 | 7774.0 |
| Ptilocichla mindanensis | B | 16 | 29.2 | | 26.3–31.0 | | Philippines 475 | 7776.0 |
| Kenopia striata | F | 3 | 19.8 | | 18.7–21.1 | | Borneo 627 | 7778.0 |
| Napothera brevicaudata | U | 10 | 19.5 | | 14.0–22.0 | | Malaysia 371 | 7784.0 |
| Napothera epilepidota | B | 5 | 16.0 | | 14.5–19.0 | | Thailand 378 | 7787.0 |
| Pnoepyga albiventer | U | 10 | 20.9 | | 17.0–24.0 | | India 5 | 7788.0 |
| Pnoepyga pusilla | U | 8 | 12.0 | | 11.0–14.0 | | India 5 | 7789.0 |
| Spelaeornis caudatus | B | 2 | 11.0 | | 10.0–12.0 | | India 5 | 7790.0 |
| Spelaeornis chocolatinus | B | 6 | | | 10.0–14.0 | | India 5 | 7794.0 |
| Neomixis tenella | B | 12 | 6.4 | 0.53 | 5.2–7.2 | | Madagascar 39 | 7797.0 |
| Neomixis viridis | U | 1 | 6.5 | | | | Madagascar 39 | 7798.0 |
| Neomixis flavoviridis | B | 3 | 9.6 | | 9.0–10.5 | | Madagascar 39 | 7800.0 |

## Body Masses of World Birds (continued)

| Species | Sex | N | Mean | Std dev | Range | Sn | Location | Number |
|---|---|---|---|---|---|---|---|---|
| Stachyris rufifrons | B | 3 | 9.7 | | 9.0–10.0 | | India 5 | 7803.0 |
| Stachyris ruficeps | U | 19 | 10.3 | | 8.0–12.0 | | India 5 | 7804.0 |
| Stachyris pyrrhops | B | 9 | | | 8.0–11.5 | | India 5 | 7805.0 |
| Stachyris chrysaea | U | 13 | 9.0 | | 7.0–10.0 | | Malaysia 371 | 7806.0 |
| Stachyris plateni | B | 23 | 8.7 | | 7.5–10.2 | | Philippines 475 | 7807.0 |
| Stachyris nigrocapitata | B | 31 | 14.1 | | 12.6–16.0 | | Philippines 475 | 7809.0 |
| Stachyris capitalis | B | 2 | 14.5 | | 14.0–15.0 | | Philippines 475 | 7810.0 |
| Stachyris nigriceps | U | 42 | 15.8 | | 13.0–20.0 | | Malaysia 371 | 7819.0 |
| Stachyris poliocephala | M | 1 | 24.0 | | | | Borneo 627 | 7820.0 |
| Stachyris maculata | M | 1 | 25.8 | | | | Borneo 627 | 7826.0 |
| Stachyris erythroptera | M | 2 | 13.0 | | 12.5–13.4 | | Borneo 627 | 7827.0 |
| Dumetia hyperythra | U | 10 | 12.9 | | 10.0–15.0 | | India 5 | 7829.0 |
| Rhopocichla atriceps | B | 3 | 16.3 | | 16.0–17.0 | | Sri Lanka 5 | 7830.0 |
| Macronous gularis | B | 9 | | | 10.0–14.0 | | India 5 | 7831.0 |
| Macronous striaticeps | B | 45 | 16.5 | | 14.2–20.0 | | Philippines 475 | 7834.0 |
| Timalia pileata | B | 2 | 16.0 | | 15.0–17.0 | | India 5 | 7837.0 |
| Chrysomma sinense hypoleucum | B | 10 | 16.1 | | 12.0–20.0 | | India 5 | 7838.0 |
| Chrysomma sinense sinense | U | 15 | 18.3 | | 15.0–21.0 | | 5 | 7838.0 |
| Turdoides nipalensis | M | 2 | 61.0 | | 58.0–64.0 | | India 5 | 7841.0 |
| Turdoides caudatus | U | 12 | 36.7 | | 30.0–40.0 | | India 5 | 7843.0 |
| Turdoides earlei | U | 2 | 47.0 | | 46.0–48.0 | | India 5 | 7844.0 |
| Turdoides longirostris | F | 1 | 35.0 | | | | India 5 | 7846.0 |

## Body Masses of World Birds (continued)

| Species | Sex | N | Mean | Std dev | Range | Sn | Location | Number |
|---|---|---|---|---|---|---|---|---|
| Turdoides malcolmi | B | 27 | 75.7 | | 63.0–92.0 | | India 5 | 7847.0 |
| Turdoides subrufus | B | | | | 57.0–78.0 | | India 5 | 7852.0 |
| Turdoides striatus orientalis | U | 10 | 66.0 | | 60.0–80.0 | | India 5 | 7853.0 |
| Turdoides reinwardtii | U | 6 | 82.0 | | 75.0–91.0 | | Ghana 228 | 7856.0 |
| Turdoides squamulatus | U | 5 | 69.0 | | 65.0–76.0 | | Somalia 691 | 7859.0 |
| Turdoides bicolor | U | | 77.3 | | | | 78 | 7862.0 |
| Turdoides plebejus | B | 7 | 67.5 | | 61.0–71.0 | | 228, 349, 680 | 7867.0 |
| Turdoides jardineii | M | 1 | 70.6 | | | | Zimbabwe | 7868.0 |
| | F | 1 | 56.3 | | | | 283 | |
| Leiothrix argentauris | U | 41 | 28.4 | | 24.0–36.0 | | Malaysia 371 | 7873.0 |
| Leiothrix lutea | B | 5 | 21.8 | | 20.0–25.0 | | India 5 | 7874.0 |
| Cutia nipalensis | M | 9 | | | 48.0–56.0 | | India | 7875.0 |
| | F | 3 | | | 40.0–46.0 | | 5 | |
| Pteruthius rufiventer | M | 7 | | | 44.0–48.0 | | India | 7876.0 |
| | F | 2 | 43.0 | | 42.0–44.0 | | 5 | |
| Pteruthius flaviscapis | B | 9 | | | 34.0–44.0 | | India 5 | 7877.0 |
| Pteruthius xanthochlorus | B | 3 | 14.3 | | 14.0–15.0 | | India 5 | 7878.0 |
| Pteruthius melanotis | U | 7 | 13.3 | | 12.0–15.0 | | Malaysia 371 | 7879.0 |
| Gampsorhynchus rufulus | M | 1 | 37.0 | | | | India | 7881.0 |
| | F | | | | | | 5 | |
| Actinodura egertoni | B | 8 | 36.0 | | 33.0–38.0 | | India 5 | 7882.0 |
| Actinodura ramsayi | B | 6 | 38.4 | | | | Thailand 378 | 7883.0 |
| Actinodura nipalensis | M | 5 | | | 44.0–48.0 | | India | 7884.0 |
| | F | 5 | | | 39.0–45.0 | | 5 | |
| Actinodura waldeni | B | 18 | | | 39.0–56.0 | | India 5 | 7885.0 |
| Minla cyanouroptera | U | 9 | 17.4 | | 16.0–19.0 | | Malaysia 371 | 7888.0 |
| Minla strigula | U | 111 | 19.2 | | 16.0–22.0 | | Malaysia 371 | 7889.0 |

## Body Masses of World Birds (continued)

| Species | Sex | N | Mean | Std dev | Range | Sn | Location | Number |
|---------|-----|---|------|---------|-------|----|----------|--------|
| Minla ignotincta | B | 6 | 14.3 | | 11.0–18.0 | | India 5 | 7890.0 |
| Alcippe chrysotis | M | 1 | 5.5 | | | | India 5 | 7891.0 |
| Alcippe cinerea | F | 2 | 11.0 | | 11.0–11.0 | | India 5 | 7893.0 |
| Alcippe castaneceps | U | 41 | 12.5 | | 11.0–20.0 | | Malaysia 371 | 7894.0 |
| Alcippe vinipectus | B | 11 | | | 11.0–13.0 | | India 5 | 7895.0 |
| Alcippe cinereiceps | M | 2 | 10.0 | | 10.0–10.0 | | India 5 | 7898.0 |
| Alcippe brunnea | M | | | | 16.0–19.0 | | India 5 | 7900.0 |
| Alcippe brunneicauda | F | 1 | 14.9 | | | | Borneo 627 | 7902.0 |
| Alcippe poioicephala | U | 14 | 20.7 | | 18.0–23.0 | | India 5 | 7903.0 |
| Alcippe morrisonia | B | 37 | 14.4 | 0.89 | 12.5–16.8 | | Thailand 378 | 7906.0 |
| Alcippe nipalensis | U | 90 | 15.8 | | 13.0–20.0 | | Malaysia 371 | 7907.0 |
| Kupeornis rufocinctus | U | 1 | 47.0 | | | | 51 | 7910.0 |
| Parophasma galinieri | U | 1 | 21.1 | | | | Ethiopia 635 | 7912.0 |
| Heterophasia annectans | B | 2 | 24.5 | | 24.2–24.8 | | Thailand 378 | 7916.0 |
| Hetcrophasia capistrata | U | 17 | 39.4 | | | | India 5 | 7917.0 |
| Heterophasia gracilis | U | | | | 34.0–42.0 | | India 5 | 7918.0 |
| Heterophasia melanoleuca | B | 11 | 32.6 | 2.22 | 30.5–36.4 | | Thailand 378 | 7919.0 |
| Heterophasia pulchella | U | | | | 35.0–47.0 | | India 5 | 7921.0 |
| Heterophasia picaoides | U | 29 | 42.4 | | 35.0–49.0 | | Malaysia 371 | 7922.0 |
| Yuhina castaniceps | M | 4 | 11.8 | | 11.0–12.0 | | India 5 | 7923.0 |
| Yuhina bakeri | U | 4 | | | 14.0–21.0 | | India 5 | 7925.0 |
| Yuhina flavicollis | B | 30 | | | 13.0–22.0 | | India 5 | 7926.0 |

## Body Masses of World Birds (continued)

| Species | Sex | N | Mean | Std dev | Range | Sn | Location | Number |
|---|---|---|---|---|---|---|---|---|
| Yuhina gularis | B | 18 | | | 18.0−24.0 | | India 5 | 7928.0 |
| Yuhina occipitalis | B | 4 | 13.0 | | 12.0−16.0 | | India 5 | 7930.0 |
| Yuhina nigrimenta | B | 6 | 9.5 | | 8.0−11.0 | | India 5 | 7932.0 |
| Yuhina xantholeuca | B | 6 | 11.8 | | 8.0−17.0 | | India 5 | 7933.0 |
| Myzornis pyrrhoura | B | 6 | | | 11.0−13.0 | | India 5 | 7934.0 |
| Oxylabes madagascariensis | M | 2 | 22.8 | | 22.5−23.2 | | Madagascar 39 | 7935.0 |
| | F | 1 | 20.7 | | | | | |
| Mystacornis crossleyi | M | 2 | 24.0 | | 22.7−25.2 | | Madagascar 39 | 7937.0 |

### ORDER: PASSERIFORMES  FAMILY: PANURIDAE

| Species | Sex | N | Mean | Std dev | Range | Sn | Location | Number |
|---|---|---|---|---|---|---|---|---|
| Panurus biarmicus | U | 50 | 15.7 | | 12.5−20.0 | S | Britain 258 | 7938.0 |
| Paradoxornis unicolor | U | 2 | 34.0 | | 32.0−36.0 | | India 5 | 7941.0 |
| Paradoxornis gularis | B | 2 | 29.0 | | 29.0−29.0 | | India 5 | 7942.0 |
| Paradoxornis guttaticollis | F | 1 | 26.0 | | | | India 5 | 7944.0 |
| Paradoxornis webbianus | U | 73 | 10.9 | 1.50 | | | Taiwan 552 | 7946.0 |
| Paradoxornis nipalensis | B | | | | 5.0−6.0 | | India 5 | 7952.0 |
| Paradoxornis ruficeps | M | 1 | 30.0 | | | | India 5 | 7956.0 |
| | F | 1 | 34.0 | | | | | |

### ORDER: PASSERIFORMES  FAMILY: RHABDORNITHIDAE

| Species | Sex | N | Mean | Std dev | Range | Sn | Location | Number |
|---|---|---|---|---|---|---|---|---|
| Rhabdornis mysticalis | B | 21 | 25.6 | | 22.4−31.5 | | Philippines 475 | 7960.0 |
| Rhabdornis inornatus | U | 54 | 39.1 | | 31.6−47.2 | | Philippines 475 | 7962.0 |

### ORDER: PASSERIFORMES  FAMILY: PARIDAE

| Species | Sex | N | Mean | Std dev | Range | Sn | Location | Number |
|---|---|---|---|---|---|---|---|---|
| Parus palustris | U | 50 | 10.6 | | 9.5−12.3 | Y | Britain 258 | 7963.0 |
| Parus lugubris | U | | 17.5 | | | | 78 | 7964.0 |
| Parus montanus | U | 50 | 10.2 | | 8.6−11.5 | Y | Britain 258 | 7965.0 |

## Body Masses of World Birds (continued)

| Species | Sex | N | Mean | Std dev | Range | Sn | Location | Number |
|---|---|---|---|---|---|---|---|---|
| Parus carolinensis | M | 18 | 10.5 | 0.72 | | | Florida, USA | 7966.0 |
| | F | 24 | 9.8 | 0.59 | | | 244 | |
| Parus atricapillus | B | 1880 | 10.8 | 1.38 | 8.2–13.6 | Y | Pennsylvania, USA | 7967.0 |
| | | | | | | | 101 | |
| Parus gambeli | M | 292 | 11.5 | 0.68 | 9.0–14.3 | | western USA | 7968.0 |
| | F | 164 | 10.1 | 0.45 | 8.2–14.5 | | 33 | |
| Parus sclateri | B | 10 | 11.0 | 0.78 | 10.1–12.5 | B | Arizona, USA | 7969.0 |
| | | | | | | | 594 | |
| Parus cinctus | B | 2 | 12.4 | | 12.0–12.8 | | Alaska, USA | 7972.0 |
| | | | | | | | 305 | |
| Parus hudsonicus | B | 84 | 9.8 | 0.81 | 7.0–12.4 | PB | Ontario, Canada | 7973.0 |
| | | | | | | | 177 | |
| Parus rufescens | B | 32 | 9.7 | 0.57 | | B | Washington, USA | 7974.0 |
| | | | | | | | 612 | |
| Parus rufonuchalis | M | 13 | | | 12.3–14.7 | | India | 7975.0 |
| | F | 8 | | | 11.4–12.4 | | 5 | |
| Parus rubidiventris | M | 16 | | | 11.6–13.1 | | India | 7976.0 |
| | F | 9 | | | 10.5–12.3 | | 5 | |
| Parus melanolophus | M | 23 | | | 8.0–9.8 | | India | 7977.0 |
| | F | 12 | | | 7.0–9.5 | | 5 | |
| Parus ater | U | 50 | 9.1 | | 7.8–10.1 | S | Britain | 7978.0 |
| | | | | | | | 258 | |
| Parus elegans | U | 5 | 12.2 | | 11.4–12.6 | | Philippines | 7980.0 |
| | | | | | | | 475 | |
| Parus cristatus | U | 1 | 10.2 | | | | | 7982.0 |
| | | | | | | | 227 | |
| Parus dichrous | M | 20 | | | 11.2–14.0 | | India | 7983.0 |
| | | | | | | | 5 | |
| Parus leucomelas | U | 10 | 16.1 | 1.20 | 13.1–17.5 | | Ghana | 7985.0 |
| | | | | | | | 228 | |
| Parus niger | M | 2 | 19.8 | | 19.7–19.8 | | Zimbabwe | 7986.0 |
| | F | 3 | 17.8 | | 17.2–18.5 | | 280, 282 | |
| Parus fasciiventer | U | 1 | 17.6 | | | | | 7993.0 |
| | | | | | | | 51 | |
| Parus cinerascens | U | 3 | 19.4 | | 18.5–20.1 | | South Africa | 7996.0 |
| | | | | | | | 256 | |
| Parus afer | U | 4 | 19.8 | | 18.0–22.4 | | | 7997.0 |
| | | | | | | | 78 | |
| Parus major caschmirensis | U | 33 | 15.0 | | 13.2–17.1 | | India | 7998.0 |
| | | | | | | | 5 | |
| Parus major | U | 50 | 19.0 | | 15.5–21.5 | S | Britain | 7998.0 |
| | | | | | | | 258 | |
| Parus monticolus | M | 24 | | | 12.0–16.8 | | India | 8000.0 |
| | F | 14 | | | 12.3–15.3 | | 5 | |

## Body Masses of World Birds (continued)

| Species | Sex | N | Mean | Std dev | Range | Sn | Location | Number |
|---|---|---|---|---|---|---|---|---|
| Parus xanthogenys | M | 11 | | | 13.7–16.9 | | India | 8002.0 |
| | F | 9 | | | 12.9–15.9 | | 5 | |
| Parus spilonotus | M | 2 | 18.8 | | 18.3–19.4 | | India | 8003.0 |
| | | | | | | | 5 | |
| Parus caeruleus | U | 50 | 13.3 | | 9.0–14.0 | B | Britain | 8005.0 |
| | | | | | | | 258 | |
| Parus cyanus | M | 2 | 12.1 | | 12.0–12.2 | | India | 8006.0 |
| | F | 1 | 13.7 | | | | 5 | |
| Parus varius | U | 11 | 17.0 | | 15.9–18.2 | | | 8008.0 |
| | | | | | | | 78 | |
| Parus wollweberi | B | 14 | 10.4 | 0.45 | 9.8–11.5 | PB | Arizona, USA | 8010.0 |
| | | | | | | | 177 | |
| Parus inornatus | M | 17 | 18.7 | | 15.4–23.1 | | California, USA | 8011.0 |
| | F | 13 | 16.3 | | 13.2–22.2 | | 229 | |
| Parus bicolor | B | 668 | 21.6 | | 17.5–26.1 | Y | Pennsylvania, USA | 8012.0 |
| | | | | | | | 336 | |
| Parus atricristatus | F | 1 | 16.0 | | | | Coahuila, Mexico | 8013.0 |
| | | | | | | | 634 | |
| Sylviparus modestus | B | 21 | | | 6.0–8.5 | | India | 8014.0 |
| | | | | | | | 5 | |
| Melanochlora sultanea | B | 4 | 37.6 | | 35.0–41.0 | | India | 8015.0 |
| | | | | | | | 5 | |

### ORDER: PASSERIFORMES                    FAMILY: REMIZIDAE

| Species | Sex | N | Mean | Std dev | Range | Sn | Location | Number |
|---|---|---|---|---|---|---|---|---|
| Remiz pendulinus | U | 14 | 9.3 | | 8.4–12.5 | | | 8016.0 |
| | | | | | | | 78 | |
| Anthoscopus caroli | B | 5 | 6.5 | | 6.2–6.9 | | Zimbabwe | 8023.0 |
| | | | | | | | 284, 285 | |
| Cephalopyrus flammiceps | U | 2 | 7.0 | | 7.0–7.0 | | India | 8026.0 |
| | | | | | | | 5 | |

### ORDER: PASSERIFORMES                    FAMILY: ALAUDIDAE

| Species | Sex | N | Mean | Std dev | Range | Sn | Location | Number |
|---|---|---|---|---|---|---|---|---|
| Mirafra passerina | U | 1 | 23.0 | | | | | 8028.0 |
| | | | | | | | 78 | |
| Mirafra javanica | U | | 23.0 | | | | | 8030.0 |
| | | | | | | | 133 | |
| Mirafra hova | M | 4 | | | 20.7–22.0 | | Madagascar | 8033.0 |
| | | | | | | | 39 | |
| Mirafra africana | M | 1 | 51.0 | | | | Zambia | 8040.0 |
| | | | | | | | 166 | |
| Mirafra rufocinnamomea | M | 1 | 26.0 | | | | Zimbabwe | 8043.0 |
| | | | | | | | 281 | |
| Mirafra apiata | U | 7 | 30.7 | | 26.0–34.2 | | South Africa | 8044.0 |
| | | | | | | | 256 | |

## Body Masses of World Birds (continued)

| Species | Sex | N | Mean | Std dev | Range | Sn | Location | Number |
|---|---|---|---|---|---|---|---|---|
| Mirafra africanoides | U | | 24.5 | | | | 78 | 8046.0 |
| Mirafra erythroptera | U | 13 | 21.3 | | 17.0–27.0 | | India 5 | 8048.0 |
| Mirafra assamica | U | 11 | 26.2 | | 20.0–30.0 | | India 5 | 8049.0 |
| Mirafra sabota | U | 3 | 24.1 | | 21.3–26.0 | | South Africa 256 | 8055.0 |
| Chersomanes albofasciata | B | 4 | 30.0 | | 27.6–32.0 | | South Africa 256 | 8066.0 |
| Eremopterix leucotis | B | 3 | 21.6 | | 20.0–23.8 | | South Africa 256 | 8067.0 |
| Eremopterix verticalis | B | 14 | 18.4 | | 16.4–21.2 | | South Africa 256 | 8069.0 |
| Eremopterix leucopareia | U | 12 | 14.5 | 0.69 | 12.6–14.5 | | Kenya 521 | 8070.0 |
| Eremopterix nigriceps | B | 4 | | | 12.0–16.0 | | Niger; Chad 119 | 8072.0 |
| Eremopterix grisea | U | 11 | 16.0 | | 14.0–18.0 | | India 5 | 8073.0 |
| Ammomanes cincturus | B | 14 | | | 14.0–23.5 | | 119 | 8074.0 |
| Ammomanes phoenicurus | B | 4 | 25.6 | | 23.0–28.0 | | India 5 | 8075.0 |
| Ammomanes deserti | M | 25 | | | 24.0–29.5 | | Algeria 119 | 8076.0 |
| | F | 16 | | | 22.0–26.0 | | | |
| Alaemon alaudipes | M | 9 | 47.2 | | 39.0–51.0 | | | 8078.0 |
| | F | 4 | 40.5 | | 30.0–47.0 | | 119 | |
| Rhamphocoris clotbey | M | 2 | 53.5 | | 52.0–55.0 | | Algeria | 8080.0 |
| | F | 2 | 45.0 | | 45.0–45.0 | | 119 | |
| Melanocorypha calandra | B | 24 | 61.6 | | 44.2–73.0 | | 119 | 8081.0 |
| Melanocorypha bimaculata | B | 20 | 54.1 | | 48.0–62.0 | | 119 | 8082.0 |
| Melanocorypha mongolica | U | 4 | 50.5 | | 24.0–60.4 | | 78 | 8084.0 |
| Melanocorypha leucoptera | B | 24 | 45.3 | | 36.0–52.0 | | USSR 119 | 8085.0 |
| Melanocorypha yeltoniensis | M | 44 | 63.7 | | 56.0–76.0 | | USSR | 8086.0 |
| | F | 20 | 56.4 | | 51.0–68.0 | | 119 | |
| Calandrella cinerea | U | 161 | 20.7 | | 16.3–24.2 | S | Morocco 8 | 8090.0 |
| Calandrella cinerea longipennis | M | 23 | 22.3 | | 19.0–25.0 | | Afghanistan | 8090.0 |
| | F | 9 | 19.9 | | 19.0–22.0 | | 5 | |

## Body Masses of World Birds (continued)

| Species | Sex | N | Mean | Std dev | Range | Sn | Location | Number |
|---|---|---|---|---|---|---|---|---|
| Calandrella acutirostris | M | 11 | | | 19.0–23.0 | | India | 8091.0 |
| | F | 7 | | | 18.0–21.0 | | 5 | |
| Calandrella rufescens | B | 23 | 24.2 | | 22.0–27.0 | B | Turkey | 8092.0 |
| | | | | | | | 119 | |
| Calandrella raytal | M | 3 | | | 18.0–19.0 | | India | 8094.0 |
| | | | | | | | 5 | |
| Chersophilus duponti | U | 33 | 39.4 | 3.20 | 32.0–47.0 | | Spain | 8104.0 |
| | | | | | | | 119 | |
| Galerida cristata magna | M | 18 | 39.0 | | 31.0–43.0 | | Afghanistan | 8105.0 |
| | F | 9 | 41.9 | | 37.0–51.0 | | 5 | |
| Galerida theklae | U | 23 | 36.8 | | 34.0–41.0 | | Spain | 8106.0 |
| | | | | | | | 119 | |
| Galerida deva | M | 5 | | | 18.0–22.0 | | India | 8108.0 |
| | | | | | | | 5 | |
| Lullula arborea | B | 961 | 26.9 | | 23.0–32.0 | | USSR | 8112.0 |
| | | | | | | | 119 | |
| Alauda arvensis | M | 102 | 42.7 | 4.70 | 32.0–51.0 | M | Britain | 8113.0 |
| | F | 286 | 37.2 | 3.60 | 29.0–47.0 | | 119 | |
| Alauda gulgula | B | 17 | 26.3 | | 24.0–30.0 | | Afghanistan | 8115.0 |
| | | | | | | | 119 | |
| Eremophila alpestris | M | 207 | 31.9 | | | Y | central USA | 8117.0 |
| | F | 93 | 30.8 | | | | 674 | |
| Eremophila alpestris flava | B | 57 | 37.0 | | 27.0–43.0 | | USSR | 8117.0 |
| | | | | | | | 119 | |
| Eremophila bilopha | M | 2 | 38.5 | | 38.0–39.0 | | Algeria | 8118.0 |
| | | | | | | | 119 | |

### ORDER: PASSERIFORMES  FAMILY: PASSERIDAE

| Species | Sex | N | Mean | Std dev | Range | Sn | Location | Number |
|---|---|---|---|---|---|---|---|---|
| Passer ammodendri | M | | 26.1 | | 25.2–27.0 | | | 8119.0 |
| | F | | | | | | 613 | |
| Passer domesticus | M | 538 | 28.0 | 1.55 | 20.0–34.0 | Y | Pennsylvania, USA | 8120.0 |
| | F | 469 | 27.4 | 2.24 | 20.1–34.5 | | 101 | |
| Passer hispaniolensis | B | 200 | 24.2 | | 18.0–28.0 | | India | 8121.0 |
| | | | | | | | 5 | |
| Passer castanopterus | U | | 18.0 | | | | | 8123.0 |
| | | | | | | | 613 | |
| Passer rutilans | M | | 19.4 | | 19.3–19.5 | | | 8124.0 |
| | F | | 17.3 | | | | 613 | |
| Passer flaveolus | B | | 19.4 | | 17.0–23.0 | | | 8125.0 |
| | | | | | | | 613 | |
| Passer moabiticus moabiticus | M | | 17.5 | | 15.0–20.0 | | | 8126.0 |
| | | | | | | | 613 | |
| Passer moabiticus yatii | B | | 15.5 | | 14.0–17.0 | | | 8126.0 |
| | | | | | | | 613 | |

## Body Masses of World Birds (continued)

| Species | Sex | N | Mean | Std dev | Range | Sn | Location | Number |
|---|---|---|---|---|---|---|---|---|
| Passer iagoensis | M | | 34.7 | | 34.0−35.8 | | | 8127.0 |
| | F | | 31.5 | | 30.6−32.0 | | 613 | |
| Passer melanurus melanurus | M | | 23.5 | | 17.4−24.6 | | | 8131.0 |
| | F | | 20.3 | | 17.3−21.0 | | 613 | |
| Passer melanurus vicinis | B | | 29.5 | | 22.0−38.0 | | | 8131.0 |
| | | | | | | | 613 | |
| Passer griseus | B | 44 | 23.9 | | 18.0−26.5 | | South Africa | 8132.0 |
| | | | | | | | 256 | |
| Passer griseus laeneni | M | | 27.0 | | 24.0−30.0 | | | 8132.1 |
| | F | | 28.5 | | 26.0−31.0 | | 613 | |
| Passer griseus luangwae | B | | 22.9 | | 22.7−23.2 | | | 8132.1 |
| | | | | | | | 613 | |
| Passer swainsonii | U | 20 | 31.6 | | 27.3−35.2 | | Ethiopia | 8133.0 |
| | | | | | | | 635 | |
| Passer simplex | B | | 19.5 | | 18.0−21.0 | | | 8137.0 |
| | | | | | | | 613 | |
| Passer montanus | B | 136 | 22.0 | 1.17 | | Y | Missouri,USA | 8138.0 |
| | | | | | | | 7 | |
| Passer luteus | B | | 14.4 | | | | | 8139.0 |
| | | | | | | | 613 | |
| Passer eminibey | U | 13 | 13.4 | 0.84 | 11.9−15.0 | | Kenya | 8141.0 |
| | | | | | | | 521 | |
| Petronia xanthocollis | B | 21 | 18.1 | | 14.0−20.0 | | | 8143.0 |
| | | | | | | | 5 | |
| Petronia superciliaris | B | 3 | 23.4 | | 21.0−26.0 | | Zimbabwe | 8144.0 |
| | | | | | | | 280, 282 | |
| Petronia dentata | B | 9 | 19.2 | | 17.0−20.0 | | Cameroon | 8145.0 |
| | | | | | | | 349 | |
| Petronia petronia | U | 5 | 31.2 | | | F | Spain | 8146.0 |
| | | | | | | | 160a | |
| Montifringilla nivalis | U | 20 | 36.9 | | 32.0−40.0 | | | 8148.0 |
| | | | | | | | 78 | |
| Montifringilla theresae | U | 40 | 24.3 | | 18.0−27.5 | | | 8154.0 |
| | | | | | | | 692 | |

### ORDER: PASSERIFORMES       FAMILY: ESTRILDIDAE

| Species | Sex | N | Mean | Std dev | Range | Sn | Location | Number |
|---|---|---|---|---|---|---|---|---|
| Parmoptila woodhousei | B | 6 | 10.3 | | 10.0−11.0 | | Liberia | 8156.0 |
| | | | | | | | 314a | |
| Nigrita fusconota | U | 1 | 9.5 | | | | Kenya | 8157.0 |
| | | | | | | | 362 | |
| Nigrita bicolor | U | 1 | 10.5 | | | | Ghana | 8158.0 |
| | | | | | | | 314a | |
| Nigrita canicapilla | B | 8 | 18.7 | | 17.0−21.0 | | Cameroon | 8160.0 |
| | | | | | | | 349 | |

## Body Masses of World Birds (continued)

| Species | Sex | N | Mean | Std dev | Range | Sn | Location | Number |
|---|---|---|---|---|---|---|---|---|
| Nesocharis shelleyi | B | 5 | 7.6 | | 7.0–9.0 | | Cameroon 349 | 8162.0 |
| Nesocharis capistrata | U | 2 | 10.4 | | 10.4–10.5 | | Ghana 228 | 8163.0 |
| Pytilia phoenicoptera | U | 5 | 14.5 | | 13.7–16.1 | | Ghana 228 | 8164.0 |
| Pytilia afra | F | 1 | 15.5 | | | | Zimbabwe 280 | 8166.0 |
| Pytilia melba | B | 29 | 13.5 | | 10.5–16.2 | | Zimbabwe 282 | 8167.0 |
| Pytilia hypogrammica | F | 3 | 14.8 | | 14.3–15.0 | | 228, 349 | 8168.0 |
| Mandingoa nitidula | B | 15 | 9.1 | 0.86 | 8.1–11.2 | B | Zimbabwe 280, 281, 284 | 8169.0 |
| Cryptospiza reichenovii | B | 45 | 13.2 | | 11.5–15.6 | | Malawi 168 | 8170.0 |
| Cryptospiza jacksoni | U | 8 | 13.0 | | | | 51 | 8172.0 |
| Spermophaga haematina | B | 15 | 22.3 | | 19.8–25.0 | | 314a, 349 | 8178.0 |
| Spermophaga ruficapilla | U | 1 | 21.8 | | | | 51 | 8179.0 |
| Clytospiza monteiri | B | 13 | 15.1 | | 14.0–16.0 | | Cameroon 349 | 8180.0 |
| Hypargos niveoguttatus | B | 25 | 14.5 | | 11.4–17.9 | | Zimbabwe 280 | 8181.0 |
| Euschistospiza dybowskii | M | 2 | 13.0 | | 12.0–14.0 | | Cameroon 349 | 8183.0 |
| Lagonosticta rufopicta | U | 13 | 9.0 | | 8.0–11.6 | | Ghana 228 | 8185.0 |
| Lagonosticta nitidula | F | 1 | 11.0 | | | | Cameroon 349 | 8186.0 |
| Lagonosticta senegala | B | 128 | 8.3 | | 7.0–10.7 | | Kenya 67 | 8187.0 |
| Lagonosticta rara | B | 12 | 10.8 | | 9.1–12.0 | | Cameroon 228, 349 | 8188.0 |
| Lagonosticta rubricata | B | 16 | 10.2 | | 8.6–11.7 | | Zimbabwe 283 | 8189.0 |
| Lagonosticta rhodopareia | B | 18 | 8.8 | | 7.1–10.1 | | Zimbabwe 280, 282, 284 | 8193.0 |
| Lagonosticta larvata | B | 7 | 9.6 | | 8.8–10.6 | | Ghana 228 | 8195.0 |
| Uraeginthus angolensis | B | 19 | 9.3 | | 6.8–12.3 | | Zimbabwe 282 | 8196.0 |

## Body Masses of World Birds (continued)

| Species | Sex | N | Mean | Std dev | Range | Sn | Location | Number |
|---|---|---|---|---|---|---|---|---|
| Uraeginthus bengalus | B | 25 | 10.3 | | 9.0–11.5 | | Kenya 67 | 8197.0 |
| Uraeginthus cyanocephala | U | 12 | 10.1 | 0.61 | 9.0–11.2 | | Kenya 521 | 8198.0 |
| Uraeginthus ianthinogaster | U | 3 | 13.1 | | 12.7–13.4 | | Kenya 521 | 8199.0 |
| Uraeginthus granatina | B | 9 | 10.0 | | 8.6–11.5 | | 256, 282 | 8200.0 |
| Estrilda caerulescens | U | 10 | 8.4 | 0.70 | 6.7–9.1 | | Ghana 228 | 8201.0 |
| Estrilda perreini | B | 2 | 7.6 | | 7.5–7.6 | | Zimbabwe 280 | 8202.0 |
| Estrilda melanotis | M | 3 | 5.9 | | 5.7–6.1 | | | 8205.0 |
| | F | 3 | 7.2 | | 7.0–7.5 | | 166, 281 | |
| Estrilda melpoda | U | 43 | 7.6 | 0.70 | 6.5–9.6 | | Ghana 228 | 8209.0 |
| Estrilda rhodopyga | U | 11 | 7.3 | | 6.1–7.9 | | Kenya 521 | 8210.0 |
| Estrilda troglodytes | U | | 6.1 | | | | 358 | 8212.0 |
| Estrilda astrild | B | 11 | 7.5 | | 6.4–9.0 | | Zimbabwe 280, 281, 283, 284 | 8213.0 |
| Estrilda nonnula | U | 3 | 8.6 | | 8.5–9.0 | | Cameroon 349 | 8215.0 |
| Estrilda atricapilla | U | 1 | 7.8 | | | | Gabon 314a | 8216.0 |
| Estrilda erythronotos | B | 5 | 8.1 | | 7.7–8.6 | | 78, 256 | 8218.0 |
| Amandava amandava | B | 4 | 9.6 | | 9.5–9.8 | | 17 | 8220.0 |
| Amandava subflava | U | 1 | 7.0 | | | | Zimbabwe 283 | 8222.0 |
| Ortygospiza atricollis | U | | 12.9 | | | | 78 | 8223.0 |
| Stagonopleura bella | U | | 15.0 | | | | 134 | 8227.0 |
| Stagonopleura guttata | U | | 19.0 | | | | Australia 194a | 8229.0 |
| Neochmia temporalis | U | | 10.9 | | | | Australia 194 | 8230.0 |
| Neochmia phaeton | U | | 10.0 | | | | 133 | 8231.0 |
| Taeniopygia guttata | U | 4 | 12.0 | | 11.5–12.7 | | captive 358 | 8234.0 |

## Body Masses of World Birds (continued)

| Species | Sex | N | Mean | Std dev | Range | Sn | Location | Number |
|---------|-----|---|------|---------|-------|-----|----------|--------|
| Taeniopygia bichenovii | U | | 10.5 | | | | Australia 194 | 8235.0 |
| Oreostruthus fuliginosus | M | 2 | 19.0 | | 18.0–20.0 | | New Guinea 216 | 8239.0 |
| Erythrura trichroa | B | 18 | 14.4 | | 11.0–18.0 | | New Guinea 154 | 8244.0 |
| Erythrura papuana | B | 17 | 19.0 | 1.85 | 17.0–24.0 | | New Guinea 154 | 8246.0 |
| Erythrura psittacea | M | 1 | 11.5 | | | | New Caledonia 506 | 8247.0 |
| Erythrura cyaneovirens | U | 49 | 13.0 | 1.00 | 10.5–15.0 | | Fiji 332 | 8249.0 |
| Lemuresthes nana | M | 2 | 8.1 | | 8.0–8.2 | | Madagascar 39 | 8253.0 |
| Lonchura cantans | M | 1 | 12.0 | | | | 17 | 8254.0 |
| Lonchura malabarica | U | 13 | 12.0 | | 10.0–14.0 | | India 5 | 8255.0 |
| Lonchura griseicapilla | U | 4 | 12.5 | | 11.6–13.5 | | Kenya 521 | 8256.0 |
| Lonchura cucullata | U | 97 | 9.2 | 0.83 | 7.7–11.8 | | Kenya 67 | 8257.0 |
| Lonchura bicolor | B | 15 | 9.9 | | 8.0–11.0 | | Cameroon 349 | 8258.0 |
| Lonchura fringilloides | B | 9 | 16.7 | | 13.0–18.9 | | 280, 282, 349 | 8260.0 |
| Lonchura striata | U | 11 | 12.3 | | 9.5–13.0 | | India 5 | 8261.0 |
| Lonchura leucogastroides | U | 3 | 11.7 | | 10.2–12.5 | | Timor 17 | 8262.0 |
| Lonchura kelaarti | U | 15 | 14.1 | | 9.5–17.0 | | India 5 | 8265.0 |
| Lonchura punctulata | B | 13 | 13.6 | | 12.0–15.0 | | India 5 | 8266.0 |
| Lonchura leucogastra | B | 7 | 11.4 | | 10.6–12.3 | | Philippines 475 | 8267.0 |
| Lonchura tristissima | B | 5 | 8.2 | | | | New Guinea 35, 217, 218 | 8268.0 |
| Lonchura malacca | B | 28 | 12.6 | | 9.8–14.4 | | Philippines 475 | 8270.0 |
| Lonchura maja | U | | 12.5 | | 11.0–14.0 | | 78 | 8273.0 |
| Lonchura grandis | M | 4 | | | 11.5–14.0 | | New Guinea 154, 217 | 8275.0 |
| | F | 3 | 14.3 | | | | | |

## Body Masses of World Birds (continued)

| Species | Sex | N | Mean | Std dev | Range | Sn | Location | Number |
|---|---|---|---|---|---|---|---|---|
| Lonchura spectabilis | B | 11 | 11.5 | | 10.3–12.7 | | New Guinea 154 | 8279.0 |
| Lonchura castaneothorax | B | 2 | 16.0 | | 16.0–16.0 | | 17 | 8284.0 |
| Padda oryzivora | B | 12 | 24.8 | 1.82 | 22.5–27.8 | | 17 | 8291.0 |
| Padda fuscata | U | 2 | 20.7 | | 19.2–22.2 | | Timor 17 | 8292.0 |
| Amadina fasciata | U | 12 | 15.4 | 1.34 | 11.6–16.7 | | Kenya 521 | 8293.0 |
| Amadina erythrocephala | U | 192 | 22.5 | | 17.5–30.0 | | 78 | 8294.0 |
| Vidua chalybeata | B | 12 | 12.5 | | 10.7–14.5 | | 228, 256, 633 | 8295.0 |
| Vidua funerea | B | 7 | 13.4 | | 11.7–15.5 | | Zimbabwe 280, 282, 284 | 8298.0 |
| Vidua fischeri | U | 1 | 13.6 | | | | Kenya 521 | 8302.0 |
| Vidua regia | U | 4 | 13.8 | | 12.6–14.9 | | 78, 256 | 8303.0 |
| Vidua macroura | B | 10 | 14.4 | | 12.6–15.4 | | 256, 284, 349 | 8304.0 |
| Vidua paradisaea | F | 1 | 22.2 | | | | Zimbabwe 282 | 8308.0 |

### ORDER: PASSERIFORMES  FAMILY: MOTACILLIDAE

| Species | Sex | N | Mean | Std dev | Range | Sn | Location | Number |
|---|---|---|---|---|---|---|---|---|
| Dendronanthus indicus | U | 15 | | | 14.0–17.0 | | India 5 | 8310.0 |
| Motacilla alba alba | U | 93 | 21.0 | 2.10 | 17.3–25.0 | F | Sweden 119 | 8311.0 |
| Motacilla alba personata | B | 21 | 24.4 | | 21.0–28.0 | | Afghanistan 119 | 8311.0 |
| Motacilla madaraspatensis | B | 4 | 30.5 | | 26.0–36.0 | | India 5 | 8314.0 |
| Motacilla aguimp | U | 55 | 27.0 | | 23.0–30.5 | | 78 | 8315.0 |
| Motacilla capensis | U | 174 | 20.8 | | 17.0–25.0 | | 78 | 8316.0 |
| Motacilla flaviventris | B | 2 | 23.4 | | 23.0–23.7 | | Madagascar 39 | 8317.0 |
| Motacilla citreola citreola | B | 25 | 20.3 | | 17.0–23.0 | B | USSR 119 | 8318.0 |
| Motacilla citreola calcarata | B | 21 | 18.0 | | 15.0–21.0 | | India 5 | 8318.0 |

## Body Masses of World Birds (continued)

| Species | Sex | N | Mean | Std dev | Range | Sn | Location | Number |
|---|---|---|---|---|---|---|---|---|
| Motacilla flava | M | 129 | 14.9 | | 12.3–20.4 | S | Morocco | 8319.0 |
| tschutschensis | F | 61 | 13.9 | | 11.2–17.9 | | 8 | |
| Motacilla flava beema | B | 20 | 16.9 | | 14.0–21.0 | | India | 8319.0 |
| | | | | | | | 5 | |
| Motacilla flava flava | B | 162 | 17.6 | | 14.0–27.0 | | Germany | 8319.0 |
| | | | | | | | 119 | |
| Motacilla cinerea | M | 87 | 18.0 | 1.40 | 15.0–22.0 | | | 8320.0 |
| | F | 26 | 17.2 | 1.40 | 14.0–20.0 | | 119 | |
| Motacilla cinerea caspica | U | 34 | 16.3 | | 13.0–20.0 | | India | 8320.0 |
| | | | | | | | 5 | |
| Motacilla clara | U | 3 | 19.9 | | 19.4–20.4 | | Kenya | 8321.0 |
| | | | | | | | 362 | |
| Macronyx capensis | U | 4 | 47.5 | | 41.3–50.8 | | South Africa | 8325.0 |
| | | | | | | | 256 | |
| Anthus lineiventris | B | 4 | 32.9 | | 30.3–34.8 | | Zimbabwe | 8332.0 |
| | | | | | | | 285 | |
| Anthus novaeseelandiae | B | 19 | 24.2 | | 22.2–26.3 | | Philippines | 8339.0 |
| | | | | | | | 475 | |
| Anthus leucophrys | B | 2 | 27.0 | | 25.0–29.0 | | Cameroon | 8340.0 |
| | | | | | | | 349 | |
| Anthus vaalensis | F | 1 | 29.7 | | | | Zimbabwe | 8341.0 |
| | | | | | | | 285 | |
| Anthus melindae | U | 6 | 21.9 | | 19.0–25.0 | | | 8343.0 |
| | | | | | | | 78 | |
| Anthus campestris | B | 44 | 23.0 | | 19.0–28.0 | | USSR | 8344.0 |
| | | | | | | | 119 | |
| Anthus godlewskii | B | 15 | 25.8 | | 22.0–28.0 | B | USSR | 8345.0 |
| | | | | | | | 119 | |
| Anthus berthelotti | B | 2 | 16.5 | | 16.0–17.0 | | Madeira Is. | 8346.0 |
| | | | | | | | 119 | |
| Anthus similis | B | 7 | 28.9 | | 27.0–31.5 | | | 8349.0 |
| | | | | | | | 119 | |
| Anthus trivialis | M | 413 | 21.7 | 1.20 | 18.0–26.0 | B | Germany | 8354.0 |
| | F | 23 | 25.1 | 1.80 | 22.0–29.0 | | 119 | |
| Anthus hodgsoni | B | 56 | 19.3 | | 18.0–26.0 | B | Mongolia | 8355.0 |
| | | | | | | | 119 | |
| Anthus gustavi | U | 10 | 19.8 | 1.20 | 17.0–21.0 | | SE USSR | 8356.0 |
| | | | | | | | 119 | |
| Anthus pratensis | U | 228 | 18.4 | 1.00 | 13.9–23.4 | | Wales | 8357.0 |
| | | | | | | | 119 | |
| Anthus cervinus | B | 53 | 20.4 | | 17.0–24.0 | B | NW USSR | 8358.0 |
| | | | | | | | 119 | |
| Anthus roseatus | B | 28 | | | 17.0–25.0 | | India | 8359.0 |
| | | | | | | | 5 | |

## Body Masses of World Birds (continued)

| Species | Sex | N | Mean | Std dev | Range | Sn | Location | Number |
|---|---|---|---|---|---|---|---|---|
| Anthus petrosus | U | 19 | 23.5 | | 19.5–27.5 | M | Germany 119 | 8360.0 |
| Anthus spinoletta | U | 100 | 23.9 | 2.10 | 19.5–24.0 | W | Germany 119 | 8361.0 |
| Anthus rubescens | M | 28 | 21.6 | | 19.2–25.5 | B | Alaska 278 | 8362.0 |
| | F | 8 | 20.1 | | 18.6–23.2 | | | |
| Anthus correndera | B | 6 | 19.9 | | 18.0–23.0 | | Brazil 38 | 8365.0 |
| Anthus spragueii | B | 20 | 25.3 | 1.87 | 22.3–29.2 | | Oklahoma, USA 231 | 8367.0 |
| Anthus furcatus | B | 29 | | | 17.5–23.0 | | Brazil 38 | 8368.0 |
| Anthus hellmayri | B | 33 | | | 16.5–21.0 | | Brazil 38 | 8369.0 |
| Anthus bogotensis | M | 2 | 26.0 | | 26.0–26.0 | | Venezuela 644 | 8370.0 |
| Anthus lutescens | B | 18 | | | 14.0–17.0 | | Brazil 38 | 8371.0 |
| Anthus lutescens parvus | B | 14 | 12.6 | | | | Panama 244 | 8371.0 |
| Anthus nattereri | M | 2 | 19.5 | | 19.0–20.0 | | Brazil 38 | 8373.0 |

### ORDER: PASSERIFORMES      FAMILY: PRUNELLIDAE

| Species | Sex | N | Mean | Std dev | Range | Sn | Location | Number |
|---|---|---|---|---|---|---|---|---|
| Prunella collaris | M | 21 | 47.1 | 0.67 | | | Japan 410 | 8375.0 |
| | F | 12 | 40.3 | 2.03 | | | | |
| Prunella collaris rufilata | B | 14 | 30.7 | | 25.0–35.0 | Y | USSR 119 | 8375.0 |
| Prunella himalayana | M | 4 | | | 25.5–29.8 | | India 5 | 8376.0 |
| | F | 2 | 25.7 | | 24.3–27.1 | | | |
| Prunella rubeculoides | M | 4 | | | 23.2–25.7 | | India 5 | 8377.0 |
| | F | 3 | | | 21.2–24.1 | | | |
| Prunella strophiata | B | 29 | | | 16.2–20.2 | Y | India 5 | 8378.0 |
| Prunella montanella | B | 12 | 17.6 | | 17.0–20.0 | | 119 | 8379.0 |
| Prunella ocularis | M | 2 | 22.5 | | 20.0–25.0 | | Turkey; Iran 119 | 8380.0 |
| Prunella fulvescens | M | 1 | 20.0 | | | | India 5 | 8382.0 |
| Prunella atrogularis | B | 15 | 18.5 | | 14.5–23.6 | F | USSR 119 | 8383.0 |
| Prunella modularis modularis | U | 527 | 19.7 | 1.90 | 14.0–26.0 | Y | Germany 119 | 8385.0 |

## Body Masses of World Birds (continued)

| Species | Sex | N | Mean | Std dev | Range | Sn | Location | Number |
|---|---|---|---|---|---|---|---|---|
| Prunella modularis occidentalis | U | 50 | 20.8 | 2.50 | 13.0–26.0 | | Britain 119 | 8385.0 |
| Prunella immaculata | U | 3 | 20.7 | | 19.0–22.5 | | India 5 | 8387.0 |

### ORDER: PASSERIFORMES                    FAMILY: PLOCEIDAE

| Species | Sex | N | Mean | Std dev | Range | Sn | Location | Number |
|---|---|---|---|---|---|---|---|---|
| Bubalornis albirostris | M | 7 | 64.5 | | 56.0–80.0 | | South Africa 256, 521 | 8388.0 |
| Dinemellia dinemelli | U | 8 | 63.9 | | 57.0–71.0 | | Kenya 521 | 8390.0 |
| Sporopipes frontalis | U | | 16.8 | | | | 522 | 8391.0 |
| Sporopipes squamifrons | B | 4 | 10.6 | | 8.5–12.6 | | South Africa 256 | 8392.0 |
| Plocepasser mahali | M | 7 | 44.9 | | 40.6–51.4 | | South Africa 256 | 8393.0 |
| | F | 6 | 41.7 | | 36.5–49.0 | | | |
| Plocepasser superciliosus | U | 4 | 34.8 | | 32.0–36.0 | | Ghana; Cameroon 228, 349 | 8394.0 |
| Pseudonigrita arnaudi | U | 15 | 20.1 | 0.89 | 18.0–21.5 | | Kenya 521 | 8398.0 |
| Philetarius socius | B | 35 | 26.7 | | 20.8–32.0 | | South Africa 256 | 8400.0 |
| Ploceus baglafecht | U | 114 | 31.5 | 2.36 | 24.5–37.5 | | Ethiopia 633 | 8404.0 |
| Ploceus luteolus | U | 8 | 12.7 | | 11.2–13.5 | | 228, 349 | 8408.0 |
| Ploceus intermedius | U | 6 | 21.2 | | 20.0–23.4 | | Kenya 521 | 8409.0 |
| Ploceus ocularis | U | 38 | 25.2 | | 24.0–28.7 | | Kenya 67 | 8410.0 |
| Ploceus nigricollis | B | 16 | 24.1 | | 21.0–31.2 | | Ghana 228 | 8411.0 |
| Ploceus melanogaster | B | 20 | 22.4 | | 20.0–25.0 | | Cameroon 349 | 8412.0 |
| Ploceus alienus | U | 11 | 22.2 | 2.25 | | | 51 | 8413.0 |
| Ploceus capensis | U | 55 | 42.4 | | 31.9–54.9 | | 78 | 8415.0 |
| Ploceus subaureus | M | 1 | 26.2 | | | | Malawi 299 | 8416.0 |
| | F | 1 | 38.0 | | | | | |
| Ploceus burnieri | M | 12 | 19.6 | | 17.0–21.3 | | Tanzania 15 | 8416.1 |
| | F | 21 | 16.6 | | 14.2–17.5 | | | |
| Ploceus xanthops | M | 1 | 47.8 | | | | Zimbabwe 282 | 8417.0 |
| | F | 1 | 34.6 | | | | | |

**Body Masses of World Birds (continued)**

| Species | Sex | N | Mean | Std dev | Range | Sn | Location | Number |
|---------|-----|---|------|---------|-------|----|----------|--------|
| Ploceus heuglini | U | 2 | 24.7 | | 23.0–26.4 | | Ghana 228 | 8425.0 |
| Ploceus velatus | M | 39 | 33.6 | 1.80 | 28.0–36.4 | | South Africa 256 | 8429.0 |
| | F | 29 | 28.5 | 1.90 | 25.5–34.1 | | | |
| Ploceus cucullatus | M | 48 | 45.2 | 2.40 | 37.5–50.0 | | Kenya 67 | 8433.0 |
| | F | 62 | 36.6 | 2.50 | 32.5–43.5 | | | |
| Ploceus spekei | U | 2 | 30.5 | | 30.0–31.0 | | Kenya 521 | 8435.0 |
| Ploceus nigerrimus | M | 6 | 38.5 | | 34.0–41.0 | | Cameroon 349 | 8437.0 |
| | F | 11 | 30.5 | | 27.0–38.0 | | | |
| Ploceus golandi | M | 5 | | | 24.0–26.0 | | Kenya 495 | 8439.0 |
| Ploceus rubiginosus | U | 10 | 28.6 | | 25.0–35.0 | | Kenya 521 | 8444.0 |
| Ploceus nelicourvi | M | 7 | 25.4 | | 23.5–27.2 | | Madagascar 39 | 8448.0 |
| | F | 5 | 22.1 | | 21.0–23.7 | | | |
| Ploceus sakalava | M | 2 | 24.4 | | 24.2–24.5 | | Madagascar 39 | 8449.0 |
| Ploceus benghalensis | B | 16 | 19.8 | | 18.0–22.0 | | India 5 | 8450.0 |
| Ploceus manyar | M | 2 | 17.4 | | 16.4–18.4 | | Thailand 378 | 8451.0 |
| Ploceus philippinus | B | 23 | 28.2 | | 24.0–32.0 | | India 5 | 8452.0 |
| Ploceus hypoxanthus | M | 1 | 18.6 | | | | Thailand 378 | 8453.0 |
| Ploceus megarhynchus | M | 3 | | | 34.0–40.0 | | India 5 | 8454.0 |
| | F | 7 | | | 30.0–34.0 | | | |
| Ploceus bicolor | B | 26 | 33.1 | 2.10 | 29.4–37.5 | | Somalia 691 | 8455.0 |
| Ploceus insignis | M | 1 | 35.0 | | | | Cameroon 349 | 8460.0 |
| Malimbus rubricollis | U | 2 | 42.2 | | 40.5–44.0 | | Kenya 362 | 8474.0 |
| Anaplectes rubriceps | M | 1 | 22.0 | | | | Ghana 228 | 8475.0 |
| Quelea erythrops | M | 6 | 17.0 | | 12.0–20.0 | | Cameroon 349 | 8478.0 |
| | F | 7 | 15.4 | | 14.0–16.0 | | | |
| Quelea quelea | M | 80 | 19.5 | 1.30 | 17.2–22.8 | | Zimbabwe 284 | 8479.0 |
| | F | 98 | 18.3 | 1.00 | 16.1–21.2 | | | |
| Foudia madagascariensis | M | 25 | 18.2 | 0.20 | | | Mauritius 152 | 8480.0 |
| | F | 13 | 16.3 | 0.36 | | | | |
| Foudia madagascariensis | M | 10 | 16.8 | | 14.9–18.3 | | Seychelles 152 | 8480.0 |
| | F | 9 | 15.2 | | 13.5–16.5 | | | |

## Body Masses of World Birds (continued)

| Species | Sex | N | Mean | Std dev | Range | Sn | Location | Number |
|---|---|---|---|---|---|---|---|---|
| Foudia eminentissima | M | | | | 24.5–26.3 | | Aldabra Is. | 8481.0 |
| | F | | | | 22.7–24.7 | | 152 | |
| Foudia omissa | B | 9 | 18.4 | | 16.0–21.0 | | Madagascar | 8482.0 |
| | | | | | | | 152 | |
| Foudia rubra | B | 17 | 17.7 | | 16.0–20.1 | | Mauritius | 8483.0 |
| | | | | | | | 152 | |
| Foudia sechellarum | B | 27 | 17.1 | | 15.5–19.1 | | Seychelles | 8484.0 |
| | | | | | | | 152 | |
| Foudia flavicans | M | 26 | 15.9 | 0.10 | 14.7–17.0 | | Rodrigues Is. | 8485.0 |
| | F | 14 | 14.8 | 0.22 | 13.5–16.0 | | 152 | |
| Euplectes afer | M | 81 | 14.7 | 2.70 | 11.2–20.5 | | Ghana | 8486.0 |
| | F | 44 | 13.9 | 1.00 | 12.2–16.2 | | 228 | |
| Euplectes hordaceus | M | 12 | 20.4 | 1.40 | 18.0–22.9 | | Kenya | 8489.0 |
| | F | 6 | 18.4 | | 17.2–19.5 | | 67 | |
| Euplectes orix | M | 30 | 17.0 | 1.00 | 14.3–18.9 | | Ghana | 8491.0 |
| | F | 20 | 15.5 | 2.20 | 11.4–21.2 | | 228 | |
| Euplectes capensis | M | 14 | 19.9 | 1.00 | 17.7–21.1 | | Zimbabwe | 8494.0 |
| | F | 15 | 16.5 | 0.90 | 15.5–19.1 | | 284 | |
| Euplcctcs axillaris | M | 69 | 26.5 | 1.40 | 23.0–29.2 | | Kenya | 8495.0 |
| | F | 84 | 20.9 | 1.50 | 18.2–24.9 | | 67 | |
| Euplectes macrourus | M | 8 | 22.9 | | 17.0–30.0 | | Cameroon | 8496.0 |
| | F | 5 | 17.2 | | 12.0–20.0 | | 349 | |
| Euplectes ardens | M | 60 | 18.3 | 1.40 | 15.6–22.0 | | Zimbabwe | 8498.0 |
| | F | 92 | 16.0 | 1.20 | 13.1–19.3 | | 284 | |
| Euplectes hartlaubi | M | 1 | 35.5 | | | | Zambia | 8499.0 |
| | | | | | | | 166 | |
| Euplectes progne | U | 95 | 35.1 | | 25.0–46.9 | | | 8501.0 |
| | | | | | | | 78 | |
| Amblyospiza albifrons | U | 6 | 35.8 | | 29.6–43.3 | | | 8504.0 |
| | | | | | | | 280, 281, 314a, 349 | |

### ORDER: PASSERIFORMES                          FAMILY: PROMERIPIDAE

| Species | Sex | N | Mean | Std dev | Range | Sn | Location | Number |
|---|---|---|---|---|---|---|---|---|
| Promerops gurneyi | M | 2 | 40.4 | | 36.6–44.3 | | South Africa | 8505.0 |
| | | | | | | | 256, 354 | |

### ORDER: PASSERIFORMES                          FAMILY: DICAEIDAE

| Species | Sex | N | Mean | Std dev | Range | Sn | Location | Number |
|---|---|---|---|---|---|---|---|---|
| Prionochilus olivaceus | B | 200 | 9.0 | | 8.4–10.7 | | Philippines | 8507.0 |
| | | | | | | | 475 | |
| Prionochilus xanthopygius | B | 3 | 9.5 | | 8.9–10.2 | | Borneo | 8511.0 |
| | | | | | | | 627 | |
| Dicaeum agile | M | 11 | 9.0 | | 7.5–11.0 | | India | 8514.0 |
| | F | 1 | 8.0 | | | | 5 | |
| Dicaeum chrysorrheum | M | 1 | 9.0 | | | | India | 8517.0 |
| | | | | | | | 5 | |

## Body Masses of World Birds (continued)

| Species | Sex | N | Mean | Std dev | Range | Sn | Location | Number |
|---------|-----|---|------|---------|-------|----|----------|--------|
| Dicaeum nigrilore | B | 20 | 11.0 | | 9.5–12.4 | | Philippines 475 | 8521.0 |
| Dicaeum anthonyi | M | 2 | 12.6 | | 12.1–13.0 | | Philippines 475 | 8522.0 |
| Dicaeum bicolor | M | 4 | 8.2 | | 7.8–8.5 | | Philippines 475 | 8523.0 |
| | F | 3 | 9.4 | | 8.6–10.0 | | | |
| Dicaeum australe | B | 26 | 8.7 | | 7.9–10.5 | | Philippines 475 | 8525.0 |
| Dicaeum trigonostigma | B | 30 | 7.1 | | 5.9–8.4 | | Philippines 475 | 8527.0 |
| Dicaeum hypoleucum | B | 38 | 8.0 | | 6.9–9.6 | | Philippines 475 | 8528.0 |
| Dicaeum erythrorhynchos | B | 17 | 6.3 | | 4.0–8.0 | | India 5 | 8529.0 |
| Dicaeum concolor | U | 15 | 6.2 | | 5.0–8.0 | | India 5 | 8530.0 |
| Dicaeum pygmaeum | B | 4 | 5.2 | | 4.5–5.6 | | Philippines 475 | 8531.0 |
| Dicaeum pectorale | B | 5 | 7.4 | | 7.0–7.8 | | New Guinea 216, 267 | 8535.0 |
| Dicaeum geelvinkianum | B | 12 | 6.8 | 0.63 | 5.3–7.5 | | New Guinea 154 | 8536.0 |
| Dicaeum eximium | B | | | | 7.0–9.5 | | New Britain 219 | 8538.0 |
| Dicaeum ignipectus | M | 16 | | | 4.0–8.0 | | India 5 | 8543.0 |
| | F | 5 | | | 5.5–6.1 | | | |
| Dicaeum hirundinaceum | M | | 8.0 | | | | Australia 194a | 8547.0 |
| Dicaeum cruentatum | M | 3 | | | 7.0–8.0 | | India 5 | 8548.0 |

### ORDER: PASSERIFORMES                FAMILY: NECTARINIIDAE

| Species | Sex | N | Mean | Std dev | Range | Sn | Location | Number |
|---------|-----|---|------|---------|-------|----|----------|--------|
| Anthreptes fraseri | U | 6 | 10.9 | | 9.7–12.3 | | Liberia; Gabon 314a | 8550.0 |
| Anthreptes simplex | M | 1 | 7.0 | | | | Borneo 627 | 8554.0 |
| Anthreptes malacensis | M | 6 | 12.2 | | 11.4–13.0 | | Philippines 475 | 8555.0 |
| | F | 3 | 10.7 | | 10.5–10.8 | | | |
| Anthreptes rhodolaema | B | 7 | 13.1 | | 12.0–15.0 | | Borneo 627 | 8556.0 |
| Anthreptes singalensis | B | 2 | 8.2 | | 8.0–8.4 | | Borneo 627 | 8557.0 |
| Anthreptes longuemarei | F | 1 | 14.0 | | | | Cameroon 349 | 8559.0 |

## Body Masses of World Birds (continued)

| Species | Sex | N | Mean | Std dev | Range | Sn | Location | Number |
|---------|-----|---|------|---------|-------|----|----------|--------|
| Anthreptes collaris | B | 23 | 7.2 | | 5.8–8.5 | | Kenya 67 | 8566.0 |
| Nectarinia olivacea | M | 28 | 11.2 | 0.68 | | | central Africa | 8572.0 |
| | F | 39 | 9.8 | 0.90 | | | 51 | |
| Nectarinia olivacea | M | 54 | 7.7 | 0.40 | 6.8–9.1 | | Somalia | 8572.0 |
| | F | 49 | 7.1 | 0.50 | 6.2–8.7 | | 691 | |
| Nectarinia violacea | U | 2 | 9.2 | | 9.2–9.3 | | 78 | 8573.0 |
| Nectarinia veroxii | U | | 10.5 | | | | 453 | 8574.0 |
| Nectarinia oritis | M | 13 | 13.0 | | 11.0–15.0 | | Cameroon | 8579.0 |
| | F | 7 | 10.7 | | 10.0–12.0 | | 349 | |
| Nectarinia alinae | M | 28 | 12.9 | 1.11 | | | | 8580.0 |
| | F | 22 | 11.5 | 1.22 | | | 51 | |
| Nectarinia verticalis | B | 12 | 14.0 | | 13.0–15.0 | | Cameroon 349 | 8581.0 |
| Nectarinia dussumieri | U | | 10.8 | | | | 453 | 8585.0 |
| Nectarinia amethystina | B | 43 | 10.4 | | 8.3–12.5 | | Zimbabwe 283 | 8588.0 |
| Nectarinia senegalensis | M | 16 | 10.3 | 0.60 | 9.6–11.9 | | Ghana | 8589.0 |
| | F | 7 | 8.4 | | 6.8–9.8 | | 228 | |
| Nectarinia zeylonica | B | 17 | | | 7.0–11.0 | | India 5 | 8592.0 |
| Nectarinia minima | B | 16 | | | 4.0–6.0 | | India 5 | 8593.0 |
| Nectarinia sperata | B | 46 | 6.4 | | 5.2–7.5 | | Philippines 475 | 8594.0 |
| Nectarinia aspasia | U | 9 | 8.0 | | | | New Guinea 35 | 8595.0 |
| Nectarinia jugularis | M | 22 | 9.2 | | 7.9–11.9 | | Philippines | 8597.0 |
| | F | 18 | 8.1 | | 6.8–10.0 | | 475 | |
| Nectarinia souimanga | M | 5 | 7.3 | | 6.5–7.7 | | Madagascar | 8600.0 |
| | F | 5 | 6.5 | | 5.5–7.5 | | 39 | |
| Nectarinia venusta | B | 31 | 6.8 | | 5.3–9.0 | | Zimbabwe 283 | 8604.0 |
| Nectarinia talatala | B | 2 | 6.4 | | 6.0–6.7 | | Zimbabwe 282, 283 | 8606.0 |
| Nectarinia bouvieri | M | 5 | 8.9 | | 8.5–10.0 | | Cameroon 349 | 8608.0 |
| Nectarinia asiatica | B | 18 | 8.1 | | 5.0–11.0 | | India 5 | 8610.0 |
| Nectarinia lotenia | B | 15 | | | 7.0–11.0 | | Sri Lanka 5 | 8612.0 |

## Body Masses of World Birds (continued)

| Species | Sex | N | Mean | Std dev | Range | Sn | Location | Number |
|---|---|---|---|---|---|---|---|---|
| Nectarinia manoensis | B | 16 | 9.5 | | 8.2–12.8 | | Zimbabwe 282, 283, 284 | 8613.0 |
| Nectarinia chalybea | M | 5 | 9.5 | | 8.4–10.2 | | Zimbabwe | 8614.0 |
| | F | 4 | 7.8 | | 7.4–8.2 | | 281 | |
| Nectarinia preussi | M | 9 | 9.5 | | 9.0–10.0 | | Cameroon 349 | 8618.0 |
| Nectarinia afra | M | 4 | 9.9 | | 8.3–13.0 | | | 8619.0 |
| | F | 3 | 8.2 | | 8.0–8.6 | | 166, 256 | |
| Nectarinia mediocris | B | 48 | 8.8 | | 7.2–10.7 | | Malawi | 8620.0 |
| | | 44 | 7.8 | | 6.3–9.9 | | 168 | |
| Nectarinia neergaardi | U | | 6.2 | | | | 453 | 8621.0 |
| Nectarinia chloropygia | B | 14 | 6.3 | | 5.0–7.0 | | 314a, 349 | 8622.0 |
| Nectarinia minulla | F | 1 | 6.5 | | | | Cameroon 349 | 8623.0 |
| Nectarinia regia | B | 9 | 6.0 | | | | 51 | 8624.0 |
| Nectarinia cuprea | M | 5 | 7.2 | | 6.8–7.6 | | Ghana | 8627.0 |
| | F | 5 | 6.1 | | 5.5–6.8 | | 228 | |
| Nectarinia fusca | M | 1 | 10.0 | | | | 78 | 8628.0 |
| Nectarinia rufipennis | M | 1 | 10.0 | | | | Tanzania | 8629.0 |
| | F | 1 | 8.7 | | | | 290 | |
| Nectarinia tacazze | U | 34 | 16.5 | 1.65 | 12.0–19.0 | | Ethiopia 633 | 8630.0 |
| Nectarinia kilimensis | U | | 16.2 | | | | 453 | 8633.0 |
| Nectarinia reichenowi | U | | 14.5 | | | | 453 | 8634.0 |
| Nectarinia famosa | B | 3 | 17.1 | | 16.2–18.3 | | South Africa 256 | 8635.0 |
| Nectarinia mariquensis | U | 4 | 11.4 | | 10.6–12.5 | | 78 | 8640.0 |
| Nectarinia bifasciata | M | 18 | 7.1 | | 6.5–8.1 | | Kenya | 8641.0 |
| | F | 11 | 6.3 | | 6.0–6.5 | | 61 | |
| Nectarinia notata | M | 1 | 14.5 | | | | | 8643.0 |
| | F | 1 | 20.0 | | | | 39, 97 | |
| Nectarinia superba | M | 1 | 18.0 | | | | Cameroon 349 | 8646.0 |
| Nectarinia pulchella | B | 19 | 6.9 | | 5.6–10.2 | | Ghana 228 | 8647.0 |
| Aethopyga boltoni | M | 5 | 8.0 | | 7.5–8.7 | | Philippines | 8650.0 |
| | F | 6 | 6.7 | | 6.1–7.4 | | 475 | |

## Body Masses of World Birds (continued)

| Species | Sex | N | Mean | Std dev | Range | Sn | Location | Number |
|---|---|---|---|---|---|---|---|---|
| Aethopyga pulcherrima | B | 46 | 6.2 | | 4.4–7.5 | | Philippines 475 | 8652.0 |
| Aethopyga shelleyi | M | 4 | 4.7 | | 3.9–5.3 | | Philippines | 8654.0 |
| | F | 2 | 3.6 | | 3.1–4.2 | | 475 | |
| Aethopyga gouldiae | M | 5 | | | 6.5–8.0 | | India | 8655.0 |
| | F | 2 | 5.0 | | 4.0–6.0 | | 5 | |
| Aethopyga nipalensis | M | 4 | 6.9 | | 6.5–8.0 | | India | 8656.0 |
| | F | 8 | | | 5.0–6.5 | | 5 | |
| Aethopyga saturata | U | 22 | 5.3 | | 3.0–7.0 | | Malaysia 371 | 8659.0 |
| Aethopyga siparaja | M | 6 | | | 6.4–7.9 | | India 5 | 8660.0 |
| Aethopyga mystacalis | M | 1 | 6.0 | | | | Borneo 627 | 8661.0 |
| Aethopyga ignicauda | M | 2 | 8.2 | | 7.5–9.0 | W | India 5 | 8662.0 |
| Arachnothera longirostra | M | 12 | 11.7 | | 10.5–13.6 | | Philippines | 8663.0 |
| | F | 7 | 10.2 | | 9.4–10.7 | | 475 | |
| Arachnothera crassirostris | F | 1 | 14.8 | | | | Borneo 627 | 8664.0 |
| Arachnothera flavigaster | F | 1 | 38.4 | | | | Borneo 627 | 8666.0 |
| Arachnothera chrysogenys | U | 2 | 23.2 | | 23.0–23.5 | | Malaysia; Borneo 371, 627 | 8667.0 |
| Arachnothera clarae | B | 17 | 30.1 | | 27.5–34.8 | | Philippines 475 | 8668.0 |
| Arachnothera affinis | B | 4 | 26.4 | | 22.2–29.2 | | Borneo 627 | 8669.0 |
| Arachnothera magna | U | 34 | 30.7 | | 23.0–35.0 | | Malaysia 371 | 8671.0 |

### ORDER: PASSERIFORMES     FAMILY: MELANOCHARITIDAE

| Species | Sex | N | Mean | Std dev | Range | Sn | Location | Number |
|---|---|---|---|---|---|---|---|---|
| Melanocharis nigra | M | 9 | 13.2 | | 11.5–14.7 | | New Guinea | 8674.0 |
| | F | 5 | 15.2 | | 14.0–16.5 | | 154 | |
| Melanocharis versteri | M | 10 | 11.9 | 1.00 | 9.7–13.7 | | New Guinea | 8676.0 |
| | F | 10 | 18.1 | 0.80 | 16.7–20.0 | | 154 | |
| Melanocharis striativentris | M | 21 | 16.1 | | 14.6–19.2 | | Australia | 8677.0 |
| | F | 11 | 19.2 | | 15.5–22.5 | | 524 | |
| Melanocharis crassirostris | U | 1 | 18.0 | | | | New Guinea 216 | 8678.0 |
| Toxorhamphus novaeguineae | M | 4 | 13.9 | | 12.0–16.0 | | New Guinea | 8679.0 |
| | F | 1 | 10.0 | | | | 218, 267 | |
| Toxorhamphus poliopterus | M | 10 | 12.4 | 1.00 | 10.5–14.3 | | New Guinea | 8680.0 |
| | F | 10 | 10.4 | 0.70 | 9.3–11.5 | | 154 | |

## Body Masses of World Birds (continued)

| Species | Sex | N | Mean | Std dev | Range | Sn | Location | Number |
|---|---|---|---|---|---|---|---|---|
| Toxorhamphus iliolophus | U | 35 | 11.0 | | | | New Guinea 35 | 8681.0 |
| Toxorhamphus pygmaeum | U | 4 | 5.4 | | 5.0–6.0 | | New Guinea 154, 218 | 8682.0 |
| **ORDER: PASSERIFORMES** | | | | | **FAMILY: PARAMYTHIIDAE** | | | |
| Oreocharis afaki | B | 8 | 21.1 | | 17.0–22.5 | | New Guinea 154, 216 | 8683.0 |
| Paramythia montium | M | 8 | 41.1 | | 38.5–43.0 | | New Guinea 216 | 8684.0 |
| | F | 6 | 42.9 | | 38.0–49.0 | | | |
| **ORDER: PASSERIFORMES** | | | | | **FAMILY: FRINGILLIDAE** | | | |
| Fringilla coelebs | M | 91 | 21.9 | 1.67 | 17.0–28.5 | | New Zealand 500 | 8685.0 |
| | F | 61 | 20.9 | 1.52 | 18.0–27.0 | | | |
| Fringilla montifringilla | U | 50 | 24.0 | | 21.8–31.0 | W | Britain 258 | 8687.0 |
| Serinus pusillus | B | 5 | | | 10.5–12.7 | | India 5 | 8688.0 |
| Serinus serinus | U | 20 | 11.2 | | 8.5–14.0 | | 78 | 8689.0 |
| Serinus canaria | U | 1 | 8.4 | | | | 227 | 8691.0 |
| Serinus citrinella | U | 1 | 12.0 | | | | 227 | 8692.0 |
| Serinus thibetanus | U | 6 | | | 10.0–11.0 | | 5 | 8693.0 |
| Serinus canicollis | B | 6 | 13.8 | | 11.3–17.5 | | 166, 256 | 8694.0 |
| Serinus citrinelloides | U | 24 | 15.0 | | 13.0–17.0 | | Ethiopia 633 | 8697.0 |
| Serinus koliensis | U | 30 | 13.6 | | 11.3–16.1 | | 78 | 8700.0 |
| Serinus atrogularis | U | 206 | 11.4 | | 8.0–14.0 | | 78 | 8706.0 |
| Serinus mozambicus | B | 71 | 10.6 | | 9.1–13.5 | | Ghana 228 | 8708.0 |
| Serinus donaldsoni | U | 3 | 24.0 | | 20.7–26.9 | | Kenya 521 | 8709.0 |
| Serinus dorsostriatus | U | | 14.4 | | | | Kenya 522 | 8711.0 |
| Serinus flaviventris | U | 68 | 16.3 | | 11.5–21.0 | | 78 | 8712.0 |
| Serinus sulphuratus | B | 10 | 19.2 | | 16.6–23.4 | | 166, 283 | 8713.0 |

## Body Masses of World Birds (continued)

| Species | Sex | N | Mean | Std dev | Range | Sn | Location | Number |
|---|---|---|---|---|---|---|---|---|
| Serinus albogularis | B | 3 | 25.9 | | 23.6–28.1 | | South Africa 256 | 8714.0 |
| Serinus gularis | B | 8 | 15.5 | | 10.5–17.9 | | Zimbabwe 281, 283 | 8717.0 |
| Serinus tristriatus | U | 125 | 15.5 | | 12.2–19.4 | | Ethiopia 634 | 8719.0 |
| Serinus striolatus | U | 35 | 22.4 | | 19.5–25.5 | | Ethiopia 633 | 8722.0 |
| Serinus burtoni | U | 3 | 29.9 | | | | 51, 349 | 8724.0 |
| Linurgus olivaceus | B | 25 | 23.2 | | 21.0–28.0 | | Cameroon 349 | 8734.0 |
| Carduelis chloris | U | 21 | 27.8 | | 25.0–31.5 | B | Britain 258 | 8736.0 |
| Carduelis sinica | U | 2 | 31.3 | | 30.8–31.8 | F | Alaska 213 | 8737.0 |
| Carduelis spinoides | B | 9 | 18.6 | | 18.5–20.8 | | 5 | 8738.0 |
| Carduelis spinus | U | 50 | 14.5 | | 11.6  18.2 | S | Britain 258 | 8741.0 |
| Carduelis pinus | B | 328 | 14.6 | 1.01 | 10.8–20.1 | | Pennsylvania, USA 101 | 8742.0 |
| Carduelis atriceps | M F | | 12.5 14.5 | | | | Chiapas, Mexico 273 | 8743.0 |
| Carduelis magellanica | U | 2 | 11.0 | | 11.0–11.0 | | 38, 111 | 8748.0 |
| Carduelis notata | B | 13 | 10.9 | | 10.0–12.2 | | Belize 510 | 8751.0 |
| Carduelis barbata | U | | 16.6 | | | | Chile 52 | 8752.0 |
| Carduelis xanthogaster | U | 10 | 12.7 | 0.90 | 10.8–13.8 | | Panama 337 | 8753.0 |
| Carduelis tristis | M F | 2178 1547 | 13.2 12.6 | 1.13 0.81 | 8.6–20.7 10.0–17.1 | Y | Pennsylvania, USA 101 | 8756.0 |
| Carduelis psaltria | B | 202 | 9.5 | | 8.0–11.5 | B | California, USA 554 | 8757.0 |
| Carduelis lawrencei | M F | 33 30 | 10.6 11.3 | 0.85 1.03 | 8.8–12.5 9.8–14.3 | B | California, USA 177 | 8758.0 |
| Carduelis dominicensis | U | | 9.0 | | | | 185 | 8759.0 |
| Carduelis carduelis | U | 50 | 15.6 | | 13.4–17.8 | B | Britain 258 | 8760.0 |
| Carduelis hornemanni | B | 54 | 12.7 | | 10.4–16.1 | | Alaska, USA 278 | 8761.0 |

## Body Masses of World Birds (continued)

| Species | Sex | N | Mean | Std dev | Range | Sn | Location | Number |
|---|---|---|---|---|---|---|---|---|
| Carduelis flammea | B | 30 | 13.0 | | 10.1–15.0 | B | Alaska, USA 278 | 8762.0 |
| Carduelis flavirostris | U | 50 | 15.4 | | 13.0–17.6 | W | Britain 258 | 8763.0 |
| Carduelis cannabina | U | 50 | 15.3 | | 14.5–21.0 | Y | Britain 258 | 8764.0 |
| Leucosticte nemoricola | B | 19 | | | 19.8–25.6 | | India 5 | 8767.0 |
| Leucosticte brandti | M | 5 | | | 26.4–28.9 | | India | 8768.0 |
| | F | 5 | | | 26.0–28.3 | | 5 | |
| Leucosticte arctoa dawsoni | B | 42 | 24.6 | 1.04 | | W | California, USA 177 | 8769.0 |
| Leucosticte arctoa atrata | B | 111 | 26.5 | | | W | Utah, USA 321 | 8769.0 |
| Leucosticte arctoa australis | B | 87 | 26.9 | | | B | 302 | 8769.0 |
| Rhodopechys sanguinea | M | 2 | 39.0 | | 34.0–44.0 | | India | 8774.0 |
| | F | 2 | 32.5 | | 32.0–33.0 | | 5 | |
| Rhodopechys mongolica | B | 20 | | | 18.0–24.0 | | 5 | 8776.0 |
| Rhodopechys obsoleta | B | 2 | 25.5 | | 25.0–26.0 | | India 5 | 8777.0 |
| Uragus sibiricus | F | 1 | 16.0 | | | | 149 | 8778.0 |
| Carpodacus nipalensis | M | 4 | | | 22.0–23.0 | | | 8780.0 |
| | F | 3 | 22.1 | | 20.0–23.5 | | 5 | |
| Carpodacus erythrinus | B | 19 | 24.1 | | 21.0–30.0 | | 5 | 8781.0 |
| Carpodacus purpureus | B | 316 | 24.9 | 1.60 | 18.1–35.3 | | Pennsylvania, USA 101 | 8782.0 |
| Carpodacus cassinii | B | 62 | 26.5 | 2.37 | 20.4–37.8 | | Arizona, USA 177 | 8783.0 |
| Carpodacus mexicanus | B | 220 | 21.4 | 1.29 | 19.0–25.5 | Y | California, USA 177 | 8784.0 |
| Carpodacus pulcherrimus | M | 13 | | | 17.5–20.0 | | India | 8785.0 |
| | F | 5 | | | 17.4–19.2 | | 5 | |
| Carpodacus rhodochrous | M | 9 | | | 17.0–20.0 | | India | 8787.0 |
| | F | 10 | | | 16.0–18.5 | | 5 | |
| Carpodacus edwardsii | F | 1 | 26.5 | | | | India 5 | 8789.0 |
| Carpodacus rhodopeplus | M | 1 | 23.0 | | | | India 5 | 8793.0 |
| Carpodacus thura | M | 14 | | | 24.0–35.6 | | India | 8794.0 |
| | F | 8 | | | 30.0–36.0 | | 5 | |

## Body Masses of World Birds (continued)

| Species | Sex | N | Mean | Std dev | Range | Sn | Location | Number |
|---|---|---|---|---|---|---|---|---|
| Carpodacus rhodochlamys | B | 7 | | | 31.0–36.0 | | India 5 | 8795.0 |
| Carpodacus rubicilla | M | 1 | 43.2 | | | | India 5 | 8797.0 |
| Carpodacus puniceus | M | 11 | | | 42.8–51.2 | | India 5 | 8798.0 |
| | F | 5 | | | 43.0–50.0 | | | |
| Pinicola enucleator | B | 17 | 56.4 | 3.24 | 52.0–62.0 | F | Canada 177 | 8801.0 |
| Pinicola subhimachalus | M | 6 | | | 42.5–48.2 | | India 5 | 8802.0 |
| | F | 4 | | | 44.0–50.0 | | | |
| Haematospiza sipahi | B | 5 | 39.5 | | 38.0–42.5 | | India 5 | 8803.0 |
| Loxia pytyopsittacus | U | 37 | 53.0 | | 44.0–58.2 | | 78 | 8804.0 |
| Loxia curvirostra | B | 36 | 36.5 | 3.71 | 29.2–44.9 | W | Arizona, USA 594 | 8806.0 |
| Loxia curvirostra | U | 638 | 40.6 | | 34.0–48.0 | PB | Britain 258 | 8806.0 |
| Loxia leucoptera | M | 16 | 27.8 | 1.30 | | W | Manitoba, Canada 544 | 8807.0 |
| | F | 15 | 25.3 | 1.70 | | | | |
| Pyrrhula nipalensis | B | 7 | 24.6 | | 20.0–29.0 | | 5, 371 | 8808.0 |
| Pyrrhula leucogenys | M | 1 | 19.0 | | | | Philippines 475 | 8809.0 |
| Pyrrhula aurantiaca | B | 11 | | | 17.0–22.0 | | India 5 | 8810.0 |
| Pyrrhula erythrocephala | M | 14 | | | 18.0–22.5 | | India 5 | 8811.0 |
| | F | 12 | | | 18.0–27.6 | | | |
| Pyrrhula erythaca | U | 1 | 20.0 | | | | India 5 | 8812.0 |
| Pyrrhula pyrrhula | U | 34 | 21.8 | | 21.0–27.0 | B | Britain 258 | 8813.0 |
| Coccothraustes coccothraustes | U | 50 | 54.0 | | 48.0–62.0 | Y | Britain 258 | 8814.0 |
| Eophona migratoria | M | 1 | 50.0 | | | S | 149 | 8815.0 |
| Eophona personata | U | | 80.0 | | | | 149 | 8816.0 |
| Mycerobas icteriodes | U | 2 | 67.0 | | | | India 467 | 8817.0 |
| Mycerobas affinis | M | 3 | | | 69.0–72.0 | | India 5 | 8818.0 |
| | F | 1 | 83.0 | | | | | |
| Mycerobas melanozanthos | B | 2 | 62.0 | | 50.0–74.0 | | India 5 | 8819.0 |

## Body Masses of World Birds (continued)

| Species | Sex | N | Mean | Std dev | Range | Sn | Location | Number |
|---------|-----|---|------|---------|-------|----|----------|--------|
| Mycerobas carnipes | M | 6 | | | 56.1–66.0 | | India | 8820.0 |
| | F | 4 | | | 54.7–59.2 | | 5 | |
| Hesperiphona vespertina | M | 852 | 60.1 | 3.24 | 38.7–86.1 | W | Pennsylvania, USA | 8821.0 |
| | F | 1157 | 58.7 | 3.83 | 43.2–73.5 | | 101 | |
| Hesperiphona abeillei | B | 4 | 48.5 | | 47.1–49.7 | | Mexico | 8822.0 |
| | | | | | | | 456 | |
| Pyrrhoplectes epauletta | B | 3 | 19.0 | | 19.0–19.0 | | India | 8823.0 |
| | | | | | | | 5 | |

### ORDER: PASSERIFORMES          FAMILY: DREPANIDIDAE

| Species | Sex | N | Mean | Std dev | Range | Sn | Location | Number |
|---------|-----|---|------|---------|-------|----|----------|--------|
| Telespiza cantans | U | 12 | 32.8 | 3.60 | 24.4–37.9 | | Hawaiian Is. | 8825.0 |
| | | | | | | | 663 | |
| Loxioides bailleui | M | 4 | 36.0 | | 33.5–40.0 | | Hawaiian Is. | 8828.0 |
| | | | | | | | 663 | |
| Pseudonestor xanthophrys | M | 1 | 20.0 | | | | Hawaiian Is. | 8832.0 |
| | | | | | | | 399 | |
| Viridonia virens | M | 144 | 14.0 | 1.20 | 11.0–15.6 | | Hawaiian Is. | 8833.0 |
| | F | 88 | 13.4 | 0.94 | 10.5–16.2 | | 637 | |
| Viridonia parva | U | 4 | 7.9 | | | | Hawaiian Is. | 8834.0 |
| | | | | | | | 357 | |
| Oreomystis bairdi | U | | 14.0 | | | | estimated | 8839.0 |
| | | | | | | | 400 | |
| Paroreomyza montana | U | | 11.0 | | | | estimated | 8841.0 |
| | | | | | | | 400 | |
| Loxops coccineus | U | | 11.0 | | | | estimated | 8845.0 |
| | | | | | | | 400 | |
| Vestiaria coccinea | U | 6 | 24.9 | | | | Hawaiian Is. | 8847.0 |
| | | | | | | | 357 | |
| Palmeria dolei | U | | 22.0 | | | | estimated | 8850.0 |
| | | | | | | | 400 | |
| Himatione sanguinea | U | 7 | 25.1 | | | | Hawaiian Is. | 8851.0 |
| | | | | | | | 357 | |

### ORDER: PASSERIFORMES          FAMILY: PARULIDAE

| Species | Sex | N | Mean | Std dev | Range | Sn | Location | Number |
|---------|-----|---|------|---------|-------|----|----------|--------|
| Peucedramus taeniatus | B | 16 | 11.0 | 0.60 | 10.1–12.1 | B | Coahuila, Mexico | 8853.0 |
| | | | | | | | 386 | |
| Vermivora pinus | B | 24 | 8.4 | 0.57 | 7.2–11.0 | | Pennsylvania, USA | 8855.0 |
| | | | | | | | 101 | |
| Vermivora chrysoptera | M | 164 | 8.7 | 0.31 | 7.2–11.2 | | Pennsylvania, USA | 8856.0 |
| | F | 90 | 8.9 | 0.22 | 7.5–11.8 | | 101 | |
| Vermivora peregrina | M | 214 | 10.2 | 0.22 | 7.3–18.4 | M | Pennsylvania, USA | 8857.0 |
| | F | 168 | 9.8 | 0.31 | 7.8–13.4 | | 101 | |
| Vermivora celata | B | 72 | 9.0 | 0.84 | 7.3–11.6 | M | California, USA | 8858.0 |
| | | | | | | | 177 | |

## Body Masses of World Birds (continued)

| Species | Sex | N | Mean | Std dev | Range | Sn | Location | Number |
|---------|-----|---|------|---------|-------|----|----------|--------|
| Vermivora ruficapilla | M | 198 | 8.9 | 0.42 | 7.0–13.9 | M | Pennsylvania, USA | 8859.0 |
|  | F | 257 | 8.6 | 0.42 | 6.7–11.1 |  | 101 |  |
| Vermivora virginiae | B | 8 | 7.8 |  | 7.0–9.0 |  | Arizona, USA | 8860.0 |
|  |  |  |  |  |  |  | 594 |  |
| Vermivora crissalis | B | 7 | 9.7 |  | 8.0–11.5 |  |  | 8861.0 |
|  |  |  |  |  |  |  | 585, 586, 590 |  |
| Vermivora luciae | B | 97 | 6.6 | 0.53 | 5.1–7.9 | B | Arizona, USA | 8862.0 |
|  |  |  |  |  |  |  | 177 |  |
| Parula americana | B | 44 | 8.6 |  | 7.1–10.2 |  | Florida, USA | 8863.0 |
|  |  |  |  |  |  |  | 190 |  |
| Parula pitiayumi | M | 10 | 7.2 | 0.38 |  |  | Panama | 8864.0 |
|  | F | 6 | 6.6 |  |  |  | 244 |  |
| Parula superciliosa | B | 2 | 9.0 |  |  |  | Mexico | 8865.0 |
|  |  |  |  |  |  |  | 368 |  |
| Parula gutturalis | B | 7 | 9.5 |  |  |  | Panama | 8866.0 |
|  |  |  |  |  |  |  | 244 |  |
| Dendroica petechia | M | 186 | 9.8 | 0.68 | 7.9–12.8 |  | Pennsylvania, USA | 8867.0 |
|  | F | 139 | 9.2 | 0.59 | 7.4–16.0 |  | 101 |  |
| Dendroica petechia cruciana | U | 12 | 12.8 | 1.39 |  |  | Puerto Rico | 8867.0 |
|  |  |  |  |  |  |  | 676 |  |
| Dendroica pensylvanica | M | 66 | 9.8 | 0.21 | 8.1–13.1 |  | Pennsylvania, USA | 8868.0 |
|  | F | 46 | 9.4 | 0.31 | 7.5–10.9 |  | 101 |  |
| Dendroica magnolia | M | 531 | 8.9 | 0.58 | 7.0–12.9 |  | Pennsylvania, USA | 8869.0 |
|  | F | 430 | 8.5 | 0.35 | 6.6–12.6 |  | 101 |  |
| Dendroica tigrina | B | 102 | 11.0 | 0.27 | 9.3–17.3 | M | Pennsylvania, USA | 8870.0 |
|  |  |  |  |  |  |  | 101 |  |
| Dendroica caerulescens | M | 89 | 10.5 |  | 8.4–12.4 | F | Florida, USA | 8871.0 |
|  | F | 124 | 9.8 |  | 8.8–12.1 |  | 190 |  |
| Dendroica coronata | M | 231 | 12.9 | 0.76 | 10.6–16.7 | M | Pennsylvania, USA | 8872.0 |
|  | F | 290 | 12.2 | 1.29 | 9.9–15.3 |  | 101 |  |
| Dendroica coronata auduboni | M | 109 | 12.3 | 0.94 | 10.0–16.0 |  | California, USA | 8872.0 |
|  | F | 79 | 11.9 | 0.86 | 10.0–14.0 |  | 177 |  |
| Dendroica nigrescens | M | 11 | 8.8 | 0.61 | 7.8–10.3 | S | California, USA | 8873.0 |
|  | F | 13 | 7.9 | 0.53 | 7.1–8.8 |  | 177 |  |
| Dendroica townsendi | M | 48 | 9.1 | 0.71 | 7.3–10.4 | S | California, USA | 8874.0 |
|  | F | 48 | 8.6 | 0.69 | 7.3–10.7 |  | 177 |  |
| Dendroica occidentalis | M | 36 | 9.5 | 0.81 | 7.7–10.7 | S | California, USA | 8875.0 |
|  | F | 18 | 8.8 | 0.78 | 8.0–11.2 |  | 177 |  |
| Dendroica virens | B | 100 | 8.8 | 0.65 | 7.7–11.3 | M | Pennsylvania, USA | 8876.0 |
|  |  |  |  |  |  |  | 177 |  |
| Dendroica chrysoparia | B | 18 | 10.2 | 0.87 | 8.7–12.1 | B | Texas, USA | 8877.0 |
|  |  |  |  |  |  |  | 473 |  |
| Dendroica fusca | M | 12 | 10.0 | 2.01 |  |  |  | 8878.0 |
|  | F | 18 | 9.5 | 0.59 |  |  | 244 |  |

## Body Masses of World Birds (continued)

| Species | Sex | N | Mean | Std dev | Range | Sn | Location | Number |
|---|---|---|---|---|---|---|---|---|
| Dendroica dominica | B | 6 | 9.4 | | 8.8 – 10.0 | | N. Carolina, USA 177 | 8879.0 |
| Dendroica graciae | B | 9 | 8.1 | | 7.5 – 9.1 | B | Nevada, USA 300 | 8880.0 |
| Dendroica adelaidae | U | 23 | 6.7 | 0.60 | 5.3 – 8.0 | | Puerto Rico 186 | 8881.0 |
| Dendroica pityophila | B | 7 | 7.9 | | 7.2 – 8.4 | | Bahamas 600 | 8882.0 |
| Dendroica pinus | B | 21 | 11.9 | 1.24 | 9.4 – 15.1 | | Minnesota, USA 177 | 8883.0 |
| Dendroica kirtlandii | M | 81 | 13.7 | 0.86 | 12.4 – 15.8 | B | Michigan, USA 647 | 8884.0 |
| | F | 32 | 13.8 | 1.10 | 12.2 – 16.0 | | | |
| Dendroica discolor | M | 149 | 8.0 | 0.87 | 6.1 – 10.1 | F | Florida, USA 421 | 8885.0 |
| | F | 110 | 7.3 | 0.88 | 65.7 – 10.8 | | | |
| Dendroica vitellina | B | 27 | 6.7 | | 6.2 – 7.5 | | Cayman Is. 432 | 8886.0 |
| Dendroica palmarum | B | 176 | 10.3 | 0.47 | 7.0 – 12.9 | M | Pennsylvania, USA 101 | 8887.0 |
| Dendroica castanea | M | 16 | 13.1 | 0.47 | 11.6 – 15.1 | M | Pennsylvania, USA 101 | 8888.0 |
| | F | 14 | 12.0 | 0.27 | 10.7 – 13.6 | | | |
| Dendroica striata | B | 170 | 13.0 | 2.27 | 9.7 – 20.9 | M | Pennsylvania, USA 177 | 8889.0 |
| Dendroica cerulea | M | 16 | 9.5 | 0.40 | 8.6 – 10.2 | B | Pennsylvania, USA 177 | 8890.0 |
| | F | 11 | 9.2 | 0.45 | 8.4 – 10.3 | | | |
| Dendroica plumbea | U | | 11.0 | | | | 185 | 8891.0 |
| Dendroica pharetra | B | 10 | 10.3 | | | | Jamaica 127, 600 | 8892.0 |
| Dendroica angelae | U | 3 | 8.4 | | 7.8 – 8.7 | | Puerto Rico 125, 317 | 8893.0 |
| Catharopeza bishopi | U | 3 | 17.2 | | 16.0 – 19.0 | | St. Vincent 317 | 8894.0 |
| Mniotilta varia | M | 24 | 11.0 | 1.36 | 8.8 – 15.2 | | Pennsylvania, USA 101 | 8895.0 |
| | F | 46 | 10.6 | 0.16 | 9.0 – 12.7 | | | |
| Setophaga ruticilla | M | 143 | 8.5 | 0.56 | 7.0 – 12.0 | | Pennsylvania, USA 101 | 8896.0 |
| | F | 170 | 8.1 | 0.69 | 6.7 – 11.2 | | | |
| Protonotaria citrea | M | 18 | 15.0 | | 13.6 – 15.8 | B | Michigan, USA 646 | 8897.0 |
| | F | 61 | 17.4 | | 13.6 – 20.0 | | | |
| Protonotaria citrea | B | 95 | 14.3 | 0.14 | | B | Tennessee, USA 463 | 8897.0 |
| Helmitheros vermivorus | B | 37 | 13.0 | 0.89 | – 15.2 | M | Jamaica 153 | 8898.0 |
| Limnothlypis swainsonii | B | 19 | 18.9 | 1.70 | 14.3 – 20.4 | F | Florida, USA 377 | 8899.0 |

## Body Masses of World Birds (continued)

| Species | Sex | N | Mean | Std dev | Range | Sn | Location | Number |
|---|---|---|---|---|---|---|---|---|
| Seiurus aurocapillus | B | 181 | 19.4 | 1.22 | 14.0–28.8 | | Pennsylvania, USA 101 | 8900.0 |
| Seiurus noveboracensis | B | 289 | 17.8 | 2.90 | 13.8–24.4 | M | Pennsylvania, USA 101 | 8901.0 |
| Seiurus motacilla | M | 39 | 19.8 | 1.13 | 17.4–22.7 | B | Pennsylvania, USA 101 | 8902.0 |
| | F | 23 | 20.8 | 1.76 | 17.7–26.0 | | | |
| Oporornis formosus | M | 96 | 14.3 | 0.59 | 12.0–20.6 | | Pennsylvania, USA 101 | 8903.0 |
| | F | 43 | 13.7 | 0.59 | 11.4–16.5 | | | |
| Oporornis agilis | B | 134 | 15.2 | 2.94 | 10.7–26.8 | F | Pennsylvania, USA 177 | 8904.0 |
| Oporornis philadelphia | M | 140 | 13.0 | 0.79 | 9.6–17.9 | M | Pennsylvania, USA 101 | 8905.0 |
| | F | 89 | 12.0 | 0.90 | 10.0–14.7 | | | |
| Oporornis tolmiei | B | 26 | 10.4 | 1.25 | 8.6–12.6 | M | Arizona, USA 177 | 8906.0 |
| Geothlypis trichas | M | 965 | 10.3 | 0.66 | 7.6–15.5 | | Pennsylvania, USA 101 | 8907.0 |
| | F | 644 | 9.9 | 0.78 | 7.6–15.3 | | | |
| Geothlypis trichas occidentalis | M | 14 | 10.1 | | 9.5–10.7 | | western USA 32 | 8907.0 |
| Geothlypis trichas sirpicola | M | 11 | 9.2 | | 8.2–10.0 | | California, USA 32 | 8907.0 |
| Geothlypis beldingi | M | 28 | 15.7 | | 13.8–17.7 | | Mexico 32 | 8908.0 |
| Geothlypis flavovelata | M | 5 | 10.9 | | 10.2–11.5 | | Mexico 584 | 8909.0 |
| Geothlypis rostrata | B | 9 | 16.1 | | 15.1–17.3 | | Bahamas 600 | 8910.0 |
| Geothlypis semiflava | M | 1 | 17.0 | | | | Ecuador 589 | 8911.0 |
| Geothlypis speciosa | B | 2 | 10.8 | | 10.0–11.6 | | Mexico City 584 | 8912.0 |
| Geothlypis nelsoni | M | 4 | 11.1 | | 10.1–11.8 | | Mexico 584 | 8913.0 |
| Geothlypis aequinoctialis | B | 13 | 13.1 | 1.36 | 11.2–15.0 | | 38, 188, 384, 441 | 8914.0 |
| Geothlypis poliocephala | B | 13 | 14.6 | 1.04 | 13.2–16.2 | | 272, 457, 510 | 8915.0 |
| Microligea palustris | U | | 13.0 | | | | 185 | 8916.0 |
| Teretistris fernandinae | B | 6 | 10.8 | | 9.0–13.8 | | Cuba 429, 496 | 8917.0 |
| Teretistris fornsi | U | 4 | 10.6 | | 10.0–11.2 | | Cuba 429, 496 | 8918.0 |
| Wilsonia citrina | M | 89 | 10.8 | 0.38 | 8.1–13.9 | M | Pennsylvania, USA 101 | 8920.0 |
| | F | 115 | 10.1 | 0.51 | 8.2–12.5 | | | |

## Body Masses of World Birds (continued)

| Species | Sex | N | Mean | Std dev | Range | Sn | Location | Number |
|---|---|---|---|---|---|---|---|---|
| Wilsonia pusilla | B | 502 | 7.7 | 0.08 | 6.3–10.5 | M | Pennsylvania, USA 101 | 8921.0 |
| Wilsonia pusilla | B | 515 | 6.9 | 0.04 | 5.4–8.9 | S | California, USA 107 | 8921.0 |
| Wilsonia canadensis | M | 309 | 10.6 | 0.29 | 8.7–13.5 | M | Pennsylvania, USA 101 | 8922.0 |
|  | F | 263 | 10.2 | 0.36 | 9.1–12.6 |  |  |  |
| Cardellina rubrifrons | B | 5 | 9.8 |  | 8.2–11.2 | B | Arizona, USA 594 | 8923.0 |
| Ergaticus ruber | B | 5 | 8.1 |  | 7.6–8.7 |  | Mexico 456 | 8924.0 |
| Ergaticus versicolor | M | 1 | 10.0 |  |  |  | Guatemala 584 | 8925.0 |
| Myioborus pictus | B | 12 | 7.9 | 1.14 | 5.9–9.6 | S | Mexico; Arizona 594 | 8926.0 |
| Myioborus miniatus | B | 37 | 9.5 |  |  |  | Panama 244 | 8927.0 |
| Myioborus torquatus | B | 6 | 10.5 |  |  |  | Panama 244, 436 | 8933.0 |
| Myioborus melanocephalus | B | 3 | 11.2 |  |  |  | Peru 193 | 8934.0 |
| Euthlypis lachrymosa | B | 10 | 15.2 | 0.63 | 14.2–16.5 |  | Mexico 548, 584, 593 | 8938.0 |
| Basileuterus fraseri | B | 18 | 11.6 |  |  |  | Peru 672 | 8939.0 |
| Basileuterus bivittatus | B | 38 | 14.6 |  |  |  | Peru 193 | 8940.0 |
| Basileuterus chrysogaster | B | 21 | 11.1 |  |  |  | Peru 193 | 8941.0 |
| Basileuterus signatus | U | 3 | 12.6 |  | 12.1–13.4 |  | Peru 193 | 8942.0 |
| Basileuterus luteoviridis | U | 19 | 16.5 | 1.40 |  |  | Peru 668 | 8943.0 |
| Basileuterus nigrocristatus | B | 37 | 13.8 |  | 11.6–17.2 |  | Ecuador 320 | 8944.0 |
| Basileuterus coronatus | B | 12 | 16.1 |  | 14.4–17.2 |  | Ecuador 320 | 8949.0 |
| Basileuterus culicivorus | U | 22 | 10.5 |  | 9.5–12.0 |  | Trinidad 188 | 8950.0 |
| Basileuterus hypoleucus | U | 1 | 9.5 |  |  |  | Brazil 441 | 8952.0 |
| Basileuterus rufifrons | U | 17 | 10.9 |  |  |  | Panama; Belize 315, 510, 611 | 8953.0 |
| Basileuterus belli | U | 2 | 10.4 |  | 10.2–10.7 |  | Mexico 456 | 8954.0 |

## Body Masses of World Birds (continued)

| Species | Sex | N | Mean | Std dev | Range | Sn | Location | Number |
|---|---|---|---|---|---|---|---|---|
| Basileuterus melanogenys | B | 9 | 11.8 | | | | Panama 244 | 8955.0 |
| Basileuterus tristriatus | U | 22 | 12.7 | 12.90 | | | Peru 668 | 8957.0 |
| Basileuterus leucoblepharus | U | 7 | 16.3 | | 14.0–21.0 | | Brazil 441 | 8958.0 |
| Basileuterus flaveolus | U | 3 | 14.5 | | 14.0–15.0 | | Brazil 440 | 8960.0 |
| Basileuterus fulvicauda | B | 12 | 14.9 | | | | 244, 315 | 8961.0 |
| Basileuterus rivularis | U | 11 | 13.5 | | 11.5–16.5 | | Brazil 441 | 8962.0 |
| Zeledonia coronata | U | | 21.0 | | | | 603a | 8963.0 |
| Icteria virens | M | 248 | 25.5 | 1.43 | 20.3–31.7 | | Pennsylvania, USA 101 | 8964.0 |
| | F | 173 | 25.1 | 1.54 | 20.2–33.8 | | | |
| Granatellus venustus | B | 12 | 10.8 | | 10.2–11.4 | | Mexico 593 | 8965.0 |
| Granatellus sallaei | B | 9 | 9.9 | | 8.8–11.0 | | Mexico 457, 584 | 8966.0 |
| Granatellus pelzelni | F | 2 | 11.2 | | 11.0–11.5 | | Brazil 226, 610a | 8967.0 |

### ORDER: PASSERIFORMES  FAMILY: EMBERIZIDAE

| Species | Sex | N | Mean | Std dev | Range | Sn | Location | Number |
|---|---|---|---|---|---|---|---|---|
| Melophus lathami | B | 23 | 21.8 | | 18.0–25.0 | | India 5 | 8970.0 |
| Emberiza citrinella | U | 50 | 26.5 | | 24.0–31.0 | B | Britain 258 | 8972.0 |
| Emberiza leucocephalos | B | 3 | 27.6 | | 27.0–29.0 | | 5 | 8973.0 |
| Emberiza cirlus | U | 1 | 23.1 | | | | 227 | 8974.0 |
| Emberiza cia | U | 11 | 19.3 | | 18.0–21.0 | | India 5 | 8976.0 |
| Emberiza cioides | M | 1 | 22.0 | | | | 149 | 8978.0 |
| Emberiza buchanani | B | 18 | 20.5 | | 17.0–22.0 | F | India 5 | 8980.0 |
| Emberiza hortulana | B | 17 | 23.8 | | 17.0–36.0 | | 8, 228, 570 | 8982.0 . |
| Emberiza stewarti | M | 6 | 17.0 | | 16.0–18.0 | S | India 5 | 8983.0 |
| | F | 11 | 15.0 | | 13.0–17.0 | | | |
| Emberiza striolata | U | 3 | 75.0 | | 73.0–79.0 | | 78 | 8985.0 |

## Body Masses of World Birds (continued)

| Species | Sex | N | Mean | Std dev | Range | Sn | Location | Number |
|---|---|---|---|---|---|---|---|---|
| Emberiza impetuani | B | 26 | 15.1 | | 13.3–17.2 | | South Africa 256 | 8986.0 |
| Emberiza tahapisi | U | 1 | 12.8 | | | | Ghana 228 | 8987.0 |
| Emberiza capensis | B | 3 | 22.4 | | 21.0–23.6 | | South Africa 256 | 8989.0 |
| Emberiza tristrami | U | | | | 17.5–20.0 | | 149 | 8991.0 |
| Emberiza fucata | M | 6 | | | 18.0–21.0 | S | India | 8992.0 |
|  | F | 7 | | | 16.0–19.0 | | 5 | |
| Emberiza pusilla | B | 10 | 13.0 | | 11.0–14.0 | S | India 5 | 8993.0 |
| Emberiza chrysophrys | U | | 20.0 | | | | 78 | 8994.0 |
| Emberiza rustica | B | 3 | 23.2 | | 19.2–27.0 | | 84, 213 | 8995.0 |
| Emberiza aureola | M | 1 | 21.7 | | | S | India 5 | 8997.0 |
| Emberiza flaviventris | M | 7 | 21.3 | | 15.1–28.4 | | Zimbabwe | 8998.0 |
|  | F | 7 | 18.2 | | 16.1–21.2 | | 280, 281, 282 | |
| Emberiza poliopleura | U | 1 | 20.6 | | | | 521 | 8999.0 |
| Emberiza affinis | U | 4 | 15.4 | | 14.5–16.0 | | Ghana 228 | 9000.0 |
| Emberiza cabanisi | U | 4 | 24.0 | | 22.3–25.5 | | 228, 284 | 9001.0 |
| Emberiza rutila | U | 5 | 17.4 | | 16.7–18.9 | | 78 | 9002.0 |
| Emberiza melanocephala | B | 15 | 29.7 | | 27.0–35.0 | | India 5 | 9003.0 |
| Emberiza bruniceps | M | 11 | | | 22.0–27.0 | | India | 9004.0 |
|  | F | 7 | | | 20.0–25.0 | | 5 | |
| Emberiza spodocephala | M | 1 | 18.0 | | | | 5 | 9006.0 |
| Emberiza pallasi | M | 5 | 12.5 | | 11.8–14.0 | | | 9008.0 |
|  | F | | 11.0 | | | | 149 | |
| Emberiza schoeniclus | U | 50 | 18.3 | | 15.0–22.0 | B | Britain 258 | 9009.0 |
| Miliaria calandra | U | 37 | 46.0 | | 38.0–55.0 | Y | Britain 258 | 9010.0 |
| Calcarius mccownii | B | 5 | 23.2 | | | | 592 | 9011.0 |
| Calcarius lapponicus | B | 68 | 27.3 | | 23.5–32.5 | | Alaska, USA 305 | 9012.0 |

## Body Masses of World Birds (continued)

| Species | Sex | N | Mean | Std dev | Range | Sn | Location | Number |
|---|---|---|---|---|---|---|---|---|
| Calcarius pictus | M | 22 | 28.5 | | 25.5–31.8 | B | Alaska, USA 278 | 9013.0 |
| | F | 7 | 24.3 | | 22.0–26.0 | | | |
| Calcarius ornatus | B | 20 | 18.9 | 2.19 | 10.8–20.9 | W | Arizona, USA 594 | 9014.0 |
| Plectrophenax nivalis | B | 35 | 42.2 | 5.48 | 34.0–56.0 | W | Alaska, USA 177 | 9015.0 |
| Plectrophenax hyperboreus | B | 11 | 54.5 | 7.06 | 38.0–62.0 | W | Alaska, USA 177 | 9016.0 |
| Calamospiza melanocorys | B | 40 | 37.6 | 3.66 | 29.5–51.5 | | Arizona, USA 177 | 9017.0 |
| Passerella iliaca | B | 711 | 32.3 | | 21.7–42.1 | W | California, USA 344 | 9018.0 |
| Melospiza melodia | M | 238 | 21.0 | 1.17 | 18.2–29.8 | Y | Pennsylvania, USA 101 | 9019.0 |
| | F | 176 | 20.5 | 1.54 | 11.9–26.1 | | | |
| Melospiza melodia maxima | B | 4 | 42.4 | | 40.2–45.7 | | Alaska, USA 213 | 9019.0 |
| Melospiza melodia gouldii | U | 39 | 20.0 | 0.89 | | | California, USA 365 | 9019.0 |
| Melospiza mclodia maxillaris | U | 27 | 20.8 | 0.91 | | | California, USA 365 | 9019.0 |
| Melospiza melodia pusillula | U | 37 | 18.6 | 0.80 | | | California, USA 365 | 9019.0 |
| Melospiza melodia samuelis | U | 17 | 18.8 | 1.19 | | | California, USA 365 | 9019.0 |
| Melospiza lincolnii | B | 360 | 17.4 | 0.48 | 10.4–24.0 | M | Pennsylvania, USA 101 | 9020.0 |
| Melospiza georgiana | B | 991 | 17.0 | 0.96 | 10.9–22.2 | M | Pennsylvania, USA 101 | 9021.0 |
| Zonotrichia capensis | M | 15 | 21.0 | 0.89 | | | Panama 244 | 9022.0 |
| | F | 11 | 20.0 | 1.63 | | | | |
| Zonotrichia querula | M | 19 | 38.8 | | 36.8–41.7 | S | Kansas, USA 696 | 9023.0 |
| | F | 19 | 33.7 | | 31.4–36.3 | | | |
| Zonotrichia leucophrys leucophrys | B | 162 | 29.4 | 1.03 | 21.6–38.5 | M | Pennsylvania, USA 101 | |
| Zonotrichia leucophrys gambelii | B | 50 | 25.5 | 1.69 | 21.0–28.5 | W | California, USA 177 | 90 |
| Zonotrichia leucophrys nuttalli | B | 50 | 32.0 | 2.18 | 27.0–35.5 | W | California, USA 177 | 90 |
| Zonotrichia leucophrys pugetensis | B | 50 | 25.3 | 1.78 | 21.4–29.1 | W | California, USA 177 | |
| Zonotrichia leucophrys oriantha | B | 50 | 28.4 | 1.91 | 23.3–33.7 | B | Oregon, USA 177 | 9024.0 |
| Zonotrichia albicollis | B | 1884 | 25.9 | 2.18 | 19.0–35.4 | | Pennsylvania, USA 101 | 9025.0 |

## Body Masses of World Birds (continued)

| Species | Sex | N | Mean | Std dev | Range | Sn | Location | Number |
|---|---|---|---|---|---|---|---|---|
| Zonotrichia atricapilla | B | 1422 | 29.8 | | 21.2–42.2 | W | California, USA 344 | 9026.0 |
| Junco vulcani | U | | 28.0 | | | | 603a | 9027.0 |
| Junco hyemalis hyemalis | M | 2819 | 20.4 | 1.21 | 14.3–26.7 | | Pennsylvania, USA 101 | 9028.0 |
| | F | 1316 | 18.8 | 0.78 | 14.3–25.1 | | | |
| Junco hyemalis mearnsi | B | 221 | 18.2 | 1.30 | 15.5–23.5 | W | Arizona, USA 177 | 9028.0 |
| Junco hyemalis dorsalis | B | 170 | 21.8 | 1.40 | 18.0–26.0 | W | Arizona, USA 177 | 9028.0 |
| Junco hyemalis caniceps | B | 40 | 19.6 | 1.10 | 18.0–23.0 | W | Arizona, USA 177 | 9028.0 |
| Junco phaeonotus | B | 57 | 20.4 | 1.06 | 18.3–22.0 | | Arizona, USA 177 | 9030.0 |
| Passerculus sandwichensis | M | 71 | 20.6 | 1.35 | | B | Manitoba, Canada 662 | 9031.0 |
| | F | 35 | 19.5 | 2.29 | | | | |
| Passerculus sandwichensis | F | 10 | 17.0 | 0.83 | 15.1–17.8 | B | California, USA 677a | 9031.0 |
| Passerculus sandwichensis princeps | B | 103 | 26.0 | | | | Sable Is., Canada 662 | 9031.0 |
| Ammodramus maritimus | M | 14 | 24.2 | | 21.9–27.4 | B | New Jersey, USA 697 | 9032.0 |
| | F | 3 | 22.3 | | 19.8–24.4 | | | |
| Ammodramus caudacutus | M | 33 | 20.7 | | 18.0–23.1 | B | New Jersey, USA 697 | 9033.0 |
| | F | 14 | 17.8 | | 15.3–19.0 | | | |
| Ammodramus leconteii | M | 26 | 13.4 | 0.64 | 12.4–15.2 | B | N. Dakota, USA 405 | 9034.0 |
| Ammodramus henslowii | B | 18 | 13.1 | | 11.1–14.9 | B | 275 | 9035.0 |
| Ammodramus bairdii | B | 21 | 17.5 | 1.34 | 15.0–20.3 | S | Arizona, USA 594 | 9036.0 |
| Ammodramus savannarum | B | 60 | 17.0 | 2.75 | 13.4–28.4 | W | Arizona, USA 594 | 9037.0 |
| Ammodramus humeralis | B | 12 | 16.8 | | 14.0–19.0 | | 38, 245, 246, 384 | 9038.0 |
| Ammodramus aurifrons | U | 1 | 14.5 | | | | Brazil 45 | 9039.0 |
| Xenospiza baileyi | M | 1 | 17.4 | | | | Mexico 584 | 9040.0 |
| Spizella arborea | B | 1785 | 20.1 | 1.59 | | | 251 | 9041.0 |
| Spizella passerina | B | 934 | 12.3 | 0.84 | 9.8–18.8 | | Pennsylvania, USA 101 | 9042.0 |
| Spizella pallida | B | 18 | 12.0 | | 9.8–14.5 | B | northern USA 41 | 9043.0 |

## Body Masses of World Birds (continued)

| Species | Sex | N | Mean | Std dev | Range | Sn | Location | Number |
|---|---|---|---|---|---|---|---|---|
| Spizella breweri | B | 83 | 10.9 | 0.70 | | B | Idaho, USA 462 | 9045.0 |
| Spizella pusilla | B | 635 | 12.5 | 1.47 | 10.2–16.5 | Y | Pennsylvania, USA 101 | 9046.0 |
| Spizella atrogularis | B | 7 | 11.9 | | 10.8–13.1 | | Mexico 586 | 9048.0 |
| Pooecetes gramineus | M | 28 | 26.5 | | | S | Michigan, USA 142 | 9049.0 |
| | F | 15 | 24.9 | | | | | |
| Chondestes grammacus | B | 49 | 29.0 | 1.94 | 24.7–33.3 | | California, USA 177 | 9050.0 |
| Amphispiza bilineata | B | 89 | 13.5 | 1.15 | 10.2–16.4 | W | Arizona, USA 177 | 9051.0 |
| Amphispiza belli | B | 166 | 19.3 | 1.20 | | B | Idaho, USA 462 | 9052.0 |
| Amphispiza quinquestriata | M | 23 | 19.3 | 1.10 | 17.1–21.7 | | | 9053.0 |
| | F | 4 | 18.8 | | 17.9–19.5 | | 689 | |
| Aimophila humeralis | B | 7 | 24.1 | | 21.9–26.0 | | Mexico 46, 584 | 9055.0 |
| Aimophila ruficauda | M | 5 | 28.2 | | 26.6–30.0 | | Mexico 548, 593 | 9056.0 |
| Aimophila sumichrasti | M | 1 | 30.3 | | | | Oaxaca, Mexico 584 | 9057.0 |
| | F | 1 | 28.6 | | | | | |
| Aimophila aestivalis | M | 12 | 20.2 | 1.42 | 18.4–22.6 | | | 9060.0 |
| | | | 19.1 | | | | 689 | |
| Aimophila botterii | B | 52 | 19.9 | 1.86 | 15.7–25.5 | B | Arizona, USA 177 | 9061.0 |
| Aimophila cassinii | B | 92 | 18.9 | 1.51 | 14.0–23.5 | Y | Arizona, USA 177 | 9062.0 |
| Aimophila carpalis | B | 34 | 15.3 | 0.97 | 13.3–17.4 | | Arizona, USA 14 | 9063.0 |
| Aimophila ruficeps | M | 59 | 19.3 | 1.48 | 16.0–23.3 | | | 9064.0 |
| | F | 39 | 18.1 | 1.31 | 15.2–20.3 | | 689 | |
| Aimophila notosticta | M | 1 | 27.1 | | | | Oaxaca, Mexico 584 | 9065.0 |
| Aimophila rufescens | B | 16 | 34.3 | | 28.2–38.9 | | Nicaragua 272 | 9066.0 |
| Torreornis inexpectata | B | 14 | 26.7 | 1.43 | 24.0–29.0 | | Cuba 429, 496 | 9067.0 |
| Oriturus superciliosus | B | 13 | 41.5 | | 36.9–53.7 | | Mexico 46, 139, 456 | 9068.0 |
| Pipilo chlorurus | B | 68 | 29.4 | 3.27 | 21.5–37.3 | | Arizona, USA 177 | 9069.0 |
| Pipilo ocai | M | 4 | 64.5 | | 61.0–68.0 | | Oaxaca, Mexico 406 | 9070.0 |
| | F | 5 | 57.6 | | 54.5–62.5 | | | |

## Body Masses of World Birds (continued)

| Species | Sex | N | Mean | Std dev | Range | Sn | Location | Number |
|---|---|---|---|---|---|---|---|---|
| Pipilo erythrophthalmus | M | 205 | 41.7 | 1.95 | 32.1–50.0 | | Pennsylvania, USA | 9071.0 |
| | F | 128 | 39.3 | 1.70 | 32.1–52.3 | | 101 | |
| Pipilo socorroensis | F | 1 | 28.1 | | | | Mexico | 9072.0 |
| | | | | | | | 593 | |
| Pipilo crissalis | M | 43 | 53.9 | | 48.6–61.2 | | California, USA | 9073.0 |
| | F | 29 | 51.8 | | 46.3–61.2 | | 137 | |
| Pipilo fuscus | B | 98 | 44.4 | 3.19 | 36.6–52.5 | Y | Arizona, USA | 9074.0 |
| | | | | | | | 177 | |
| Pipilo aberti | M | 37 | 47.1 | 3.29 | 40.0–54.1 | B | Arizona, USA | 9075.0 |
| | F | 25 | 44.8 | 2.90 | 39.5–51.0 | | 177 | |
| Pipilo albicollis | B | 8 | 46.5 | | 42.0–50.0 | | Mexico | 9076.0 |
| | | | | | | | 137 | |
| Melozone kieneri | B | 11 | 38.7 | | 35.0–41.0 | | Mexico | 9077.0 |
| | | | | | | | 548 | |
| Melozone biarcuatum | U | | 28.0 | | | | | 9078.0 |
| | | | | | | | 603a | |
| Melozone leucotis | U | 2 | 42.8 | | | | Costa Rica | 9079.0 |
| | | | | | | | 315 | |
| Arremon aurantiirostris | U | 40 | 34.5 | 2.30 | | | Costa Rica | 9080.0 |
| | | | | | | | 315 | |
| Arremon schlegeli | U | 4 | 26.7 | | 25.1–28.0 | | Venezuela | 9081.0 |
| | | | | | | | 626 | |
| Arremon taciturnus | U | 42 | 24.8 | | 22.0–27.8 | | Brazil | 9082.0 |
| | | | | | | | 437 | |
| Arremon abeillei | B | 10 | 25.9 | | | | Peru | 9083.0 |
| | | | | | | | 672 | |
| Arremon flavirostris | B | 9 | 28.5 | | 27.5–30.3 | | | 9084.0 |
| | | | | | | | 111, 441 | |
| Arremonops rufivirgatus | B | 14 | 23.6 | 2.39 | 18.8–29.5 | Y | | 9085.0 |
| | | | | | | | 177 | |
| Arremonops chloronototus | B | 11 | 27.3 | 2.30 | 24.2–31.2 | | Belize | 9087.0 |
| | | | | | | | 510 | |
| Arremonops conirostris | M | 7 | 42.3 | | | | Panama | 9088.0 |
| | F | 10 | 37.2 | 2.18 | | | 244 | |
| Atlapetes albinucha | B | 4 | 32.4 | | 31.0–35.0 | | Mexico | 9089.0 |
| | | | | | | | 584 | |
| Atlapetes gutturalis | U | 10 | 34.6 | 1.90 | 31.7–37.2 | | Panama | 9090.0 |
| | | | | | | | 337 | |
| Atlapetes pallidinucha | B | 12 | 35.5 | | 31.0–40.0 | | Peru | 9091.0 |
| | | | | | | | 451 | |
| Atlapetes rufinucha | U | 1 | 25.7 | | | | Peru | 9092.0 |
| | | | | | | | 668 | |
| Atlapetes pileatus | U | 12 | 24.0 | 2.30 | 21.5–27.5 | | Oaxaca, Mexico | 9094.0 |
| | | | | | | | 406 | |

## Body Masses of World Birds (continued)

| Species | Sex | N | Mean | Std dev | Range | Sn | Location | Number |
|---------|-----|---|------|---------|-------|----|----------|--------|
| Atlapetes tricolor | U | 1 | 29.5 | | | | Peru 668 | 9098.0 |
| Atlapetes leucopterus | B | 7 | 19.3 | | | | Peru 672 | 9103.0 |
| Atlapetes brunneinucha | U | 14 | 46.6 | 3.29 | | | Peru 668 | 9111.0 |
| Atlapetes virenticeps | M | 1 | 41.0 | | | | 487 | 9112.0 |
| Atlapetes atricapillus | U | 9 | 44.6 | | 40.0–49.0 | | Panama 50 | 9113.0 |
| Atlapetes torquatus | B | 12 | 40.9 | | | | Panama; Peru 244, 668, 672 | 9114.0 |
| Pezopetes capitalis | B | 2 | 55.8 | | 55.8–55.9 | | Panama 244 | 9115.0 |
| Pselliophorus tibialis | U | 10 | 30.0 | 2.10 | 26.8–34.8 | | Panama 337 | 9116.0 |
| Lysurus crassirostris | M | 1 | 40.5 | | | | East Ecuador 589 | 9118.0 |
| | F | 1 | 37.0 | | | | | |
| Lysurus castaneiceps | M | 1 | 37.2 | | | | Peru 584 | 9119.0 |
| Gubernatrix cristata | M | 1 | 46.0 | | | | Brazil 38 | 9120.0 |
| Paroaria coronata | M | 2 | 43.0 | | 42.0–44.0 | | Brazil 38 | 9121.0 |
| Paroaria dominicana | U | | 33.2 | | | | Brazil 522 | 9122.0 |
| Paroaria gularis | U | 7 | 22.0 | | 20.8–23.8 | | Venezuela 625 | 9123.0 |
| Paroaria capitata | U | 3 | 22.4 | | 22.0–23.2 | | Argentina 111 | 9125.0 |
| Coereba flaveola | B | 136 | 9.4 | 0.80 | 7.4–12.5 | W | Puerto Rico 186 | 9126.0 |
| Conirostrum speciosum | U | | 8.4 | | | | Brazil 564 | 9127.0 |
| Conirostrum leucogenys | B | 4 | 7.0 | | 7.0–7.0 | | 384, 497 | 9128.0 |
| Conirostrum bicolor | U | 2 | 10.5 | | 10.5–10.5 | | Trinidad 188 | 9129.0 |
| Conirostrum cinereum | B | 2 | 9.5 | | 9.0–10.0 | | Peru 668 | 9131.0 |
| Conirostrum ferrugineiventre | U | 3 | 11.5 | | 11.0–12.0 | | Peru 537 | 9133.0 |
| Conirostrum sitticolor | U | 2 | 10.4 | | | | Peru 668 | 9135.0 |

## Body Masses of World Birds (continued)

| Species | Sex | N | Mean | Std dev | Range | Sn | Location | Number |
|---|---|---|---|---|---|---|---|---|
| Conirostrum albifrons | U | 1 | 10.0 | | | | Peru 668 | 9136.0 |
| Oreomanes fraseri | U | 25 | 25.0 | | 22.0–27.0 | | 279 | 9137.0 |
| Orchesticus abeillei | U | 1 | 31.5 | | | | Brazil 279 | 9138.0 |
| Schistoclamys ruficapillus | U | 3 | 29.0 | | 26.3–38.2 | | 279 | 9139.0 |
| Schistochlamys melanopis | U | 22 | 33.0 | | 29.0–40.0 | | 279 | 9140.0 |
| Neothraupis fasciata | U | 3 | 30.0 | | 29.0–32.0 | | 279 | 9141.0 |
| Cypsnagra hirundinacea | U | 7 | 29.0 | | 25.0–34.0 | | 279 | 9142.0 |
| Conothraupis speculigera | U | 9 | 25.0 | | 23.0–28.0 | | 279 | 9143.0 |
| Lamprospiza melanoleuca | U | 7 | 34.0 | | 31.0–42.0 | | 279 | 9145.0 |
| Cissopis leveriana | U | 10 | 76.0 | | 73.8–86.0 | | 279 | 9146.0 |
| Chlorornis riefferii | U | 37 | 53.0 | | 42.0–59.0 | | 279 | 9147.0 |
| Compsothraupis loricata | M | 1 | 72.5 | | | | 279 | 9148.0 |
| Sericossypha albocristata | U | 17 | 114.0 | | 95.0–125.0 | | 279 | 9149.0 |
| Nesospingus speculiferus | B | 18 | 35.9 | 2.42 | 31.0–39.7 | | Puerto Rico 431 | 9150.0 |
| Chlorospingus ophthalmicus bolivianus | U | 27 | 16.0 | | 13.3–18.0 | | 279 | 9151.0 |
| Chlorospingus ophthalmicus hiaticolus | U | 28 | 22.0 | | 16.0–22.5 | | 279 | 9151.0 |
| Chlorospingus tacarcunae | M | 13 | 18.9 | | 18.2–20.0 | | Panama 50 | 9152.0 |
| Chlorospingus inornatus | U | 14 | 28.0 | | 20.2–36.0 | | 279 | 9153.0 |
| Chlorospingus semifuscus | U | 6 | 19.0 | | 17.2–23.0 | | 279 | 9154.0 |
| Chlorospingus pileatus | U | 50 | 21.0 | | 16.0–24.1 | | 279 | 9155.0 |
| Chlorospingus zeledoni | M | 1 | 20.3 | | | | Costa Rica 279 | 9155.1 |
| Chlorospingus parvirostris | U | 15 | 24.0 | | 17.5–28.5 | | 279 | 9156.0 |

**Body Masses of World Birds (continued)**

| Species | Sex | N | Mean | Std dev | Range | Sn | Location | Number |
|---|---|---|---|---|---|---|---|---|
| Chlorospingus flavigularis | U | 8 | 28.0 | | 25.0–30.5 | | 279 | 9157.0 |
| Chlorospingus flavovirens | U | 1 | 25.0 | | | | 279 | 9158.0 |
| Chlorospingus canigularis | U | 12 | 18.0 | | 14.5–20.8 | | 279 | 9159.0 |
| Cnemoscopus rubrirostris | U | 32 | 18.0 | | 13.0–22.5 | | 279 | 9160.0 |
| Hemispingus atropileus | U | 60 | 22.0 | | 18.0–26.0 | | 279 | 9161.0 |
| Hemispingus calophrys | U | 34 | 17.0 | | 14.5–20.5 | | 279 | 9162.0 |
| Hemispingus superciliaris | U | 41 | 14.0 | | 11.4–17.2 | | 279 | 9164.0 |
| Hemispingus frontalis | U | 53 | 17.0 | | 14.0–20.0 | | 279 | 9166.0 |
| Hemispingus melanotis | U | 52 | 16.0 | | 13.0–21.8 | | 279 | 9167.0 |
| Hemispingus rufosuperciliaris | U | 21 | 29.0 | | 26.0–33.0 | | 279 | 9169 0 |
| Hemispingus verticalis | M | 4 | 13.3 | | 12.0–14.0 | | | 9170.0 |
| | F | 2 | 12.8 | | 12.5–13.0 | 451 | | |
| Hemispingus xanthophthalmus | U | 30 | 12.0 | | 10.3–15.0 | | 279 | 9171.0 |
| Hemispingus trifasciatus | U | 31 | 14.0 | | 12.0–16.0 | | 279 | 9172.0 |
| Pyrrhocoma ruficeps | U | 3 | 15.0 | | | | 279 | 9173.0 |
| Thlypopsis ornata | U | 23 | 12.0 | | 9.8–14.9 | | 279 | 9175.0 |
| Thlypopsis pectoralis | U | 8 | 15.0 | | 14.0–17.0 | | 279 | 9176.0 |
| Thlypopsis sordida | U | 29 | 17.0 | | 14.0–19.0 | | 279 | 9177.0 |
| Thlypopsis inornata | U | 8 | 15.0 | | 14.0–17.0 | | 279 | 9178.0 |
| Thlypopsis ruficeps | U | 16 | 11.0 | | 9.6–13.1 | | 279 | 9179.0 |
| Hemithraupis guira | U | 22 | 12.0 | | 9.5–14.0 | | 279 | 9180.0 |
| Hemithraupis ruficapilla | U | 2 | 12.0 | | 11.0–13.0 | | 279, 440 | 9181.0 |
| Hemithraupis flavicollis | U | 7 | 13.0 | | 11.0–15.0 | | 279 | 9182.0 |

## Body Masses of World Birds (continued)

| Species | Sex | N | Mean | Std dev | Range | Sn | Location | Number |
|---|---|---|---|---|---|---|---|---|
| Chrysothlypis chrysomelas | U | 6 | 13.0 | | 11.0–15.0 | | 279 | 9183.0 |
| Chrysothlypis salmoni | M | 1 | 12.0 | | | | 279 | 9184.0 |
| Nemosia pileata | U | 25 | 16.0 | | 12.0–20.7 | | 279 | 9185.0 |
| Phaenicophilus palmarum | M | 2 | 32.0 | | 32.0–32.0 | | Dominica | 9187.0 |
| | F | 4 | 27.2 | | 24.0–32.0 | | 600 | |
| Phaenicophilus poliocephalus | U | | 31.0 | | | | 185 | 9188.0 |
| Calyptophilus frugivorus | U | | 30.0 | | | | 185 | 9189.0 |
| Rhodinocichla rosea | U | 14 | 48.0 | | 43.0–51.8 | | 279 | 9190.0 |
| Mitrospingus cassinii | U | 12 | 40.4 | 2.90 | | | Costa Rica 315 | 9191.0 |
| Chlorothraupis carmioli | U | 23 | 39.0 | 2.48 | 34.5–43.4 | | Panama 50 | 9193.0 |
| Chlorothraupis olivacea | U | 4 | 39.0 | | 36.0–41.0 | | 279 | 9194.0 |
| Chlorothraupis stolzmanni | F | 1 | 40.3 | | | | 279 | 9195.0 |
| Eucometis penicillata | U | 52 | 27.0 | | 22.5–35.0 | | 279 | 9197.0 |
| Lanio fulvus | U | 20 | 24.0 | | 19.0–30.0 | | 279 | 9198.0 |
| Lanio versicolor | U | 18 | 17.0 | | 13.0–20.0 | | 279 | 9199.0 |
| Lanio aurantius | U | 16 | 35.0 | | 29.8–45.0 | | 279 | 9200.0 |
| Lanio leucothorax | U | | 40.0 | | | | 603a | 9201.0 |
| Creurgops verticalis | U | 6 | 24.0 | | 21.0–27.0 | | 279 | 9202.0 |
| Creurgops dentata | U | 6 | 19.0 | | 16.0–19.7 | | 279 | 9203.0 |
| Heterospingus rubrifrons | U | 2 | 38.0 | | 36.0–40.0 | | 279 | 9204.0 |
| Tachyphonus cristatus | U | 9 | 19.0 | | 16.6–23.0 | | 279 | 9206.0 |
| Tachyphonus rufiventer | U | 9 | 19.0 | | 15.5–19.8 | | 279 | 9207.0 |
| Tachyphonus surinamus | U | 37 | 19.0 | | 15.0–23.0 | | 279 | 9208.0 |

## Body Masses of World Birds (continued)

| Species | Sex | N | Mean | Std dev | Range | Sn | Location | Number |
|---|---|---|---|---|---|---|---|---|
| Tachyphonus luctuosus | U | 27 | 13.0 | | 11.5–15.0 | | 279 | 9209.0 |
| Tachyphonus delatrii | U | 11 | 20.0 | | 16.2–23.4 | | Panama 50 | 9210.0 |
| Tachyphonus coronatus | U | 33 | 29.3 | | 26.0–33.1 | | Brazil 441 | 9211.0 |
| Tachyphonus rufus | U | 172 | 34.4 | | 25.7–42.5 | | 279 | 9212.0 |
| Tachyphonus phoenicius | U | 43 | 21.0 | | 17.0–25.0 | | 279 | 9213.0 |
| Trichothraupis melanops | U | 35 | 24.3 | | 21.7–29.5 | | Brazil 441 | 9214.0 |
| Habia rubica | M | 79 | 34.0 | | 27.7–42.9 | | | 9215.0 |
| | F | 47 | 31.0 | | 22.5–37.0 | | 279 | |
| Habia fuscicauda | M | 86 | 41.0 | | 32.9–46.5 | | | 9216.0 |
| | F | 62 | 36.0 | | 29.0–44.0 | | 279 | |
| Habia gutturalis | M | 9 | 37.4 | | 32.9–41.6 | | Yucatan, Mexico | 9217.0 |
| | F | 5 | 33.4 | | 30.8–37.7 | | 457 | |
| Habia atrimaxillaris | M | 1 | 48.9 | | | | Costa Rica 279 | 9218.0 |
| Piranga bidentata bidentata | B | 9 | 34.7 | | 33.3–39.4 | | Mexico 593 | 9220.0 |
| Piranga bidentata flammea | B | 18 | 37.5 | 4.09 | 32.6–48.4 | | Mexico 593 | 9220.0 |
| Piranga flava | B | 36 | 38.0 | 4.16 | 23.2–47.4 | B | Mexico 584 | 9221.0 |
| Piranga rubra | B | 30 | 28.2 | 3.18 | | S | Louisiana, USA 503 | 9222.0 |
| Piranga roseogularis | U | 21 | 24.0 | | 20.7–29.9 | | 279 | 9223.0 |
| Piranga olivacea | U | 218 | 28.6 | 0.22 | 17.5–35.2 | | Pennsylvania, USA 101 | 9224.0 |
| Piranga ludoviciana | B | 90 | 28.1 | | 22.5–34.5 | S | California, USA 557 | 9225.0 |
| Piranga leucoptera | U | 58 | 16.0 | | 13.3–20.0 | | 279 | 9226.0 |
| Piranga erythrocephala | U | 3 | 21.8 | | 19.9–24.5 | | 279, 593 | 9227.0 |
| Piranga rubriceps | U | 7 | 35.0 | | 28.0–40.0 | | 279 | 9228.0 |
| Calochaetes coccineus | U | 4 | 46.0 | | 42.0–49.0 | | Peru 450 | 9229.0 |
| Ramphocelus sanguinolentus | U | 27 | 41.0 | | 34.7–48.1 | | 279 | 9230.0 |

## Body Masses of World Birds (continued)

| Species | Sex | N | Mean | Std dev | Range | Sn | Location | Number |
|---|---|---|---|---|---|---|---|---|
| Ramphocelus nigrogularis | U | 10 | 31.0 | | 27.0–36.0 | | 279 | 9231.0 |
| Ramphocelus dimidiatus | U | 34 | 28.0 | | 23.8–34.0 | | 279 | 9232.0 |
| Ramphocelus melanogaster | M | 1 | 25.0 | | | | Peru 279 | 9233.0 |
| Ramphocelus carbo | U | 261 | 28.0 | | 23.5–37.5 | | 279 | 9234.0 |
| Ramphocelus carbo atrosericeus | U | 21 | 24.0 | | 21.5–27.0 | | 279 | 9234.0 |
| Ramphocelus bresilius | U | 8 | 32.9 | | 27.9–35.5 | | Brazil 441 | 9235.0 |
| Ramphocelus passerinii | U | 37 | 32.0 | | 25.5–37.0 | | 279 | 9236.0 |
| Ramphocelus flammigerus | M | 13 | 33.0 | | 29.6–35.6 | | 279 | 9237.0 |
| Spindalis zena zena | B | 14 | 21.1 | | 17.0–24.5 | | Bahamas 600 | 9238.0 |
| Spindalis zena benedicti | U | 12 | 30.0 | | 26.9–35.2 | | Mexico 279 | 9238.0 |
| Spindalis zena dominicensis | U | 5 | 31.0 | | 29.1–33.2 | | 279 | 9238.0 |
| Spindalis zena nigricephala | U | 16 | 43.0 | | 42.1–47.2 | | Jamaica 279 | 9238.0 |
| Thraupis episcopus | U | 181 | 35.0 | | 27.0–45.0 | | 279 | 9239.0 |
| Thraupis glaucocolpa | U | 3 | 33.7 | | 31.0–36.5 | | Venezuela 626 | 9240.0 |
| Thraupis sayaca | U | 27 | 32.0 | | 27.9–34.4 | | 279 | 9241.0 |
| Thraupis cyanoptera | U | 5 | 44.0 | | 41.0–46.0 | | 279 | 9242.0 |
| Thraupis ornata | F | 1 | 33.0 | | | | Brazil 610a | 9243.0 |
| Thraupis abbas | U | 42 | 45.0 | | 38.0–55.0 | | 279 | 9244.0 |
| Thraupis palmarum | U | 82 | 39.0 | | 27.0–48.0 | | 279 | 9245.0 |
| Thraupis cyanocephala | U | 53 | 36.0 | | 27.0–47.0 | | 279 | 9246.0 |
| Thraupis bonariensis | U | 41 | 36.0 | | 28.2–46.5 | | 279 | 9247.0 |
| Cyanicterus cyanicterus | M | 1 | 34.0 | | | | 279 | 9248.0 |

## Body Masses of World Birds (continued)

| Species | Sex | N | Mean | Std dev | Range | Sn | Location | Number |
|---|---|---|---|---|---|---|---|---|
| Bangsia arcaei | U | 2 | 37.2 | | 34.6–39.9 | | | 9249.0 |
| | | | | | | | 50, 50a | |
| Buthraupis montana | U | 35 | 96.0 | | 72.0–116.0 | | | 9254.0 |
| | | | | | | | 279 | |
| Buthraupis montana montana | U | 17 | 79.0 | | 69.0–91.0 | | | 9254.0 |
| | | | | | | | 279 | |
| Buthraupis eximia | U | 14 | 63.0 | | 50.0–70.0 | | | 9255.0 |
| | | | | | | | 279 | |
| Buthraupis aureodorsalis | U | 10 | 85.0 | | 75.0–94.0 | | | 9256.0 |
| | | | | | | | 279 | |
| Buthraupis wetmorei | U | 2 | 62.5 | | 62.0–63.0 | | | 9257.0 |
| | | | | | | | 279 | |
| Wetmorethraupis sterrhopteron | U | 3 | 55.0 | | 54.0–56.0 | | | 9258.0 |
| | | | | | | | 279 | |
| Anisognathus melanogenys | M | 1 | 41.0 | | | | Columbia | 9259.0 |
| | | | | | | | 279 | |
| Anisognathus lacrymosus | U | 80 | 31.0 | | 25.0–38.0 | | | 9260.0 |
| | | | | | | | 279 | |
| Anisognathus igniventris | U | 98 | 34.0 | | 23.7–41.2 | | | 9261.0 |
| | | | | | | | 279 | |
| Anisognathus flavinucha | U | 53 | 42.0 | | 33.0–56.0 | | | 9262.0 |
| | | | | | | | 279 | |
| Stephanophorus diadematus | U | 4 | 41.6 | | 39.9–42.3 | | Brazil | 9264.0 |
| | | | | | | | 441 | |
| Iridosornis analis | U | 63 | 26.0 | | 20.0–29.0 | | | 9266.0 |
| | | | | | | | 279 | |
| Iridosornis jelskii | U | 45 | 20.0 | | 16.0–26.0 | | | 9267.0 |
| | | | | | | | 279 | |
| Iridosornis rufivertex | U | 14 | 23.0 | | 18.0–27.9 | | | 9268.0 |
| | | | | | | | 279 | |
| Iridosornis reinhardti | U | 73 | 24.0 | | 19.5–28.0 | | | 9269.0 |
| | | | | | | | 279 | |
| Dubusia taeniata | U | 30 | 37.0 | | 31.0–45.0 | | | 9270.0 |
| | | | | | | | 279 | |
| Delothraupis castaneoventris | U | 27 | 28.0 | | 24.6–33.0 24.6–33.0 | | | 9271.0 |
| | | | | | | | 279 | |
| Pipraeidea melanonota | U | 14 | 21.0 | | 18.0–25.2 | | | 9272.0 |
| | | | | | | | 279 | |
| Euphonia jamaica | U | 3 | 17.0 | | | | Jamaica | 9273.0 |
| | | | | | | | 125a | |
| Euphonia plumbea | U | 4 | 8.9 | | 8.7–9.5 | | | 9274.0 |
| | | | | | | | 279, 610a | |
| Euphonia affinis | U | 28 | 10.0 | | 8.5–12.8 | | | 9275.0 |
| | | | | | | | 279 | |

## Body Masses of World Birds (continued)

| Species | Sex | N | Mean | Std dev | Range | Sn | Location | Number |
|---------|-----|---|------|---------|-------|-----|----------|--------|
| Euphonia luteicapilla | U | 5 | 13.0 | | 11.4–14.5 | | 279 | 9276.0 |
| Euphonia chlorotica | U | 16 | 11.0 | | 8.0–14.3 | | 279 | 9277.0 |
| Euphonia trinitatis | U | 4 | 11.0 | | 8.8–14.0 | | 279 | 9278.0 |
| Euphonia concinna | U | 3 | 10.3 | | 9.0–12.0 | | Columbia 384 | 9279.0 |
| Euphonia finschi | U | | | | 10.0–11.0 | | 279 | 9281.0 |
| Euphonia violacea | U | 87 | 15.0 | | 12.5–17.0 | | 279 | 9282.0 |
| Euphonia laniirostris | U | 28 | 15.0 | | 13.0–16.5 | | 279 | 9283.0 |
| Euphonia hirundinacea | U | 50 | 14.0 | | 11.6–17.8 | | 279 | 9284.0 |
| Euphonia chalybea | U | 4 | 19.0 | | 18.0–20.0 | | 279 | 9285.0 |
| Euphonia elegantissima | U | 28 | 15.0 | | 13.1–17.0 | | 279 | 9286.0 |
| Euphonia musica | U | 6 | 13.0 | | 12.4–14.4 | | 279 | 9287.0 |
| Euphonia imitans | U | 2 | 14.0 | | | | 603a | 9289.0 |
| Euphonia fulvicrissa | U | 9 | 11.0 | | 10.1–13.0 | | 279 | 9290.0 |
| Euphonia gouldi | U | 20 | 14.0 | | 10.9–16.0 | | 279 | 9291.0 |
| Euphonia chrysopasta | U | 11 | 14.0 | | 11.0–16.2 | | 279 | 9292.0 |
| Euphonia mesochrysa | U | 9 | 13.0 | | 12.0–15.0 | | 279 | 9293.0 |
| Euphonia minuta | U | 13 | 10.0 | | 7.9–11.5 | | 279 | 9294.0 |
| Euphonia anneae | B | 3 | 14.9 | | 14.4–15.4 | | Panama 50, 611 | 9295.0 |
| Euphonia xanthogaster | U | 107 | 13.0 | | 9.0–16.0 | | 279 | 9296.0 |
| Euphonia rufiventris | U | 10 | 14.0 | | 13.0–18.0 | | 279 | 9297.0 |
| Euphonia cayennensis | U | 3 | 14.2 | | 11.8–16.0 | | 279, 437 | 9298.0 |
| Euphonia pectoralis | B | 10 | 14.4 | | | | 38, 200, 622 | 9299.0 |

## Body Masses of World Birds (continued)

| Species | Sex | N | Mean | Std dev | Range | Sn | Location | Number |
|---|---|---|---|---|---|---|---|---|
| Chlorophonia flavirostris | M | 1 | 11.0 | | | | 279 | 9300.0 |
| Chlorophonia cyanea | U | 13 | 14.0 | | 11.0–15.0 | | 279 | 9301.0 |
| Chlorophonia pyrrhophrys | U | 5 | 17.0 | | 16.0–18.0 | | 279 | 9302.0 |
| Chlorophonia occipitalis | B | 9 | 25.8 | | 23.0–28.3 | | 584 | 9303.0 |
| Chlorophonia callophrys | B | 6 | 25.8 | | | | Panama 244 | 9304.0 |
| Chlorochrysa phoenicotis | M | 1 | 22.0 | | | | 279 | 9305.0 |
| Chlorochrysa calliparaea | U | 18 | 17.0 | | 14.9–21.5 | | 279 | 9306.0 |
| Chlorochrysa nitidissima | B | 8 | 18.6 | | 17.3–21.6 | | Columbia 387 | 9307.0 |
| Tangara inornata | U | 13 | 18.0 | | 16.4–19.1 | | 279 | 9308.0 |
| Tangara palmeri | U | 10 | 32.3 | | | | 279 | 9310.0 |
| Tangara chilensis | U | 17 | 23.0 | | 17.0–27.0 | | 279 | 9311.0 |
| Tangara chilensis paradisea | U | 4 | 17.0 | | 16.0–17.0 | | 279 | 9311.0 |
| Tangara seledon | B | 8 | 19.0 | | | | Paraguay 200 | 9313.0 |
| Tangara cyanocephala | U | 16 | 18.0 | | 16.0–21.6 | | Brazil 441 | 9314.0 |
| Tangara desmaresti | B | 2 | 19.8 | | 19.1–20.5 | | Brazil 279, 610a | 9315.0 |
| Tangara cyanoventris | U | 2 | 16.5 | | 16.0–17.0 | | Brazil 440 | 9316.0 |
| Tangara schrankii | U | 35 | 19.0 | | 14.0–23.0 | | 279 | 9318.0 |
| Tangara florida | B | 3 | 19.3 | | 18.5–20.5 | | Panama 497 | 9319.0 |
| Tangara arthus | U | 21 | 22.0 | | 18.7–27.5 | | 279 | 9320.0 |
| Tangara icterocephala | U | 61 | 22.0 | | 17.7–24.7 | | 279 | 9321.0 |
| Tangara xanthocephala | U | 33 | 19.0 | | 15.0–23.6 | | 279 | 9322.0 |
| Tangara chrysotis | U | 4 | 24.0 | | 23.0–25.5 | | 279 | 9323.0 |

## Body Masses of World Birds (continued)

| Species | Sex | N | Mean | Std dev | Range | Sn | Location | Number |
|---|---|---|---|---|---|---|---|---|
| Tangara parzudakii | U | 21 | 28.0 | | 25.0–31.0 | | 279 | 9324.0 |
| Tangara xanthogastra | U | 17 | 15.0 | | 13.0–18.0 | | 279 | 9325.0 |
| Tangara punctata | U | 17 | 15.0 | | 13.0–17.0 | | 279 | 9326.0 |
| Tangara guttata | U | 22 | 18.4 | | 15.0–20.5 | | Trinidad 188 | 9327.0 |
| Tangara varia | U | 1 | 10.0 | | | | 277 | 9328.0 |
| Tangara rufigula | F | 1 | 19.0 | | | | 279 | 9329.0 |
| Tangara gyrola | U | 168 | 21.0 | | 17.5–26.5 | | 279 | 9330.0 |
| Tangara lavinia | U | | 24.0 | | | | 603a | 9331.0 |
| Tangara cayana | U | 30 | 18.0 | | 15.2–22.5 | | 279 | 9332.0 |
| Tangara cucullata | U | | 20.0 | | | | 185 | 9333.0 |
| Tangara preciosa | U | 3 | 23.0 | | 22.0–24.0 | | 279 | 9335.0 |
| Tangara vitriolina | U | 15 | 23.0 | | 18.4–26.8 | | 279 | 9336.0 |
| Tangara meyerdeschauenseei | B | 2 | 26.0 | | 25.4–26.5 | | Peru 539 | 9337.0 |
| Tangara ruficervix | U | 15 | 19.0 | | 16.0–22.2 | | 279 | 9339.0 |
| Tangara labradorides | U | 14 | 15.0 | | 13.0–16.4 | | 279 | 9340.0 |
| Tangara cyanotis | U | 12 | 15.0 | | 12.5–17.0 | | 279 | 9341.0 |
| Tangara cyanicollis | U | 19 | 17.0 | | 14.0–18.8 | | 279 | 9342.0 |
| Tangara larvata | U | 35 | 20.0 | | 17.1–23.9 | | 279 | 9343.0 |
| Tangara nigrocincta | U | 8 | 17.0 | | 15.0–17.8 | | 279 | 9344.0 |
| Tangara fucosa | U | 7 | 21.0 | | 18.0–23.0 | | 279 | 9346.0 |
| Tangara nigroviridis | U | 34 | 17.0 | | 14.0–19.5 | | 279 | 9347.0 |
| Tangara vassorii | U | 63 | 18.0 | | 15.0–21.0 | | 279 | 9348.0 |

## Body Masses of World Birds (continued)

| Species | Sex | N | Mean | Std dev | Range | Sn | Location | Number |
|---|---|---|---|---|---|---|---|---|
| Tangara heinei | U | 4 | 21.0 | | 19.8–22.8 | | 279 | 9349.0 |
| Tangara phillipsi | B | 4 | 19.3 | | 16.8–20.8 | | Peru 225 | 9350.0 |
| Tangara viridicollis | U | 23 | 21.0 | | 18.1–24.0 | | 279 | 9351.0 |
| Tangara argyrofenges | U | 5 | 19.0 | | 18.0–20.0 | | 279 | 9352.0 |
| Tangara cyanoptera | U | 6 | 23.0 | | 21.5–24.5 | | 279 | 9353.0 |
| Tangara velia | U | 11 | 21.0 | | 19.0–23.0 | | 279 | 9354.0 |
| Tangara callophrys | F | 1 | 26.5 | | | | Peru 193 | 9355.0 |
| Tangara mexicana | B | 13 | 20.9 | | 18.0–23.5 | | Trinidad 576 | 9355.1 |
| Iridophanes pulcherrima | U | 6 | 15.0 | | 14.0–16.0 | | 279 | 9356.0 |
| Dacnis albiventris | U | 2 | 11.2 | | 11.0–11.5 | | 279 | 9358.0 |
| Dacnis lineata | U | 20 | 11.0 | | 9.5–13.0 | | 279 | 9359.0 |
| Dacnis flaviventer | U | 3 | 13.0 | | 12.0–14.0 | | 279 | 9360.0 |
| Dacnis nigripes | U | 10 | 14.0 | | 11.0–15.5 | | 279 | 9361.0 |
| Dacnis venusta | B | 43 | 16.1 | | | | Panama 244 | 9362.0 |
| Dacnis cayana | U | 61 | 13.0 | | 10.0–15.5 | | 279 | 9363.0 |
| Chlorophanes spiza | U | 108 | 19.0 | | 14.0–23.0 | | 279 | 9366.0 |
| Cyanerpes nitidus | U | 5 | 9.0 | | 8.0–10.2 | | 279 | 9367.0 |
| Cyanerpes lucidus | B | 6 | 11.4 | | | | Panama 244 | 9368.0 |
| Cyanerpes caeruleus | U | 53 | 12.0 | | 7.8–14.0 | | 279 | 9369.0 |
| Cyanerpes cyaneus | U | 86 | 14.0 | | 11.0–18.3 | | 279 | 9370.0 |
| Xenodacnis parina | B | 11 | 11.5 | | 10.0–12.0 | | 279 | 9371.0 |
| Tersina viridis | U | 30 | 29.0 | | 25.9–35.0 | | 279 | 9372.0 |

## Body Masses of World Birds (continued)

| Species | Sex | N | Mean | Std dev | Range | Sn | Location | Number |
|---------|-----|---|------|---------|-------|----|----------|--------|
| Catamblyrhynchus diadema | U | 15 | 14.3 | 1.70 | | | Venezuela 261 | 9373.0 |
| Nephelornis oneilli | M | 15 | 17.5 | 0.94 | 16.0–19.0 | | Peru 353 | 9376.0 |
| | F | 13 | 14.9 | 0.68 | 13.5–15.5 | | | |
| Saltatricula multicolor | U | | 22.4 | | | | Argentina 522 | 9379.0 |
| Coryphospingus pileatus | U | 9 | 15.8 | | 13.8–19.0 | | 384, 625 | 9380.0 |
| Coryphospingus cucullatus | B | 32 | 14.9 | | | | Paraguay 200 | 9381.0 |
| Rhodospingus cruentus | B | 2 | 10.8 | | 10.5–11.0 | | Peru 672 | 9382.0 |
| Phrygilus gayi | M | 9 | 22.0 | | | | Chile 363a | 9385.0 |
| | F | 5 | 19.0 | | | | | |
| Phrygilus fruticeti | B | 13 | 38.8 | | 35.1–41.8 | | Argentina 411 | 9387.0 |
| Phrygilus unicolor | B | 7 | 21.2 | | | | 644, 668 | 9388.0 |
| Phrygilus plebejus | M | 1 | 13.6 | | | | Peru 672 | 9391.0 |
| Phrygilus alaudinus | U | | 18.0 | | | | Chile 292 | 9393.0 |
| Haplospiza rustica | U | 13 | 15.6 | 1.02 | | | Peru 668 | 9396.0 |
| Haplospiza unicolor | B | 7 | 16.1 | | 14.0–21.5 | | Brazil 38, 441, 610a | 9397.0 |
| Acanthidops bairdii | U | | 16.0 | | | | 603a | 9398.0 |
| Donacospiza albifrons | M | 1 | 16.5 | | | | Brazil 38 | 9401.0 |
| | F | 4 | | | 14.0–16.5 | | | |
| Diuca diuca | U | | 31.0 | | | | Chile 292 | 9406.0 |
| Idiopsar brachyurus | M | 1 | 43.0 | | | | Bolivia 643 | 9407.0 |
| Incaspiza ortizi | M | 10 | 35.0 | | 29.0–38.0 | | Peru 451 | 9412.0 |
| | F | 7 | 31.0 | | 28.0–33.0 | | | |
| Poospiza thoracica | B | 3 | 11.9 | | 11.0–12.8 | | Brazil 38, 441 | 9415.0 |
| Poospiza ornata | U | | 11.8 | | | | Argentina 522 | 9419.0 |
| Poospiza nigrorufa | U | 4 | 19.4 | | 17.0–20.7 | | Argentina 111 | 9421.0 |
| Poospiza lateralis | U | 8 | 19.9 | | 18.5–21.0 | | Brazil 441 | 9423.0 |

## Body Masses of World Birds (continued)

| Species | Sex | N | Mean | Std dev | Range | Sn | Location | Number |
|---|---|---|---|---|---|---|---|---|
| Poospiza torquata | U | | 10.3 | | | | Argentina 522 | 9429.0 |
| Poospiza melanoleuca | U | 4 | 14.8 | | 14.0–15.0 | | Argentina 111 | 9430.0 |
| Poospiza cinerea | B | 9 | 11.2 | | 8.4–13.0 | | Paraguay 610 | 9431.0 |
| Sicalis flaveola | U | 21 | 19.7 | 1.50 | 17.0–23.4 | | Venezuela 625 | 9440.0 |
| Sicalis luteola | U | | 16.0 | | | | Chile 52 | 9441.0 |
| Emberizoides herbicola herbicola | B | 9 | 28.0 | | 26.0–31.1 | | 182 | 9445.0 |
| Emberizoides herbicola sphenurus | B | 6 | 25.3 | | 23.0–28.0 | | Surinam 182 | 9445.0 |
| Emberizoides ypiranganus | B | 5 | 20.1 | | 18.5–22.0 | | 182 | 9447.0 |
| Embernagra platensis | M<br>F | 3<br>1 | <br>42.0 | | 51.0–59.0 | | Brazil 38 | 9448.0 |
| Volatinia jacarina | B | 41 | 9.7 | | 8.0–12.0 | | Trinidad 576 | 9450.0 |
| Sporophila schistacea | B | 9 | 12.6 | | 12.0–13.0 | | Brazil 610a | 9453.0 |
| Sporophila intermedia | B | 15 | 12.1 | | 11.0–13.5 | | Venezuela 625 | 9454.0 |
| Sporophila plumbea | B | 3 | 11.6 | | 9.7–13.0 | | Surinam; Brazil 38, 246 | 9455.0 |
| Sporophila americana | B | 11 | 10.7 | | 9.5–12.5 | | 87, 245, 311 | 9456.0 |
| Sporophila americana aurita | U | 14 | 11.4 | 1.10 | | | Costa Rica 315 | 9456.0 |
| Sporophila americana aurita | U | 30 | 9.9 | 1.00 | | | Panama 315 | 9456.0 |
| Sporophila torqueola | B | 41 | 8.7 | 1.04 | 6.3–12.0 | B | Mexico 584 | 9457.0 |
| Sporophila collaris | B | 5 | | | 13.0–14.0 | | Brazil 38 | 9458.0 |
| Sporophila bouvronides | M | 1 | 9.6 | | | | Venezuela 626 | 9459.0 |
| Sporophila lineola | B | 26 | 9.7 | | 7.5–12.0 | | Trinidad 576 | 9460.0 |
| Sporophila luctuosa | U | 4 | 12.5 | | | | Peru 668 | 9461.0 |
| Sporophila nigricollis | B | 12 | 9.6 | | 8.5–11.2 | | Venezuela 625 | 9462.0 |

## Body Masses of World Birds (continued)

| Species | Sex | N | Mean | Std dev | Range | Sn | Location | Number |
|---|---|---|---|---|---|---|---|---|
| Sporophila caerulescens | U | 8 | 11.0 | | 8.9–15.0 | | Brazil 38, 441 | 9465.0 |
| Sporophila albogularis | U | | 9.7 | | | | Brazil 522 | 9466.0 |
| Sporophila leucoptera | M | 3 | 15.5 | | 15.0–16.0 | | Paraguay 610 | 9467.0 |
| Sporophila bouvreuil | M F | 1 | 8.5 | | | | Brazil 38 | 9471.0 |
| Sporophila minuta | B | 19 | 7.8 | | 7.0–9.0 | | Trinidad 576 | 9472.0 |
| Sporophila hypoxantha | B | 3 | 9.5 | | 9.0–10.5 | | Brazil 38 | 9473.0 |
| Sporophila palustris | M F | 1 1 | 9.0 7.5 | | | | Brazil 38 | 9475.0 |
| Sporophila castaneiventris | U | 7 | 7.7 | | | | 161, 246, 668 | 9479.0 |
| Sporophila melanogaster | B | 3 | 9.3 | | 8.0–10.5 | | Brazil 38 | 9480.0 |
| Sporophila telasco | M | 1 | 8.4 | | | | Peru 672 | 9481.0 |
| Oryzoborus crassirostris | F | 1 | 19.8 | | | | Venezuela 625 | 9484.0 |
| Oryzoborus maximiliani | M | 1 | 25.0 | | | | Peru 193 | 9486.0 |
| Oryzoborus angolensis | U | 24 | 12.3 | 0.90 | | | Panama 315 | 9487.0 |
| Amaurospiza concolor | U | 1 | 13.0 | | | | 603a | 9489.0 |
| Amaurospiza moesta | B | 4 | 13.4 | | 13.0–14.0 | | Brazil 38 | 9490.0 |
| Melopyrrha nigra | B | 19 | 10.9 | | 14.5–18.5 | | Cayman Is. 432 | 9491.0 |
| Catamenia analis | F | 1 | 16.6 | | | | Ecuador 320 | 9493.0 |
| Catamenia inornata | U | 16 | 13.4 | | 12.3–14.5 | | Ecuador 320 | 9494.0 |
| Tiaris obscura | U | 29 | 11.2 | 0.84 | | | Peru 668 | 9496.0 |
| Tiaris canora | U | | 10.0 | | | | 185 | 9497.0 |
| Tiaris olivacea | B | 21 | 8.9 | | | | Panama 244 | 9498.0 |
| Tiaris bicolor | U | 59 | 9.7 | 0.70 | 7.8–11.2 | W | Puerto Rico 186 | 9499.0 |

## Body Masses of World Birds (continued)

| Species | Sex | N | Mean | Std dev | Range | Sn | Location | Number |
|---|---|---|---|---|---|---|---|---|
| Tiaris fuliginosa | B | 24 | 13.3 | | 11.0–16.0 | | Trinidad 576 | 9500.0 |
| Loxipasser anoxanthus | B | 12 | 11.4 | | 10.5–12.5 | | Jamaica 600 | 9501.0 |
| Loxigilla portoricensis | M | 12 | 34.8 | 2.25 | 31.1–39.1 | | Puerto Rico 431 | 9502.0 |
| | F | 8 | 29.2 | | 23.4–36.7 | | | |
| Loxigilla violacea | M | 2 | 21.6 | | 20.8–22.5 | | Bahamas 600 | 9503.0 |
| | F | 5 | 19.2 | | 18.0–20.5 | | | |
| Loxigilla noctis | U | 81 | 18.4 | 1.57 | | | Dominica 471 | 9504.0 |
| Diglossa baritula | U | 11 | 8.0 | | 6.0–9.4 | | 279 | 9505.0 |
| Diglossa plumbea | U | 4 | 9.9 | | 9.3–10.1 | | 279, 315 | 9506.0 |
| Diglossa sittoides | U | 13 | 9.0 | | 7.0–10.7 | | 279 | 9507.0 |
| Diglossa albilatera | U | 26 | 10.0 | | 8.2–12.0 | | 279 | 9509.0 |
| Diglossa lafresnayii | U | 105 | 16.0 | | 12.0–19.2 | | 279 | 9511.0 |
| Diglossa carbonaria carbonaria | U | 22 | 11.0 | | 9.0–14.8 | | 279 | 9516.0 |
| Diglossa carbonaria brunneiventris | U | 72 | 12.0 | | 9.8–15.0 | | 279 | 9516.0 |
| Diglossa duidae | U | 10 | 16.0 | | 13.8–16.5 | | 279 | 9517.0 |
| Diglossopis glauca | U | 37 | 12.0 | | 9.5–13.0 | | 279 | 9520.0 |
| Diglossopis caerulescens | U | 27 | 13.0 | | 10.1–16.0 | | 279 | 9521.0 |
| Diglossopis cyanea | U | 180 | 17.0 | | 12.0–22.5 | | 279 | 9522.0 |
| Euneornis campestris | U | 260 | 16.0 | | 13.2–19.2 | | 279 | 9523.0 |
| Melanospiza richardsoni | B | 6 | 20.9 | | 18.0–23.0 | | St. Lucia 60, 590, 631 | 9524.0 |
| Geospiza magnirostris | U | 24 | 34.5 | | 27.0–39.0 | | Galapagos Is. 61 | 9525.0 |
| Geospiza fortis | U | 59 | 24.1 | | 18.0–31.0 | | Galapagos Is. 61 | 9526.0 |
| Geospiza fuliginosa | U | 33 | 14.5 | | 12.0–17.0 | | Galapagos Is. 61 | 9527.0 |
| Geospiza difficilis | U | | 12.3 | | | | Galapagos Is. 1 | 9528.0 |

## Body Masses of World Birds (continued)

| Species | Sex | N | Mean | Std dev | Range | Sn | Location | Number |
|---|---|---|---|---|---|---|---|---|
| Geospiza scandens | U | 3 | 22.6 | | 20.0–25.0 | | Galapagos Is. 61 | 9530.0 |
| Geospiza conirostris | U | | 27.6 | | | | Galapagos Is. 1 | 9531.0 |
| Camarhynchus crassirostris | U | 33 | 35.5 | | 29.0–40.0 | | Galapagos Is. 61 | 9532.0 |
| Camarhynchus psittacula | U | 42 | 18.7 | | 15.0–21.0 | | Galapagos Is. 61 | 9533.0 |
| Camarhynchus parvulus | U | 47 | 13.1 | | 10.0–17.0 | | Galapagos Is. 61 | 9535.0 |
| Camarhynchus pallidus | U | 16 | 23.1 | | 20.0–31.0 | | Galapagos Is. 61 | 9536.0 |
| Certhidea olivacea | U | 14 | 9.5 | | 8.0–12.0 | | Galapagos Is. 61 | 9538.0 |
| Pinaroloxias inornata | U | | 12.5 | | | | Cocos Is. 603a | 9539.0 |
| Spiza americana | M | 18 | 29.3 | | | B | Great Plains, USA | 9540.0 |
| | F | 13 | 24.6 | | | | 674 | |
| Pheucticus chrysopeplus | B | 3 | 62.7 | | 54.0–77.6 | | Peru; Mexico 593, 668, 672 | 9541.0 |
| Pheucticus tibialis | B | 12 | 62.3 | | | | Panama 244 | 9543.0 |
| Pheucticus aureoventris | U | 1 | 65.5 | | | | Peru 668 | 9544.0 |
| Pheucticus ludovicianus | B | 277 | 45.6 | 0.40 | 35.4–65.0 | | Pennsylvania, USA 101 | 9545.0 |
| Pheucticus melanocephalus | M | 18 | 41.8 | | 35.0–46.0 | S | California, USA | 9546.0 |
| | F | 15 | 42.2 | | 37.0–48.8 | | 558 | |
| Cardinalis cardinalis | M | 591 | 45.4 | 4.29 | 33.7–63.2 | Y | Pennsylvania, USA | 9547.0 |
| | F | 517 | 43.9 | 4.53 | 33.6–64.9 | | 101 | |
| Cardinalis sinuatus | M | 81 | 36.7 | 2.49 | 31.6–44.0 | Y | Arizona, USA | 9549.0 |
| | F | 61 | 34.3 | 1.72 | 29.7–38.4 | | 177 | |
| Caryothraustes poliogaster | B | 11 | 41.8 | 3.15 | 34.9–46.7 | | Belize; Mexico 457, 510 | 9550.0 |
| Caryothraustes canadensis | B | 6 | 34.5 | | 31.0–36.0 | | 246, 497, 610a | 9551.0 |
| Caryothraustes humeralis | F | 1 | 37.0 | | | | Brazil 610a | 9552.0 |
| Pitylus grossus | U | 5 | 49.0 | | | | Panama; Brazil 38, 50, 315, 437 | 9555.0 |
| Saltator atriceps | B | 17 | 79.7 | | 68.2–91.2 | | Yucatan, Mexico 457 | 9557.0 |
| Saltator maximus | M | 17 | 46.2 | 2.96 | | | Panama | 9558.0 |
| | F | 13 | 49.1 | 1.73 | | | 244 | |

## Body Masses of World Birds (continued)

| Species | Sex | N | Mean | Std dev | Range | Sn | Location | Number |
|---|---|---|---|---|---|---|---|---|
| Saltator atripennis | B | 2 | 56.4 | | 55.1–57.7 | | Columbia 387 | 9559.0 |
| Saltator coerulescens | U | 21 | 54.9 | 5.01 | 49.0–67.0 | | Venezuela 625 | 9560.0 |
| Saltator similis | U | 10 | 47.7 | | 43.5–53.0 | | Brazil 38, 440, 441 | 9561.0 |
| Saltator orenocensis | U | 2 | 35.6 | | 34.5–36.8 | | Venezuela 625 | 9562.0 |
| Saltator aurantiirostris | M | 4 | | | 53.5–58.0 | | Brazil | 9564.0 |
| | F | 3 | | | 50.0–60.0 | | 38 | |
| Saltator maxillosus | B | 4 | 50.8 | | 48.0–54.0 | | Brazil 38 | 9565.0 |
| Saltator cinctus | B | 7 | 48.6 | | 43.0–53.0 | | Peru 435 | 9566.0 |
| Saltator albicollis | U | 25 | 36.9 | | 30.0–44.0 | | Trinidad 188 | 9569.0 |
| Cyanoloxia glaucocaerulea | U | 4 | | | 16.0–19.5 | | Brazil 38, 441 | 9570.0 |
| Cyanocompsa cyanoides | U | 44 | 32.5 | | | | 315 | 9571.0 |
| Cyanocompsa parellina | B | 11 | 15.0 | | 13.4–16.8 | | Yucatan, Mexico 457 | 9572.0 |
| Cyanocompsa brissonii | B | 5 | 20.6 | | 18.9–22.5 | | Venezuela 625 | 9573.0 |
| Guiraca caerulea | M | 10 | 29.3 | 1.42 | 27.0–31.4 | B | N. Carolina,USA | 9574.0 |
| | F | 5 | 27.5 | | 26.1–29.8 | | 177 | |
| Passerina amoena | M | 58 | 16.0 | | 13.0–19.5 | S | California, USA | 9575.0 |
| | F | 25 | 15.0 | | 12.7–16.9 | | 555 | |
| Passerina cyanea | M | 464 | 14.9 | 1.39 | 12.3–21.4 | | Pennsylvania, USA | 9576.0 |
| | F | 339 | 14.1 | 1.41 | 11.2–18.6 | | 101 | |
| Passerina versicolor | B | 10 | 11.8 | 0.79 | 10.0–12.7 | | Sinaloa, Mexico 594 | 9577.0 |
| Passerina ciris | M | 116 | 16.1 | | 13.3–19.0 | | Florida, USA | 9578.0 |
| | F | 131 | 15.0 | | 12.9–19.0 | | 190 | |
| Passerina rositae | B | 2 | 20.0 | | 19.5–20.5 | | Chiapas, Mexico 584 | 9579.0 |
| Passerina leclancherii | B | 7 | 14.0 | | 12.0–15.1 | | Mexico 584 | 9580.0 |

## ORDER: PASSERIFORMES          FAMILY: ICTERIDAE

| Species | Sex | N | Mean | Std dev | Range | Sn | Location | Number |
|---|---|---|---|---|---|---|---|---|
| Psarocolius oseryi | M | 5 | 190.0 | | | | Peru | 9582.0 |
| | F | 4 | 99.8 | | | | 335 | |
| Psarocolius decumanus | M | 7 | 319.0 | | | | | 9583.0 |
| | F | 8 | 152.0 | | 124.0–170.0 | | 87, 188, 244, 246 | |

## Body Masses of World Birds (continued)

| Species | Sex | N | Mean | Std dev | Range | Sn | Location | Number |
|---|---|---|---|---|---|---|---|---|
| Psarocolius viridis | M | 2 | 410.0 | | 360.0–460.0 | | Fr. Guiana | 9584.0 |
| | F | 1 | 215.0 | | | | 161 | |
| Psarocolius atrovirens | F | 1 | 152.0 | | | | Peru | 9585.0 |
| | | | | | | | 193 | |
| Psarocolius angustifrons | M | 1 | 278.0 | | | | Venezuela | 9586.0 |
| | F | 4 | 177.0 | | 167.0–184.0 | | 451 | |
| Psarocolius waglcri | M | 9 | 214.0 | | | | Panama | 9587.0 |
| | F | 7 | 113.0 | | | | 244 | |
| Gymnostinops montezuma | M | 4 | 423.0 | | 353.0–528.0 | | Belize; Panama | 9589.0 |
| | F | 7 | 225.0 | | 198.0–254.0 | | 87, 510, 611 | |
| Gymnostinops yuracares | U | | 360.0 | | | | Peru | 9591.1 |
| Cacicus cela | M | 8 | 105.0 | | 81.0–121.0 | | | 9593.0 |
| | F | 10 | 77.9 | | 67.0–110.0 | | 87, 161, 245, 246, 611 | |
| Cacicus haemorrhous | M | 10 | 102.0 | | 93.0–108.0 | | | 9594.0 |
| | F | 4 | 68.0 | | 62.0–74.0 | | 38, 161, 200, 245, 441 | |
| Cacicus uropygialis | M | 4 | 69.3 | | 62.5–71.7 | | Panama | 9595.0 |
| | F | 5 | 54.0 | | 50.0–57.6 | | 611 | |
| Cacicus chrysopterus | M | 1 | 39.0 | | | | Brazil | 9597.0 |
| | F | 2 | 32.8 | | 31.5–34.0 | | 38 | |
| Cacicus solitarius | M | 5 | 90.1 | | | | Peru | 9600.0 |
| | F | 2 | 80.0 | | 75.0–85.0 | | 193 | |
| Cacicus melanicterus | M | 4 | 89.9 | | 70.0–107.0 | | Mexico | 9601.0 |
| | F | 2 | 69.7 | | 68.6–70.9 | | 458, 593 | |
| Amblycercus holosericeus | M | 13 | 71.0 | 4.36 | | | Panama | 9602.0 |
| | F | 4 | 56.4 | | | | 244 | |
| Icterus chrysocephalus | B | 4 | 41.2 | | 38.0–44.0 | | Surinam | 9603.0 |
| | | | | | | | 245, 246 | |
| Icterus cayanensis | U | | 39.0 | | | | Peru | 9604.0 |
| | | | | | | | 623 | |
| Icterus chrysater | B | 16 | 53.6 | | 46.2–61.5 | | | 9605.0 |
| | | | | | | | 272, 384, 510, 611 | |
| Icterus nigrogularis | B | 16 | 40.2 | | 31.0–50.0 | | | 9606.0 |
| | | | | | | | 188, 245, 246, 625 | |
| Icterus leucopteryx | B | 5 | 39.5 | | | | | 9607.0 |
| | | | | | | | 125a, 511, 600 | |
| Icterus auratus | B | 9 | 32.1 | | 26.7–36.1 | | Yucatan, Mexico | 9608.0 |
| | | | | | | | 323, 457 | |
| Icterus mesomelas | B | 14 | 39.9 | | 33.8–47.7 | | Belize; Mexico | 9609.0 |
| | | | | | | | 457, 510, 584 | |
| Icterus auricapillus | M | 2 | 32.2 | | 31.0–33.5 | | | 9610.0 |
| | | | | | | | 384, 611 | |

## Body Masses of World Birds (continued)

| Species | Sex | N | Mean | Std dev | Range | Sn | Location | Number |
|---|---|---|---|---|---|---|---|---|
| Icterus pectoralis | M | 3 | 44.5 | | 42.5–46.9 | | W. Mexico 593 | 9612.0 |
| | F | | | | | | | |
| Icterus gularis yucatanensis | M | 7 | 59.1 | | 56.4–64.3 | B | Yucatan, Mexico 457 | 9613.0 |
| | F | 3 | 51.5 | | 47.4–53.6 | | | |
| Icterus pustulatus | B | 10 | 37.0 | 2.53 | 32.1–41.9 | | Sonora, Mexico 594 | 9614.0 |
| Icterus cucullatus | B | 56 | 24.3 | 2.02 | 20.6–33.2 | B | Arizona, USA 177 | 9615.0 |
| Icterus icterus | U | 19 | 72.2 | 6.10 | 60.9–83.8 | W | Puerto Rico 186 | 9616.0 |
| Icterus galbula galbula | M | 57 | 34.3 | 1.62 | 22.3–41.5 | S | Pennsylvania, USA 101 | 9618.0 |
| | F | 59 | 33.2 | 2.44 | 28.1–41.3 | | | |
| Icterus galbula bullockii | B | 33 | 33.6 | 2.26 | 29.0–38.0 | | California, USA 177 | 9618.0 |
| Icterus spurius | B | 45 | 19.6 | 1.77 | 16.0–25.1 | | Texas, USA 177 | 9619.0 |
| Icterus spurius fuertesi | M | 2 | 19.0 | | 18.8–19.2 | | Mexico 590 | 9619.0 |
| Icterus dominicensis prosthemelas | M | 9 | 29.5 | | 26.2–32.1 | | Belize 510 | 9620.0 |
| | F | 6 | 26.5 | | 22.9–291 | | | |
| Icterus dominicensis portoricensis | M | 4 | 42.7 | | 41.5–44.9 | | Puerto Rico 431 | 9620.0 |
| Icterus wagleri | B | 10 | 42.4 | 4.18 | 36.7–50.0 | | Mexico 59, 548, 593 | 9621.0 |
| Icterus laudabilis | U | | 35.0 | | | | 185 | 9622.0 |
| Icterus bonana | U | | 35.0 | | | | 185 | 9623.0 |
| Icterus oberi | U | | 35.0 | | | | 185 | 9624.0 |
| Icterus graduacauda | B | 25 | 42.2 | 4.87 | 31.0–52.0 | B | Mexico 584 | 9625.0 |
| Icterus parisorum | B | 39 | 37.4 | 2.31 | 32.1–41.0 | | Arizona, USA 177 | 9627.0 |
| Nesopsar nigerrimus | U | | 39.0 | | | | 185 | 9628.0 |
| Gymnomystax mexicanus | U | 1 | 93.0 | | | | Venezuela 625 | 9629.0 |
| Xanthocephalus xanthocephalus | M | 3 | 79.7 | | 72.5–85.5 | | Oregon, USA 177 | 9630.0 |
| | F | 10 | 49.3 | 4.00 | 42.4–56.0 | | | |
| Agelaius flavus | U | | 43.0 | | | | Brazil 564 | 9631.0 |
| Agelaius xanthophthalmus | U | 24 | 45.1 | | | | Peru 193 | 9632.0 |

## Body Masses of World Birds (continued)

| Species | Sex | N | Mean | Std dev | Range | Sn | Location | Number |
|---|---|---|---|---|---|---|---|---|
| Agelaius thilius | M | 1 | 30.0 | | | | Brazil 38 | 9633.0 |
| Agelaius cyanopus | M | 1 | 43.0 | | | | | 9634.0 |
| | F | 1 | 32.2 | | | | 38, 111 | |
| Agelaius phoeniceus | M | 290 | 63.6 | 4.43 | 52.9–81.1 | | Pennsylvania, USA | 9635.0 |
| | F | 249 | 41.5 | 2.74 | 29.0–55.0 | | 101 | |
| Agelaius tricolor | M | 52 | 68.2 | 3.64 | 60.0–79.0 | B | California, USA | 9636.0 |
| | F | 5 | 49.2 | | 46.0–54.5 | | 177 | |
| Agelaius icterocephalus | M | 27 | 35.4 | 2.00 | 31.4–40.0 | | Trinidad | 9637.0 |
| | F | 17 | 26.6 | | 24.0–31.0 | | 576 | |
| Agelaius humeralis | M | 1 | 37.0 | | | | Cuba | 9638.0 |
| | F | 4 | 29.2 | | | | 429 | |
| Agelaius xanthomus | B | 8 | 38.4 | | | | Puerto Rico 676 | 9639.0 |
| Agelaius ruficapillus | M | | | | 39.0–44.0 | | Brazil | 9640.0 |
| | F | 2 | 32.0 | | 32.0–32.0 | | 38 | |
| Leistes militaris | M | 9 | 46.2 | | | | Panama | 9641.0 |
| | F | 4 | 35.5 | | | | 244 | |
| Leistes superciliaris | M | 2 | 53.0 | | 53.0–53.0 | | Brazil | 9642.0 |
| | F | 1 | 39.5 | | | | 38 | |
| Sturnella loyca | U | | 113.0 | | | | Chile 292 | 9645.0 |
| Sturnella magna | M | 20 | 102.0 | 11.20 | | | Florida, USA | 9646.0 |
| | F | 9 | 76.0 | | | | 243 | |
| Sturnella neglecta | M | 51 | 112.0 | | | B | Great Plains, USA | 9647.0 |
| | F | 32 | 89.4 | | | | 674 | |
| Pseudoleistes guirahuro | M | 1 | 98.0 | | | | Brazil 38 | 9648.0 |
| Pseudoleistes virescens | M | 1 | 88.0 | | | | Brazil | 9649.0 |
| | F | 1 | 64.0 | | | | 38 | |
| Amblyramphus holosericeus | M | 1 | 86.0 | | | | Brazil | 9650.0 |
| | F | 1 | 75.5 | | | | 38 | |
| Curaeus curaeus | U | | 90.0 | | | | Chile 292 | 9652.0 |
| Gnorimopsar chopi | M | 2 | 79.5 | | 79.0–80.0 | | Brazil | 9654.0 |
| | F | 3 | | | 75.0–84.0 | | 38, 610a | |
| Dives dives | B | 4 | 96.2 | | 83.4–102.0 | | Belize; Mexico 139, 510 | 9660.0 |
| Quiscalus mexicanus | M | 16 | 191.0 | 22.80 | 157.0–234.0 | B | Arizona, USA | 9662.0 |
| | F | 12 | 107.0 | 11.40 | 96.0–140.0 | | 594 | |
| Quiscalus major | M | 149 | 214.0 | | 175.0–253.0 | Y | | 9663.0 |
| | F | 48 | 119.0 | | 1-02.0–132.0 | | 545 | |
| Quiscalus quiscula | M | 197 | 127.0 | | | W | Tennessee, USA | 9666.0 |
| | F | 135 | 100.0 | | | | 164 | |

## Body Masses of World Birds (continued)

| Species | Sex | N | Mean | Std dev | Range | Sn | Location | Number |
|---|---|---|---|---|---|---|---|---|
| Quiscalus niger | M | 10 | 75.1 | 7.10 | 67.5–88.0 | | Cayman Is. | 9667.0 |
| caymanensis | F | 11 | 57.4 | 4.30 | 51.0–64.0 | | 432 | |
| Quiscalus niger | M | 89 | 86.4 | 5.75 | | | Puerto Rico | 9667.0 |
| | F | 95 | 61.7 | 4.97 | | | 676 | |
| Quiscalus lugubris | M | 5 | 74.2 | | 70.0–80.0 | | Martinique | 9668.0 |
| | F | 8 | 54.8 | | 49.0–58.0 | | 535 | |
| Euphagus carolinus | M | 91 | 64.3 | 2.38 | 45.9–80.4 | M | Pennsylvania, USA | 9669.0 |
| | F | 105 | 55.2 | 3.59 | 47.0–76.5 | | 101 | |
| Euphagus cyanocephalus | M | 19 | 67.2 | 3.20 | 60.0–73.0 | | Oregon, USA | 9670.0 |
| | F | 15 | 58.1 | 4.90 | 50.6–67.0 | | 177 | |
| Molothrus badius | B | 5 | 44.5 | | 41.0–50.0 | | | 9671.0 |
| | | | | | | | 38, 111 | |
| Molothrus rufoaxillaris | M | 4 | | | 56.0–65.0 | | Brazil | 9672.0 |
| | F | 2 | 47.5 | | 38.0–57.0 | | 38 | |
| Molothrus bonariensis | M | 479 | 38.7 | 1.09 | | | Puerto Rico | 9673.0 |
| | F | 670 | 31.9 | 1.04 | | | 676 | |
| Molothrus aeneus | M | 17 | 66.7 | 4.42 | 59.6–72.0 | | Mexico | 9674.0 |
| | F | 7 | 57.4 | | 39.6–65.5 | | 585 | |
| Molothrus ater | M | 757 | 49.0 | 1.77 | 32.4–58.0 | | Pennsylvania, USA | 9675.0 |
| | F | 692 | 38.8 | 1.93 | 30.5–51.2 | | 101 | |
| Scaphidura oryzivora | M | 6 | 219.0 | | 174.0–242.0 | | | 9676.0 |
| | F | 2 | 162.0 | | 156.0–167.0 | | 87, 161, 510 | |
| Dolichonyx orizivorus | M | 22 | 47.0 | | 28.5–56.3 | F | Delaware, USA | 9677.0 |
| | F | 5 | 37.1 | | 26.5–44.3 | | 375 | |

## PART II

## BODY MASSES AND COMPOSITION OF MIGRANT BIRDS IN THE EASTERN UNITED STATES

# INTRODUCTION

## Eugene P. Odum

In the 1960s, my students, colleagues, and I undertook an extensive study of lipid storage and use by migrating birds. As part of these studies, we compiled an extensive data set on avian body mass and composition, collected mostly from birds killed during migration. Although we have published numerous papers from this study (e.g., Connell et al. 1960, Odum 1960, Odum et al. 1961, Marshall and Baker 1964, Hicks 1967), the full data set has never been published. Since the data are relatively unique, and still requested by researchers, I present them in the following table.

A majority of the bird specimens that were analyzed in this study were spring and fall nocturnal migrants that were killed in collision with a television tower located in Leon County, Florida, between Thomasville, Georgia, and Tallahassee, Florida, near the Gulf coast. During the fall of 1956, Herbert Stoddard searched the grounds of the TV tower at dawn each morning from August to December, collecting the birds that had been killed the previous night. All specimens were immediately frozen, and then transported to the University of Georgia where the study was conducted.

Data for some of the species were collected from other locations. A large series was collected by L. D. Caldwell from a TV tower in Michigan (Caldwell et al. 1964). W. H. Drury donated a fall series of Blackpoll Warblers (*Dendroica striata*) from Massachusetts, all in premigratory and very fat condition. Chimney Swifts (*Chaetura pelagica*) were collected from a chimney roost in Athens, Georgia. Wintering and premigratory Savannah Sparrows (*Passerculus sandwichensis*) and fall premigratory hummingbirds were collected at the Savannah River Ecology Laboratory, Aiken County, South Carolina. Some individuals of other species were collected from TV towers in the southeastern United States, however, most specimens for these species came from the Leon County tower.

Upon receipt at the University of Georgia, the specimens were classified according to species, age and sex, and then weighed and measured. This first mass is referred to as the "wet mass." Aging was done by removing the skin on the top of the head and examining the degree of skull ossification. Specimens were sexed by plumage or gonadal inspection. After weighing, the birds were placed in a vacuum oven, which was set at a constant 40°C, and allowed to dry for 36–108 h, depending on the size of the specimen. When the bird was completely dry, it was weighed again, producing the "dry mass" (Marshall and Baker 1964).

To estimate the amount of fat in each specimen, each bird was then ground in a blender or meat chopper containing about 5 cm of 95% ethyl alcohol (Connell et al. 1960). The pulverized mixture was boiled in a beaker placed in a water bath, and then strained through a very fine mesh sieve. This boiling procedure was repeated 3 times with petroleum ether substituted for the alcohol, after which the residue was dried for 12 h. The residue was weighed again, and the difference between this last measurement and the dry mass was the amount of fat removed by the petroleum ether. The "fat-free mass" was calculated by subtracting the amount of fat from the wet mass. An alternate fat extraction method using chloroform was used on some specimens.

The nonfat dry residue was then burned in a muffle furnace at 600°C until burning eliminated all organic matter, reducing the residue to the ash content of the specimen (Odum et al. 1965). The "ash-free mass" was calculated by subtracting the mass of ash from the fat-free mass. This step was not done on all specimens.

The following table reports wet mass, dry mass, fat-free mass, and ash-free mass for 43 species. For each age and sex class, I report the sample size, mean, standard deviation, and range. Birds in their first calendar year (immatures) are classified as HY (hatching year) in the table. Adult birds are classified as AHY (after hatching year). In addition, I have included statistics for the entire sample for each species, including individual birds that were of unknown sex or age class. The addition of these individuals to the total sample results in slightly larger sample sizes in the total sample than can be summed from the individual age and sex classes.

I would like to express my special appreciation to Shirley Marshall who managed my lipid extraction laboratory with great efficiency during the period when most of these data were obtained.

## Body Mass and Composition of North American Migrant Birds

|  | N | Mean | S.D. | Min | Max |
|---|---|---|---|---|---|
| Porzana carolina |  |  |  |  |  |
| Male HY |  |  |  |  |  |
| wet | 6 | 71.14 | 9.85 | 59.40 | 84.13 |
| dry | 6 | 32.34 | 5.35 | 27.57 | 40.60 |
| fat-free | 6 | 57.44 | 8.78 | 45.17 | 64.78 |
| Male AHY |  |  |  |  |  |
| wet | 4 | 68.56 |  | 58.21 | 73.20 |
| dry | 4 | 31.34 |  | 20.89 | 38.96 |
| fat-free | 4 | 54.84 |  | 50.84 | 61.91 |
| Female HY |  |  |  |  |  |
| wet | 8 | 66.90 | 13.13 | 48.67 | 89.87 |
| dry | 8 | 35.68 | 9.58 | 24.41 | 49.21 |
| fat-free | 8 | 47.42 | 6.95 | 36.70 | 61.24 |
| Female AHY |  |  |  |  |  |
| wet | 10 | 65.49 | 8.81 | 52.70 | 78.10 |
| dry | 10 | 31.09 | 9.13 | 17.50 | 44.21 |
| fat-free | 10 | 49.99 | 2.30 | 46.20 | 54.89 |
| All individuals |  |  |  |  |  |
| wet | 28 | 67.54 | 9.94 | 48.67 | 89.87 |
| dry | 28 | 32.70 | 8.32 | 17.50 | 49.21 |
| fat-free | 28 | 51.55 | 6.83 | 36.70 | 64.78 |
| Cuculus americanus |  |  |  |  |  |
| Male HY |  |  |  |  |  |
| wet | 12 | 72.09 | 10.88 | 51.60 | 86.34 |
| dry | 12 | 41.25 | 10.34 | 22.03 | 55.68 |
| fat-free | 12 | 46.67 | 2.93 | 42.16 | 52.08 |
| ash-free | 11 | 44.71 | 2.85 | 40.49 | 49.40 |
| Male AHY |  |  |  |  |  |
| wet | 19 | 77.28 | 11.31 | 54.97 | 88.91 |
| dry | 19 | 46.50 | 13.03 | 22.20 | 59.07 |
| fat-free | 19 | 46.04 | 4.29 | 37.00 | 55.88 |
| ash-free | 17 | 44.14 | 4.36 | 35.51 | 53.84 |
| Female HY |  |  |  |  |  |
| wet | 13 | 68.56 | 14.62 | 50.50 | 89.23 |
| dry | 13 | 36.64 | 13.97 | 19.70 | 59.04 |
| fat-free | 13 | 45.19 | 6.27 | 31.56 | 52.29 |
| ash-free | 9 | 46.23 | 2.81 | 43.06 | 50.24 |
| Female AHY |  |  |  |  |  |
| wet | 14 | 75.07 | 15.64 | 50.34 | 93.50 |
| dry | 14 | 41.13 | 16.25 | 17.27 | 62.02 |
| fat-free | 14 | 49.84 | 3.38 | 42.62 | 55.13 |
| ash-free | 14 | 47.88 | 3.27 | 40.95 | 53.22 |

## Body Mass and Composition of North American Migrant Birds (continued)

| | N | Mean | S.D. | Min | Max |
|---|---|---|---|---|---|
| **All individuals** | | | | | |
| wet | 77 | 71.51 | 13.49 | 49.25 | 93.50 |
| dry | 77 | 39.71 | 13.88 | 17.27 | 62.02 |
| fat-free | 77 | 46.80 | 4.49 | 31.56 | 55.88 |
| ash-free | 63 | 45.56 | 3.86 | 35.51 | 53.84 |
| **Chaetura pelagica** | | | | | |
| Male HY | | | | | |
| wet | 13 | 30.37 | 1.23 | 28.00 | 32.41 |
| dry | 13 | 16.42 | 1.37 | 13.10 | 18.62 |
| fat-free | 13 | 19.71 | 0.84 | 18.03 | 21.00 |
| Female HY | | | | | |
| wet | 8 | 30.34 | 1.29 | 27.87 | 31.65 |
| dry | 8 | 15.96 | 0.88 | 14.89 | 17.59 |
| fat-free | 8 | 20.08 | 1.39 | 17.64 | 22.22 |
| All individuals | | | | | |
| wet | 21 | 30.36 | 1.22 | 27.87 | 32.41 |
| dry | 21 | 16.24 | 1.20 | 13.10 | 18.62 |
| fat-free | 21 | 19.85 | 1.06 | 17.64 | 22.22 |
| **Archilochus colubris** | | | | | |
| Male AHY | | | | | |
| wet | 9 | 3.50 | 0.77 | 2.38 | 4.99 |
| dry | 9 | 1.64 | 0.62 | 1.12 | 2.93 |
| fat-free | 9 | 2.50 | 0.32 | 1.84 | 2.89 |
| Female HY | | | | | |
| wet | 5 | 3.42 | 0.52 | 2.76 | 4.18 |
| dry | 5 | 1.44 | 0.40 | 1.07 | 1.96 |
| fat-free | 5 | 2.71 | 0.30 | 2.33 | 3.00 |
| ash-free | 1 | 2.27 | | | |
| Female AHY | | | | | |
| wet | 7 | 3.99 | 0.98 | 2.91 | 5.65 |
| dry | 7 | 1.96 | 0.79 | 1.21 | 3.33 |
| fat-free | 7 | 2.72 | 0.42 | 2.25 | 3.37 |
| All individuals | | | | | |
| wet | 21 | 3.64 | 0.80 | 2.38 | 5.65 |
| dry | 21 | 1.70 | 0.64 | 1.07 | 3.33 |
| fat-free | 21 | 2.62 | 0.35 | 1.84 | 3.37 |
| ash-free | 1 | 2.27 | | | |
| **Vireo griseus** | | | | | |
| Male HY | | | | | |
| wet | 34 | 10.12 | 0.77 | 8.28 | 11.99 |
| dry | 34 | 4.31 | 0.57 | 3.29 | 5.52 |
| fat-free | 34 | 8.46 | 0.67 | 7.06 | 10.02 |
| Female HY | | | | | |
| wet | 52 | 9.98 | 0.80 | 7.56 | 11.59 |
| dry | 52 | 4.36 | 0.65 | 3.16 | 5.84 |
| fat-free | 52 | 8.18 | 0.62 | 5.99 | 9.66 |
| All individuals | | | | | |
| wet | 86 | 10.04 | 0.79 | 7.56 | 11.99 |
| dry | 86 | 4.34 | 0.62 | 3.16 | 5.84 |
| fat-free | 86 | 8.29 | 0.65 | 5.99 | 10.02 |
| **Vireo philadelphicus** | | | | | |
| Male HY | | | | | |
| wet | 2 | 14.58 | | 13.95 | 15.20 |

## Body Mass and Composition of North American Migrant Birds (continued)

|  | N | Mean | S.D. | Min | Max |
|---|---|---|---|---|---|
| dry | 2 | 6.56 | | 6.36 | 6.76 |
| fat-free | 2 | 10.64 | | 9.76 | 11.51 |
| **Male AHY** | | | | | |
| wet | 2 | 11.58 | | 10.95 | 12.21 |
| dry | 2 | 5.09 | | 4.95 | 5.23 |
| fat-free | 2 | 9.04 | | 8.45 | 9.62 |
| **Female HY** | | | | | |
| wet | 1 | 14.51 | | | |
| dry | 1 | 7.30 | | | |
| fat-free | 1 | 9.85 | | | |
| **Female AHY** | | | | | |
| wet | 1 | 13.75 | | | |
| dry | 1 | 6.59 | | | |
| fat-free | 1 | 10.07 | | | |
| **All individuals** | | | | | |
| wet | 6 | 13.43 | 1.57 | 10.95 | 15.20 |
| dry | 6 | 6.20 | 0.92 | 4.95 | 7.30 |
| fat-free | 6 | 9.88 | 0.98 | 8.45 | 11.51 |
| | | | | | |
| *Vireo olivaceus* | | | | | |
| **Male HY** | | | | | |
| wet | 39 | 20.25 | 3.60 | 15.47 | 27.46 |
| dry | 39 | 10.03 | 3.37 | 5.50 | 16.36 |
| fat-free | 39 | 14.84 | 1.19 | 12.78 | 17.16 |
| ash-free | 11 | 13.87 | 0.68 | 12.66 | 14.93 |
| **Male AHY** | | | | | |
| wet | 113 | 19.32 | 2.99 | 13.93 | 26.37 |
| dry | 113 | 8.90 | 2.62 | 4.93 | 15.12 |
| fat-free | 113 | 14.99 | 1.33 | 11.13 | 19.79 |
| ash-free | 12 | 14.67 | 1.94 | 12.34 | 19.28 |
| **Female HY** | | | | | |
| wet | 55 | 18.65 | 2.62 | 14.12 | 24.78 |
| dry | 55 | 8.67 | 2.40 | 5.15 | 14.57 |
| fat-free | 55 | 14.54 | 1.19 | 11.64 | 17.11 |
| ash-free | 23 | 13.56 | 0.87 | 12.53 | 16.12 |
| **Female AHY** | | | | | |
| wet | 111 | 18.07 | 2.76 | 13.95 | 26.20 |
| dry | 111 | 8.42 | 2.34 | 5.10 | 15.29 |
| fat-free | 111 | 14.09 | 1.26 | 11.38 | 18.03 |
| ash-free | 9 | 13.36 | 1.15 | 11.93 | 15.23 |
| **All individuals** | | | | | |
| wet | 323 | 18.88 | 3.01 | 13.93 | 27.46 |
| dry | 323 | 8.82 | 2.62 | 4.93 | 16.36 |
| fat-free | 323 | 14.59 | 1.31 | 11.13 | 19.79 |
| ash-free | 57 | 13.88 | 1.26 | 11.93 | 19.28 |
| | | | | | |
| *Catharus fuscescens* | | | | | |
| **Male HY** | | | | | |
| wet | 41 | 42.83 | 4.02 | 32.09 | 53.56 |
| dry | 41 | 24.93 | 4.06 | 12.38 | 33.48 |
| fat-free | 41 | 27.75 | 3.85 | 22.54 | 38.46 |
| **Male AHY** | | | | | |
| wet | 3 | 43.28 | | 42.40 | 44.32 |
| dry | 3 | 24.84 | | 24.51 | 25.50 |
| fat-free | 3 | 33.26 | | 32.88 | 33.55 |
| **Female HY** | | | | | |
| wet | 40 | 39.59 | 4.41 | 28.52 | 48.58 |

**Body Mass and Composition of North American Migrant Birds (continued)**

| | N | Mean | S.D. | Min | Max |
|---|---|---|---|---|---|
| dry | 40 | 22.42 | 4.66 | 11.59 | 30.01 |
| fat-free | 40 | 25.27 | 2.65 | 21.24 | 34.33 |
| Female AHY | | | | | |
| wet | 6 | 39.72 | 2.54 | 36.74 | 43.40 |
| dry | 6 | 23.92 | 3.27 | 20.46 | 29.20 |
| fat-free | 6 | 26.75 | 4.27 | 22.91 | 34.33 |
| All individuals | | | | | |
| wet | 100 | 41.15 | 4.26 | 28.52 | 53.56 |
| dry | 100 | 23.68 | 4.29 | 11.59 | 33.48 |
| fat-free | 100 | 26.66 | 3.61 | 21.24 | 38.46 |
| | | | | | |
| Catharus minimus | | | | | |
| Male HY | | | | | |
| wet | 43 | 31.29 | 4.88 | 23.90 | 47.14 |
| dry | 43 | 13.88 | 5.68 | | 29.51 |
| fat-free | 43 | 25.25 | 2.94 | 20.09 | 34.93 |
| ash-free | 42 | 24.21 | 2.95 | 19.21 | 33.90 |
| Male AHY | | | | | |
| wet | 11 | 38.38 | 5.82 | 28.45 | 47.91 |
| dry | 11 | 19.53 | 5.92 | 10.56 | 29.15 |
| fat-free | 11 | 27.30 | 2.36 | 24.38 | 31.01 |
| ash-free | 11 | 26.26 | 2.31 | 23.31 | 29.88 |
| Female HY | | | | | |
| wet | 59 | 29.99 | 3.60 | 24.66 | 42.17 |
| dry | 59 | 13.00 | 4.10 | | 24.29 |
| fat-free | 59 | 24.42 | 4.08 | 14.46 | 41.84 |
| ash-free | 58 | 23.63 | 3.86 | 19.00 | 40.74 |
| Female AHY | | | | | |
| wet | 15 | 35.58 | 5.92 | 27.17 | 45.20 |
| dry | 15 | 17.35 | 5.72 | 9.63 | 25.87 |
| fat-free | 15 | 26.12 | 1.94 | 22.78 | 30.49 |
| ash-free | 12 | 25.26 | 2.09 | 21.86 | 29.49 |
| All individuals | | | | | |
| wet | 133 | 32.08 | 5.43 | 23.90 | 47.91 |
| dry | 133 | 14.62 | 5.54 | | 29.51 |
| fat-free | 133 | 25.20 | 3.45 | 14.46 | 41.84 |
| ash-frcc | 127 | 24.28 | 3.36 | 19.00 | 40.74 |
| | | | | | |
| Catharus ustulatus | | | | | |
| Male HY | | | | | |
| wet | 94 | 32.40 | 7.59 | 22.70 | 53.41 |
| dry | 94 | 15.61 | 6.62 | 7.12 | 34.96 |
| fat-free | 94 | 24.33 | 1.90 | 20.58 | 29.57 |
| ash-free | 88 | 23.20 | 1.85 | 19.72 | 28.39 |
| Male AHY | | | | | |
| wet | 52 | 36.52 | 7.87 | 22.00 | 49.51 |
| dry | 52 | 18.58 | 7.05 | 7.14 | 30.28 |
| fat-free | 52 | 25.53 | 2.10 | 20.79 | 28.95 |
| ash-free | 39 | 24.26 | 2.18 | 19.89 | 27.93 |
| Female HY | | | | | |
| wet | 68 | 28.74 | 6.45 | 20.56 | 45.45 |
| dry | 68 | 13.17 | 5.44 | 6.99 | 26.83 |
| fat-free | 68 | 22.54 | 1.89 | 19.56 | 27.70 |
| ash-free | 61 | 21.39 | 1.72 | 18.77 | 26.39 |
| Female AHY | | | | | |
| wet | 50 | 34.76 | 6.79 | 20.90 | 44.60 |
| dry | 50 | 17.78 | 6.00 | 7.00 | 27.74 |

## Body Mass and Composition of North American Migrant Birds (continued)

|  | N | Mean | S.D. | Min | Max |
|---|---|---|---|---|---|
| fat-free | 50 | 24.32 | 2.31 | 18.99 | 28.52 |
| ash-free | 38 | 23.11 | 2.35 | 18.18 | 27.18 |
| All individuals |  |  |  |  |  |
| wet | 299 | 32.66 | 7.51 | 20.56 | 53.41 |
| dry | 299 | 15.85 | 6.51 | 6.99 | 34.96 |
| fat-free | 299 | 24.18 | 2.20 | 18.99 | 29.57 |
| ash-free | 239 | 22.96 | 2.18 | 18.18 | 28.39 |
| **Catharus guttatus** |  |  |  |  |  |
| Male HY |  |  |  |  |  |
| wet | 2 | 30.03 |  | 27.58 | 32.48 |
| dry | 2 | 11.74 |  | 10.58 | 12.90 |
| fat-free | 2 | 26.77 |  | 24.84 | 28.70 |
| ash-free | 2 | 25.74 |  | 23.87 | 27.62 |
| Female HY |  |  |  |  |  |
| wet | 6 | 28.58 | 0.84 | 27.09 | 29.44 |
| dry | 6 | 11.38 | 1.00 | 9.95 | 13.03 |
| fat-free | 6 | 25.53 | 1.68 | 23.12 | 27.18 |
| ash-free | 6 | 24.57 | 1.63 | 22.24 | 26.15 |
| All individuals |  |  |  |  |  |
| wet | 12 | 29.13 | 1.55 | 27.09 | 32.48 |
| dry | 12 | 11.82 | 1.07 | 9.95 | 13.17 |
| fat-free | 12 | 25.62 | 1.60 | 23.12 | 28.70 |
| ash-free | 12 | 24.66 | 1.55 | 22.24 | 27.62 |
| **Catharus (Hylocichla) mustelina** |  |  |  |  |  |
| Male HY |  |  |  |  |  |
| wet | 24 | 59.87 | 5.93 | 48.66 | 73.98 |
| dry | 24 | 29.84 | 6.70 | 17.24 | 43.34 |
| fat-free | 24 | 43.32 | 2.46 | 38.57 | 49.09 |
| ash-free | 22 | 41.58 | 2.52 | 36.85 | 47.21 |
| Male AHY |  |  |  |  |  |
| wet | 19 | 59.08 | 6.61 | 47.65 | 72.14 |
| dry | 19 | 30.07 | 8.39 | 19.23 | 48.65 |
| fat-free | 19 | 42.06 | 4.02 | 34.31 | 50.13 |
| ash-free | 15 | 41.47 | 3.34 | 36.04 | 48.28 |
| Female HY |  |  |  |  |  |
| wet | 18 | 54.03 | 6.63 | 44.06 | 65.76 |
| dry | 18 | 25.86 | 6.33 | 16.76 | 36.38 |
| fat-free | 18 | 40.85 | 3.26 | 34.51 | 48.06 |
| ash-free | 16 | 39.53 | 3.33 | 32.89 | 46.36 |
| Female AHY |  |  |  |  |  |
| wet | 28 | 55.68 | 8.36 | 41.94 | 70.86 |
| dry | 28 | 26.60 | 7.96 | 16.00 | 39.91 |
| fat-free | 28 | 41.89 | 3.19 | 35.85 | 49.77 |
| ash-free | 26 | 40.48 | 3.10 | 34.52 | 48.00 |
| All individuals |  |  |  |  |  |
| wet | 105 | 56.92 | 7.03 | 41.94 | 73.98 |
| dry | 105 | 27.70 | 7.33 | 16.00 | 48.65 |
| fat-free | 105 | 42.21 | 3.32 | 34.13 | 50.13 |
| ash-free | 95 | 40.84 | 3.15 | 32.89 | 48.28 |
| **Dumetella carolinensis** |  |  |  |  |  |
| Male HY |  |  |  |  |  |
| wet | 20 | 36.89 | 2.66 | 32.52 | 41.90 |
| dry | 20 | 15.23 | 1.98 | 11.57 | 19.43 |
| fat-free | 20 | 31.62 | 2.66 | 27.15 | 35.20 |

## Body Mass and Composition of North American Migrant Birds (continued)

| | N | Mean | S.D. | Min | Max |
|---|---|---|---|---|---|
| ash-free | 20 | 30.38 | 2.62 | 25.96 | 33.82 |
| **Male AHY** | | | | | |
| wet | 20 | 37.50 | 2.83 | 32.53 | 42.32 |
| dry | 20 | 14.88 | 1.62 | 11.63 | 17.35 |
| fat-free | 20 | 32.28 | 2.18 | 28.76 | 36.78 |
| ash-free | 20 | 31.08 | 2.13 | 27.70 | 35.54 |
| **Female HY** | | | | | |
| wet | 37 | 36.37 | 3.81 | 30.29 | 50.14 |
| dry | 37 | 14.74 | 3.37 | 11.44 | 29.71 |
| fat-free | 37 | 31.36 | 1.89 | 27.15 | 34.37 |
| ash-free | 37 | 30.13 | 1.84 | 26.02 | 33.04 |
| **Female AHY** | | | | | |
| wet | 20 | 38.62 | 3.99 | 33.91 | 50.51 |
| dry | 20 | 16.25 | 3.77 | 12.24 | 27.33 |
| fat-free | 20 | 32.01 | 1.83 | 26.90 | 35.64 |
| ash-free | 20 | 30.80 | 1.77 | 25.89 | 34.28 |
| **All individuals** | | | | | |
| wet | 104 | 37.17 | 3.42 | 30.29 | 50.51 |
| dry | 104 | 15.13 | 2.87 | 11.44 | 29.71 |
| fat-free | 104 | 31.80 | 2.13 | 26.90 | 36.78 |
| ash-free | 104 | 30.58 | 2.08 | 25.89 | 35.54 |
| | | | | | |
| *Cistothorus palustris* | | | | | |
| **Male HY** | | | | | |
| wet | 42 | 10.43 | 1.05 | 7.46 | 12.76 |
| dry | 42 | 3.78 | 0.64 | 2.90 | 5.99 |
| fat-free | 42 | 9.28 | 0.77 | 6.48 | 11.13 |
| ash-free | 27 | 8.76 | 0.71 | 6.19 | 9.89 |
| **Male AHY** | | | | | |
| wet | 23 | 9.98 | 1.28 | 8.07 | 13.21 |
| dry | 23 | 4.01 | 1.06 | 2.57 | 6.60 |
| fat-free | 23 | 8.43 | 0.72 | 7.17 | 9.61 |
| ash-free | 12 | 8.09 | 0.66 | 6.89 | 8.79 |
| **Female HY** | | | | | |
| wet | 30 | 9.47 | 1.14 | 7.04 | 11.21 |
| dry | 30 | 3.64 | 0.82 | 2.34 | 5.55 |
| fat-free | 30 | 8.16 | 0.89 | 6.59 | 9.89 |
| ash-free | 14 | 8.04 | 0.79 | 6.86 | 9.48 |
| **Female AHY** | | | | | |
| wet | 22 | 8.85 | 0.93 | 7.41 | 10.45 |
| dry | 22 | 3.37 | 0.49 | 2.51 | 4.64 |
| fat-free | 22 | 7.73 | 0.83 | 6.36 | 8.78 |
| ash-free | 17 | 7.31 | 0.84 | 6.09 | 8.40 |
| **All individuals** | | | | | |
| wet | 127 | 9.77 | 1.25 | 7.04 | 13.21 |
| dry | 127 | 3.72 | 0.81 | 2.34 | 6.60 |
| fat-free | 127 | 8.48 | 1.01 | 6.36 | 11.13 |
| ash-free | 77 | 8.09 | 0.96 | 6.09 | 9.89 |
| | | | | | |
| *Passer domesticus* | | | | | |
| **Male HY** | | | | | |
| wet | 8 | 27.22 | 3.31 | 21.00 | 32.55 |
| dry | 8 | 10.21 | 2.38 | 7.29 | 15.39 |
| fat-free | 8 | 25.04 | 2.50 | 20.34 | 27.75 |
| **Male AHY** | | | | | |
| wet | 5 | 25.58 | 3.47 | 21.35 | 29.36 |
| dry | 5 | 9.32 | 2.49 | 7.07 | 12.41 |

**Body Mass and Composition of North American Migrant Birds (continued)**

|  | N | Mean | S.D. | Min | Max |
|---|---|---|---|---|---|
| fat-free | 5 | 24.15 | 2.74 | 20.73 | 27.61 |
| Female HY |  |  |  |  |  |
| wet | 14 | 24.82 | 2.80 | 19.99 | 31.01 |
| dry | 14 | 8.73 | 1.78 | 6.99 | 13.22 |
| fat-free | 14 | 23.00 | 2.42 | 19.45 | 28.32 |
| Female AHY |  |  |  |  |  |
| wet | 5 | 23.65 | 3.80 | 19.78 | 28.82 |
| dry | 5 | 9.73 | 3.61 | 6.86 | 15.67 |
| fat-free | 5 | 21.42 | 1.81 | 19.13 | 23.89 |
| All individuals |  |  |  |  |  |
| wet | 32 | 25.36 | 3.27 | 19.78 | 32.55 |
| dry | 32 | 9.35 | 2.34 | 6.86 | 15.67 |
| fat-free | 32 | 23.44 | 2.60 | 19.13 | 28.32 |
| *Vermivora peregrina* |  |  |  |  |  |
| Male HY |  |  |  |  |  |
| wet | 13 | 13.26 | 1.02 | 11.72 | 14.71 |
| dry | 13 | 7.78 | 0.71 | 6.59 | 8.82 |
| fat-free | 13 | 7.99 | 0.55 | 6.92 | 8.99 |
| ash-free | 3 | 7.27 |  | 6.65 | 7.63 |
| Male AHY |  |  |  |  |  |
| wet | 17 | 13.17 | 1.26 | 10.52 | 15.37 |
| dry | 17 | 7.79 | 1.26 | 4.43 | 9.93 |
| fat-free | 17 | 7.97 | 0.49 | 6.84 | 8.81 |
| Female HY |  |  |  |  |  |
| wet | 27 | 12.86 | 1.25 | 10.36 | 15.83 |
| dry | 27 | 7.71 | 1.15 | 5.71 | 10.72 |
| fat-free | 27 | 7.70 | 0.45 | 6.75 | 8.94 |
| ash-free | 7 | 7.11 | 0.43 | 6.52 | 7.56 |
| Female AHY |  |  |  |  |  |
| wet | 36 | 12.63 | 1.15 | 9.25 | 14.15 |
| dry | 36 | 7.25 | 1.24 | 3.22 | 8.80 |
| fat-free | 36 | 7.87 | 0.50 | 6.73 | 9.35 |
| All individuals |  |  |  |  |  |
| wet | 95 | 12.90 | 1.18 | 9.25 | 15.83 |
| dry | 95 | 7.57 | 1.16 | 3.22 | 10.72 |
| fat-free | 95 | 7.85 | 0.49 | 6.73 | 9.35 |
| ash-free | 12 | 7.19 | 0.41 | 6.52 | 7.63 |
| *Parula americana* |  |  |  |  |  |
| Male HY |  |  |  |  |  |
| wet | 75 | 8.09 | 1.23 | 5.51 | 11.17 |
| dry | 75 | 3.87 | 1.20 | 2.26 | 7.07 |
| fat-free | 75 | 6.22 | 0.46 | 4.88 | 7.24 |
| ash-free | 56 | 6.11 | 0.37 | 5.13 | 6.96 |
| Male AHY |  |  |  |  |  |
| wet | 54 | 8.07 | 1.20 | 5.73 | 10.64 |
| dry | 54 | 3.87 | 1.17 | 2.20 | 6.30 |
| fat-free | 54 | 6.20 | 0.38 | 5.28 | 7.11 |
| ash-free | 54 | 5.97 | 0.37 | 5.09 | 6.86 |
| Female HY |  |  |  |  |  |
| wet | 96 | 6.92 | 1.03 | 5.56 | 10.08 |
| dry | 96 | 3.10 | 0.98 | 2.11 | 5.56 |
| fat-free | 96 | 5.58 | 0.41 | 4.40 | 6.78 |
| ash-free | 49 | 5.52 | 0.39 | 4.20 | 6.52 |
| Female AHY |  |  |  |  |  |
| wet | 39 | 7.45 | 1.08 | 5.81 | 10.14 |

## Body Mass and Composition of North American Migrant Birds (continued)

| | N | Mean | S.D. | Min | Max |
|---|---|---|---|---|---|
| dry | 39 | 3.50 | 1.15 | 2.33 | 6.11 |
| fat-free | 39 | 5.82 | 0.34 | 5.03 | 6.56 |
| ash-free | 39 | 5.60 | 0.32 | 4.85 | 6.31 |
| All individuals | | | | | |
| wet | 304 | 7.60 | 1.25 | 5.51 | 11.17 |
| dry | 304 | 3.56 | 1.17 | 2.11 | 7.07 |
| fat-free | 304 | 5.93 | 0.49 | 4.40 | 7.24 |
| ash-free | 236 | 5.82 | 0.43 | 4.20 | 6.96 |
| | | | | | |
| **Dendroica pensylvanica** | | | | | |
| Male HY | | | | | |
| wet | 11 | 13.22 | 0.71 | 12.08 | 13.99 |
| dry | 11 | 7.85 | 0.65 | 6.65 | 8.69 |
| fat-free | 11 | 8.03 | 0.39 | 7.30 | 8.53 |
| Male AHY | | | | | |
| wet | 11 | 12.92 | 1.40 | 9.21 | 14.46 |
| dry | 11 | 7.17 | 1.30 | 3.69 | 8.89 |
| fat-free | 11 | 8.33 | 0.40 | 7.41 | 8.78 |
| Female HY | | | | | |
| wet | 20 | 12.17 | 1.20 | 9.23 | 13.72 |
| dry | 20 | 6.89 | 1.27 | 3.44 | 8.36 |
| fat-free | 20 | 7.84 | 0.43 | 6.97 | 8.44 |
| Female AHY | | | | | |
| wet | 26 | 13.04 | 0.92 | 11.16 | 15.31 |
| dry | 26 | 7.61 | 0.74 | 6.01 | 9.41 |
| fat-free | 26 | 7.97 | 0.42 | 7.17 | 9.01 |
| All individuals | | | | | |
| wet | 73 | 12.79 | 1.22 | 9.05 | 15.57 |
| dry | 73 | 7.35 | 1.16 | 3.16 | 9.41 |
| fat-free | 73 | 8.01 | 0.45 | 6.97 | 9.07 |
| | | | | | |
| **Dendroica magnolia** | | | | | |
| Male HY | | | | | |
| wet | 21 | 9.68 | 1.55 | 7.43 | 12.51 |
| dry | 21 | 4.88 | 1.51 | 2.89 | 7.22 |
| fat-free | 21 | 6.95 | 0.47 | 6.00 | 7.70 |
| ash-free | 19 | 6.74 | 0.42 | 6.07 | 7.42 |
| Male AHY | | | | | |
| wet | 2 | 9.31 | | 8.30 | 10.32 |
| dry | 2 | 4.38 | | 3.04 | 5.72 |
| fat-free | 2 | 7.06 | | 6.66 | 7.45 |
| ash-free | 2 | 6.80 | | 6.40 | 7.20 |
| Female HY | | | | | |
| wet | 9 | 9.56 | 1.37 | 7.49 | 11.25 |
| dry | 9 | 5.02 | 1.38 | 2.78 | 6.33 |
| fat-free | 9 | 6.68 | 0.35 | 6.31 | 7.44 |
| ash-free | 6 | 6.38 | 0.22 | 6.07 | 6.54 |
| Female AHY | | | | | |
| wet | 3 | 9.69 | | 9.30 | 10.04 |
| dry | 3 | 4.92 | | 4.58 | 5.19 |
| fat-free | 3 | 7.35 | | 6.84 | 8.32 |
| All individuals | | | | | |
| wet | 35 | 9.63 | 1.39 | 7.43 | 12.51 |
| dry | 35 | 4.89 | 1.39 | 2.78 | 7.22 |
| fat-free | 35 | 6.92 | 0.49 | 6.00 | 8.32 |
| ash-free | 27 | 6.67 | 0.41 | 6.07 | 7.42 |

## Body Mass and Composition of North American Migrant Birds (continued)

| | N | Mean | S.D. | Min | Max |
|---|---|---|---|---|---|
| Dendroica coronata | | | | | |
| Male HY | | | | | |
| wet | 25 | 11.75 | 0.89 | 10.16 | 13.77 |
| dry | 25 | 4.38 | 0.43 | 3.67 | 5.34 |
| fat-free | 25 | 10.56 | 0.82 | 9.18 | 12.13 |
| Male AHY | | | | | |
| wet | 1 | 14.59 | | | |
| dry | 1 | 7.37 | | | |
| fat-free | 1 | 10.43 | | | |
| ash-free | 1 | 10.10 | | | |
| Female HY | | | | | |
| wet | 91 | 11.97 | 1.35 | 8.63 | 14.99 |
| dry | 91 | 5.16 | 1.04 | 3.54 | 8.08 |
| fat-free | 91 | 10.00 | 0.84 | 6.99 | 12.32 |
| All individuals | | | | | |
| wet | 120 | 11.94 | 1.28 | 8.63 | 14.99 |
| dry | 120 | 5.00 | 1.01 | 3.54 | 8.08 |
| fat-free | 120 | 10.13 | 0.86 | 6.99 | 12.32 |
| ash-free | 1 | 10.10 | | | |
| | | | | | |
| Dendroica fusca | | | | | |
| Male HY | | | | | |
| wet | 6 | 14.01 | 0.86 | 12.83 | 15.16 |
| dry | 6 | 8.33 | 0.49 | 7.84 | 9.08 |
| fat-free | 6 | 8.34 | 0.70 | 7.45 | 9.46 |
| ash-free | 6 | 8.01 | 0.68 | 7.12 | 9.08 |
| Male AHY | | | | | |
| wet | 12 | 13.31 | 1.61 | 10.35 | 16.41 |
| dry | 12 | 7.91 | 1.60 | 4.42 | 10.34 |
| fat-free | 12 | 7.86 | 0.41 | 7.37 | 8.72 |
| ash-free | 12 | 7.56 | 0.39 | 7.08 | 8.37 |
| Female HY | | | | | |
| wet | 17 | 12.46 | 1.25 | 10.19 | 14.66 |
| dry | 17 | 7.50 | 1.22 | 5.66 | 10.69 |
| fat-free | 17 | 7.30 | 0.47 | 6.47 | 8.08 |
| ash-free | 17 | 7.02 | 0.46 | 6.16 | 7.76 |
| Female AHY | | | | | |
| wet | 9 | 13.03 | 0.62 | 11.88 | 13.90 |
| dry | 9 | 7.82 | 0.38 | 7.16 | 8.27 |
| fat-free | 9 | 7.60 | 0.63 | 6.83 | 8.37 |
| ash-free | 9 | 7.31 | 0.61 | 6.55 | 8.07 |
| All individuals | | | | | |
| wet | 52 | 12.98 | 1.28 | 10.19 | 16.41 |
| dry | 52 | 7.81 | 1.13 | 4.42 | 10.69 |
| fat-free | 52 | 7.58 | 0.61 | 6.47 | 9.46 |
| ash-free | 51 | 7.30 | 0.59 | 6.16 | 9.08 |
| | | | | | |
| Dendroica dominica | | | | | |
| Male HY | | | | | |
| wet | 4 | 9.71 | | 9.31 | 10.27 |
| dry | 4 | 3.98 | | 3.58 | 4.36 |
| fat-free | 4 | 8.20 | | 7.88 | 8.47 |
| ash-free | 4 | 7.86 | | 7.57 | 8.10 |
| Female HY | | | | | |
| wet | 9 | 9.16 | 0.69 | 7.96 | 10.12 |
| dry | 9 | 3.68 | 0.45 | 3.05 | 4.37 |
| fat-free | 9 | 7.93 | 0.44 | 7.18 | 8.37 |

## Body Mass and Composition of North American Migrant Birds (continued)

| | N | Mean | S.D. | Min | Max |
|---|---|---|---|---|---|
| ash-free | 9 | 7.61 | 0.41 | 6.87 | 7.98 |
| **Female AHY** | | | | | |
| wet | 1 | 15.26 | | | |
| dry | 1 | 8.79 | | | |
| fat-free | 1 | 9.37 | | | |
| **All individuals** | | | | | |
| wet | 14 | 9.76 | 1.70 | 7.96 | 15.26 |
| dry | 14 | 4.13 | 1.41 | 3.05 | 8.79 |
| fat-free | 14 | 8.11 | 0.53 | 7.18 | 9.37 |
| ash-free | 13 | 7.69 | 0.38 | 6.87 | 8.10 |
| | | | | | |
| **Dendroica palmarum** | | | | | |
| **Male HY** | | | | | |
| wet | 30 | 10.43 | 1.23 | 8.27 | 12.51 |
| dry | 30 | 4.60 | 1.19 | 3.25 | 7.02 |
| fat-free | 30 | 8.69 | 0.47 | 7.46 | 9.65 |
| ash-free | 29 | 8.36 | 0.46 | 7.19 | 9.29 |
| **Male AHY** | | | | | |
| wet | 14 | 11.34 | 1.09 | 9.68 | 13.60 |
| dry | 14 | 5.48 | 1.02 | 4.16 | 7.62 |
| fat-free | 14 | 8.73 | 0.43 | 7.88 | 9.67 |
| ash-free | 14 | 8.42 | 0.42 | 7.55 | 9.32 |
| **Female HY** | | | | | |
| wet | 34 | 9.76 | 1.37 | 7.75 | 12.90 |
| dry | 34 | 4.24 | 1.27 | 2.70 | 6.91 |
| fat-free | 34 | 8.13 | 0.54 | 7.29 | 9.36 |
| ash-free | 34 | 7.84 | 0.52 | 7.00 | 9.01 |
| **Female AHY** | | | | | |
| wet | 4 | 9.53 | 0.64 | 8.78 | 10.22 |
| dry | 4 | 4.23 | 0.65 | 3.55 | 5.11 |
| fat-free | 4 | 7.92 | 0.21 | 7.71 | 8.19 |
| ash-free | 4 | 7.64 | 0.23 | 7.43 | 7.94 |
| **All individuals** | | | | | |
| wet | 100 | 10.25 | 1.29 | 7.75 | 13.60 |
| dry | 100 | 4.62 | 1.18 | 2.70 | 7.62 |
| fat-free | 100 | 8.38 | 0.58 | 7.10 | 9.67 |
| ash-free | 99 | 8.07 | 0.56 | 6.82 | 9.32 |
| | | | | | |
| **Dendroica castanea** | | | | | |
| **Male HY** | | | | | |
| wet | 2 | 14.52 | | 12.99 | 16.04 |
| dry | 2 | 8.44 | | 6.78 | 10.11 |
| fat-free | 2 | 9.56 | | 9.42 | 9.69 |
| **Male AHY** | | | | | |
| wet | 3 | 16.84 | | 16.50 | 17.46 |
| dry | 3 | 8.87 | | 7.55 | 10.19 |
| fat-free | 3 | 11.24 | | 10.90 | 11.79 |
| **Female HY** | | | | | |
| wet | 2 | 14.89 | | 14.56 | 15.22 |
| dry | 2 | 8.09 | | 7.77 | 8.41 |
| fat-free | 2 | 9.92 | | 9.81 | 10.03 |
| **Female AHY** | | | | | |
| wet | 6 | 14.60 | 1.16 | 12.80 | 16.10 |
| dry | 6 | 7.71 | 0.96 | 6.28 | 8.95 |
| fat-free | 6 | 10.10 | 0.63 | 9.24 | 10.96 |
| **All individuals** | | | | | |
| wet | 14 | 14.99 | 1.47 | 12.80 | 17.46 |

## Body Mass and Composition of North American Migrant Birds (continued)

| | N | Mean | S.D. | Min | Max |
|---|---|---|---|---|---|
| dry | 14 | 8.15 | 1.13 | 6.28 | 10.19 |
| fat-free | 14 | 10.17 | 0.79 | 9.03 | 11.79 |
| **Dendroica striata** | | | | | |
| Male HY | | | | | |
| wet | 6 | 16.20 | 3.85 | 11.13 | 19.97 |
| dry | 6 | 8.73 | 4.02 | 3.87 | 12.21 |
| fat-free | 6 | 10.77 | 0.89 | 9.29 | 11.84 |
| ash-free | 5 | 10.66 | 0.54 | 9.96 | 11.38 |
| Male AHY | | | | | |
| wet | 7 | 13.90 | 5.33 | 9.34 | 24.08 |
| dry | 7 | 6.88 | 4.79 | 3.22 | 15.87 |
| fat-free | 7 | 10.11 | 0.90 | 8.82 | 11.80 |
| ash-free | 5 | 9.66 | 0.29 | 9.19 | 9.95 |
| Female HY | | | | | |
| wet | 3 | 15.98 | | 11.92 | 18.14 |
| dry | 3 | 8.81 | | 4.66 | 11.07 |
| fat-free | 3 | 10.36 | | 10.20 | 10.52 |
| ash-free | 3 | 9.93 | | 9.66 | 10.14 |
| Female AHY | | | | | |
| wet | 3 | 11.30 | | 10.96 | 11.59 |
| dry | 3 | 4.30 | | 4.04 | 4.67 |
| fat-free | 3 | 9.99 | | 9.71 | 10.25 |
| ash-free | 3 | 9.64 | | 9.36 | 9.91 |
| All individuals | | | | | |
| wet | 19 | 14.54 | 4.26 | 9.34 | 24.08 |
| dry | 19 | 7.36 | 4.03 | 3.22 | 15.87 |
| fat-free | 19 | 10.34 | 0.78 | 8.82 | 11.84 |
| ash-free | 16 | 10.02 | 0.57 | 9.19 | 11.38 |
| **Setophaga ruticilla** | | | | | |
| Male HY | | | | | |
| wet | 23 | 9.89 | 1.19 | 7.11 | 11.83 |
| dry | 23 | 5.41 | 0.91 | 3.03 | 6.64 |
| fat-free | 23 | 6.66 | 0.53 | 5.50 | 7.94 |
| ash-free | 23 | 6.42 | 0.52 | 5.19 | 7.67 |
| Male AHY | | | | | |
| wet | 27 | 8.81 | 1.20 | 6.99 | 11.42 |
| dry | 27 | 4.30 | 1.26 | 2.51 | 6.95 |
| fat-free | 27 | 6.62 | 0.37 | 5.93 | 7.35 |
| ash-free | 26 | 6.39 | 0.37 | 5.71 | 7.10 |
| Female HY | | | | | |
| wet | 18 | 8.93 | 1.50 | 6.69 | 10.85 |
| dry | 18 | 4.57 | 1.37 | 2.52 | 6.34 |
| fat-free | 18 | 6.39 | 0.33 | 5.59 | 7.00 |
| ash-free | 18 | 6.15 | 0.31 | 5.37 | 6.75 |
| Female AHY | | | | | |
| wet | 15 | 8.91 | 1.08 | 7.00 | 10.79 |
| dry | 15 | 4.54 | 1.10 | 2.64 | 6.47 |
| fat-free | 15 | 6.38 | 0.23 | 6.07 | 6.90 |
| ash-free | 15 | 6.16 | 0.22 | 5.85 | 6.68 |
| All individuals | | | | | |
| wet | 102 | 9.11 | 1.30 | 6.60 | 11.83 |
| dry | 102 | 4.71 | 1.22 | 2.48 | 6.95 |
| fat-free | 102 | 6.49 | 0.42 | 5.50 | 7.94 |
| ash-free | 101 | 6.26 | 0.41 | 5.19 | 7.67 |

## Body Mass and Composition of North American Migrant Birds (continued)

| | N | Mean | S.D. | Min | Max |
|---|---|---|---|---|---|
| **Protonotaria citrea** | | | | | |
| *Male HY* | | | | | |
| wet | 47 | 13.62 | 1.31 | 11.90 | 18.64 |
| dry | 47 | 6.21 | 1.16 | 4.86 | 11.56 |
| fat-free | 47 | 11.12 | 0.67 | 10.04 | 12.75 |
| ash-free | 46 | 10.65 | 0.65 | 9.61 | 12.24 |
| *Male AHY* | | | | | |
| wet | 2 | 17.66 | | 17.09 | 18.23 |
| dry | 2 | 10.6 | | 8.96 | 11.17 |
| fat-free | 2 | 11.36 | | 10.64 | 12.07 |
| ash-free | 2 | 10.92 | | 10.23 | 11.61 |
| *Female HY* | | | | | |
| wet | 20 | 13.77 | 1.35 | 11.39 | 16.78 |
| dry | 20 | 6.57 | 1.32 | 4.63 | 9.90 |
| fat-free | 20 | 10.68 | 0.52 | 9.68 | 11.86 |
| ash-free | 19 | 10.25 | 0.50 | 9.27 | 11.40 |
| *Female AHY* | | | | | |
| wet | 3 | 17.61 | | 17.12 | 18.13 |
| dry | 3 | 10.19 | | 9.46 | 11.45 |
| fat-free | 3 | 11.10 | | 10.23 | 11.63 |
| ash-free | 3 | 10.66 | | 9.78 | 11.22 |
| *All individuals* | | | | | |
| wet | 72 | 13.94 | 1.62 | 11.39 | 18.64 |
| dry | 72 | 6.58 | 1.55 | 4.63 | 11.56 |
| fat-free | 72 | 11.01 | 0.66 | 9.68 | 12.75 |
| ash-free | 70 | 10.55 | 0.64 | 9.27 | 12.24 |
| | | | | | |
| **Helmitheros vermivorus** | | | | | |
| *Male HY* | | | | | |
| wet | 6 | 16.71 | 2.65 | 11.38 | 18.30 |
| dry | 6 | 9.30 | 2.49 | 4.40 | 10.85 |
| fat-free | 6 | 11.16 | 0.67 | 10.25 | 12.03 |
| ash-free | 3 | 10.27 | | 9.80 | 10.97 |
| *Female HY* | | | | | |
| wet | 6 | 16.43 | 2.51 | 11.62 | 18.47 |
| dry | 6 | 9.00 | 2.11 | 4.99 | 10.78 |
| fat-free | 6 | 11.10 | 0.62 | 10.17 | 11.89 |
| ash-free | 5 | 10.47 | 0.54 | 9.71 | 11.11 |
| *Female AHY* | | | | | |
| wet | 6 | 15.57 | 2.54 | 11.13 | 18.22 |
| dry | 6 | 8.60 | 2.24 | 4.98 | 10.55 |
| fat-free | 6 | 10.58 | 0.73 | 9.58 | 11.54 |
| ash-free | 5 | 9.98 | 0.69 | 9.15 | 11.05 |
| *All individuals* | | | | | |
| wet | 28 | 15.63 | 2.40 | 11.01 | 18.47 |
| dry | 28 | 8.47 | 2.18 | 4.40 | 10.85 |
| fat-free | 28 | 10.79 | 0.78 | 9.58 | 12.46 |
| ash-free | 20 | 10.11 | 0.62 | 9.15 | 11.11 |
| | | | | | |
| **Seiurus aurocapillus** | | | | | |
| *Male HY* | | | | | |
| wet | 10 | 23.92 | 2.08 | 20.94 | 26.68 |
| dry | 10 | 13.12 | 1.41 | 10.68 | 14.81 |
| fat-free | 10 | 15.78 | 0.96 | 14.65 | 17.21 |
| ash-free | 8 | 15.33 | 0.93 | 14.04 | 16.54 |

**Body Mass and Composition of North American Migrant Birds (continued)**

|  | N | Mean | S.D. | Min | Max |
|---|---|---|---|---|---|
| **Male AHY** |  |  |  |  |  |
| wet | 2 | 22.86 |  | 22.50 | 23.23 |
| dry | 2 | 11.96 |  | 11.66 | 12.27 |
| fat-free | 2 | 15.72 |  | 15.62 | 15.82 |
| ash-free | 2 | 15.12 |  | 15.04 | 15.19 |
| **Female HY** |  |  |  |  |  |
| wet | 9 | 21.90 | 1.38 | 20.74 | 24.85 |
| dry | 9 | 11.31 | 0.95 | 10.61 | 13.50 |
| fat-free | 9 | 15.57 | 0.69 | 14.98 | 16.45 |
| ash-free | 4 | 15.41 |  | 14.48 | 15.82 |
| **Female AHY** |  |  |  |  |  |
| wet | 11 | 22.05 | 0.59 | 20.82 | 22.99 |
| dry | 11 | 11.72 | 0.47 | 11.09 | 12.51 |
| fat-free | 11 | 15.05 | 0.77 | 13.89 | 16.43 |
| ash-free | 2 | 13.91 |  | 13.30 | 14.53 |
| **All individuals** |  |  |  |  |  |
| wet | 33 | 22.68 | 1.63 | 20.74 | 26.68 |
| dry | 33 | 12.04 | 1.19 | 10.61 | 14.81 |
| fat-free | 33 | 15.52 | 0.87 | 13.89 | 17.21 |
| ash-free | 17 | 15.23 | 0.92 | 13.30 | 16.57 |
| **Seiurus noveboracensis** |  |  |  |  |  |
| **Male HY** |  |  |  |  |  |
| wet | 7 | 17.17 | 4.13 | 13.22 | 25.69 |
| dry | 7 | 8.55 | 3.23 | 6.04 | 15.39 |
| fat-free | 7 | 12.94 | 1.71 | 9.94 | 14.95 |
| ash-free | 5 | 12.09 | 1.95 | 9.41 | 14.38 |
| **Male AHY** |  |  |  |  |  |
| wet | 7 | 21.86 | 3.91 | 14.06 | 25.49 |
| dry | 7 | 11.53 | 2.93 | 5.80 | 14.13 |
| fat-free | 7 | 14.91 | 1.41 | 12.24 | 16.55 |
| ash-free | 6 | 14.37 | 1.53 | 11.69 | 15.94 |
| **Female HY** |  |  |  |  |  |
| wet | 15 | 20.30 | 3.72 | 13.17 | 24.25 |
| dry | 15 | 11.17 | 2.90 | 6.12 | 14.32 |
| fat-free | 15 | 13.67 | 1.72 | 8.76 | 15.80 |
| ash-free | 15 | 13.12 | 1.69 | 8.27 | 14.18 |
| **Female AHY** |  |  |  |  |  |
| wet | 33 | 20.77 | 2.34 | 14.92 | 24.56 |
| dry | 33 | 11.55 | 2.37 | 5.44 | 14.72 |
| fat-free | 33 | 13.61 | 1.11 | 11.71 | 17.00 |
| ash-free | 31 | 13.07 | 1.12 | 11.18 | 16.43 |
| **All individuals** |  |  |  |  |  |
| wet | 89 | 20.37 | 3.41 | 13.17 | 25.69 |
| dry | 89 | 11.13 | 2.96 | 4.94 | 15.49 |
| fat-free | 89 | 13.68 | 1.30 | 8.76 | 17.00 |
| ash-free | 82 | 13.14 | 1.32 | 8.27 | 16.43 |
| **Oporornis formosus** |  |  |  |  |  |
| **Male HY** |  |  |  |  |  |
| wet | 69 | 16.69 | 2.64 | 10.89 | 20.31 |
| dry | 69 | 8.95 | 2.70 | 4.21 | 16.28 |
| fat-free | 69 | 11.52 | 1.16 | 5.29 | 13.43 |
| ash-free | 65 | 11.11 | 1.13 | 4.76 | 12.89 |
| **Male AHY** |  |  |  |  |  |
| wet | 10 | 15.68 | 2.71 | 12.12 | 19.84 |
| dry | 10 | 7.77 | 2.70 | 4.52 | 11.62 |

## Body Mass and Composition of North American Migrant Birds (continued)

| | N | Mean | S.D. | Min | Max |
|---|---|---|---|---|---|
| fat-free | 10 | 11.48 | 0.51 | 10.66 | 12.27 |
| ash-free | 10 | 11.04 | 0.52 | 10.21 | 11.88 |
| Female HY | | | | | |
| wet | 48 | 16.24 | 2.42 | 11.82 | 19.77 |
| dry | 48 | 8.60 | 2.37 | 4.25 | 11.91 |
| fat-free | 48 | 11.16 | 0.82 | 9.19 | 12.84 |
| ash-free | 48 | 10.72 | 0.80 | 8.77 | 12.31 |
| Female AHY | | | | | |
| wet | 10 | 16.58 | 2.60 | 12.00 | 19.30 |
| dry | 10 | 8.76 | 2.41 | 4.34 | 10.80 |
| fat-free | 10 | 11.38 | 0.52 | 10.46 | 12.26 |
| ash-free | 10 | 10.91 | 0.50 | 9.99 | 11.68 |
| All individuals | | | | | |
| wet | 156 | 16.46 | 2.54 | 10.83 | 20.31 |
| dry | 156 | 8.76 | 2.53 | 4.21 | 16.28 |
| fat-free | 156 | 11.36 | 0.96 | 5.29 | 13.43 |
| ash-free | 151 | 10.93 | 0.93 | 4.76 | 12.89 |
| | | | | | |
| Geothlypis trichas | | | | | |
| Male HY | | | | | |
| wet | 9 | 11.41 | 1.37 | 9.98 | 13.67 |
| dry | 9 | 5.35 | 1.35 | 3.75 | 6.90 |
| fat-free | 9 | 8.67 | 0.54 | 7.90 | 9.46 |
| Male AHY | | | | | |
| wet | 8 | 9.94 | 0.30 | 9.61 | 10.35 |
| dry | 8 | 4.34 | 0.31 | 3.98 | 4.68 |
| fat-free | 8 | 8.07 | 0.42 | 7.55 | 8.65 |
| All individuals | | | | | |
| wet | 19 | 10.78 | 1.19 | 9.61 | 13.67 |
| dry | 19 | 4.95 | 1.06 | 3.75 | 6.90 |
| fat-free | 19 | 8.36 | 0.54 | 7.55 | 9.46 |
| | | | | | |
| Wilsonia citrina | | | | | |
| Male HY | | | | | |
| wet | 74 | 10.59 | 1.99 | 8.17 | 14.99 |
| dry | 74 | 4.84 | 1.79 | 3.24 | 9.01 |
| fat-free | 74 | 8.41 | 0.82 | 4.06 | 9.94 |
| ash-free | 37 | 8.46 | 0.94 | 3.75 | 9.57 |
| Male AHY | | | | | |
| wet | 8 | 12.76 | 2.80 | 9.93 | 16.42 |
| dry | 8 | 6.30 | 2.44 | 3.84 | 9.46 |
| fat-free | 8 | 9.22 | 0.79 | 8.54 | 10.76 |
| ash-free | 8 | 8.98 | 0.77 | 8.24 | 10.37 |
| Female HY | | | | | |
| wet | 52 | 9.26 | 1.53 | 7.46 | 13.87 |
| dry | 52 | 4.04 | 1.34 | 2.79 | 8.74 |
| fat-free | 52 | 7.66 | 0.61 | 6.14 | 9.26 |
| ash-free | 12 | 7.92 | 0.60 | 6.82 | 8.92 |
| All individuals | | | | | |
| wet | 153 | 10.29 | 2.02 | 7.46 | 16.42 |
| dry | 153 | 4.69 | 1.73 | 2.79 | 9.46 |
| fat-free | 153 | 8.20 | 0.84 | 4.06 | 10.76 |
| ash-free | 74 | 8.35 | 0.84 | 3.75 | 10.37 |
| | | | | | |
| Melospiza georgiana | | | | | |
| Male HY | | | | | |
| wet | 4 | 17.22 | | 16.20 | 18.06 |

## Body Mass and Composition of North American Migrant Birds (continued)

|  | N | Mean | S.D. | Min | Max |
|---|---|---|---|---|---|
| dry | 4 | 7.35 |  | 6.70 | 7.81 |
| fat-free | 4 | 14.76 |  | 13.99 | 15.79 |
| ash-free | 4 | 14.17 |  | 13.41 | 15.21 |
| Female HY |  |  |  |  |  |
| wet | 3 | 14.87 |  | 13.72 | 15.93 |
| dry | 3 | 6.04 |  | 5.35 | 6.71 |
| fat-free | 3 | 12.97 |  | 12.23 | 13.63 |
| ash-free | 3 | 12.48 |  | 11.76 | 13.11 |
| All individuals |  |  |  |  |  |
| wet | 7 | 16.22 | 1.54 | 13.72 | 18.06 |
| dry | 7 | 6.79 | 0.88 | 5.35 | 7.81 |
| fat-free | 7 | 13.99 | 1.21 | 12.23 | 15.79 |
| ash-free | 7 | 13.45 | 1.16 | 11.76 | 15.21 |
| *Zonotrichia albicollis* |  |  |  |  |  |
| Male HY |  |  |  |  |  |
| wet | 7 | 25.41 | 2.58 | 23.28 | 29.27 |
| dry | 7 | 10.42 | 1.86 | 8.46 | 13.92 |
| fat-free | 7 | 21.85 | 1.50 | 20.11 | 24.69 |
| ash-free | 2 | 22.76 |  | 21.37 | 24.14 |
| Male AHY |  |  |  |  |  |
| wet | 5 | 24.54 | 2.04 | 21.80 | 27.14 |
| dry | 5 | 9.44 | 1.79 | 7.36 | 12.29 |
| fat-free | 5 | 21.99 | 1.16 | 20.49 | 23.20 |
| Female HY |  |  |  |  |  |
| wet | 26 | 23.42 | 1.65 | 20.90 | 28.13 |
| dry | 26 | 9.18 | 1.06 | 7.53 | 12.02 |
| fat-free | 26 | 20.78 | 1.01 | 19.04 | 23.23 |
| ash-free | 5 | 21.15 | 1.02 | 20.32 | 22.31 |
| Female AHY |  |  |  |  |  |
| wet | 2 | 21.94 |  | 21.57 | 22.32 |
| dry | 2 | 8.30 |  | 7.72 | 8.88 |
| fat-free | 2 | 19.50 |  | 18.86 | 20.14 |
| All individuals |  |  |  |  |  |
| wet | 40 | 23.84 | 2.00 | 20.90 | 29.27 |
| dry | 40 | 9.39 | 1.37 | 7.36 | 13.92 |
| fat-free | 40 | 21.06 | 1.25 | 18.86 | 24.69 |
| ash-free | 7 | 21.61 | 1.39 | 20.32 | 24.14 |
| *Passerculus sandwichensis* |  |  |  |  |  |
| Male HY |  |  |  |  |  |
| wet | 55 | 18.39 | 1.64 | 15.00 | 23.30 |
| dry | 55 | 7.09 | 0.92 | 5.47 | 9.30 |
| fat-free | 55 | 16.54 | 1.31 | 13.69 | 20.14 |
| Male AHY |  |  |  |  |  |
| wet | 75 | 17.48 | 1.43 | 14.44 | 22.60 |
| dry | 75 | 6.70 | 0.91 | 5.28 | 10.72 |
| fat-free | 75 | 15.81 | 1.22 | 13.12 | 20.34 |
| Female HY |  |  |  |  |  |
| wet | 51 | 17.11 | 1.34 | 14.00 | 20.65 |
| dry | 51 | 6.36 | 0.76 | 5.12 | 8.14 |
| fat-free | 51 | 15.51 | 1.20 | 12.61 | 18.40 |
| Female AHY |  |  |  |  |  |
| wet | 130 | 16.39 | 1.23 | 13.54 | 20.30 |
| dry | 130 | 6.39 | 0.93 | 4.99 | 9.94 |
| fat-free | 130 | 14.64 | 1.07 | 12.00 | 16.96 |

## Body Mass and Composition of North American Migrant Birds (continued)

|  | N | Mean | S.D. | Min | Max |
|---|---|---|---|---|---|
| All individuals | | | | | |
| wet | 311 | 17.12 | 1.55 | 13.54 | 23.30 |
| dry | 311 | 6.58 | 0.93 | 4.99 | 10.72 |
| fat-free | 311 | 15.40 | 1.37 | 12.00 | 20.34 |
| Ammodramus maritimus | | | | | |
| Male HY | | | | | |
| wet | 22 | 22.85 | 1.51 | 19.65 | 25.56 |
| dry | 22 | 8.23 | 0.68 | 7.08 | 9.68 |
| fat-free | 22 | 20.46 | 1.46 | 17.15 | 22.84 |
| ash-free | 16 | 19.43 | 1.57 | 16.49 | 21.89 |
| Male AHY | | | | | |
| wet | 4 | 20.62 | | 18.75 | 23.00 |
| dry | 4 | 7.35 | | 6.95 | 8.19 |
| fat-free | 4 | 18.61 | | 16.96 | 20.68 |
| ash-free | 4 | 17.84 | | 16.24 | 19.86 |
| Female HY | | | | | |
| wet | 16 | 21.12 | 1.38 | 19.20 | 24.25 |
| dry | 16 | 7.72 | 0.51 | 6.96 | 8.74 |
| fat-free | 16 | 18.83 | 1.33 | 16.55 | 21.77 |
| ash-free | 16 | 18.03 | 1.30 | 15.85 | 20.99 |
| Female AHY | | | | | |
| wet | 6 | 20.87 | 1.63 | 18.47 | 23.00 |
| dry | 6 | 7.31 | 0.73 | 6.41 | 8.32 |
| fat-free | 6 | 18.81 | 1.53 | 16.69 | 20.91 |
| ash-free | 4 | 17.71 | | 15.99 | 19.98 |
| All individuals | | | | | |
| wet | 50 | 21.81 | 1.73 | 18.47 | 25.56 |
| dry | 50 | 7.84 | 0.72 | 6.41 | 9.68 |
| fat-free | 50 | 19.55 | 1.60 | 16.55 | 22.84 |
| ash-free | 42 | 18.54 | 1.58 | 15.85 | 21.89 |
| Piranga rubra | | | | | |
| Male HY | | | | | |
| wet | 47 | 34.96 | 6.27 | 25.50 | 46.58 |
| dry | 47 | 17.89 | 6.81 | 10.04 | 28.63 |
| fat-free | 47 | 25.29 | 2.10 | 19.40 | 30.82 |
| ash-free | 32 | 24.63 | 2.13 | 18.43 | 29.75 |
| Male AHY | | | | | |
| wet | 8 | 39.22 | 3.42 | 35.82 | 44.84 |
| dry | 8 | 23.51 | 2.80 | 20.01 | 26.84 |
| fat-free | 8 | 23.45 | 1.68 | 20.78 | 25.79 |
| ash-free | 1 | 21.71 | | | |
| Female HY | | | | | |
| wet | 23 | 36.12 | 5.68 | 27.36 | 47.63 |
| dry | 23 | 19.90 | 6.23 | 10.51 | 28.34 |
| fat-free | 23 | 24.37 | 2.02 | 20.01 | 27.70 |
| ash-free | 9 | 24.41 | 1.14 | 22.81 | 26.09 |
| Female AHY | | | | | |
| wet | 20 | 39.44 | 4.81 | 31.04 | 45.66 |
| dry | 20 | 23.06 | 5.06 | 12.18 | 28.85 |
| fat-free | 20 | 24.33 | 2.58 | 16.05 | 27.88 |
| ash-free | 4 | 25.56 | | 23.58 | 26.94 |
| All individuals | | | | | |
| wet | 98 | 36.50 | 5.91 | 25.50 | 47.63 |
| dry | 98 | 19.88 | 0.44 | 10.04 | 28.85 |

## Body Mass and Composition of North American Migrant Birds (continued)

|  | N | Mean | S.D. | Min | Max |
|---|---|---|---|---|---|
| fat-free | 98 | 24.73 | 2.21 | 16.05 | 30.82 |
| ash-free | 46 | 24.60 | 1.95 | 18.43 | 29.75 |
| **Piranga olivacea** |  |  |  |  |  |
| Male HY |  |  |  |  |  |
| wet | 23 | 39.45 | 3.01 | 31.28 | 44.12 |
| dry | 23 | 23.48 | 2.39 | 18.55 | 28.41 |
| fat-free | 23 | 24.10 | 2.12 | 19.28 | 27.98 |
| ash-free | 14 | 23.38 | 2.38 | 18.43 | 26.95 |
| Male AHY |  |  |  |  |  |
| wet | 11 | 40.19 | 5.29 | 30.43 | 44.85 |
| dry | 11 | 23.85 | 6.66 | 10.05 | 30.57 |
| fat-free | 11 | 24.56 | 2.36 | 19.28 | 28.59 |
| ash-free | 5 | 15.07 | 1.69 | 23.55 | 27.66 |
| Female HY |  |  |  |  |  |
| wet | 19 | 39.10 | 4.73 | 24.50 | 44.05 |
| dry | 19 | 23.12 | 4.72 | 8.34 | 29.28 |
| fat-free | 19 | 24.13 | 1.62 | 21.44 | 27.59 |
| ash-free | 11 | 23.42 | 1.79 | 20.97 | 26.59 |
| Female AHY |  |  |  |  |  |
| wet | 19 | 37.00 | 5.60 | 25.46 | 46.80 |
| dry | 19 | 19.71 | 6.76 | 8.41 | 29.72 |
| fat-free | 19 | 25.19 | 2.68 | 20.99 | 31.41 |
| ash-free | 10 | 25.00 | 2.71 | 20.81 | 30.34 |
| All individuals |  |  |  |  |  |
| wet | 74 | 38.68 | 4.73 | 24.50 | 46.80 |
| dry | 74 | 22.27 | 5.42 | 8.34 | 30.57 |
| fat-free | 74 | 24.50 | 2.21 | 19.28 | 31.41 |
| ash-free | 40 | 24.01 | 2.31 | 18.43 | 30.34 |
| **Passerina cyanea** |  |  |  |  |  |
| Male HY |  |  |  |  |  |
| wet | 23 | 15.25 | 1.78 | 12.89 | 19.61 |
| dry | 23 | 6.65 | 1.80 | 4.99 | 10.99 |
| fat-free | 23 | 12.68 | 0.91 | 10.47 | 14.27 |
| ash-free | 11 | 11.72 | 0.79 | 10.04 | 12.78 |
| Male AHY |  |  |  |  |  |
| wet | 56 | 15.58 | 2.07 | 12.92 | 22.18 |
| dry | 56 | 6.89 | 2.12 | 4.62 | 13.05 |
| fat-free | 56 | 12.70 | 0.78 | 11.01 | 14.52 |
| ash-free | 31 | 11.88 | 0.67 | 10.54 | 13.16 |
| Female HY |  |  |  |  |  |
| wet | 15 | 16.04 | 2.88 | 13.31 | 21.36 |
| dry | 15 | 7.84 | 2.91 | 5.08 | 12.36 |
| fat-free | 15 | 11.94 | 0.87 | 9.62 | 13.12 |
| ash-free | 14 | 11.41 | 0.84 | 9.19 | 12.63 |
| Female AHY |  |  |  |  |  |
| wet | 59 | 15.24 | 1.98 | 11.91 | 20.63 |
| dry | 59 | 6.97 | 1.99 | 4.38 | 12.29 |
| fat-free | 59 | 12.00 | 0.97 | 8.13 | 14.81 |
| ash-free | 42 | 11.40 | 1.06 | 7.72 | 14.44 |
| All individuals |  |  |  |  |  |
| wet | 155 | 15.45 | 2.07 | 11.91 | 22.18 |
| dry | 155 | 6.99 | 2.11 | 4.38 | 13.05 |
| fat-free | 155 | 12.34 | 0.95 | 8.13 | 14.81 |
| ash-free | 100 | 11.59 | 0.90 | 7.72 | 14.44 |

**Body Mass and Composition of North American Migrant Birds (continued)**

| | N | Mean | S.D. | Min | Max |
|---|---|---|---|---|---|
| Icterus spurius | | | | | |
| Male HY | | | | | |
| wet | 2 | 22.75 | | 21.20 | 24.30 |
| dry | 2 | 9.70 | | 8.80 | 10.60 |
| fat-free | 2 | 19.00 | | 18.10 | 19.90 |
| ash-free | 2 | 18.23 | | 17.34 | 19.12 |
| Male AHY | | | | | |
| wet | 1 | 25.50 | | | |
| dry | 1 | 12.00 | | | |
| fat-free | 1 | 19.60 | | | |
| ash-free | 1 | 18.78 | | | |
| Female HY | | | | | |
| wet | 5 | 21.58 | 2.18 | 20.20 | 25.40 |
| dry | 5 | 9.04 | 1.40 | 8.20 | 11.50 |
| fat-free | 5 | 18.22 | 1.09 | 17.30 | 20.00 |
| ash-free | 5 | 17.44 | 1.10 | 16.53 | 19.24 |
| Female AHY | | | | | |
| wet | 3 | 22.90 | | 21.30 | 24.60 |
| dry | 3 | 10.47 | | 9.30 | 11.60 |
| fat-free | 3 | 18.20 | | 16.30 | 20.50 |
| ash-free | 3 | 17.46 | | 15.66 | 19.65 |
| All individuals | | | | | |
| wet | 11 | 22.51 | 2.07 | 20.20 | 25.50 |
| dry | 11 | 9.82 | 1.45 | 8.20 | 12.00 |
| fat-free | 11 | 18.48 | 1.33 | 16.30 | 20.50 |
| ash-free | 11 | 17.71 | 1.30 | 15.66 | 19.65 |
| | | | | | |
| Agelaius phoeniceus | | | | | |
| Male Unknown Age | | | | | |
| wet | 2 | 58.76 | | 53.80 | 63.72 |
| dry | 2 | 22.37 | | 22.25 | 22.49 |
| fat-free | 2 | 54.24 | | 48.15 | 60.34 |
| Female Unknown Age | | | | | |
| wet | 10 | 42.72 | 2.51 | 39.33 | 47.17 |
| dry | 10 | 17.29 | 1.59 | 14.97 | 19.97 |
| fat-free | 10 | 37.73 | 2.62 | 33.91 | 43.38 |
| All individuals | | | | | |
| wet | 12 | 45.39 | 6.98 | 39.33 | 63.72 |
| dry | 12 | 18.14 | 2.44 | 14.97 | 22.49 |
| fat-free | 12 | 40.48 | 7.33 | 33.91 | 60.34 |
| | | | | | |
| Dolichonyx orizivorus | | | | | |
| Male HY | | | | | |
| wet | 88 | 35.00 | 7.32 | 19.04 | 51.23 |
| dry | 88 | 19.12 | 5.94 | 10.46 | 33.71 |
| fat-free | 88 | 24.23 | 3.30 | 13.13 | 30.16 |
| ash-free | 77 | 22.92 | 3.41 | 11.92 | 28.80 |
| Male AHY | | | | | |
| wet | 19 | 44.29 | 5.29 | 35.18 | 54.10 |
| dry | 19 | 26.19 | 4.48 | 18.59 | 34.45 |
| fat-free | 19 | 26.82 | 2.62 | 22.22 | 31.80 |
| ash-free | 11 | 26.28 | 2.49 | 21.32 | 30.71 |
| Female HY | | | | | |
| wet | 52 | 27.91 | 5.75 | 19.62 | 40.29 |
| dry | 52 | 14.17 | 4.62 | 7.53 | 23.97 |
| fat-free | 52 | 20.56 | 2.51 | 16.54 | 27.63 |

**Body Mass and Composition of North American Migrant Birds (continued)**

|               | N   | Mean  | S.D. | Min   | Max   |
|---------------|-----|-------|------|-------|-------|
| ash-free      | 41  | 19.80 | 2.50 | 15.68 | 26.49 |
| Female AHY    |     |       |      |       |       |
| wet           | 18  | 37.93 | 5.01 | 30.20 | 49.20 |
| dry           | 18  | 22.08 | 4.16 | 15.00 | 30.71 |
| fat-free      | 18  | 23.46 | 2.55 | 19.86 | 29.05 |
| ash-free      | 6   | 23.36 | 2.93 | 18.97 | 28.05 |
| All individuals |   |       |      |       |       |
| wet           | 178 | 34.20 | 8.09 | 19.04 | 54.10 |
| dry           | 178 | 18.73 | 6.36 | 7.53  | 34.45 |
| fat-free      | 178 | 23.33 | 3.54 | 13.13 | 31.80 |
| ash-free      | 136 | 22.24 | 3.57 | 11.92 | 30.71 |

# PART III

# LITERATURE CITED

1. Abbott, I., Abbott, L.K., and Grant, P.R. 1977. Comparative ecology of Galapagos ground finches (*Geospiza* Gould): evaluation of the importance of floristic diversity and interspecific competition. Ecol. Monogr. 47:151–184.

2. Abdulali, H. 1965. Notes on Indian birds 3 - The Alpine Swift, *Apus melba* (Linnaeus), with a description of one new race. J. Bombay Nat. Hist Soc. 62:153–159.

3. Abrams, R.W., and Underhill, L.G. 1985. Relationships of pelagic seabirds with the southern ocean environment assessed by correspondence analysis. Auk 103:221–225.

4. Ainley, D.G., and Emison, W. 1972. Sexual size dimorphism in Adelie Penguins. Ibis 114:267–271.

5. Ali, S., and Ripley, S.D. 1968–1974. Handbook of the birds of India and Pakistan. Oxford Univ. Press, Oxford. Volumes 1–10.

6. Anderson, A.H., and Anderson, A. 1973. The Cactus Wren. Univ. Arizona Press, Tucson, AZ.

7. Anderson, T.R. 1978. Population studies of European sparrow in North America. Occas. Papers Mus. Nat. Hist., Univ. Kansas 70.

8. Ash, J.S. 1969. Spring weights of trans-Saharan migrants in Morocco. Ibis 111:1–10.

9. Ash, J.S. 1973. *Luscinia megarhynchus* and *L. luscinia* in Ethiopia. Ibis 115:267–269.

10. Ashmole, N.P. 1962. The Black Noddy *Anous tenuirostris* on Ascension Island part 1: general biology. Ibis 103b:235–273.

11. Astheimer, L.B., and Grau, C.R. 1990. A comparison of yolk growth rates in seabird eggs. Ibis 132:380–394.

12. Atwood, J. Unpubl. data.

13. Atwood, J.L. 1979. Body weights of the Santa Cruz Island Scrub Jay. N. Am. Bird Bander 4:148–153.

14. Austin, G.T., and Ricklefs, R.E. 1977. Growth and development of the Rufous-winged Sparrow (*Aimophila carpalis*). Condor 79:37–50.

15. Baker, N.E., and Baker, E.M. 1990. A new species of weaver from Tanzania. Bull. Br. Ornithol. Club 110:51–58.

16. Baltz, D.M., and Morejohn, G.V. 1977. Food habits and niche overlap of seabirds wintering on Monterey Bay, California. Auk 94:526–543.

17. Baptista, L.F. Unpubl. data.

18. Baptista, L.F., Boarman, W.J., and Kandianidis, P. 1983. Behavior and taxonomic status of Grayson's Dove. Auk 100:907–919.

18a. Barlow, J.C., Dick, J.A., Baldwin, D.H., and Davis, R.A. 1969. New records of birds from British Honduras. Ibis 111:399–402.

19. Barlow, J.C., and Nash, S.V. 1985. Behavior and nesting biology of the St. Andrew Vireo. Wilson Bull. 97:265–272.

20. Barrett, V.A., and Vyse, E.R. 1982. Comparative genetics of three Trumpeter Swan populations. Auk 99:103–108.

21. Bateman, G.C., and Balda, R.P. 1973. Growth, development and food habits of young Pinon Jays. Auk 90:39–61.

22. Bates, J.M., Garvin, M.C., Schmitt, D.C., and Schmitt, C.G. 1989. Notes on bird distribution in northeastern Dpto. Santa Cruz, Bolivia, with 15 species new to Bolivia. Bull. Br. Ornithol. Club 109:236–244.

23. Bauer, K.M., and Glutz von Blotzheim, U.N. (eds.). 1966. Handbuch der Vogel Mitteleuropas. Akademische Verlagsgesellschaft. Wiesbaden, Germany.

24. Baumel, J.J. 1953. Individual variation in the White-necked Raven. Condor 55:26–32.

25. Baumel, J.J. 1957. Individual variation in the Fish Crow *Corvus ossifragus*. Auk 74:73–78.

26. Beck, T.D., and Braun, C.E. 1978. Weights of Colorado Sage Grouse. Condor 80:241–243.

27. Becking, J.H. 1971. The breeding of *Collocalia gigas*. Ibis 113:330–334.

28. Becking, J.H. 1975. The ultrastructure of the avian eggshell. Ibis 117:143–151.

29. Bedard, J. 1969. Adaptive radiation in Alcidae. Ibis 111:189–198.

30. Beehler, B. 1978. Notes on the mountain birds of New Ireland. Emu 78:65–70.

31. Beehler, B. 1983. Frugivory and polygamy in birds of paradise. Auk 100:1–12.

32. Behle, W.H. 1950. Clines in the yellow-throats of western North America. Condor 52:193–219.

33. Behle, W.H. 1956. A systematic review of the Mountain Chickadee. Condor 58:51–70.

34. Bell, H.L. 1981. Information on New Guinean kingfishers, Alcedinidae. Ibis 123:51–61.

35. Bell, H.L. 1982. A bird community of lowland rainforest in New Guinea. 1. Composition and density of the avifauna. Emu 82:24–41.

36. Belopol'skii, L.O. 1957. [Ecology of sea colony birds of the Barents Sea.] Translated by: Israel Program for Scientific Translations, Jerusalem.

37. Belton, W. 1984. Birds of Rio Grande do Sul, Brazil. Part 1. Rheidae through Furnariidae. Bull. Am. Mus. Nat. Hist. 178:371–631.

38. Belton, W. 1985. Birds of Rio Grande do Sul, Brazil. Part 2. Formicariidae through Corvidae. Bull. Am. Mus. Nat. Hist. 180:1–241.

39. Benson, C.W., Colebrook-Robjent, J.F.R., and Williams, A. 1976. Contribution a L'Ornithologie de Mada-gascar. L'Oiseau et La Revue Francaise D'Ornithologie 46:103–134.

40. Benson, C.W., and Penny, M.J. 1971. The land birds of Aldabra. Phil. Trans. Royal Soc. London B 260:417–527.

41. Bent, A.C. 1968. Life histories of North American cardinals, grosbeaks, buntings, towhees, finches, sparrows, and allies. U.S. Nat. Mus. Bull. 237.

42. Berruti, A. 1979. The breeding biologies of the Sooty Albatrosses *Phoebetria fusca* and *P. palpebrata*. Emu 79:161–175.

43. Best, H., and Bellingham, P. 1990. Feeding ecology of North Island Kokako *Callaeas cinerea wilsoni* in relation to its conservation. Paper presented at the 20th International Ornithological Congress, Christchurch, New Zealand.

44. Biermann, G.C., and Sealy, S.G. 1985. Seasonal dynamics of body mass of insectivorous passerines breeding on the forested dune ridge, Delta Marsh, Manitoba. Can. J. Zool. 63:1675–1682.

45. Bierregaard, R.O. 1988. Morphological data from understory birds in terre firme forest in the Central Amazonian Basin. Rev. Brasil. Biol. 48:169–178.

46. Binford, L.C. 1968. A preliminary survey of the avifauna of the Mexican state of Oaxaca. Ph.D. dissertation, Louisiana State Univ., Baton Rouge.

47. Binford, L.C. 1985. Re-evaluation of the "hybrid" hummingbird *Cynanthus sordidus* X *C. latirostris* from Mexico. Condor 87:148–150.

48. Binford, L. C. 1989. A distributional survey of the birds of the Mexican state of Oaxaca. Ornithol. Monogr. 43.

49. Birt-Friesen, V.L., Montevecchi, W.A., Cairns, D.K., and Macko, S.A. 1989. Activity-specific metabolic rates of free-living Northern Gannets and other seabirds. Ecology 70:357–367.

50. Blake, J. Unpubl. data.

50a. Blake, J., and Loiselle, B.A. Unpubl. data.

51. Blake, J.G., Loiselle, B.A., and Vande Weghe, J.P. 1990. Weights and measurements of some Central African birds. Le Gerfaut 80:3–11.

52. Blondel, J., Vuilleumier, F., Marcus, L.F., and Terouanne, E. 1984. Is there ecomorphological convergence among Mediterranean bird communities of Chile, California, and France? Evol. Biol. 18:141–213.

53. Bloom, P.H. 1973. Seasonal variation in body weight of Sparrow Hawks in California. Western Bird Bander pp. 17–19.

54. Bohl, W.H. 1957. Chukars in New Mexico. New Mexico Dept. Game & Fish Bull. 6.

55. Boles, W.E. 1988. The robins and flycatchers of Australia. Angus & Robertson Publ. North Ryde, NSW, Australia.

56. Bond, A.B., Wilson, K.-J., and Diamond, J. 1991. Sexual dimorphism in the Kea *Nestor notabilis*. Emu 91:12–19.

57. Bosque, C., and Lentino, M. 1987. The nest, eggs, and young of the White-whiskered Spinetail (*Synallaxis* [*Poecilurus*] *candei*). Wilson Bull. 99:104–106.

58. Bowers, D.E. 1960. Correlation of variation in the Wrentit with environmental gradients. Condor 62:91–120.

59. Bowers, R.K. Unpubl. data.

60. Bowman, R., and Gullege, J.L. Unpubl. data.

61. Bowman, R.I. 1961. Morphological differentiation and adaption in the Galapagos finches. Univ. Cal. Publ. Zool. 58:1–326.

62. Boyd, H., and Maltby, L.S. 1980. Weights and primary growth of Brent Geese *Branta bernicla* moulting in the Queen Elizabeth Islands, N.W.T., Canada. Ornis Scand. 11:135–141.

63. Braun, C.E. 1976. Banding worksheet for western birds - Band-tailed Pigeon. W. Bird Banding Assn.

64. Bretagnolle, V., Zotier, R., and Jouventin, P. 1990. Comparative population biology of four prions (genus *Pachyptila*) from the Indian Ocean and consequences for their taxonomic status. Auk 107:305–316.

65. Britton, P.L. 1967. Weights of the Carmine Bee-eater *Merops nubicoides*. Ibis 109:606–614.

65a. Britton, P.L. 1969. Weights of the Pennant-winged Nightjar. Bull. Br. Ornithol. Club 89:21–23.

66. Britton, P.L. 1970. Some non-passerine bird weights from East Africa. Bull. Br. Ornithol. Club 90:142–144; 90:152–154.

67. Britton, P.L. 1977. Weights of birds in western and coastal Kenya: a comparison. Scopus 1:70–73.

68. Britton, P.L. 1978. Seasonality, density, and diversity of birds of a papyrus swamp in western Kenya. Ibis 120:450–466.

69. Britton, P.L., and Dowsett, R.J. 1969. More weights of the Carmine Bee-eater. Bull. Br. Ornithol. Club 89:85–86.

70. Brooke, M. 1978. Weights and measurements of the Manx Shearwater *Puffinus puffinus*. J. Zool. (London) 186:359–374.

71. Brooke, M. 1983. Ecological segregation of woodcreepers (Dendrocolaptidae) in the state of Rio de Janeiro, Brazil. Ibis 125:562–567.

72. Brooke, R.K. 1969. *Apus berliozi*, its races and siblings. Bull. Br. Ornithol. Club 89:11–16.
73. Brooke, R.K. 1971. Taxonomic notes on some lesser known *Apus* swifts. Bull. Br. Ornithol. Club 91:33–36.
73a. Brooke, R.K. 1971. The eastern and southern populations of the Mottled Spinetail. Bull. Br. Ornithol. Club 91:134–135.
74. Brooke, R.K. 1971. Geographical variation in the Alpine Swift *Apus (Tachymarptis) melba* (Aves: Apodidae). Durban Mus. Novitates 9:131–143.
74a. Brooke, R.K. 1971. Geographical variation in the Little Swift *Apus affinus* (Aves: Apodidae). Durban Mus. Novitates 9:93–103.
75. Brooke, R.K. 1971. Taxonomic and distributional notes on the African Chaeturini. Bull. Br. Ornithol. Club 91:76–79.
76. Brooker, M.G. 1988. Some aspects of the biology and conservation of the Thick-billed Grasswren *Amytornis textilis* in the Shark Bay area, Western Australia. Corella 12:101–108.
77. Brooker, M.G., and Brooker, L.C. 1989. The comparative breeding behavior of two sympatric cuckoos, Horsfield's Bronze-Cuckoo *Chrysococcyx basalis* and the Shining Bronze-Cuckoo *C. lucidus*, in western Australia: a new model for the evolution of egg morphology & host specificity in avian brood parasites. Ibis 131:528–547.
78. Brough, T. 1983. Average weights of birds. Avian Bird Unit, Worplesdon Laboratory, Worplesdon, U.K.
79. Brown, J.H., and Bowers, M.A. 1985. Community organization in hummingbirds: relationships between morphology and ecology. Auk 102:251–269.
80. Brown, L. 1976. Eagles of the world. Universe Books, New York.
81. Brown, L., and Amadon, D. 1968. Eagles, hawks and falcons of the world. McGraw-Hill Publ. Co., New York. Vols. 1 & 2.
82. Brown, L.H., Urban, E.K., and Newman, K. 1982. The birds of Africa. Academic Press, New York. Vol. 1.
83. Brown, R.G., Barker, S.P., Gaskins, D.E., and Sandeman, M.R. 1981. The foods of Great and Sooty Shearwaters *Puffinus gravis* and *P. griseus* in eastern Canadian waters. Ibis 123:19–30.
84. Browne, K., and Browne, E. 1956. An analysis of the weights of birds trapped on Skokholm. British Birds 49:241–257.
85. Bump, G., Darrow, R.W., Edminster, F.C., and Crissey, W.F. 1947. The Ruffed Grouse: life history, propagation, management. New York State Conserv. Dept.
86. Burton, P.J.K. 1973. Non-passerine bird weights from Panama and Columbia. Bull. Br. Ornithol. Club 93:116–118.
87. Burton, P.J.K. 1975. Passerine bird weights from Panama and Columbia, with some notes on 'soft-part' colours. Bull. Br. Ornithol. Club 95:82–86.
88. Byrd, G.V., Trapp, J.L., and Gibson, D.D. 1978. New information on Asiatic birds in the Aleutian Islands, Alaska. Condor 80:309–315.
89. Cairns, D.K. 1987. The ecology and energetics of chick provisioning by Black Guillemots. Condor 89:627–635.
90. Calder, W.A. 1984. Size, function, and life history. Harvard Univ. Press, Cambridge, MA.
91. Calder, W.A., Waser, N.M., Hiebert, S.M., Inouye, D.W., and Miller, S. 1983. Site-fidelity, longevity, and population dynamics of Broad-tailed hummingbirds: a ten year study. Oecologia 56:359–364.
91a. Caldwell, L.D., Odum, E.P., and Marshall, S.G. 1964. Comparison of fat levels in migrating birds killed at a central Michigan and a Florida Gulf Coast television tower. Wilson Bull. 75:428–434.
92. Cape May Raptor Banding Project. Cape May Bird Observatory, Cape May, NJ, USA. Unpubl. data.
93. Capparella, A.P., and Lanyon, S.M. 1985. Biochemical and morphologic analyses of the sympatric, neotropical, sibling species *Mionectes macconnellii* and *M. oleagineus*. Ornithol. Monogr. 36:347–355.
94. Cardiff, S.W., and Remsen, J.V. 1981. Three bird species new to Bolivia. Bull. Br. Ornithol. Club 101:304–305.
95. Chan, K., Ford, H.A., and Ambrose, S.J. 1990. Ecophysiological adaptations of the Eastern Spinebill *Acanthorhynchus tenuirostris* to a high altitudinal winter environment. Emu 90:119–122.
96. Reference deleted.
97. Cheke, A.S., and Diamond, A.W. 1986. Birds on Moheli and Grande Comore (Comoro Islands) in February 1975. Bull. Br. Ornithol. Club 106:138–148.
98. Cherel, Y., and LeMaho, Y. 1988. Changes in body mass and plasma metabolites during short-term fasting in the King Penguin. Condor 90:257–258.
99. Clark, G.A. 1979. Body weights of birds: a review. Condor 81:193–202.
100. Clements, J.F. 1991. Birds of the world: a checklist. Ibis Publishing Co., Vista, CA.
101. Clench, M.H., and Leberman, R.C. 1978. Weights of 151 species of Pennsylvania birds analyzed by month, age and sex. Bull. Carnegie Mus. Nat. Hist. 5.
101a. Cohn-Haft, M. Unpubl. data.
102. Colbourne, R., and Kleinpaste, R. 1983. A banding study of North Island Brown Kiwis in an exotic forest. Notornis 30:109–124.

103. Collins, C.T. Unpubl. data.

104. Collins, C.T. 1968. The comparative biology of two species of swifts in Trinidad, West Indies. Bull. Florida State Mus. 11:257–320.

105. Collins, C.T. 1972. Weights of some birds of north-central Venezuela. Bull. Br. Ornithol. Club 92:151–153.

106. Collins, C.T. 1972. Banding worksheet for western birds-Violet-green Swallow. W. Bird Banding Assoc.

106a. Collins, C.T. 1980. Notes on the food of the Horus Swift *Apus horus* in Kenya. Scopus 4:10–13.

107. Collins, C.T., and Bradley, R.A. 1971. Analysis of body weights of spring migrants in southern California. Western Bird Bander pp. 38–40.

108. Collins, C.T., and Brooke, R.K. 1976. A review of the swifts of the genus *Hirundapus* (Aves: Apodidae). Nat. Hist Mus. Los Angeles Co. Contributions in Science 282:1–22.

109. Colwell, M.A. and Oring, L.W. 1988. Breeding biology of Wilson's Phalarope in southcentral Saskatchewan. Wilson Bull. 100:567–582.

110. Colwell, R.K. 1989. Hummingbirds of the Juan Fernandez Islands: natural history, evolution and population status. Ibis 131:548–566.

111. Contreras, J.R. 1979. Bird weights from northeastern Argentina. Bull. Br. Ornithol. Club 99:21–24.

111a. Connell, C.E., Odum, E.P., and Kale, H. 1960. Fat-free weights of birds. Auk 77:1–9.

112. Contreras, J.R. 1986. Notas sobre el peso de aves Argentinas. V. Historia Natural 6:100.

113. Cooch, F.G., Stirret, G.M., and Boyce, G.F. 1960. Autumn weights of Blue Geese (*Chen caerulescens*). Auk 77:460–465.

114. Craig, R.J. 1990. Foraging behavior and microhabitat use of two species of white-eyes (Zosteropidae) on Saipan, Micronesia. Auk 107:500–505.

115. Cramp, S., and Simmons, K.E.L. 1977. Handbook of the birds of Europe, the Middle East and North Africa. Oxford Univ. Press, Oxford. Vol. 1.

116. Cramp, S., and Simmons, K.E.L. 1980. Handbook of the birds of Europe, the Middle East and North Africa. Oxford Univ. Press, Oxford. Vol. 2.

117. Cramp, S., and Simmons, K.E.L. 1983. Handbook of the birds of Europe, the Middle East and North Africa. Academic Press, New York. Vol. 3.

118. Cramp, S., and Simmons, K.E.L. 1985. Handbook of the birds of Europe, the Middle East and North Africa. Oxford Univ. Press, Oxford, UK. Vol. 4.

119. Cramp, S., and Simmons, K.E.L. 1988. Handbook of the birds of Europe, the Middle East and North Africa. Oxford Univ. Press, Oxford, UK. Vol. 5.

119a. Crick, H.Q.P., and Fry, C.H. 1986. Effects of helpers on parental condition in Red-throated Bee-eaters (*Merops bullocki*). J. Anim. Ecol. 55:893–905.

120. Crome, F.H.J., and Rushton, D.K. 1975. Development of plumage in the Plains-wanderer. Emu 75:181–184.

121. Crossin, R.S. 1967. The breeding biology of the Tufted Jay. Proc. West. Found. Vert. Zool. 1:265–299.

122. Crossin, R.S. 1974. The storm petrels (Hydrobatidae). In King, W.B. (ed.). Pelagic studies of seabirds in the central and eastern Pacific Ocean. Smithsonian Cont. Zool. 158.

123. Croxall, J.P. 1977. Feeding behaviour and ecology of New Guinean rainforest insectivorous passerines. Ibis 119:113–146.

124. Croxall, J.P., and Furse, J.E 1980. Food of the Chinstrap Penguins *Pygoscelis antarctica* and Macaroni Penguins *Eudyptes chrysolophus* at Elephant Island group, South Shetland Islands. Ibis 122:237–245.

125. Cruz, A. Unpubl. data.

125a. Cruz, A. 1974. Feeding assemblages of Jamaican birds. Condor 76:103–107.

126. Cruz, A. 1976. Distribution, ecology, and breeding biology of the Rufous-throated Solitaire in Jamaica. Auk 93:39–45.

127. Cruz, A. 1980. Feeding ecology of the Black-whiskered Vireo and associated gleaning birds in Jamaica. Wilson Bull. 92:40–52.

128. Cruz, A., and Johnston, D.W. 1984. Ecology of the West Indian Red-bellied Woodpecker on Grand Cayman: distribution and foraging. Wilson Bull. 96:366–379.

129. Cunningham, J.B. 1984. Differentiating the sexes of the Brown Creeper. Notornis 31:19–22.

130. Curry, R.L., and Grant, P.R. 1990. Galapagos mockingbirds: territorial cooperative breeding in a climactically variable environment. In Stacey, P.B., and Koenig, W.D. (eds.). Cooperative breeding in birds: long-term studies of ecology and behavior. Cambridge Univ. Press, Cambridge, England pp. 289-332.

131. Czechura, G.V., and Debus, S.J.S. 1985. The Black Falcon *Falco subniger*: a summary of information and comparison with the Brown Falcon, *Falco berigora*. Austral. Bird Watcher 11:80–91.

132. Czechura, G.V., and Debus, S.J.S. 1985. The Grey Falcon *Falco hypoleucos*: a summary of information. Austral. Bird Watcher 11:9–16.

133. Czechura, G.V., and Debus, S.J.S. 1986. The Australian hobby *Falco longipennis*: a review. Austral. Bird Watcher 11:185–207.

134. Czechura, G.V., Debus, S.J.S., and Mooney, N.J. 1987. The Collared Sparrowhawk *Accipiter cirrocephalus*: a review and comparison with the Brown Goshawk *Accipiter fasciatus*. Austral. Bird Watcher 12:335–362.

135. Darrieu, C.A. 1986. Estudios sobre la avifauna de corrientes. III. Nuevos registros de aves passeriformes (Dendrocolaptidae, Furnariidae, Formicaridae, Cotingidae y Pipridae) y consideraciones sobre su distribucion geographica. Historia Natural 6:93–99.

136. Darrieu, C.A. 1987. Estudios sobre la avifauna de corrientes IV. Nuevos registros de aves (Passeriformes, Tyrannidae) y consideraciones sobre su distribucion geographica. Neotropica 33:29–35.

136a. da Silva, J.M.C., Lima, M. de F.C., and Marceliano, M.L.V. 1990. Pesos de aves de duas localidades na Amazonia Oriental. Ararajuba 1:99–104.

137. Davis, J. 1951. Distribution and variation of the brown towhees. Univ. Cal. Publ. Zool. 52:1–120.

138. Davis, T.J. 1986. Distribution and natural history of some birds from the departments of San Martin and Amazonaz, northern Peru. Condor 88:50–56.

139. Davis, W.B. 1945. Notes on Veracruzan birds. Auk 62:272–286.

140. Dawn, P. 1981. Breeding of the Banded and Masked Lapwings in southern Victoria. Emu 81:121–127.

141. Dawson, T.J., Read, D., Russell, E.M., and Herd, R.M. 1984. Seasonal variation in daily activity patterns, water relations and diet of Emus. Emu 84:93–102.

142. Dawson, W.R., and Evans, F.C. 1960. Relation of growth and development to temperature regulation in nestling Vesper Sparrow. Condor 62:329–340.

143. Day, R.H., and Byrd, G.V. 1989. Food habits of the Whiskered Auklet at Buldir Island, Alaska. Condor 91:65–72.

144. Debus, S.J. 1984. Weights and measurements - Little Eagle. In Data exchange. Corella 8:52.

145. Debus, S.J.S. 1984. Biology of the Little Eagle on the Northern Tablelands of New South Wales. Emu 84:87–92.

146. Debus, S.J.S. 1984. A re-appraisal of the dimensions of male Forest Ravens. Corella 8:19–20.

147. Debus, S.J.S., and Czechura, G.V. 1988. The Red Goshawk *Erythrotriorchis radiatus*: a review. Austral. Bird Watcher 12:175–194.

148. Delannoy, C.A., and Cruz, A. 1988. Breeding biology of the Puerto Rican Sharp-shinned Hawk (*Accipiter striatus venator*). Auk 105:649–663.

149. Dement'ev, G.P., Gladkov, N.A., and Spangenberg, E.P. 1951. [Birds of the Soviet Union.] Translated by: Israel Program for Scientific Translations, Jerusalem.

150. Diamond, A.W. 1975. The biology of tropicbirds at Aldabra Atoll, Indian Ocean. Auk 92:16–39.

151. Diamond, A.W. 1975. Biology and behavior of frigatebirds *Fregata* spp. on Aldabra Atoll. Ibis 117:302–323.

152. Diamond, A.W. 1987. Studies of Mascarene Island birds. Cambridge Univ. Press, Cambridge.

153. Diamond, A.W., Lack, P., and Smith, R.W. 1977. Weights and fat condition of some migrant warblers in Jamaica. Wilson Bull. 89:456–466.

154. Diamond, J.M. 1972. Avifauna of the eastern highlands of New Guinea. Publ. Nuttall Ornithol. Club 12.

155. Diamond, J.M. 1981. Distribution, habits and nest of *Chenorhamphus grayi*, a malurid endemic to New Guinea. Emu 81:87–100.

156. Diamond, J.M. 1983. *Melampitta gigantea*: possible relation between feather structure and underground roosting habits. Condor 85:89–91.

157. Diamond, J.M. 1985. New distributional records and taxa from the outlying mountain ranges of New Guinea. Emu 85:65–91.

158. Diamond, J.M. 1987. Flocks of brown and black New Guinean birds: a bicolored mixed-species foraging association. Emu 87:201–211.

159. Diamond, J.M. 1989. A new subspecies of the Island Thrush *Turdus poliocephalus* from Tolokiawa Island in the Bismark Archipeligo. Emu 89:58–60.

160. Diamond, J.M., Pimm, S.L., Gilpin, M.E., and LeCroy, M. 1989. Rapid evolution of character displacement in myzomelid honeyeaters. Am. Nat. 134:675–708.

160a. Diaz, M. 1990. Interspecific patterns of seed selection among granivorous passerines: effects of seed size, seed nutritive value and bird morphology. Ibis 132:467–476.

161. Dick, J.A., McGillivray, W.B., and Brooks, D.J. 1984. A list of birds and their weights from Saul, French Guiana. Wilson Bull. 96:347–365.

162. Dickerman, R.W., Parkes, K.C., and Bell, J. 1982. Notes on the plumages of the Boat-billed Heron. Living Bird 19:115–120.

163. Diebold, E. Unpubl. weights of captive birds at Milwaukee County Zoo, Milwaukee, Wisconsin.

164. Dolbeer, R.A., Wornecki, P.P., Stickley, A.R., and White, S.B. 1978. Agricultural impact of a winter population of blackbirds and Starlings. Wilson Bull. 90:31–44.

165. Dostine, P.L., and Morton, S.R. 1989. Food of the Black-winged Stilt *Himantopus himantopus* in the Alligator River region, Northern Territory. Emu 89:250–253.

166. Dowsett, R.J. 1970. A collection of birds from Nyiaka Plateau, Zambia. Bull. Br. Ornithol. Club 90:49–53.

167. Dowsett, R.J. 1981. Breeding and other observations of the Slaty Egret *Egretta vinaceigula*. Bull. Br. Ornithol. Club 101:323–327.

168. Dowsett, R.J. 1983. Sexual size dimorphism in some montane forest passerines from south-central Africa. Bull. Br. Ornithol. Club 103:59–64.
169. Dowsett, R.J., and Fry, C.H. 1971. Weight losses of trans-Saharan migrants. Ibis 113:531–533.
170. Dowsett, R.J., and Stjernstedt, R. 1979. The *Bradypterus cinnamomeus-marie* complex in central Africa. Bull. Br. Ornithol. Club 99:86–94.
171. Duffy, D.C. 1980. Patterns of piracy by Peruvian seabirds: a depth hypothesis. Ibis 122:521–525.
172. Duffy, D.C. 1987. Ecological implications of intercolony size-variation in Jackass Penguins. Ostrich 58:54–57.
173. Duffy, D.C., Siegfried, W.R., and Jackson, S. 1987. Seabirds as consumers in the southern Benguela Region. S. Afr. J. Mar. Sci. 5:771–790.
174. Dunn, P.O., May, T.A., McCollough, M.A., and Howe, M.A. 1988. Length of stay and fat content of migrant Semipalmated Sandpipers in eastern Maine. Condor 90:824–835.
175. Dunnet, G.M. 1985. Pycroft's Petrel in the breeding season at Hen and Chickens Islands. Notornis 32:5–21.
176. Dunning, J.B. Unpubl. data.
177. Dunning, J.B. 1984. Body weights of 686 species of North American birds. Assn. W. Bird Banding Assn. Monogr. 1.
178. Dunning, J.B. 1985. Owl weights in the literature: a review. Raptor Res. 19:113–121.
179. Dunning, J.B. 1988. Yellow-footed Gull kills Eared Grebe. Colonial Waterbirds 11:117–118.
180. Earhart, C.M., and Johnson, N.K. 1970. Size dimorphism and food habits of North American owls. Condor 72:251–264.
181. Edwards, T.C., and Kochert, M.N. 1986. Use of body weight and length of footpad as predictors of sex in Golden Eagles. J. Field Ornithol. 57:317–319.
182. Eisenmann, E., and Short, L.L. 1982. Systematics of the avian genus *Emberizoides* (Emberizidae). Am. Mus. Novitates 2740:1–21.
183. Erard, C. 1978. A new race of *Parisoma lugens* from the highlands of Bale, Ethiopia. Bull. Br. Ornithol. Club 98:43–50.
184. Estbergs, J.A., and Braithwaite, R.W. 1985. The diet of the Rufous Owl *Ninox rufa* near Cooinda in the Northern Territory. Emu 85:202–205.
185. Faaborg, J. 1985. Ecological constraints on West Indian bird distributions. Ornithol. Monogr. 36:169–197.
186. Faaborg, J., and Winters, J.E. 1979. Winter returns and longevity and weights of Puerto Rican birds. Bird-Banding 50:216–223.
186a. Farabaugh, S.M. 1982. The ecological and social significance of duetting. In Kroodsma, D.E., and Miller, E.H. (eds.). Acoustic communication in birds. Academic Press, New York pp. 85-124.
187. Feinsinger, P. 1976. Organization of a tropical guild of nectivorous birds. Ecol. Monogr. 46:257–291.
188. ffrench, R. 1973. A guide to the birds of Trinidad and Tobago. Livingston Publ., Wynnewood, PA.
189. Finlayson, J.C. 1981. Seasonal distribution, weights and fat of passerine migrants at Gibralter. Ibis 123:88–95.
190. Fisk, E.J. 1979. Fall and winter birds near Homestead, Florida. Bird-Banding 50:224–243, 297–303.
191. Fitzpatrick, J.W., and O'Neill, J.P. 1979. A new tody-tyrant from northern Peru. Auk 96:443–447.
192. Fitzpatrick, J.W., and Willard, D.E. 1990. *Cercomacra manu*, a new species of antbird from southwestern Amazonia. Auk 107:239–245.
193. Fitzpatrick, J.W., Willard, D.E., and Stotz, D.F. Weights and wing measurements of birds from southeastern Peru. Unpubl. manuscript.
194. Ford, H.A., and Bell, H. 1981. Density of birds in eucalypt woodland affected to varying degrees by dieback. Emu 81:202–208.
194a. Ford, H.A., Noske, S., and Bridges, L. 1986. Foraging of birds in eucalypt woodland in north-eastern New South Wales. Emu 86:168–179.
195. Ford, J. 1979. A new subspecies of Grey Butcherbird from the Kimberly, Western Australia. Emu 79:191–194.
196. Ford, J. 1979. Subspeciation, hybridization and relationship in the Little Shrike-Thrush *Colluricincla megarhyncha* of Australia and New Guinea. Emu 79:195–210.
197. Ford, J. 1983. Evolutionary and ecological relationships between quail-thrushes. Emu 83:152–172.
198. Ford, J., and Parker, S.A. 1973. First record of *Acanthiza robustirostris* in Queensland. Emu 73:27.
199. Ford, J., and Parker, S.A. 1974. Distribution and taxonomy of some birds from south-western Queensland. Emu 74:177–194.
200. Foster, M.S. 1987. Feeding methods and efficiencies of selected frugivorous birds. Condor 89:566–580.
200a. Francis, C.M. Unpubl. data.
201. Francis, C.M. 1987. Hatching asynchrony and egg size variation in White-bellied Swiftlets (*Collocalia esculenta*). M.S. thesis, Queens University.
202. Freeman, S., and Jackson, W.M. 1990. Univariate metrics are not adequate to measure avian body size. Auk 107:69–74.

203. Friedmann, H. 1955. The honey-guides. Smithsonian Inst. Press, Washington, D.C.

204. Frings, H., and Frings, M. 1961. Some biometric studies on the albatrosses of Midway Atoll. Condor 63:304–312.

205. Frith, C.B., and Frith, D.W. 1985. Parental care and investment in the Tooth-billed Bowerbird *Scenopoeetes dentirostris* (Ptilonorhynchidae). Austral. Bird Watcher 11:103–113.

206. Frith, C.B., and Frith, D.W. 1990. Nesting biology and relationships of the Lesser Melampitta *Melampitta lugubris*. Emu 90:65–73.

207. Fry, C.H., Keith, S., and Urban, E.K. 1988. The birds of Africa. Academic Press, New York. Vol. 3.

208. Fugler, S.R., Hunter, S., Newton, I.P., and Steele, W.K. 1987. Breeding biology of Blue Petrels *Halobaena caerulea* at the Prince Edwards Islands. Emu 87:103–110.

209. Galbraith, I.C.J., and Parker, S.A. 1969. The Atherton Scrub-Wren *Sericornis keri* Mathews: a neglected Australian species. Emu 69:212–232.

210. Garrido, O.H. 1983. A new subspecies of *Caprimulgus cubanensis* (Aves: Caprimulgidae) from the Isle of Pines, Cuba. Auk 100:988–991.

211. Gaymer, R., Blackman, R.A.A., Dawson, P.G., Penny, M., and Penny, C.M. 1969. The endemic birds of Seychelles. Ibis 111:157–176.

212. Gayou, D.C. 1985. Body weights of south Texas Green Jays. Bull. Texas Ornithol. Soc. 18:30–31.

213. Gibson, D.D. 1981. Migrant birds at Shemya Island, Aleutian Islands, Alaska. Condor 83:65–66.

214. Gill, B.J., and Dow, D.D. 1983. Morphology and development of nesting Gray-crowned and Hall's Babblers. Emu 83:41–43.

215. Gill, B.J., and Veitch, C.R. 1990. Measurements of bush birds on Little Barrier Island, New Zealand. Notornis 37:141–145.

216. Gilliard, E.T., and LeCroy, M. 1961. Birds of the Victor Emanuel and Hindenburg Mountains, New Guinea. Bull. Am. Mus. Nat. Hist. 123:1–86.

217. Gilliard, E.T., and LeCroy, M. 1966. Birds of the Middle Sepik region, New Guinea. Bull. Am. Mus. Nat. Hist. 132:247–275.

218. Gilliard, E.T., and LeCroy, M. 1967. Annotated list of birds of the Adelbert Mountains, New Guinea. Bull. Am. Mus. Nat. Hist. 138:51–81.

219. Gilliard, E.T., and LeCroy, M. 1967. Results of the 1958–1959 Gilliard New Britain Expedition. Bull. Am. Mus. Nat. Hist. 135:175–216.

220. Goldstein, D. Unpubl. data.

221. Goodman, S.M., and Gonzales, P.C. 1989. Notes on Philippine birds, 12. Seven species new to Catanduanes Island. Bull. Br. Ornithol. Club 109:48–50.

222. Goodman, S.M., and Schulenberg, T.S. 1991. The rediscovery of the Red-tailed Newtonia *Newtonia fanovanae* in south-eastern Madagascar with notes on the natural history of the genus Newtonia. Bird Conservation Internat. 1:33–45.

222a. Gorski, L.J. 1969. Systematics and ecology of sibling species of Traill's Flycatcher. Ph.D. dissertation, Univ. Connecticut, Storrs, CT.

223. Grant-Mackie, J.A. 1980. Acquisition of a specimen of the New Caledonian Kagu (Cagou). Notornis 27:292–293.

224. Graves, G.R., and Restrepo, D.U. 1989. A new allopatric taxon in the *Hapalopsittaca amazonina* (Psittacidae) superspecies from Columbia. Wilson Bull. 101:369–376.

225. Graves, G.R., and Weske, J.S. 1987. *Tangara phillipsi*, a new species of tanager from the Cerros del Sira, eastern Peru. Wilson Bull. 99:1–6.

226. Graves, G.R., and Zusi, R. 1990. Avian body weights from the lower Rio Xingu, Brazil. Bull. Br. Ornithol. Club 110:20–25.

227. Greenewalt, C.H. 1962. Dimensional relationships for flying animals. Smithson. Misc. Coll. 144.

228. Greig-Smith, P.W., and Davidson, N.C. 1977. Weights of West African savanna birds. Bull. Br. Ornithol. Club 97:96–99.

229. Grinnell, J., Dixon, J., and Linsdale, J.M. 1930. Vertebrate natural history of a section of Northern California through the Lassen Peak region. Univ. Cal. Publ. Zool. 35.

230. Gross, A.O. 1937. Birds of the Bowdoin-MacMillan Arctic Expedition 1934. Auk 54:12–42.

231. Grzybowski, J.A. 1980. Ecological relationships among grassland birds during winter. Ph.D. dissertation, Univ. Oklahoma, Norman, OK.

232. Hackett, S.J., and Rosenberg, K.V. 1990. Comparison of phenotypic and genetic differentiation in South American antwrens (Formicariidae). Auk 107:473–489.

233. Hamilton, R.B. 1975. Comparative behavior of the American Avocet and the Black-necked Stilt (Recurvirostridae). Ornithol. Monogr. 17.

234. Hanson, H.C., and Kossack, C.W. 1957. Weight and body-fat relationships of Mourning Doves in Illinois. J. Wildl. Manage. 21:169–181.

235. Hardy, J.W. 1983. Weights and measurements - Variegated Fairy-Wren. Corella 7:48.

236. Harrington-Tweit, B. 1979. A seabird die-off on the Washington coast in mid-winter 1976. W. Birds 10:49–56.
237. Harris, M.P. 1969. Food as a factor controlling the breeding of *Puffinus lherminieri*. Ibis 111:139–156.
238. Harris, M.P. 1969. The biology of storm petrels in the Galapagos Islands. Proc. Cal. Acad. Sci. 37:95–166.
239. Harris, M.P. 1970. Breeding ecology of the Swallow-tailed Gull, *Creagrus furcatus*. Auk p 43 87:215–243.
240. Harris, M.P. 1979. Measurements and weights of British Puffins. Bird Study 26:179–186.
241. Harris, S.W. 1974. Status, chronology and ecology of nesting storm petrels in northwestern California. Condor 76:249–261.
242. Harrison, C.S., Hiola, T.S., and Seki, M.P. 1983. Hawaiian seabird feeding ecology. Wildlife Monogr. 85.
243. Hartman, F.A. 1955. Heart weight in birds. Condor 57:221–238.
244. Hartman, F.A. 1961. Locomotor mechanisms of birds. Smithsonian Misc. Coll. 142.
245. Haverschmidt, F. 1948. Bird weights from Surinam. Wilson Bull. 60:230–239.
246. Haverschmidt, F. 1952. More bird weights from Surinam. Wilson Bull. 64:234–241.
247. Haverschmidt, F. 1968. Birds of Surinam. Oliver & Boyd Co., London.
248. Haverschmidt, F. 1972. *Pachyramphus surinamus* nesting in Surinam. Ibis 114:393–395.
248a. Hawke, D. 1989. Gould's Petrel from Dunedin City. Notornis 36:189–190.
249. Hay, R. Unpubl. data.
250. Heath, S. Unpubl. data.
251. Helms, C.W. 1963. Tentative field estimates of metabolism in buntings. Auk 80:318–334.
252. Henny, C.J., Carter, J.L., and Carter, B.J. 1981. A review of Bufflehead sex and age criteria with notes on weights. Wildfowl 32:117–122.
253. Henny, C.J., and VanCamp, L.F. 1979. Annual weight cycle in wild Screech Owls. Auk 96:795–796.
254. Herd, R.M. 1985. Anatomy and histology of the gut of the Emu *Dromaius novaehollandiae*. Emu 85:43–46.
255. Herholdt, J.J. 1988. Bird weights from the Orange Free State (Part I: Non-passerines). Safring News 17:3–14.
256. Herholdt, J.J. 1988. Bird weights from the Orange Free State (Part II: Passerines). Safring News 17:43–57.
257. Herlugson, C.J. 1983. Growth of nestling Mountain Bluebirds. J. Field Ornithol. 54:259–265.
258. Hickling, R. (ed.). 1983. Enjoying ornithology. T & AD Poyser, Ltd., Staffordshire, UK.
258a. Hicks, D.L. 1967. Adipose tissue composition and cell size in fall migratory thrushes (Turdidae). Condor 69:387–399.
259. Hicks, L.E. 1934. Individual and sexual variations in the European Starling. Bird-Banding 5:103–118.
260. Hilty, S.L. 1977. *Chlorospingus flavovirens* rediscovered, with notes on other Pacific Columbian and Cauca Valley birds. Auk 94:44–49.
261. Hilty, S.L., Parker, T.A., and Silliman, J. 1979. Observations on Plush-capped Finches in the Andes with a description of the juvenile and immature plumages. Wilson Bull. 91:145–148.
262. Hitchcock, W.B., and McKean, J.L. 1969. Square-tailed Kite in the Northern Territory. Emu 69:115.
263. Hodgdon, K.Y. 1979. Operation Horned Grebe. N. Amer. Bird Bander 4:110.
264. Höglund, J., Kålås, J.A., and Løfaldli, L. 1990. Sexual dimorphism in the lekking Great Snipe. Ornis Scand. 21:1–6.
265. Hogstad, O. 1976. Sexual dimorphism and divergence in winter foraging behavior of three-toed woodpeckers *Picoides tridactylus*. Ibis 118:41–50.
266. Holmes, R.T., and Recher, H.F. 1986. Determinants of guild structure in forest bird communities: an intercontinental comparison. Condor 88:427–439.
267. Hoogerwerf, A. 1971. On a collection of birds from the Vogelkop near Manokwari, north-western New Guinea. Emu 71:1–12; 71:73–83.
268. Horne, R.S.C. 1985. Diet of Royal and Rockhopper Penguins at MacQuarie Island. Emu 85:150–156.
269. Houston, D.C. 1976. Breeding of the White-backed and Ruppell's Griffon Vultures, *Gyps africanus* and *G. rueppellii*. Ibis 118:14–40.
269a. Howe, H.F. 1977. Bird activity and seed dispersal of a tropical wet forest tree. Ecology 58:539–550.
270. Howe, H.F., and Vande Kerckhove, G.A. 1981. Removal of wild nutmeg (*Virola surinamensis*) crops by birds. Ecology 62:1093–1106.
271. Howe, M.A. 1980. Problems with wing tags: evidence of harm to Willets. J. Field Ornithol. 51:72–73.
272. Howell, T.R. 1972. Birds of the lowland pine savannah of northeastern Nicaragua. Condor 74:316–340.
273. Hubbard, J.P. 1967. Notes on some Chiapas birds. Wilson Bull. 79:236.
274. Hubbell, P. Unpubl. data
275. Hyde, A.S. 1939. The life history of Henslow's Sparrow, *Passerherbulus henslowi* (Audubon). Univ. Mich. Mus. Zool. Misc. Publ. 41.
276. Imber, M.J. 1976. Breeding biology of the Grey-faced Petrel *Pterodroma macroptera gouldi*. Ibis 118:51–64.
277. Inigo, E.E., Ramos, M., and Gonzalez, F. 1987. Two recent records of neotropical eagles in southern Veracruz, Mexico. Condor 89:671–672.
278. Irving, L. 1960. Birds of Anaktuvuk Pass, Kobuk, and O  Crow. U.S. Nat. Mus. Bull. 217.
279. Isler, M.L., and Isler, P.R. 1987. The tanagers. Smithsonia  Inst. Press, Washington, D.C.

280. Jackson, H.D. 1972. Avifaunal survey of the Umtali Municipal Area. I. The Muneni River collection: a comparison of samples from riparian forest and miombo woodland. Arnoldia Rhod. 6:1–10.

281. Jackson, H.D. 1976. Avifaunal survey of the Umtali Municipal Area. II. The cecil Kop collection: a comparison of samples from montane grassland, montane thicket, and montane forest. Arnoldia Rhod. 8:1–11.

282. Jackson, H.D. 1986. Avifaunal survey of the Mutare Municipal Area. III. The Gimboki collection: a comparison of samples from riparian thicket, miombo woodland on sandflats and miombo woodland on rocky slopes. Arnoldia Zimbabwe 9:325–332.

283. Jackson, H.D. 1987. Avifaunal survey of the Mutare Municipal Area. IV. The Matika collection:a comparison of samples from grassland, thicket, woodland and riparian forest. Arnoldia Zimbabwe 9:353–360.

284. Jackson, H.D. 1987. Avifaunal survey of the Mutare Municipal Area. V. The Birkley South collection: a comparison of samples from grassland, woodland and thicket. Arnoldia Zimbabwe 9:361–367.

285. Jackson, H.D. 1989. Weights of birds collected in the Mutare Municipal Area, Zimbabwe. Bull. Br. Ornithol. Club 109:100–106.

286. Jehl, J. Unpubl. data.

287. Jehl, J.R. 1974. The near-shore avifauna of the Middle American west coast. Auk 91:681–699.

288. Jehl, J.R. 1987. Geographic variation and evolution in the California Gull (*Larus californicus*). Auk 104:421–428.

289. Jehl, J.R., and Murray, B.G. Jr. 1986. The evolution of normal and reverse sexual size dimorphism in shorebirds and other birds. Current Ornithol. 3:1–86.

290. Jenkins, J.M. 1983. The native forest birds of Guam. Ornithol. Monogr. 31.

291. Jensen, F.P. 1983. A new species of sunbird from Tanzania. Ibis 125:447–449.

292. Jimenez, J.E., and Jaksic, F.M. 1989. Biology of the Austral Pygmy-Owl. Wilson Bull. 101:377–389.

293. Jimenez, J.E., and Jaksic, F.M. 1989. Behavioral ecology of Grey Eagle-Buzzards, *Geranoaetus melanoleucus*, in central Chile. Condor 91:913–921.

293a. Johnsgard, P.A. 1973. Grouse and quail of North America. Univ. Nebraska Press, Lincoln, NE.

294. Johnsgard, P.A. 1981. The plovers, sandpipers, and snipes of the world. Univ. Nebraska Press, Lincoln, NE.

295. Johnsgard, P.A. 1983. The hummingbirds of North America. Smithsonian Inst. Press, Washington, D.C.

296. Johnsgard, P.A. 1983. Cranes of the world. Indiana Univ. Press, Bloomington, IN.

297. Johnsgard, P.A. 1986. The pheasants of the world. Oxford Univ. Press, Oxford.

298. Johnsgard, P.A. 1988. The quails, partridges, and francolins of the world. Oxford Univ. Press, Oxford.

299. Johnson, D.W. 1985. Weight, moult, and breeding condition of some Malawi birds. Ostrich 56:216–217.

300. Johnson, N.K. 1965. The breeding avifaunas of the Sheep and Spring Ranges in southern Nevada. Condor 67:93–124.

301. Johnson, N.K., and Jones, R.E. 1990. Geographic differentiation and distribution of the Peruvian Screech-Owl. Wilson Bull. 102:199–212.

302. Johnson, R.E. 1977. Seasonal variation in the genus *Leucosticte* in North America. Condor 79:76–86.

303. Johnson, T.B., and Hilty, S. 1976. Notes on the Sickle-winged Guan in Columbia. Auk 93:194–195.

304. Johnson, T.H., and Temple, S.A. 1990. A global review of island endemic birds. Ibis 132:167–180.

305. Johnston, D.W. 1963. Heart weights of some Alaskan birds. Wilson Bull. 75:435–446.

306. Johnston, D.W., and Williamson, F.S. 1960. Heart weights of North American crows and ravens. Wilson Bull. 72:248–252.

307. Johnston, R.F., and Johnson, R.F. 1989. Nonrandom mating in feral pigeons. Condor 91:23–29.

308. Jolly, J. Unpubl. data.

309. Jouventin, P., Martinez, J., and Roux, J.P. 1989. Breeding biology and current status of the Amsterdam Island Albatross *Diomedea amsterdamensis*. Ibis 131:171–182.

310. Jouventin, P., Mougin, J.-L., Stahl, J.-C., and Weimerskirch, H. 1985. Comparative biology of the burrowing petrels of the Crozet Islands. Notornis 32:157–220.

311. Junge, G.C.A., and Mees, G.F. 1961. The avifauna of Trinidad and Tobago. Zoologische Verhandelingen 37:1–172.

312. Kagarise, C.M. 1979. Breeding biology of the Wilson's phalarope in North Dakota. Bird-Banding 50:12–22.

313. Kahl, M.P. Unpubl. compilation.

314. Karr, J.R. 1971. Structure of avian communities in selected Panama and Illinois habitats. Ecol. Monogr. 41:207–233.

314a. Karr, J.R. 1976. Weights of African birds. Bull. Br. Ornithol. Club 96:92–96.

315. Karr, J.R., Willson, M.F., and Moriarty, D.J. 1978. Weights of some Central American birds. Brenesia 14–15:249–257.

315a. Kemp, A.C. 1989. Estimation of biological indices for little-known African owls. In Meyburg, B.-U., and Chancellor, R.D. (eds.). Raptors in the modern world. World Working Group on Birds of Prey and Owls, Berlin, Germany pp. 441-449.

316. Kepler, C.B. 1969. Breeding biology of the Blue-faced Booby *Sula dactylatra personata* on Green Island, Kure Atoll. Publ. Nuttall Ornithol. Club.

317. Kepler, C.B., and Parkes, K.C. 1972. A new species of warbler (Parulidae) from Puerto Rico. Auk 89:1–19.
318. Kerlinger, P., and Lein, M. 1988. Causes of mortality, fat condition, and weights of wintering Snowy Owls. J. Field Ornithol. 59:7–12.
319. King, J.R. 1989. Notes on the birds of the Rio Mazan Valley, Azuay Province, Ecuador, with special reference to *Leptosittaca branickii*, *Hapalopsittaca amazonica pyrrhops*, and*Metallura baroni*. Bull. Br. Ornithol. Club 109:140–147.
320. King, J.R. 1991. Body weights of some Ecuadorean birds. Bull. Br. Ornithol. Club 111:46–49.
321. King, J.R., and Wales, E.E. 1965. Photoperiodic regulation of testicular metamorphosis and fat deposition in three taxa of rosy finches. Physiol. Zool. 38:49–68.
322. Kitson, A.R. 1980. *Larus relictus* - a review. Bull. Br. Ornithol. Club 100:178–185.
323. Klaas, E.E. 1968. Summer birds from the Yucatan Peninsula, Mexico. Univ. Kansas Mus. Nat. Hist. Publ. 17:579–611.
324. Klomp, N.I., and Wooller, R.D. 1988. The size of Little Penguins *Eudyptula minor*, on Penguin Island, Western Australia. Rec. West. Aust. Mus. 14:211–215.
325. Koenig, W.D. 1980. Variation and age determination in a population of Acorn Woodpeckers. J. Field Ornithol. 51:10–16.
326. Koford, C.B. 1953. The California Condor. Nat. Audubon Soc. Res. Rep. 4.
327. Komen, J. 1991. Energy requirements of nestling Cape Vultures. Condor 93:153–158.
328. Kushlan, J.A. 1977. Sexual size dimorphism in the White Ibis. Wilson Bull. 89:92–98.
329. LaBastille, A. 1974. Ecology and management of the Atitlan Grebe, Lake Atitlan, Guatemala. Wildl. Monogr. 37:1–66.
330. Lambert, F.R. 1989. Pigeons as seed predators and dispersers of figs in a Malaysian lowland forest. Ibis 131:521–527.
331. Langham, N.P.E. 1983. Growth strategies in marine terns. Studies in Avian Biol. 8:73–83.
332. Langham, N.P.E. 1987. Morphometrics and moult in Fijian passerines. New Zealand J. Zool. 14:463–475.
333. Lanyon, S.M., Stotz, D.F., and Willard, D.E. 1990. *Clytoctantes atrogularis*, a new species of antbird from western Brazil. Wilson Bull. 102:571–580.
334. Lanyon, W.E. 1961. Specific limits and distribution of Ash–throated and Nutting Flycatcher. Condor 63:421–449.
335. Leak, J., and Robinson, S.K. 1989. Notes on the social behavior and mating system of the Casqued Oropendula. Wilson Bull. 101:134–137.
336. Leberman, R.D. 1973. A study of Tufted Titmouse weights. E. Bird Banding Assn. News 36:34–38.
337. Leck, C.F. 1975. Weights of migrants and resident birds in Panama. Bird-Banding 46:201–206.
338. LeCroy, M. 1981. The genus *Paradisaea* display and evolution. Am. Mus. Novitates 2714.
339. LeCroy, M., and LeCroy, S. 1974. Growth and fledging in the Common Tern (*Sterna hirundo*). Bird-Banding 45:326–340.
340. Lehmann, V.W. 1941. Attwater's Prairie Chicken: its life history and management. North Am. Fauna 57.
341. Leopold, A.S., and McCabe, R.A. 1957. Natural history of the Montezuma Quail in Mexico. Condor 59:3–26.
342. Lessells, C.M., and Ovenden, G.N. 1989. Heritability of wing length and weight in European Bee-eaters (*Merops apiaster*). Condor 91:210–214.
343. Ligon, J.D. 1968. Sexual differences in foraging behavior in two species of *Dendrocopus* woodpeckers. Auk 85:203–215.
344. Linsdale, J.M., and Sumner, E.L. 1934. Winter weights of Golden-crowned and Fox Sparrows. Condor 36:107–112.
345. Livezey, B.C. 1989. Phylogenetic relationships and incipient flightlessness of the extinct Auckland Islands Merganser. Wilson Bull. 101:410–435.
346. Livezey, B.C. 1990. Evolutionary morphology of flightlessness in the Auckland Islands Teal. Condor 92:639–673.
347. Long, J.L. 1984. Weights and Measurements - Western Rosella. In Data exchange. Corella 8:28.
348. Longmore, N.W., and Boles, W.E. 1983. Description and systematics of the Eungella Honeyeater *Meliphaga hindwoodi*, a new species of honeyeater from central eastern Queensland, Australia. Emu 83:59–65.
349. Louette, M., and Prevost, J. 1987. Passereaux collectes par J. Prevost au Cameroun. Malimbus 9:83–96.
350. Lougheed, S.C., Arnold, T.W., and Bailey, R.C. 1991. Measurement error of external and skeletal variables in birds and its effects on principal components. Auk 108:432–436.
351. Lowe, K.W., Clark, A., and Clarke, R.A. 1985. Body measurements, plumage and moult of the Sacred Ibis in South Africa. Ostrich 56:111–116.
352. Lowery, G.H., and O'Neill, J.P. 1969. A new species of antpitta from Peru and a revision of the subfamily Grallariinae. Auk 86:1–12.
353. Lowery, G.H., and Tallman, D.A. 1976. A new genus and species of nine-primaried oscine of uncertain affinities from Peru. Auk 93:415–429.

353a. Lyles, A. Unpubl. data from captive birds at New York Zoological Society, Bronx, New York.

354. MacLean, G.L. 1985. Robert's Birds of Southern Africa. John Voeloker Bird Book Fund, Cape Town, South Africa.

355. MacLean, S.F. 1974. Lemming bones as a source of calcium for arctic sandpipers (*Calidris* spp.). Ibis 116:552–557.

356. MacLean, S.F., and Holmes, R.T. 1971. Bill lengths, wintering areas, and taxonomy of North American Dunlins *Calidris alpina*. Auk 88:893–901.

357. MacMillen, R.E. 1981. Nonconformance of standard metabolic rate with body mass in Hawaiian Honeycreepers. Oecologia 49:340–343.

358. MacMillen, R.E. 1990. Water economy of granivorous birds: a predictive model. Condor 92:379–394.

359. Maher, W.J. 1974. Ecology of Pomarine, Parasitic and Long-tailed Jaegers in northern Alaska. Pacific Coast Avifauna 37.

360. Maher, W.J. 1986. Growth and development of the Brown-backed Honeyeater *Ramsayornis modestus* in North Queensland. Emu 86:245–248.

361. Majumdar, N. 1984. On a collection of birds from Bastar district, Madhya Pradesh. Records Zool.Survey of India Occas. Papers 59:1–54.

362. Mann, C.F. 1985. An avifaunal study in Kakamega Forest, Kenya, with particular reference to species diversity, weight and moult. Ostrich 56:236–262.

363. Marchant, S., and Higgins, P. (ed.). 1990. The handbook of Australian, New Zealand and Antarctic birds. Oxford Univ. Press, Oxford, UK. Vol. 1.

363a. Marin A.M., Kiff, L.F., and Pena G., L. 1989. Notes on Chilean birds, with descriptions of two new subspecies. Bull. Br. Ornithol. Club 109:66–82.

364. Marion, W.R. 1977. Growth and development of the Plain Chachalaca in south Texas. Wilson Bull. 89:47–56.

365. Marshall, J.T. 1948. Ecological races of Song Sparrows in the San Francisco Bay Region. Condor 50:233–256.

366. Marshall, J.T. 1978. Systematics of smaller Asian night birds based on voice. Ornithol. Monogr. 25.

366a. Marshall, S.G., and Baker, V.B. 1964. The role of fat in bird migration. Oriole 29:35–39.

367. Marti, C.D. 1990. Sex and age dimorphism in the Barn Owl and a test of mate choice. Auk 107:246–254.

368. Martin, P.S., Robins, C.R., and Heed, W.B. 1954. Birds and biogeography of the Sierra de Tamaulipas, an isolated pine-oak habitat. Wilson Bull. 66:38–57.

369. Mason, I.J., and Wolfe, T.O. 1975. First record of the Marsh Crake for the Northern Territory. Emu 75:235.

370. Maxson, S.J., and Bernstein, N.P. 1982. Kleptoparasitism by South Polar Skuas on Blue-eyed Shags in Antarctica. Wilson Bull. 94:269–281.

371. McClure, H.E. 1964. Avian bionomics in Malaya: 1. The avifauna above 5000 feet altitude at Mount Brinchang, Pahang. Bird-Banding 35:141–183.

372. McKean, J.L., and Lewis, J.H. 1971. Record of the Fulmar Prion off southern New South Wales. Emu 71:141.

373. McLandress, M.R., and Raveling, D.G. 1981. Changes in diet and body composition of Canada Geese before spring migration. Auk 98:65–79.

374. McNicholl, M.K. Unpubl. data.

375. Mcanley, B. 1967. Aging and sexing blackbirds, Bobolinks and Starlings. Patuxent Wildl. Res. Center. Spec. Report.

376. Meanley, B. 1969. Natural history of the King Rail. North Am. Fauna 67.

377. Meanley, B. 1971. Natural history of the Swainson's Warbler. North Am. Fauna 69.

378. Melville, D.S., and Round, P.D. 1984. Weights and gonad condition of some Thai birds. Bull. Br. Ornithol. Club 104:127–138.

379. Mendelssohn, H., Yom-Tov, Y., and Safriel, U. 1975. Hume's Tawny-Owl *Strix butleri* in the Judean, Negev and Sinai deserts. Ibis 117:110–111.

380. Merritt, R.E., Bell, P.A., and Laboudallon, V. 1986. Breeding biology of the Seychelles Black Parrot (*Coracopsis nigra barklyi*). Wilson Bull. 98:160–163.

381. Merton, D.V., Morris, R.B., and Atkinson, I.A.E. 1984. Lek behaviour in a parrot: the Kakapo *Strigops habroptilus* of New Zealand. Ibis 126:277–283.

382. Mewaldt, L.R. 1952. Reproduction and moult in Clark's Nutcracker, *Nucifraga columbiana* Wilson. Ph.D. dissertation, Washington State Univ., Pullman, WA.

383. Milledge, D.R. 1983. The Marbled Frogmouth *Podargus ocellatus plumiferus* in the Nightcap Range, Northeastern NSW. Emu 83:43–44.

384. Miller, A.H. 1947. The tropical avifauna of the upper Magdelena Valley, Columbia. Auk 64:351–381.

385. Miller, A.H. 1952. Supplementary data on the tropical avifauna of the arid upper Magdalena Valley of Columbia. Auk 69:450–457.

386. Miller, A.H. 1955. The avifauna of the Sierra del Carmen of Coahuila, Mexico. Condor 57:154–178.

387. Miller, A.H. 1963. Seasonal activity and ecology of the avifauna of an American equatorial cloud forest. Univ. Cal. Publ. Zool. 66:1–73.

388. Miskelly, C.M. 1990. Breeding systems of New Zealand Snipe *Coenocorypha aucklandica* and Chatham Island Snipe *C. pusilla*: are they food limited? Ibis 132:366–379.
389. Montague, T.L. 1984. The food of Antarctic Petrels (*Thalassoica anarctica*). Emu 84:244–245.
390. Moorhouse, R.J. Unpubl. data.
391. Moreau, R.E. 1969. Comparative weights of some Trans-Saharan migrants at intermediate points. Ibis 111:621–629.
392. Moreau, R.E., and Dolp, R.M. 1970. Fat, water, weights and wing-lengths of autumn migrants in transit on the northwest coast of Egypt. Ibis 112:209–228.
393. Morris, A.K. 1971. Distribution of the Black Bittern in New South Wales. Emu 71:175–176.
394. Morris, A.K. 1971. The Red-chested Quail in New South Wales, Australia. Emu 71:178–180.
395. Morrison, P. 1962. Modification of body temperature by activity in Brazilian hummingbirds. Condor 64:315–323.
396. Morse, R.A., and Laigo, F.M. 1969. The Philippines Spine-tail Swift, *Chaetura dubia* McGreggor, as a honey bee predator. Philippine Ent. 1:138–143.
397. Morton, E.S. 1977. Intratropical migration in the Yellow-green Vireo and Piratic Flycatcher. Auk 94:97–106.
398. Moser, T.D., and Rusch, D.H. 1988. Indices of structural size and condition of Canada Geese. J. Wildl. Manage. 52:202–208.
399. Mountainspring, S. 1987. Ecology, behavior and conservation of the Maui Parrotbill. Condor 89:24–39.
400. Mountainspring, S., and Scott, J.M. 1985. Interspecific competition among Hawaiian forest birds. Ecol. Monogr. 55:219–239.
401. Mueller, H.C., Berger, D.D., and Allez, G. 1976. Age and sex variation in the size of Goshawks. Bird-Banding 47:310–318.
402. Mueller, H.C., Berger, D.D., and Allez, G. 1979. Age and sex differences in the size of Sharp-shinned Hawks. Bird-Banding 50:34–44.
403. Mueller, H.C., Berger, D.D., and Allez, G. 1981. Age, sex, and seasonal differences in size of Cooper's Hawks. J. Field Ornithol. 52:112–126.
404. Murata, K. Unpubl. data from captive birds at Kobe Oji Zoo, Kobe, Japan.
405. Murray, B.G. 1969. A comparative study of the LeConte's and Sharp-tailed sparrows. Auk 86:199–231.
406. Murray, B.G., and Hardy, J.W. 1981. Behavior and ecology of four syntopic species of finches in Mexico. Z. Tierpsychol. 57:57–72.
407. Murray, B.G., and Jehl, J.R. 1964. Weights of autumn migrants from coastal New Jersey. Bird-Banding 35:253–263.
408. Murray, K.G., Winnett-Murray, K., Eppley, Z.A., Hunt, G.L., and Schwartz, D.B. 1983. Breeding biology of the Xanthus' Murrelet. Condor 85:12–21.
409. Naik, S., and Naik, R.M. 1966. Studies on the House Swift, *Apus affinis* (G.E. Gray) VI. Body weight. Pavo 4:54–91.
410. Nakamura, M. 1990. Cloacal protuberance and copulatory behavior of the Alpine Accentor (*Prunella collaris*). Auk 107:284–295.
411. Navas, J.R., and Bó, N.A. 1986. Notas sobre una coleccion de aves del Parque Nacional Lihue Calel, La Pampa, Argentina. El Hornero 12:250–261.
412. Nelson, A.L., and Martin, A.C. 1953. Gamebird weights. J. Wildlife Manage. 17:36–42.
413. Nelson, B. 1978. The Sulidae. Oxford Univ. Press, Oxford.
414. Nesbit, I.C. 1981. Biological characteristics of the Roseate Tern *Sterna dougallii*. Office of Endangered Species, U.S. Fish & Wildl. Serv., Washington, D.C.
415. Nesbitt, S.A., Gilbert, D.T., and Barbour, D.B. 1976. Capturing and banding Limpkins in Florida. Bird-Banding 47:164–165.
416. Neufeldt, I., and Ivanov, A.I. 1960. Some notes of the biology of the Needle-tailed Swift in Siberia. British Birds 53:433–435.
417. New Zealand Dept. Conservation. Unpubl. data.
417a. Newmark, W.D. 1991. Tropical forest fragmentation and the local extinction of understory birds in the eastern Usambara Mountains, Tanzania. Conservation Biol. 5:67–78.
418. Newton, I., Marquiss, M., and Village, A. 1983. Weights, breeding, and survival in European Sparrowhawks. Auk 100:344–354.
419. Nichols, J.D., and Haramis, G.M. 1980. Sex-specific differences in winter distribution patterns of Canvasbacks. Condor 82:406–416.
420. Nisbet, I.C.T. 1968. Weights of birds caught at night at a Malayan radio tower. Ibis 110:352–354.
421. Nolan, V. 1978. The ecology and behavior of the Prairie Warbler *Dendroica discolor*. Ornithol. Monogr. 26.
422. Norris, R.A. 1958. Comparative biosystematics and life history of the nuthatches *Sitta pygmaea* and *Sitta pusilla*. Univ. Cal. Publ. Zool. 56:119–300.
423. Noske, R.A. 1979. Coexistence of three species of treecreepers in north-eastern New South Wales. Emu 79:120–128.

423a. Odum, E.P. 1960. Lipid deposition in nocturnal migrant birds. Proc. Internat. Ornithol. Congress 12:563–576.

423b. Odum, E.P., Connell, C.E., and Stoddard, H.L. 1961. Flight energy and estimated flight ranges of some migratory birds. Auk 78:515–527.

423c. Odum, E.P., Marshall, S.G., and Marples, T.G. 1965. The caloric content of migrating birds. Ecology 46:901–904.

424. Olsen, P.D., Debus, S.J.S., Czechura, G.V., and Mooney, N.J. 1990. Comparative feeding ecology of the Grey Goshawk *Accipiter novaehollandiae* and Brown Goshawk *Accipiter fasciatus*. Austral. Bird Watcher 13:178–191.

425. Olsen, P.D., and Olsen, J. 1985. A natural hybridization of the Brown Goshawk *Accipiter fasciatus* and Gray Goshawk *Accipiter novaehollandiae* in Australia, and a comparison of the two species. Emu 85:250–257.

426. Olsen, P.D., and Olsen, J. 1986. Distribution, status, movements and breeding of the Grey Falcon *Falco hypoleucos*. Emu 86:47–51.

427. Olsen, P.D., and Olsen, J. 1987. Movements and measurements of the Australian Kestrel *Falco cenchroides*. Emu 87:35–41.

428. Olsen, P.D., and Stokes, T. 1989. State of knowledge of the Christmas Island Hawk-Owl *Ninox squampila natalis*. In Meyburg, B.-U., and Chancellor, R.D. (eds.). Raptors in the modern world. Proc. Third World Conference on Birds of Prey and Owls, World Working Group on Birds of Prey and Owl, Berlin, Germany.

429. Olson, S.L. 1985. Weights of some Cuban birds. Bull. Br. Ornithol. Club 105:68–69.

430. Reference deleted.

431. Olson, S.L., and Angle, J.P. 1977. Weights of some Puerto Rican birds. Bull. Br. Ornithol. Club 97:105–107.

432. Olson, S.L., James, H.L., and Meister, C.A. 1981. Winter field notes and specimen weights of Cayman Island birds. Bull. Br. Ornithol. Club 101:339–346.

433. O'Neill, J.P., and Graves, G.R. 1977. A new genus and species of owl (Aves: Strigidae) from Peru. Auk 94:409–416.

434. O'Neill, J.P., Munn, C.A., and Franke J., I. 1991. *Nannopsittaca dachilleae*, a new species of parrotlet from eastern Peru. Auk 108:225–229.

435. O'Neill, J.P., and Schulenberg, T.S. 1979. Notes on the Masked Saltator, *Saltator cinctus*, in Peru. Auk 79:610–613.

436. Oniki, Y. 1972. Some temperatures of Panamaniam birds. Condor 74:209–214.

437. Oniki, Y. 1974. Some temperatures of birds of Belem, Brazil. Acta Amazonica 4:63–68.

438. Oniki, Y. 1975. Temperatures of some Puerto Rican birds, with notes of low temperatures in todies. Condor 77:344.

439. Oniki, Y. 1978. Weights, digestive tracts and gonadal conditions of some Amazonian birds. Rev. Brasil. Biol. 38:679–681.

440. Oniki, Y. 1980. Weights and cloacal temperatures of some birds of Minas Gerais, Brazil. Rev. Brasil. Biol. 40:1–4.

441. Oniki, Y. 1981. Weights, cloacal temperatures, plumage and molt condition of birds in the state of Sao Paulo. Rev. Brasil. Biol. 41:451–460.

441a. Ottenwalder, J., and Vargas, T. 1979. Nueva localidad para el Diablotin en la Republica Dominicana. Natur. Postal. Univer. Aut. de Santo Domingo 36.

442. Otto, J.E. 1983. Breeding ecology of the Pied-billed Grebe (*Podilymbus podiceps* (Linnaeus)) on Rush Lake, Winnebago County, Wisconsin. M.S. thesis, Univ. Wisconsin-Oshkosh, Oshkosh, WI.

443. Owen, M., and Cook, W.A. 1977. Variations in body weight, wing length and condition of Mallard *Anas platyrhynchos platyrhynchos* and their relationship to environmental changes. J. Zool. (London) 183:377–395.

444. Owen, M., and Ogilvie, M.A. 1979. Wing molt and weights of Barnacle Geese in Spitsbergen. Condor 81:42–52.

445. Palmer, R.S. (ed.). 1962. Handbook of North American birds. Vol. 1. Yale University Press, New Haven, CT.

446. Palmer, R.S. (ed.). 1976. Handbook of North American birds. Vols. 2 & 3. Yale University Press, New Haven, CT.

447. Palmer, R.S. (ed.) 1988. Handbook of North American birds. Vols. 4 & 5. Yale University Press, New Haven, CT.

448. Parker, T.A. 1981. Distribution and biology of the White-cheeked Cotinga *Zaratornis stresemanni*, a high Andean frugivore. Bull. Br. Ornithol. Club 101:256–265.

449. Parker, T.A. 1984. Notes on the behavior of *Ramphotrigon* flycatchers. Auk 101:186–188.

450. Parker, T.A., and Parker, S.A. 1982. Behavioral and distributional notes on some unusual birds of a lower montane cloud forest in Peru. Bull. Br. Ornithol. Club 102:63–70.

451. Parker, T.A., Schulenberg, T.S., Graves, G.R., and Braun, M.J. 1985. The avifauna of the Huancabamba region, northern Peru. Ornithol. Monogr. 36:169–197.

452. Parmelee, D.F., and MacDonald, S.D. 1960. The birds of west-central Ellesmere Island and adjacent areas. Nat. Mus. Canada Bull. 169.

453. Paton, D.C., and Collins, B.G. 1989. Bills and tongues of nectar-feeding birds: a review of morphology, function and performance, with intercontinental comparisons. Austral. J. Ecol. 14:473–506.

454. Paton, D.C., and Ford, H.A. 1976. Pollination by birds of native plants in South Australia. Emu 77:73–85.

455. Payne, R.B. 1984. Sexual selection, lek and arena behavior, and sexual size dimorphism in birds. Ornithol. Monogr. 33. p. 43.

456. Paynter, R.A. 1952. Birds from Popocatepetl and Ixtaccihuatl, Mexico. Auk 69:293–301.

457. Paynter, R.A. 1955. The ornithogeography of the Yucatan Peninsula. Bull. Peabody Mus. Nat. Hist. 9:1–347.

458. Paynter, R.A. 1956. Avifauna of the Jorulla Region, Michoacan, Mexico. Postilla 25:1–10.

459. Pearson, D.J. 1970. Weights of Red-backed Shrikes, on autumn passage in Uganda. Ibis 112:114–115.

460. Pearson, D.J., and Backhurst, G.C. 1976. The southward migration of palaearctic birds over Ngulia, Kenya. Ibis 118:78–105.

461. Pearson, D.L. 1977. Ecological relationships of small antbirds in Amazonian bird communities. Auk 94:283–292.

462. Peterson, K.L., Best, L.B., and Winter, B.M. 1986. Growth of nestling Sage Sparrows and Brewer's Sparrows. Wilson Bull. 98:535–546.

463. Petit, L. Unpubl. data.

464. Pierce, R.J. 1984. Plumage, morphology and hybridization of New Zealand stilts *Himantopus* ssp. Notornis 31:106–130.

464a. Piersma, T., and Davidson, N.C. 1991. Confusions of mass and size. Auk 108:441–444.

465. Power, D.M. 1980. Evolution of land birds on the California Islands. In Power, D.M. (eds.). The California Islands: proceedings of a multidisciplinary symposium. Santa Barbara Mus. Nat. Hist.

466. Price, T. 1979. The seasonality and occurrence of birds in the Eastern Ghats of Andhra Pradesh. J. Bombay Nat. Hist. Soc. 76:379–422.

467. Price, T., and Jamdar, N. 1990. The breeding birds of Overa Wildlife Sanctuary, Kashmir. J. Bombay Natural Hist. Soc. 87:1–15.

468. Prince, P.A., Ricketts, C., and Thomas, G. 1981. Weight loss in incubating albatrosses and its implications for the energy and food requirements. Condor 83:238–242.

469. Prinzinger, R., Kruger, K., and Schuchman, K.L. 1981. Metabolism-weight relationship in p43 17 hummingbird species at different temperatures during day and night. Experientia 37:1307–1309.

470. Prum, R.O., and Johnson, A.E. 1987. Display behavior, foraging ecology, and systematics of the Golden-winged Manakin (*Masius chrysopterus*). Wilson Bull. 99:521–539.

471. Prys-Jones, R.P. 1982. Molt and weight of some land-birds on Dominica, West Indies. J. Field Ornithol. 53:352–362.

472. Puerta, M.L., Alonso, J.C., Huecas, V., Alonso, J.A., Abelenda, M., and Munoz-Pulido, R. 1990. Hematology and blood chemistry of wintering Common Cranes. Condor 92:210–214.

473. Pulich, W.M. 1976. The Golden-cheeked Warbler. Texas Parks & Wildl. Dept., Austin, TX.

474. Quinn, J.S. 1990. Sexual size dimorphism and parental care patterns in a monomorphic and a dimorphic larid. Auk 107:260–274.

475. Rand, A.L., and Rabor, D.S. 1960. Birds of the Philippine Islands: Siquijor, Mount Malindang, Bohol, and Samar. Fieldiana: Zoology 35:222–441.

476. Rao, P., and Muralidharan, S. 1990. Weight of White-necked Stork *Ciconia episcopus*. J. Bombay Nat. Hist. Soc. 87:139.

477. Ratti, J.T., Timm, D.E., and Robards, F.C. 1977. Weights and measurements of Vancouver Canada Geese. Bird-Banding 48:354–357.

478. Raveling, D.G. 1968. Weights of *Branta canadensis interior* during winter. J. Wildl. Manage. 32:412–414.

479. Raveling, D.G. 1978. Morphology of the Cackling Canada Goose. J. Wildl. Manage. 42:897–900.

480. Recher, H.F. 1989. Foraging segregation of Australian warblers (Acanthizidae) in open forest near Sydney, NSW. Emu 89:204–215.

480a. Recher, H.F., Holmes, R.T., Schulz, M., Shields, J., and Kavanagh, R. 1985. Foraging patterns of breeding birds in eucalypt forest and woodland of southeastern Australia. Austral. J. Ecol. 10:399–419.

481. Recher, H.F., Davis, W.E., and Holmes, R.T. 1987. Ecology of the Brown and Striated Thornbills in front of south-eastern New South Wales, with comments on forest management. Emu 87:1–13.

482. Reid, N. 1989. Dispersal of mistletoes by honeyeaters and flowerpeckers: components of seed dispersal quality. Ecology 70:137–145.

483. Remsen, J.V. 1977. Five bird species new to Columbia. Auk 94:363.

484. Remsen, J.V., and Traylor, M.A. 1983. Additions to the avifauna of Bolivia, Part II. Condor 85:95–98.

485. Reynolds, C.M. 1972. Mute Swan weights in relation to breeding. Wildfowl 23:111–118.

486. Richdale, L.E. 1964. Notes on the Mottled Petrel *Pterodroma inexpectata* and other petrels. Ibis 106:110–114.

487. Ricklefs, R.E. Unpubl. compilation of weight data.

487a. Ricklefs, R.E. 1968. Weight recession in nestling birds. Auk 85:30–35.

488. Ricklefs, R.E. 1983. Some considerations on the reproductive energetics of pelagic seabirds. Studies in Avian Biol. 8:84–94.

489. Ricklefs, R.E., Bruning, D.F., and Archibald, G.W. 1986. Growth rates of cranes reared in captivity. Auk 103:125–134.

490. Rinke, D. 1986. Notes on the avifauna of Niuafo'ou Island, Kingdom of Tonga. Emu 86:82–86.

491. Rinke, D. 1987. The avifauna of 'Eua and its offshore islet Kalau, Kingdom of Tonga. Emu 87:26–34.

492. Rinke, D. 1989. The reproductive biology of the Red Shining Parrot *Prosopeia tabuensis* on the island of 'Eua, Kingdom of Tonga. Ibis 131:238–249.

493. Ripley, S.D. 1950. A small collection of birds from Argentine Tierra del Fuega. Postilla 3:1–11.

494. Ripley, S.D., and Beehler, B.M. 1987. New evidence for sympatry in the sibling species *Caprimulgus atripennis* Jerdon and *C. macrurus* Horsfield. Bull. Br. Ornithol. Club 107:47–49.

495. Ripley, S.D., and Bond, G.M. 1971. Systematic notes on a collection of birds from Kenya. Smithsonian Cont. Zool. 111:1–21.

496. Ripley, S.D., and Watson, G.E. 1956. Cuban bird notes. Postilla 26:1–6.

496a. Rising, J.D., and Somers, K.M. 1989. The measurement of overall body size in birds. Auk 106:666–674.

497. Robbins, M.B., Parker, T.A., and Allen, S.E. 1985. The avifauna of Cerro Pirre, Darien, eastern Panama. Ornithol. Monogr. 36:198–232.

498. Robbins, M.B., and Ridgely, R.S. 1991. *Sipia rosenbergi* (Formicariidae) is a synonym of *Myrmeciza* [*laemosticta*] *nigricauda*, with comments on the validity of the genus *Sipia*. Bull. Br. Ornithol. Club 111:11–18.

499. Robertson, H.A., and Dennison, M.D. 1984. Sexual dimorphism of the Chatham Island Warbler *Gerygone albofrontata*. Emu 84:103–107.

500. Robertson, H.A., Whitaker, A.H., and Fitzgerald, B.M. 1983. Morphometrics of forest birds in the Orongorongo Valley, Wellington, New Zealand. New Zealand J. Zool. 10:87–98.

501. Robertson, J.S., and Woodall, P.F. 1987. Survival of Brown Honeyeaters in south-east Queensland. Emu 87:137–142.

502. Rodrigues de los Santos, M., and Rubio Garcia, J.C. 1986. Biology and biometry of the Pallid Swift (*Apus pallidus*) in southern Spain. Gerfaut 76:19–30.

503. Rogers, D.T., and Odum, E.P. 1966. A study of autumnal postmigrant weights and vernal fattening of North American migrant in the tropics. Wilson Bull. 78:415–433.

504. Rohwer, F.C. 1988. Inter- and intraspecific relationships between egg size and clutch size in waterfowl. Auk 105:161–176.

505. Rohwer, S., and Butler, J. 1977. Ground foraging and rapid molt in the Chuck-will's-widow. Wilson Bull. 89:165–166.

506. Ross, C.A. 1988. Weights of some New Caledonian birds. Bull. Brit. Ornithol. Club 108:91–93.

507. Round, P.D., and Swan, R.L. 1977. Aspects of the breeding of Cory's Shearwater *Calonectris diomedea* in Crete. Ibis 119:350–353.

508. Rowley, I. 1978. Communal activities among White-winged Choughs *Corcorax melanorhamphus*. Ibis 120:178–197.

509. Russell, S.M. Unpubl. data.

510. Russell, S.M. 1964. A distributional study of the birds of British Honduras. Ornithol. Monogr. 1.

511. Russell, S.M., Barlow, J.C., and Lamm, D.W. 1979. Status of some birds on Isla San Andres and Isla Providencia, Columbia. Condor 81:98–100.

512. Russell, S.M., and Lamm, D.W. 1978. Species of Formicariidae new to Columbia. Auk 95:421.

513. Ryan, P.G., and Moloney, C.L. 1991. Tristan Thrushes kill adult White-bellied Storm Petrels. Wilson Bull. 103:130–132.

514. Ryan, P.G., Watkins, B.P., and Siegfried, W.R. 1989. Morphometrics, metabolic rate and body temperature of the smallest flightless bird: the Inaccessible Island Rail. Condor 91:465–467.

515. Ryan, P.G., and Watlans, B.B. 1989. Snow Petrel breeding biology at an inland site in continental Antarctica. Colonial Waterbirds 12:176–184.

516. Ryder, J.P. 1978. Sexing Ring-billed Gulls externally. Bird-Banding 49:218–222.

517. Sanderson, G.C. 1977. Management of migratory shore and upland game birds in North America. Internat. Assn. Fish & Wildl. Agencies. University Nebraska Press, Lincoln.

518. Saunders, D.A. 1978. Measurements of the Little Corella from Kununurra, Western Australia. Emu 78:37–39.

519. Saunders, D.A. 1979. Distribution and taxonomy of the White-tailed and Yellow-tailed Black-Cockatoos *Calypturhynchus* spp. Emu 79:215–229.

520. Saunders, D.A., Smith, G.T., and Campbell, N.A. 1984. The relationship between body weight, egg weight, incubation period, nestling period and nest site in the Psittaciformes, Falconiformes, Strigiformes and Columbiformes. Aust. J. Zool. 32:57–65.

521. Schluter, D. Unpubl. data.

522. Schluter, D., and Repasky, R. 1991. Worldwide limitation of finch densities by food and other factors. Ecology 72:1763–1774.

523. Schmitt, C.G., and Cole, D.C. 1981. First records of Black-legged Seriema (*Chunga burmeisteri*) in Bolivia. Condor 83:182–183.

524. Schodde, R. 1978. The identify of five type-specimens of New Guinean birds. Emu 78:1–6.

525. Schodde, R. 1982. The Fairy-wrens. Lansdowne Editions, Melbourne, Australia.

526. Schodde, R. 1984. First specimens of Campbell's Fairy-Wren, *Malurus campbelli*, from New Guinea. Emu 84:249–250.

527. Schodde, R., and Christidis, L. 1987. Genetic differentiation and subspeciation in the Grey Grasswren (*Amytornis barbatus*) (Maluridae). Emu 87:188–192.

528. Schodde, R., and Mason, I.J. 1979. Revision of the Zitting Cisticola *Cisticola juncidis* (Rafinesque) in Australia, with description of a new subspecies. Emu 79:49–53.

529. Schodde, R., and Mason, I.J. 1980. Nocturnal birds of Australia. Lansdowne Editions, Melbourne, Australia

530. Schodde, R., Mason, I.J., and McKean, J.L. 1979. A new subspecies of *Philemon buceroides* from Arnhem Land. Emu 79:24–30.

531. Schodde, R., and McKean, J.L. 1973. Distribution, taxonomy and evolution of the Gardener Bowerbirds *Amblyornis* spp. in eastern New Guinea with descriptions of two new subspecies. Emu 73:51–60.

532. Schramm, M. 1983. The breeding biology of the petrels *Pterodroma macroptera*, *P. brevirostris* and *P. mollis* at Marion Island. Emu 83:75–81.

533. Schreiber, E.A., and Schreiber, R.W. 1988. Great Frigatebird size dimorphism on two Central Pacific atolls. Condor 90:90–99.

534. Schreiber, R.W., and Schreiber, E.A. 1979. Notes on measurements, mortality, molt, and gonad condition in Florida west coast Laughing Gulls. Fla. Field Nat. 7:19–23.

535. Schreiber, R.W., and Schreiber, E.A. 1984. Mensural and moult data for some birds of Martinique, French West Indies. Bull. Br. Ornithol. Club 104:62–69.

536. Schuchmann, K. 1979. Notes on sexual dimorphism and the nest of the Greenish Puffleg *Haplophaedia aurelia caucensis*. Bull. Br. Ornithol. Club 99:59–61.

537. Schulenberg, T.S. 1985. An intergeneric hybrid conebill (*Conirostrum* x *Oreomanes*) from Peru. Ornithol. Monogr. 36:390–395.

538. Schulenberg, T.S. 1987. Observations on two rare birds, *Upucerthia albigula* and *Conirostrum tamarugense*, from the Andes of southwestern Peru. Condor 89:654–658.

539. Schulenberg, T.S., and Binford, L.C. 1985. A new species of tanager (Emberizidae: Thraupidae, *Tangara*) from southern Peru. Wilson Bull. 97:413–421.

540. Scott, J.M., Hoffman, W., Ainley, D., and Zeillemaker, C.F. 1974. Range expansion and activity patterns in Rhinoceros Auklets. Western Birds 5:13–20.

541. Scott, P., and The Wildfowl Trust. 1972. The swans. Wildfowl Trust, London.

542. Sealy, S.G. 1975. Egg size of murrelets. Condor 77:500–501.

543. Sealy, S.G. 1976. Biology of nesting Ancient Murrelets. Condor 78:294–306.

544. Sealy, S.G., Sexton, D.A., and Collins, K.M. 1980. Observations of a White-winged Crossbill invasion of southeastern Manitoba. Wilson Bull. 92:114–116.

545. Selander, R.K. 1958. Age determination and molt in the Boat-tailed Grackle. Condor 60:355–376.

546. Selander, R.K. 1966. Sexual dimorphism and differential niche utilization in birds. Condor 68:113–151.

547. Selander, R.K., and Baker, J.K. 1957. The Cave Swallow in Texas. Condor 59:345–363.

548. Selander, R.K., and Giller, D.R. 1959. The avifauna of the Barranca del Oblatos, Jalisco, Mexico. Condor 61:210–222.

549. Servat, G., and Pearson, D.L. 1991. Natural history notes and records for seven poorly-known bird species from Amazonian Peru. Bull. Br. Ornithol. Club 111:92–95.

550. Serventy, D.L., and Whittell, H.M. 1976. Birds of Western Australia. Univ. West Australia Press, Perth, W. Australia.

551. Seton, D., and Lavery, H.J. 1974. Nest and eggs of the Atherton Scrubwren *Sericornis keri*. Emu 74:53.

552. Severinghaus, L.L. 1987. Social behavior of the Vinous-throated Parrotbill during the non-breeding season. Bull. Inst. Zool., Academia Sinica 26:231–244.

553. Shannon, P. Unpubl. weights from captive birds at Audubon Zool. Garden, New Orleans, Louisiana.

554. Sheppard, J. 1972. Banding worksheet for western birds - Lesser Goldfinch. W. Bird Banding Assn.

555. Sheppard, J. 1972. Banding worksheet for western birds - Lazuli Bunting. W. Bird Banding Assn.

556. Sheppard, J. 1973. An initial study of LeConte's Thrasher (*Toxostoma lecontei*). MA thesis, California State Univ., Long Beach, CA.

557. Sheppard, J., and Collins, C.T. 1971. Banding worksheet for western birds - Western Tanager. W. Bird Banding Assn.

558. Sheppard, J., and Collins, C.T. 1971. Banding worksheet for western birds - Black-headed Grosbeak. W. Bird Banding Assn.

559. Short, L.L. 1969. Observations of three sympatric species of tapaculos (Rhinocryptidae) in Argentina. Ibis 111:239–240.

560. Short, L.L. 1982. Woodpeckers of the world. Delaware Mus. Nat. Hist. Monogr. 4. Greenville, DE pp. 1–676.

561. Short, L.L., and Banks, R.C. 1965. Notes on birds of northwestern Baja California. Trans. San Diego Soc. Nat. Hist. 14:41–52.

562. Short, L.L., and Morony, J.J. 1969. Notes on some birds of central Peru. Bull. Br. Ornithol. Club 89:11–115.

562a. Sibley, C.G., and Monroe, B.L. 1990. Distribution and taxonomy of birds of the world. Yale Univ. Press, New Haven, CT.

562b. Sibley, C.G., and Ahlquist, J.E. 1991. Phylogeny and classification of birds: a study in molecular evolution. Yale Univ. Press, New Haven, CT.

563. Sick, H. 1979. Notes on some Brazilian birds. Bull. Br. Ornithol. Club 99:115–120.

564. Sick, H. 1984. Ornitologia Brasileira. Linha Grafica Editora.

565. Sick, H. 1991. Distribution and subspeciation of the Biscutate Swift *Streptoprocne biscutata*. Bull. Br. Ornithol. Club 111:38–40.

566. Simons, T.R. 1985. Biology and behavior of the endangered Hawaiian Dark-rumped Petrel. Condor 87:229–245.

567. Skeel, M.A. 1982. Sex determination of adult Whimbrels. J. Field Ornithol. 53:414–416.

568. Smith, G.T. 1987. Observations on the biology of the Western Bristlebird *Dasyornis longirostris*. Emu 87:111–118.

569. Smith, G.T. 1991. Ecology of the Western Whipbird *Psophodes nigrogularis* in Western Australia. Emu 91:145–157.

570. Smith, V.W. 1966. Autumn and spring weights of some Palaearctic migrants in central Nigeria. Ibis 108:492–512.

570a. Smithe, F.B. 1966. The birds of Tikal. Nat. Hist. Press, Garden City, New York.

571. Smithe, F.B., and Paynter, R.A. 1963. Birds of Tikal, Guatemala. Harvard Univ. Mus. Comp. Zool. Bull. 128:245–324.

572. Snow, B.K. 1966. Observations on the behaviour and ecology of the Flightless Cormorant *Nannopterum harrisi*. Ibis 108:265–280.

573. Snow, B.K. 1981. Relationships between hermit hummingbirds and their food plants in eastern Ecuador. Bull. Br. Ornithol. Club 101:387–396.

574. Snow, D.W. 1975. *Laniisoma elegans* in Peru. Auk 92:583–584.

575. Snow, D.W. 1982. The cotingas. Cornell Univ. Press, Cornell, New York.

576. Snow, D.W., and Snow, B.K. 1963. Weights and wing-lengths of some Trinidad birds. Zoologica 48:1–12.

577. Snyder, N.F.R., and Wiley, J.W. 1976. Sexual size dimorphism in hawks and owls of North America. Ornithol. Monogr. 20.

578. Snyder, N.F.R., Wiley, J.W., and Kepler, C.B. 1987. The parrots of Luqillo: natural history and conservation of the Puerto Rican Parrot. W. Foundation Vert. Zool., Los Angeles.

579. Sokal, R.R., and Rohlf, F.J. 1973. Introduction to biostatistics. W.H. Freeman and Co., San Francisco.

580. Somadikarta, S. 1968. The Giant Swiftlet, *Collocalia gigas* Hartert and Butler. Auk 85:549–559.

581. Sorrie, B.A. 1977. Banding worksheet for western birds - Varied Thrush. W. Bird Banding Assn.

582. Sparks, J., and Soper, T. 1967. Penguins. Facts on File Publ., New York.

583. Specimens from California State University. at Sacramento, CA.

583a. Specimens from the Los Angeles County Museum, Los Angeles.

584. Specimens from Louisiana State University, Baton Rouge, LA.

585. Specimens from Texas Cooperative Wildl. Collections, Texas A & M University, College Station, TX.

586. Specimens from the Delaware Museum of Natural History, Dover, DE.

586a. Specimens from the Field Museum of Natural History, Chicago.

587. Specimens from the Museum of Vertebrate Zoology, University of California, Berkeley.

588. Specimens from the Natural History Museum (Bulawayo, Zimbabwe).

589. Specimens from the Philadelphia Academy of Natural Sciences, Philadelphia, PA.

590. Specimens from the Royal Ontario Museum, Toronto, Ontario, Canada.

590a. Specimens from the University of Oklahoma, Norman, OK.

591. Specimens from the San Diego Museum of Natural History, San Diego, CA.

592. Specimens from the University of Minnesota, Minneapolis, MN.

593. Specimens from the Cowan Vertebrate Museum, University of British Columbia, Vancouver, British Columbia, Canada.

594. Specimens from the University of Arizona, Tucson.

594a. Specimens from the Western Foundation of Vertebrate Zoology, Los Angeles.

595. Specimens from the University of Washington, Seattle, WA.

596. Specimens from the Yale Peabody Museum, New Haven, CT.

596a. Specimens from the British Museum of Natural History, London.

597. Specimens from the New Zealand National Museum, Private Bag, New Zealand.

597a. Specimens from the U.S. National Museum, Washington, D.C.

598. Specimens from the Philadelphia Zoological Gardens, Philadelphia, PA.
598a. Specimens from the University of Alaska, Fairbanks, AK.
599. Spring, L. 1971. A comparison of functional and morphological adaptations in the Common Murre (*Uria aalge*) and Thick-billed Murre (*Uria lomvia*). Condor 73:1–27.
600. Steadman, D.W., Olson, S.L., Barber, J.C., Meister, C.A., and Melville, M.E. 1980. Weights of some West Indian birds. Bull. Br. Ornithol. Club 100:155–158.
601. Stiles, F.G. 1978. Possible specialization for hummingbird-hunting in the Tiny Hawk. Auk 95:550–553.
602. Stiles, F.G. 1983. Systematics of the southern forms of *Selasphorus* (Trochilidae). Auk 100:311–325.
603. Stiles, F.G. 1983. The taxonomy of *Microcerculus* wrens (Troglodytidae) in Central America. Wilson Bull. 95:169–183.
603a. Stiles, F.G. and Stiles, A.F. 1989 A guide to the birds of Costa Rica. Cornell Univ. Press, Ithaca, New York.
604. Stokes, T. 1980. Notes on the landbirds of New Caledonia. Emu 80:81–86.
605. Stonehouse, B. 1970. Geographic variation in Gentoo Penguins *Pygoscelis papua*. Ibis 112:52–57.
606. Stoner, D. 1936. Studies on the Bank Swallow *Riparia riparia riparia* (Linnaeus) in the Oneida Lake region. Roosevelt Wildl. Annals 4:127–233.
607. Storer, R.W. 1982. A hybrid between the Hooded and Silver Grebes (*Podiceps gallardoi* and *P. occipitalis*). Auk 99:168–169.
608. Storer, R.W. 1982. The Hooded Grebe on Laguna de los Escarchados: ecology and behavior. Living Bird 19:51–67.
609. Storer, R.W. 1987. Morphology and relationships of the Hoary-headed Grebe and the New Zealand Dabchick. Emu 87:150–158.
610. Storer, R.W. 1989. Notes on Paraguayan birds. Occas. Pap. Mus. Zool. Univ. Mich. 719:1–21.
610a. Stotz, D. Unpubl. data.
611. Strauch, J.G. 1977. Further bird weights from Panama. Bull. Br. Ornithol. Club 97:61–65.
611a. Strong, A.M., and Bancroft, G.T. Unpubl. data.
612. Sturman, W.A. 1968. The foraging ecology of *Parus atricapillus* and *P. rufescens* in the breeding season, with comparisons with other species of *Parus*. Condor 70:309–322.
613. Summers-Smith, J.D. 1988. The sparrows: a study of the genus *Passer*. T & AD Poyser, Inc., Staffordshire, UK.
613a. Sutton, R. Unpubl. data.
614. Swales, M. 1965. The seabirds of Gough Island. Ibis 107:17–42.
615. Sykes, P.W. 1975. Caribbean Coot collected in southern Florida. Fla. Field Nat. 3:25–27.
616. Tanner, J.T. 1942. The Ivory-billed Woodpecker. Nat. Audubon Soc. Res. Report 1.
617. Tarburton, M.K. 1986. Breeding of the White-rumped Swiftlet in Fiji. Emu 86:214–227.
618. Tashian, R.E. 1952. Some birds from the Palenque region of northeastern Chiapas, Mexico. Auk 69:60–66.
619. Teixeira, D.M. 1987. Notas sobre o "Gravatazeiro," *Rhopornis ardesiaca* (Wied, 1831) (Aves, Formicariidae). Rev. Brasil. Biol. 47:409–414.
620. Teixeira, D.M., and Luigi, G. 1989. Notas sobre *Cranioleuca semicinerea* (Reichenbach 1853) (Aves: Furnariidae). Rev. Brasil. Biol. 49:605–613.
621. Teixeira, D.M., Nacinovic, J.B., and Luigi, G. 1989. Notes on some birds of northeastern Brazil. Bull. Br. Ornithol. Club 109:152–157.
622. Teixeira, D.M., Nacinovic, J.B., and Tavares, M.S. 1986. Notes on some birds of northeastern Brazil. Bull. Br. Ornithol. Club 106:70–74.
623. Terborgh, J., Robinson, S.K., Parker, T.A., Munn, C.A., and Pierpont, N. 1990. Structure and organization of an Amazonian forest bird community. Ecol. Monogr. 60:213–238.
623a. Tershy, B.R., and Breese, D. Unpubl. data.
624. Thiollay, J.-M. 1988. Comparative foraging success of insectivorous birds in tropical and temperate forests: ecological implications. Oikos 53:17–30.
625. Thomas, B.T. 1982. Weights of some Venezuelan birds. Bull. Br. Ornithol. Club 102:48–52.
626. Thomas, B.T. 1990. Additional weights of Venezuelan birds. Bull. Br. Ornithol. Club 110:48–51.
627. Thompson, M.C. 1966. Birds from North Borneo. Univ. Kansas Publ. Mus. Nat. Hist. 17:377–433.
628. Threlfall, W., and Mahoney, S.P. 1980. The use of measurements in sexing Common Murres from Newfoundland. Wilson Bull. 92:266–268.
629. Tomkins, R.J. 1984. Some aspects of the morphology of Wandering Albatrosses on MacQuarie Island. Emu 84:29–32.
630. Tomlinson, R.E. 1975. Weights and wing lengths of wild Sonoran Masked Bobwhites during fall and winter. Wilson Bull. 87:180–186.
631. Trail, P., and Baptista, L.P. Unpubl. data.
632. Turner, A., and Rose, C. 1989. A handbook to the swallows and martins of the world. Christopher Helm, London.

633. Tyler, S. 1979. Bird ringing in an Addis Ababa garden. Scopus 3:1–8.

634. Urban, E.K. 1959. Birds from Coahuila, Mexico. Univ. Kansas Publ., Mus. Nat. Hist. 11:443–516.

635. Urban, E.K. 1975. Weights and longevity of some birds from Addis Ababa, Ethiopia. Bull. Br. Ornithol. Club 95:96–98.

636. Urban, E.K., Fry, C.H., and Keith, S. 1986. The birds of Africa. Academic Press, New York. Vol. 2.

637. van Riper, C. III. 1987. Breeding ecology of the Hawaii Common Amakihi. Condor 89:85–102.

638. Verbeek, N.A. 1976. Banding worksheet for western birds - Yellow-billed Magpie. W. Bird Banding Assn.

639. Vermeer, K., and Cullen, L. 1982. Growth comparison of a plankton- and a fish-eating alcid. Murrelet 63:34–39.

640. Vernon, D.P. 1977. The first live Australian specimen of the Westland Petrel (*Procellaria westlandica*). Austral. Bird Watcher 7:44–46.

641. Vernon, D.P., and McKean, J.L. 1972. A specimen of the Herald Petrel from Queensland. Emu 72:115.

642. Voous, K.H. 1969. Predation potential in birds of prey from Surinam. Ardea 57:117–148.

643. Vuilleumier, F. 1969. Field notes on some birds from the Bolivian Andes. Ibis 111:599–608.

644. Vuilleumier, F., and Ewert, D.N. 1978. The distribution of birds in Venezuelan paramo. Bull. Am. Mus. Nat. Hist. 162:49–90.

645. Walkinshaw, L.H. 1939. The Yellow Rail in Michigan. Auk 56:227–237.

646. Walkinshaw, L.H. 1953. Life-history of the Prothonotary Warbler. Wilson Bull. 65:152–168.

647. Walkinshaw, L.H. 1983. Kirtland's Warbler: the natural history of an endangered species. Cranbrook Inst. Science, Bloomfield Hills, MI.

648. Wallace, G.E. Unpubl. data (in litt. from M.K. McNicholl).

649. Wallace, G.J. 1965. Studies on neotropical thrushes in Columbia. Publ. Mus., Michigan St. Univ., Biol. Series 3:1–48.

650. Wallace, M.P., and Temple, S.A. 1987. Competitive interactions within and between species in a guild of avian scavengers. Auk 104:290–295.

651. Walsberg, G.E. 1977. Ecology and energetics of contrasting social systems in *Phainopepla nitens* (Aves: Ptilogonatidae). Univ. Cal. Publ. Zool. 108.

652. Walsh, J.F., and Grimes, L.C. 1981. Observations on some Palaearctic land birds in Ghana. Bull. Br. Ornithol. Club 101:327–333.

653. Walters, P.M. 1981. Notes on the body weight and molt of the Elf Owl (*Micrathene whitneyi*) in southeastern Arizona. N. Amer. Bird Bander 6:104–105.

653a. Watling, D. 1986. Notes on the Collared Petrel *Pterodroma (leucoptera) brevipes*. Bull. Br. Ornithol. Club 106:63–70.

654. Wanless, S., Burger, A.E., and Harris, M.P. 1991. Diving depths of Shags *Phalacrocorax aristotelis* breeding on the Isle of May. Ibis 133:37–42.

655. Ward, D. 1990. Incubation temperatures and behavior of Crowned, Black-winged, and Lesser Black-winged Plovers. Auk 107:10–17.

656. Warham, J. 1958. The nesting of the shearwater *Puffinus carneipes*. Auk 75:1–14.

657. Warham, J. 1971. Body temperatures of petrels. Condor 73:214–219.

658. Warham, J. 1974. The Fiordland Crested Penguin *Eudyptes pachyrhynchus*. Ibis 116:1–27.

659. Warner, D.W. 1959. The song, nest, eggs, and young of the Long-tailed Partridge. Wilson Bull. 71:307–312.

660. Watanuki, Y. 1986. Moonlight avoidance behavior in Leach's Storm-Petrels as a defense against Slaty-backed Gull. Auk 103:14–22.

661. Watling, D. 1986. Notes on the Collared Petrel *Pterodroma (leucoptera) brevipes*. Bull. Br. Ornithol. Club 106:63–70.

662. Weatherhead, P.J. 1980. Sexual dimorphism in two Savannah Sparrow populations. Can. J. Zool. 58:412–415.

663. Weathers, W.W., and van Riper, C. 1982. Temperature regulation in two endangered Hawaiian honeycreepers: the Palila (*Psittirostra bailleui*) and the Laysan Finch (*Psittirostra cantans*). Auk 99:667–675.

664. Weeden, R.B. 1979. Relative heart size in Alaskan Tetraonidae. Auk 96:306–318.

665. Weimerskirch, H., Zotier, R., and Jouventin, P. 1989. The avifauna of the Kerguelen Islands. Emu 89:15–29.

666. Weller, M.W. 1975. Ecology and behavior of the South Georgia Pintail *Anas g. georgica*. Ibis 117:217–231.

667. Weller, M.W. 1980. The island waterfowl. Iowa State Univ. Press, Ames, IA.

668. Weske, J.S. 1972. The distribution of the avifauna in the Apurimac Valley of Peru with respect to environmental gradients, habitat, and related species. Ph.D. dissertation, Univ. Oklahoma, Norman, OK.

669. Weske, J.S., and Terborgh, J.W. 1981. *Otus marshalli*, a new species of screech-owl from Peru. Auk 98:1–7.

670. West, G.C., Weeden, R.B., Irving, L., and Peyton, L.J. 1970. Geographic variation in body size and weight of Willow Ptarmigan. Arctic 23:240–253.

671. Westerkov, K. 1960. Birds of Campbell Island. New Zealand Wildl. Publ. 61.

672. Wiedenfeld, D.A., Schulenberg, T.S., and Robbins, M.B. 1985. Birds of a tropical deciduous forest in extreme northwestern Peru. Ornithol. Monogr. 36:305–316.
673. Wiens, J.A., and Dyer, M.I. 1975. Rangeland avifaunas: their composition, energetics, and role in the ecosystem. In Smith, D.R. (ed.). Symposium on management of forest and range habitats for nongame birds. USDA Forest Service Gen. Tech. Rep. WO-1, Washington, D.C. pp. 176-182.
674. Wiens, J.A., and Rotenberry, J.T. 1980. Patterns of morphology and ecology in grassland and shrubsteppe bird populations. Ecol. Monogr. 50:287–308.
675. Wilcox, L. 1959. A twenty year banding study of the Piping Plover. Auk 76:129–152.
676. Wiley, J.W. 1986. Growth of Shiny Cowbird and host chicks. Wilson Bull. 98:126–131.
677. Williams, G.R. 1960. The birds of the Pitcairn Islands, central South Pacific Ocean. Ibis 102:58–70.
677a. Williams, J.B. and Nagy, K.A. 1985. Daily energy expenditure by female Savannah Sparrows feeding nestlings. Auk 102:187–190.
678. Willis, E.O. 1982. The behavior of Black-banded Woodcreepers (*Dendrocolaptes picumnus*). Condor 84:272–285.
679. Willis, E.O. 1982. The behavior of Scale-backed Antbirds. Wilson Bull. 94:447–462.
680. Wilson, R.T., and Ball, D.M. 1979. Morphometry, wing-loading and food of western Darfur birds. Bull. Br. Ornithol. Club 99:15–20.
681. Wilson, R.T., and Lewis, J.G. 1977. Observations on the Speckled Pigeon *Columba guinea* in Tigrai, Ethiopia. Ibis 119:195–198.
681a. Wingate, D.B. 1972. First successful hand-rearing of an abandoned Bermuda Petrel chick. Ibis 114:97–101.
682. Wingham, E.J. 1984. Breeding biology of the Australasian Gannet *Morus serratus* (Gray) at Motu Karamarama, Hauraki Gulf, New Zealand. II. Breeding success and chick growth. Emu 84:211–224.
683. Winterstein, S.R., and Raitt, R.J. 1983. Nestling growth and development and the breeding ecology of the Beechey Jay. Wilson Bull. 95:256–268.
684. Wishart, R.A. 1979. Indices of structural size and condition of American Wigeon (*Anas americana*). Can. J. Zool. 57:2369–2374.
685. Withers, P.C., Forbes, R.B., and Hedrick, M.S. 1987. Metabolic, water and thermal relations of the Chilean Tinamou. Condor 89:424–426.
686. Woinarski, J.C.Z. 1987. Notes on the status and ecology of the Red-lored Whistler *Pachycephalarufogularis*. Emu 87:224–236.
687. Woinarski, J.C.Z., and Bulman, C. 1985. Ecology and breeding biology of the Forty-spotted Pardalote and other pardalotes on North Bruny Island. Emu 85:106–120.
688. Woinarski, J.C.Z., Dorward, D.F., and Cullen, J.M. 1983. Variation in the *Pardalotus striatus* complex in south-eastern Australia. Emu 83:82–93.
689. Wolf, L.L. 1977. Species relationships in the avian genus *Aimophila*. Ornithol. Monogr. 23.
690. Wong, M. 1986. Trophic organization of understory birds in a Malaysian dipterocarp forest. Auk 103:100–116.
691. Wood, B. 1989. Biometrics, iris and bill colouration, and moult of Somali forest birds. Bull. Br. Ornithol. Club 109:11–22.
692. Wood, B., Madge, S.C., and Waller, C.S. 1978. Description, moult and measurements of *Montifringilla theresae*. Bull. Br. Ornithol. Club 98:55–59.
693. Woodall, P.F. 1984. Kleptoparasitism in Hardheads and Pacific Black Ducks, including size-related differences. Emu 84:65–70.
694. Woods, J.G. Unpubl. data.
695. Wooler, R.D., and Calver, M.C. 1981. Diet of three insectivorous birds on Barrow Island, Washington. Emu 81:48–50.
696. Woolfenden, G.E. 1955. Spring molt of the Harris Sparrow. Wilson Bull. 67:212–213.
697. Woolfenden, G.E. 1956. Comparative breeding behavior of *Ammospiza caudacuta* and *A. maritima*. Univ. Kansas Publ. Mus. Nat. Hist. 10:45–75.
698. Woolfenden, G.E. 1973. Nesting and survival in a population of Florida Scrub Jays. Living Bird 12:25–49.
699. Wyndham, E. 1979. Diurnal cycle, behaviour and social organization of the Budgerigar, *Melopsittacus undulatus*. Emu 80:25–33.
700. Yanez, J.L., Nunez, H., and Fasic, F.M. 1982. Food habits and weight of Chimango Caracaras in central Chile. Auk 99:170–171.
701. Zammuto, R.M., and Franks, E.C. 1979. Trapping flocks of Chimney Swifts in Illinois. Bird-Banding 50:201–209.
702. Zwickel, F.C., Brigham, J.H., and Buss, I.O. 1966. Autumn weights of Blue Grouse in north-central Washington 1954–1963. Condor 68:488–496.
703. Zwickel, F.C., and Brigham, J.H. 1974. Autumnal weights of Spruce Grouse in north-central Washington. J. Wildl. Manage. 38:315–319.

# INDEX

## A

Abeillia, 110
Abroscopus, 248
Aburria, 40
Acanthagenys, 187
Acanthidops, 301
Acanthisitta, 143
Acanthisittidae, 143
Acanthiza, 181, 182
Acanthorhynchus, 187
Accipiter, 33, 34
Accipitridae, 30–37
Aceros, 126
Acestrura, 118
Acridotheres, 219
Acrocephalus, 244, 245
Acropternis, 160
Acryllium, 41
Actenoides, 122
Actinodura, 254
Actophilornis, 57
Adelomyia, 114
Aechmophorus, 10
Aegithalidae, 232
Aegithalos, 232
Aegithina, 202
Aegithinidae, 202
Aegolius, 99
Aegotheles, 100
Aegothelidae, 100
Aegypius, 32
Aenigmatolimnas, 53
Aeronautes, 106
Aethia, 71
Aethopyga, 273, 274
Afropavo, 48
Agamia, 26
Agapornis, 83
Agelaius, 308, 309, 321–332
Aglaeactis, 115
Aglaiocercus, 117
Agriocharis, 49
Agriornis, 174
Ailuroedus, 179
Aimophila, 288
Aix, 21
Ajaia, 28
Alaemon, 259
Alauda, 260
Alaudidae, 258–260
Alca, 71
Alcedinidae, 120–122
Alcedo, 120
Alcidae, 71–72
Alcippe, 255
Aleadryas, 190

Alectoris, 42
Alectroenas, 79
Alectura, 41
Alethe, 216
Alisterus, 81
Alle, 71
Alophoixus, 238
Alopochen, 20
Amadina, 265
Amalocichla, 181
Amandava, 263
Amaurolimnas, 52
Amaurornis, 53
Amaurospiza, 303
Amazilia, 112, 113
Amazona, 87
Amazonetta, 21
Amblycercus, 307
Amblyornis, 179
Amblyospiza, 270
Amblyramphus, 309
Ammodramus, 287, 329
Ammomanes, 259
Ammoperdix, 41
Ampeliceps, 220
Ampelioides, 161
Ampelion, 160
Amphispiza, 288
Amytornis, 180
Anabacerthia, 150
Anairetes, 169
Anaplectes, 269
Anarhynchus, 65
Anas, 21–23
Anastomus, 29
Anatidae, 17–25
Ancistrops, 150
Andigena, 133
Androdon, 108
Andropadus, 237
Anhima, 17
Anhimidae, 17
Anhinga, 17
Anhingidae, 17
Anisognathus, 296
Anodorhynchus, 84
Anous, 70
Anser, 19
Anseranas, 17
Anseriformes, 17–25
Anthochaera, 187
Anthornis, 187
Anthoscopus, 258
Anthracoceros, 126
Anthracothorax, 109
Anthreptes, 271, 272
Anthropoides, 55